Mathematicians under the Nazis

Mathematicians under the Nazis

SANFORD L. SEGAL

PRINCETON UNIVERSITY PRESS

PRINCETON AND OXFORD

Published by Princeton University Press, 41 William Street, Princeton, New Jersey 08540
In the United Kingdom: Princeton University Press, 3 Market Place, Woodstock, Oxfordshire OX20 1SY

LIBRARY OF CONGRESS CATALOGING-IN-PUBLICATION DATA

Segal, Sanford L., 1937–
Mathematicians under the Nazis / Sanford L. Segal.
p. cm.
Includes bibliographical references and index.
ISBN 0-691-00451-X (alk. paper)
1. Mathematicians—Germany—History—20th century.
2. Mathematics—Germany—History—20th century. I. Title.
QA28 .S44 2003
510'.943'09043—dc21 2002070399

British Library Cataloging-in-Publication Data is available

This book has been composed in Berkeley

Printed on acid-free paper. ∞

www.pupress.princeton.edu

Printed in the United States of America

10 9 8 7 6 5 4 3 2 1

TO THE MEMORY OF

JAMES B. COLTON II

AND

NORMAN OLIVER BROWN,

BOTH CLASSICISTS,

BOTH HUMANISTS OF BROAD PERSPECTIVE,

WHO TAUGHT ME WHAT HISTORY WAS

AND ABOUT THE MEANING OF FACTS.

"A teacher affects eternity, he can never tell where his influence stops."

A mathematician, like a painter or a poet, is a maker of patterns. If his patterns are more permanent than theirs, it is because they are made with *ideas*. A painter makes patterns with shapes and colours, a poet with words. . . . A mathematician, on the other hand, [unlike the poet] has no material to work with but ideas, and so his patterns are likely to last longer, since ideas wear less with time than words.

—G. H. HARDY *A Mathematician's Apology* (1940)

Und hat Form nicht zweierlei Gesicht? Ist sie nicht sittlich und unsittlich zugleich—sittlich als Ergebnis und Ausdruck der Zucht, unsittlich aber und selbst widersittlich sofern sie von Natur eine moralische Gleichgultigkeit in sich schleisst, ja wesentlich bestrebt ist, das Moralische unter ihr stolzes und unumschranktes Szepter zu beugen.

—THOMAS MANN, *Der Tod in Venedig*

And has not form two aspects? Is it not moral and immoral at once; moral in so far as it is the expression and result of discipline, immoral—yes, actually hostile to morality—in that of its very essence it is indifferent to good and evil, and deliberately concerned to make the moral world stoop beneath its proud and undivided sceptre?

—THOMAS MANN, *Death in Venice* (trans. H. T. Lowe Porter)

CONTENTS

PREFACE

WHEN still a graduate student, I discovered by chance and to my fascinated horror that a prominent and distinguished German mathematician had been, *as a mathematician*, a propagandist for Nazi ideology. At that time, I thought that at some time I would want to investigate this further, to try to understand whether he was simply a freak or, as superficially appeared, the leader of a group of like-minded academic colleagues. I have always had an interest in history, and over the years this developed into a growing amateur interest in the history of science. Sporadic concentration on the subject broadened my original interest to a question of the situation of mathematics under the Nazi regime. An Alexander von Humboldt Fellowship in 1988 enabled me to spend time doing archival research in Germany, as well as talk to a number of mathematicians. Since then, my previous sporadic work has been more concentrated, including several trips to Germany, and this book is the result.

From the first, I wanted to write a book that was both scholarly, in the sense of documentation and understanding, and also readable by a wider public, as many scholarly works unfortunately are not. In particular, it should be a book that does not require any advanced knowledge about mathematics. Sensitive (perhaps hypersensitive) to the general but perverse social view of mathematicians as disembodied intellects, it also seemed appropriate to explain enough about the nature of mathematics to make its interaction with ideology more interesting and meaningful. From another internal standpoint, this also seemed necessary. To write about the history of mathematics in a period so soaked in ideology and political pressure seemed meaningless unless both a mathematical-historical and a general historical context were provided. Thus this book is an attempt at a kind of social history of mathematics and the community of mathematicians in the period. At the same time, it is *not* a history of mathematics as a scientific discipline or of mathematical results achieved during the period in question. Information about actual progress in mathematics in Germany during the Nazi period is available in the seven volumes of the FIAT review of German science devoted to mathematics and edited by Wilhelm Süss and Alwin Walther (for applied mathematics). Rather, this is a contribution to the history of the academic community of mathematicians as practitioners of an academic discipline in a time of intense political and ideological pressure.

Being a kind of social history also means that the people who figure in it are not just the mathematically famous, but also those who are mathematically now forgotten. Not only are the Behnkes and Blaschkes, Hasses and Teichmüllers important, but also the Weyrichs and Weinels, Wegners and Torniers. It also means considering mathematicians as more than just members of a certain academic profession. It is easy to refute the naive idea that mathematicians, being "analytically trained" and educated to examine the validity of hypotheses, ought not to have been prone to Nazi fellow-traveling—especially since mathematics (unlike biology, anthropology, history, or architecture) was not a subject that would seem to lend itself to ideologizing. Mathematicians need to be seen in the social context of academics in Germany, as well as in that of the history of their profession. Indeed, there were psychologists like E. R. Jaensch and mathematicians like Ludwig Bieberbach who saw mathematics as a fruitful area in which Nazi ideology could be exemplified and play a significant role. There were mathematicians like E. A. Weiss who saw mathematics as the ideal discipline for training those who would be good Nazis.

Mathematics also has its own rhetoric, a rhetoric of absolute certainty and inviolate validity. The Nazis were fond of such rhetoric. As early as 1923, the mathematician and early Nazi Theodor Vahlen spoke of mathematics as a mirror of the races that proves the presence of racial qualities in intellectual activities with mathematical, and therefore incontrovertible, certainty. A frequent Nazi argumentation (despite the irrational appeals to blood, soil, and feeling) involved what purported to be the drawing of unremitting logical conclusions from premises. If Jews were literally (and not just figuratively) poisonous parasites on the body politic, then the only solution was to expel, suppress, or, finally, exterminate them. Nazi intellectuals also practiced a kind of scientific reductionism. Present-day scientific reductionism involves the reduction of biology to chemistry, both of these to physics, which ultimately (in one frequent contemporary interpretation) finds its meaning in mathematical formulas. For the Nazis, the fundamental science was instead biology. All arguments had eventually to become biological ones. This, along with the wave of irrationalist thought and rhetoric in which Germany was awash, was accurately perceived as something of a threat to mathematics' "right to life."

Of course, even when dealing with such a restricted subject as mathematics, and even if a broader context were not employed, it would be impossible in a book of any reasonable size to cover all events of possible interest. Nevertheless, though I acknowledge the undoubted variations that existed here or there, the events selected seem representative of the interactions of Nazi politics and ideology with mathematics.

Throughout this book, the emphasis is on mathematics and mathematicians—the only exceptions being parts of chapter 3, which deals with the German academic crisis, and parts of chapter 5, which deals with life as an academic mathematician under National Socialism. Chapter 1 poses a question—Why study mathematics and mathematicians under the Nazis?—and attempts to answer it.

There is no attempt in this book to fit the narrative to any Procrustean bed of underlying interpretation. Grand theories always seem to neglect detail, especially somewhat contradictory detail, for all their fascinating and valid perceptions of harmonies and similarities. They also often relegate human beings (and human actions create history) to marionettes controlled by higher forces. It is potentially more accurate, it seems, to allow larger interpretations of a set of events to grow out of smaller, individually prescribed ones. This is done in full knowledge of the historian's dilemma: while necessarily striving for full accuracy and completeness, necessarily that is unachievable. Mistakes will be made (though one hopes not too many); some important facts will be overlooked or omitted through ignorance or underestimation. Nevertheless, the serious historical attempt provides one facet to add to those of others in obtaining a fuller view.

In this light, chapters 2 and 3 deal with two independent and concomitant crises that form the socio-historical basis for mathematicians in Germany in the 1930s. Chapter 2 discusses the crisis in mathematics. At the end of the nineteenth century, the nature of mathematical activity began to change, and a crisis in its logical foundations loomed. This latter, however, really took hold only after World War I. Already during the Weimar years, these issues had occasionally become confused with political ones, and those mathematicians interested in a *Deutsche Mathematik*, a mathematics ideologized in conformity with National Socialism, reflected these mathematical differences in some of their argumentation. Not all mathematicians who fellow-traveled with the Nazis believed in the possible distinction of an "Aryan mathematics," but several did. In addition, these changes and crises were initially within German mathematics, with German mathematicians playing a leading role. If we are to understand the interactions between politics, ideology, and mathematics in the Nazi period, it is essential to understand the nature of the mathematics of the time. For readers who have no mathematical training, this may be the most difficult chapter, but every effort has been made to be clear and to avoid indulgence in mathematical technicalities, since this book is especially intended for such readers.

Chapter 3 is a delineation of the crisis in the German professoriat also dating back to the turn of the century and well described by Fritz Ringer and others. To say that the professional crises in mathematics and the academic crises facing the German professoriat intersected to influence attitudes toward the Hitler regime is too simplistic to be the whole truth. Nevertheless, the social sense of crisis afflicting German academe was not one for which there is any evidence that mathematicians stood apart. Like other academics, Weimar mathematicians seem to have been predominantly national conservatives. They felt as declassed as their colleagues, as uncomfortable with the more open society and "Western" ideas of Weimar. Furthermore, as a group, they were not noticeably less anti-Semitic than their other non-Jewish colleagues, though notable exceptions existed among both mathematicians and other academics. In short, to understand the attitudes of mathematicians, they have to be set in the matrix of their colleagues. This chapter, then, has broader scope than the others, but nevertheless presents mathematically relevant material as well.

The lengthy chapter 4 is devoted to three detailed case studies of the interaction between politics, ideology, and mathematics under the Nazis. These are independent of one another and are not presented in chronological order, partly to emphasize them as particular extreme instances of a general situation. Two of these studies deal with applied (or "applicable") mathematics, the last (though temporally earliest) and longest with "pure" mathematics. One of these stories is from mid-war, one from 1938–40, and one from 1934, when the Nazis were completing their consolidation of power. These have been presented in great detail because, while "the devil is in the details," it is only through detail that we can come to a better approximation of the truth. Besides the bureaucratic manipulations and political in-fighting, the real and asserted motivations are fascinating, and I hope to have made them so for the reader. The circumstances of these three analyzed snapshots of mathematical life under the Nazis are sufficiently different to make them complementary rather than redundant. While chronological order has been eschewed, this does not mean that it has no importance. For example, in 1934, the question of who was a true National Socialist obviously would play a greater role than at mid-war. Two of the studies in this chapter have essentially not appeared previously in the literature, while the third and longest has been published only in a sharply abbreviated form.

Chapter 5 explores everyday life as an academic mathematician under the Nazis. As with chapter 3, the scope of this chapter is somewhat broader, for while the details as they affected mathematicians are of interest, mathematicians themselves were not singled out as a peculiar species of academic. Thus, in addition to issues like mathematical pedagogy and the training of future mathematicians, what it meant to be an academic living in those times is adumbrated.

Chapter 6 looks at mathematical institutions and their reaction at critical junctures to political and ideological pressures. Here the sorts of questions faced by mathematical journals, by the German Mathematical Society (*Deutsche Mathematiker Vereinigung*) and the national organization devoted to mathematical pedagogy (*Mathematische Reichsverband*), as well as the creation of the mathematical research center at Oberwolfach, are discussed. There is also a brief description of the history and progress of the "mathematics working group" in the Sachsenhausen concentration camp. Here, too, is a brief discussion of the status of "applied mathematics" in Nazi Germany. Rather than only simple recitals of facts, however, biographical and contextual information about the actors involved are included in an attempt to allow a more layered, more accurate, and more comprehensive picture to emerge for the reader. This has made the chapter rather long, but, I hope, far more interesting than it otherwise would have been.

Chapter 7 examines Ludwig Bieberbach and the *Deutsche Mathematik* movement. While the movement is mentioned from time to time earlier, its physical expression as a journal stems directly from an incident discussed in chapter 6. Inverting our usual notions of a universalistic science and mathematics, the movement attempted to discern different ethnic styles in the doing and teaching of mathematics, while not denying the validity of any individual mathematical fact, by whomever established. Indeed, its supporters argued that the universal validity of a mathematical fact is precisely what makes mathematics most suitable for discerning such different styles. Ludwig Bieberbach was once accounted perhaps the most brilliant mathematician of his generation, and his important supporter, the mathematician and Nazi bureaucrat Theodor Vahlen, was certainly scientifically competent. This adds some piquancy to the justifications, development, and fate of the movement with which Bieberbach was so intimately associated.

Throughout the preceding chapters, biographical material is scattered as seems appropriate. I am convinced that personality plays an important role in history: people make history. Awareness of personality seems crucial in giving a reasonable approximation of events: the actors are important to understanding actions. Chapter 8 deals with some of these actors, both non-Jews and Jews. Some of these have appeared in earlier chapters, sometimes even with brief biographical mention, but the kind of biographical detail present in this chapter, as an aid to deeper understanding, would have been out of place and too digressive earlier. A number of the actors treated in chapter 8 have not appeared earlier, but their stories as given add important nuances to matters already discussed. The biographical detail in this material varies widely as seems appropriate.

For the reader's convenience (and to improve digestibility), chapters 4–8 are broken into subsections, as indicated in the table of contents. These are also used for cross-referencing purposes throughout the book.

Two other features of this book attempt to add to the picture of the weave of academic mathematical life under the Nazis. One is a use of footnotes for frequent cross-referencing possibilities for readers so interested. This is an attempt at emphasizing the multiple layers of interaction and meaning in the linear format of writing. The other is a frequent use of direct quotation (in translation). Nuances of language are frequently missed in summary. This is especially true in the Nazi period in Germany, when sometimes the language has to be heard or seen to be believed, and sometimes the language was given peculiar meaning. Also, the same passage might well be honestly cited by different historians with different biases (and we all have biases), making different emphases to different effect. If the attempt to enhance the reader's understanding is to be successful, then the individual reader must also be enabled to make an individual judgment, despite the fact that translation and quotation are harder and longer than summary.

Having said what this book contains, it is perhaps well to say some of the things it does not contain. There is no discussion of the expulsion of Jews and others from academic posts except as this is important to other topics. The literature contains ample discussions of the general academic situation, beginning with the famous list in the *Manchester Guardian* and E. Y. Hartshorne's book on the subject (*The German Universities and National Socialism* [London: Allen and Unwin, 1937]), and continuing to the present day. Mathematicians, in particular, are discussed in an article (in German), "Fachverband-Institut-Staat," by Norbert Schappacher and Martin Kneser (1990). The reception in the United States of specifically German emigré mathematicians has been discussed by Nathan Reingold in an article in *Annals of Science* 38 (1981). Similarly, though the organization of mathematical activity within concentration camps is discussed, the Holocaust of the Jews (and others) is not mentioned except as incidental to other matters. This book contains no lengthy histories of mathematics departments during the Nazi period—to the author's knowledge, there are five of these published (all in German), only one of which attempts comprehensiveness. There is an excellent discussion of Göttingen by Norbert Schappacher (1987 and n.d.); a brief discussion of Hamburg by Christoph Maass (1991); a discussion about Berlin contained in the papers of Reinhard Siegmund-Schultze (1984b, 1984c, 1988, 1989); a discussion of Frankfurt concentrating on the three Jews, Max Dehn, Paul Epstein, and Ernst Hellinger, by their colleague, Carl Ludwig Siegel (1966, 3: 462–474); and what can be inferred about Münster from the memoirs of Heinrich Behnke (1978 and n.d.). Naturally enough, these are drawn upon and cited, but readers interested in such departmental histories should consult them.

A word is in order about translations in the text. All translations from French or German, unless otherwise indicated, are mine. Translators from other languages are indicated. Where a standard English edition of a foreign-language work exists, I have usually chosen to use it. An attempt has been made especially for the translations from German to hew closely to the original while being recognizably English. The Nazi language had its own flavor and peculiarities that need to be heard. The Viscount d'Abernon is reputed to have said that many calamities of German diplomacy were answerable to the fact that different turns of speech used by the German Foreign Office were not translatable into English or French, and when carried over into those languages, gave a completely unintended and misunderstood meaning. One trusts that such calamities have not happened, at least not frequently, here. A question also arises as to how to treat German academic titles. Because the organization of the typical American university is different from the German one of that time, I have kept the administrative titles *Kurator*, *Rektor*, *Dekan*, spelled in this German way. I have also retained the term *Ordinarius*, as such a person was traditionally far more powerful and important than a "full professor." I have used "associate professor" and *ausserordentliche Professor* interchangeably. The *Assistent* in a German academic department was far more powerful than an assistant in an American one would be. For example, an *Assistent* might actually temporarily run a department, as Otto Neugebauer or Franz Rellich did at Göttingen. Nevertheless, the German and English words have been used interchangeably as seemed appropriate. I have avoided, wherever possible, the baroque German variants on positions like *ausserplanmässige nicht beamtete ausserordentliche Professor* (roughly an associate professor without civil-service status unforeseen in the number of regular positions in the department). *Habilitation* has remained habilitation. I have also retained *Herr* instead of replacing it with "Mister." Occasionally, where it seemed it might be of interest, I have given the original German or French words as well as the translation. In German, *Geist* can mean "spirit" or "intellect" (as well as various other things), and I have tried to select the translation that seemed to me most appropriate in context. Similar remarks go for other words, especially the German distinction between "understanding" and "reason" (*Verstand* and *Vernunft*). The word "Nazi" is used frequently throughout, not as a slur, but because it was and is a common and convenient abbreviation referring to the *National Sozialistische Deutschen Arbeiter Partei*, its members, its policies, or its doctrines. The alternative "National Socialist" is not quite as accurate, and sometimes sounds rather clumsy when used repeatedly, though it also finds frequent appropriate use.

The noun *Anschauung* and its various grammatical forms are difficult to translate, and play a considerable role at various points in the text. Literally, etymologically, *Anschauung* is a "looking at" and intends a sort of descriptive intuition of the physically concrete. "Intuition" and "intuitive" are used here as translations, but one should not forget that such "intuition" always has a physically concrete referent.

Another word of some difficulty is *Volk* and its adjective *völkisch*. *Volk*, which can mean "nation," "people," or even its English cognate "folk," refers to the body of the people as a sort of mystical unity, especially in its use by the Nazis and some of their precursors and sympathizers. The historian George Mosse even invented the English neologism "volkish" to render it. In this book an attempt is made at translation (with German indications if necessary) or direct use of *völkisch*.

In writing a book about the Nazi period in Germany, a great danger is underestimating the complexity of a given situation. When evil is present, it is tempting to see things only in starkly contrasted black and white. This is especially true when the other pole of the book is mathematics, with its reputation for clarity, if also for difficulty. I hope to have avoided such pitfalls, and to have presented as accurately as possible the situation, currents of thought, and varying attitudes of mathematicians in Nazi Germany, as well as the interaction of that pressure-cooker of politics and ideology with their mathematics. This attempt to understand an undeniably complex situation partially accounts also for the detail of chapters 4, 6, and 7, and the broader scope of chapters 3 and 5. Let me emphasize here that the purpose of this book is neither to wash dirty linen publicly and assign assessments of guilt, nor to fit the events described into some preconceived social structure of mathematics or science. Rather, it is to describe the historical situation and development of the mathematics profession in Nazi Germany and its interactions with the state, allowing conclusions to emerge therefrom. Such a study, however, requires attention to nuance and respect for detail rather than mere summary.

That said, however, there is no way to begin, but without further preamble, like a Brahms quartet to begin.

ACKNOWLEDGMENTS

I AM grateful to the Alexander von Humboldt Foundation for a grant that turned sporadic interest in this subject into the initial substantial work for this book. I am also grateful to the American Philosophical Society for providing travel assistance.

I am indebted to the following institutions and their personnel for material used in the preparation of this book: Bundesarchiv Koblenz; Berlin Document Center; MIT Archives; New York Public Library; Harvard University Archives; Stanford University Libraries; Mathematisches Institut Göttingen; Niedersächsisches Staats-und-Universitätsbibliothek, Göttingen; Library of Congress; National Archives; the archives at the Universities of Aachen, Bonn, Freiburg, Freie Universität Berlin, Göttingen, Heidelberg, Hamburg, Jena, Münster, Munich, and Tübingen; the Courant Institute, New York University; the Einstein Archives, Princeton; the Archives of American Mathematics, Austin, Texas.

The author is indebted to the following people for information used in this book. Some of these people are now deceased; yet their assistance should be acknowledged anyway: Natascha Artin Brunswick, Martin Barner, Heinrich Begehr, Marianne Bernstein-Wiener, Liliane Beaulieu, Elsbeth Bredendiek, Egbert Brieskorn, Werner Burau, Nina Courant, Andreas Defant, Klaus Floret, Ulrich Hunger, Niels Jacob, Erich Kähler, Manfred Knebusch, Martin and Jutta Kneser, Matthias Landau, Gerald Liebenau, Christoph Maass, Herbert Mehrtens, Alexander Ostrowski, Volker Peckhaus, Holger Petersson, Margarete Petersson, Karin Reich, Volker Remmert, Norbert Schappacher, Karl-Heinz Schlote, Julius Schoeps, Rotraut Stanik, Karl Stein, Irmgard Süss, Peter Thullen, Horst Tietz, Magda Tisza, Renate Tobies, Heinrich Wefelscheid, K. H. Weise, Hermann Weisert, Klaus-Werner Wiegmann, and Hans Zassenhaus. This list may not be complete, and I apologize to anyone inadvertently overlooked, and assure them of my thankfulness for their help.

I wish to thank Beata Smarcynzska for a detailed survey of the contents of four German mathematical journals during the years 1933–45. A summary of this survey is in the appendix. I also wish to thank Elizaveta Pachepsky for help in composing the bibliography, Sherry Wert for her gratefully received efforts in improving the text, and Joan Robinson for her help with the index.

I am especially grateful to the typists who prepared various parts of this book: Roberta Colon, Frances Crawford, Ann Joy, and Marie Parsons.

I wish to thank preliminary readers of all or substantial parts of this book, and their suggestions: Bonnie Gold, Jonathan Liebenau, Adam Segal-Isaacson, Klaus-Werner Wiegmann, and the anonymous readers for Princeton University Press. Needless to say, however, any error of substance or diction still remaining is entirely the author's responsibility.

Finally, I wish to acknowledge the support of my wife Rima through the many years of this book's gestation. Though lively like Beatrice, she has nevertheless endured not only a husband with a beard, but also the effort he has put into this book.

ABBREVIATIONS

AVA	*Aerodynamische Versuchsanstalt* (Aerodynamic Experimental Station)
BAK	Bundesarchiv Koblenz
BDC	Berlin Document Center
BI	Interview of Ludwig Bieberbach by Herbert Mehrtens, Sept. 21, 1981
BL	Ludwig Bieberbach correspondence, Niedersächsische Staats- und Universitätsbibliothek, Göttingen
DDR	East Germany
DFG	*Deutsche Forschungsgemeinschaft* (Association for German Research)
DM	*Deutsche Mathematik* (journal)
DMV	*Deutsche Mathematiker Vereingung* (German Mathematical Society)
DNVP	*Deutschnationale Volkspartei*
GAMM	*Gesellschaft für Angewandte Mathematik und Mechanik* (Society for Applied Mathematics and Mechanics)
GS	Papers of Gabor Szegö, Stanford University Archives
HDT	Helmut Hasse's letters to Harold Davenport, in the archives at Trinity College, Cambridge, Davenport folder
HK	Personal papers of Hellmuth Kneser, in the possession of his son, Martin Kneser
HNMA	Erich Hecke *Nachlass*, *Mathematische Annalen*; in the possession of the Petersson family
JDMV	*Jahresbericht der Deutschen Mathematiker Vereinigung* (*Journal of the German Mathematical Society*)
MI	The papers of the Göttingen Mathematical Institute
MR	*Mathematische Reichsverband*
NMT	Nuremburg Military Tribunal
NSDAP	*National Sozialistische Deutschen Arbeiter Partei*
NSLB	*National Sozialistisches Lehrerbund* (Union of National Socialist Teachers)
NTM Schriftenreihe	*NTM Schriftenreihe für Geschichte der Naturwissenschaften, Technik, und Medizin*
PAB	Personalakten Blaschke, Hamburg Staatsarchiv
PAS	Personalakten Wilhelm Süss, Universität Freiburg

Poggendorf	J. C. Poggendorf, *Biographisch-Literarisches Handwörterbuch der Exakten Naturwissenschaften.* Sächsischen Akademie der Wissenschaften zu Leipzig. Margot Köstler, editor-in-chief. 7 vols. to date. Berlin: Akademie Verlag.
RFR	*Reichsforschungrat* (National Council on Research)
RLM	*Reichsluftfahrtministerium* (air ministry)
RM	*Reichsmark* (unit of currency)
RSHA	*Reichssicherheitshauptamt* (National Security Office)
SA	*Sturmabteilung* (stormtroopers)
SD	*Sicherheitsdienst* (Security Service)
SS	*Schutzstaffel* (elite guard; separated from the SA after June 30, 1934)
UAG	Archiv der Universität Göttingen
VP	Oswald Veblen papers, Library of Congress
WM	E. A. Weiss, "Wozu Mathematik," pamphlet (Bonn, 1933)
WVHA	*Wirtschaftsverwaltungshauptamt* (office for economic administration)
ZAMM	*Zeitschrift für Angewandte Mathematik und Mechanik* (Journal of Applied Mathematics and Mechanics)
ZN	Ernst Zermelo *Nachlass*, Universität Freiburg
ZWB	*Zentrale für wissenschaftliches Berichtwesen* (Air Minister's Central Office for Scientific Reporting)

Mathematicians under the Nazis

CHAPTER ONE

Why Mathematics?

MATHEMATICS under the Nazi regime in Germany? This seems at first glance a matter of no real interest. What could the abstract language of science have to say to the ideology that oppressed Germany and pillaged Europe for twelve long years? At most, perhaps, unseemly (or seemly) anecdotes about who behaved badly (or well) might be offered. While such biographical material, when properly evaluated to sift out gossip and rumor, is of interest—history is made by human beings, and their actions affect others and signify attitudes—there is much more to mathematics and how it was affected under Nazi rule. Indeed, there are several areas of interaction between promulgated Nazi attitudes and the life and work of mathematicians. Thus this book is an attempt at a particular investigation of the relationships between so-called pure (natural) science and the extra-scientific culture. That there should be strong cultural connections between the technological applications of pure science (including herein the social applications of biological theory) and various aspects of the Industrial Revolution is obvious. Social Darwinism, and similar influences of science on social thought and action, have been frequently studied. It is not at all clear at the outset, however, that theoretical science and the contemporary cultural ambience have much to do with one another. Belief in this nonconnection is strengthened by the image of science proceeding *in vacuo*, so to speak, according to its own stringent rules of logic: the scientific method. In the past thirty years, however, this naive assumption of the autonomy of scientific development has begun to be critically examined.[1]

A general investigation of this topic is impossible, even if the conclusion were indeed the total divorce of theoretical science from other aspects of culture. Hence the proposal to study one particular microcosm: the relationship between mathematics and the intensity of the Nazi *Weltanschauung* (or "worldview") in Germany. Although 1939 is a convenient dividing line in the history of Hitler's Reich, nonetheless the prewar Nazi period must also be viewed as a culmination; the Germany of those years was prepared during the Weimar Republic, and both the cultural and scientific problems that will concern us have their origins at the turn of the century. World War I symbolized the conclusion of an era whose end had already come. Similarly, World War II was a continuation of what had gone before, and a terminal date of 1939 is even more artificial and will not be adhered to.

[1] One of the earliest examples is Paul Forman, "Weimar Culture, Causality, and Quantum Theory, 1918–1927," *Historical Studies in the Physical Sciences* 3 (1971): 1–16; and by the same author, "Scientific Internationalism and the Weimar Physicists: The Ideology and Its Manipulation in Germany after World War I," *Isis* 64 (1973): 150–180.

The concentration on mathematics may perhaps need some justification. At first glance, a straw man has been set up—after all, what could be more culture-free than mathematics, with its strict logic, its axiomatic procedures, and its guarantee that a true theorem is forever true. Disputes might arise about the validity of a theorem in certain situations: whether all the hypotheses had been explicitly stated; whether in fact the logical chain purporting to lead to a certain conclusion did in fact do so; and similar technical matters; but the notion of mathematical truth is often taken as synonymous with eternal truth. Nor is this only a contemporary notion, as the well-known apocryphal incident involving Euler and Diderot at the court of Catherine the Great, or the Platonic attitude toward mathematics, indicate.[2] Furthermore, there is the "unreasonable effectiveness" of mathematics in its application to the physical and social scientific world. Even so-called applied mathematics, concerning which Carl Runge[3] remarked that it was merely pure mathematics applied to astronomy, physics, chemistry, biology, and the like, proceeds by abstracting what is hypothesized as essential in a problem, solving a corresponding mathematical problem, and reinterpreting the mathematical results in an "applied" fashion.[4]

Mathematics also has a notion of strict causality: if *A*, then *B*. It is true that the standards of rigor, the logical criteria used to determine whether or not a proof is valid, that is, to determine whether or not *B* truly follows from *A*, have changed over time; nevertheless, the notion that it is conceivable that *B* can be shown always to follow from *A* is central to mathematics. As the prominent American mathematician E. H. Moore remarked, "Sufficient unto the day is the rigor thereof."[5] Both the necessary process of abstraction and the idea of mathematical causality separate mathematics from more mundane areas. Somewhat paradoxically, perhaps, they are also partly responsible for the great power of mathematics in application. Mathematical abstraction and mathematical causality seem to elevate mathematics above the sphere of the larger culture.

[2] Diderot is supposed to have challenged Euler to prove the existence of God, and Euler to have replied, "Monsieur! $(a + b^n)/n$, donc Dieu existe; répondez!" The mathematical expression attributed to Euler is, of course, nonsense. For the mythical aspect of the story, see R. J. Gillings, "The So-Called Euler-Diderot Incident," *American Mathematical Monthly* (1954): 77–80.

[3] Carl Runge was a professor at Göttingen, and a leading "applied mathematician" of the first part of the century. The remark is attributed to Runge by Heinrich Tietze in *Famous Problems of Mathematics* (New York: Graylock, 1963).

[4] For a discussion of the role of mathematics in application, see Eugene P. Wigner, "The Unreasonable Effectiveness of Mathematics in the Natural Sciences," *Communications in Pure and Applied Mathematics* 13 (1960): 1–14. Wigner writes: "The mathematical formulation of the physicists' often crude experience leads in an uncanny number of cases to an amazingly accurate description of a large class of phenomena. This shows . . . that it is in a very real sense the correct language . . . , fundamentally, we do not know why our theories work so well, hence their accuracy may not prove their truth and consistency. . . . The miracle of the appropriateness of the language of mathematics for the formulation of the laws of physics is a wonderful gift which we neither understand nor deserve. We should be grateful for it" (pp. 8, 14).

[5] Quoted by Marvin Minsky in a lecture on computer science given at the annual meeting of the American Mathematical Society in New Orleans, Louisiana, 1965; also cited by E. T. Bell, *The Development of Mathematics* (1940), 503.

Twin popular illusions incorrectly elaborate upon this view and make mathematics seem even more remote from the general culture. The first of these is that the doing of mathematics is only a matter of calculation, or, more sophisticatedly, of logical step-by-step progress from one eternal truth to another via intermediate truths. This view is enhanced by the way mathematicians publicly present the results of their investigations: exactly as such logical progressions. In fact, however, the discovery of mathematics, as opposed to the presentation of it, is more like the reconnoitering of some unknown land. Various probes in various directions each contribute to the forming of a network of logical connections, often even unconsciously.[6] The realization of this network, the a posteriori checking for logical flaws, and the orderly presentation of the results, do not reflect the process of mathematical creativity, whatever that may be, and however ill it is understood.

The second illusion is that all that counts for a mathematician is to distinguish the correct from the incorrect. Correctness is indeed the sine qua non of mathematics, but aesthetic considerations are of great importance.

Among the various aesthetic factors influencing mathematical activity are economy of presentation, and the logic (inevitability) of often unexpected conclusions. While correctness is indeed *the* mathematical essential, some correct proofs are preferable to others. Proofs should be as clear and transparent as possible (to those cognizant of the prerequisite knowledge). A good notation, a good arrangement of the steps in a proof, are essential, not only to aid the desired clarity, but also because, by indicating fundamentals in the problem area, they actually incline toward new results. Clarity, arrangement, and logical progression of thought leading to an unexpected conclusion are well illustrated in an incident concerning no less a personage than the philosopher Thomas Hobbes:[7]

> He was 40 yeares old before he looked on Geometry; which happened accidentally. Being in a Gentleman's Library, Euclid's Elements lay open, and 'twas the 47 *El. libri* I.[8] He read the Proposition. By G——, sayd he (he would now and then sweare an emphaticall Oath by way of emphasis) *this is impossible*! So he reads the Demonstration of it, which referred him back to such a Proposition; which proposition he read. That referred him back to another, which he also read. *Et sic deinceps* [and so on] that at last he was demonstratively convinced of that truth. This made him in love with Geometry.

[6] Several personal examples can be found in J. E. Littlewood, *A Mathematician's Miscellany* (1953); reissued by Cambridge University Press (Cambridge, 1986), with other added material by Littlewood and a foreword (ed. Béla Bollobás) as *Littlewood's Miscellany*. See also Jacques Hadamard, *The Psychology of Invention in the Mathematical Field* (1949).

[7] John Aubrey, *Brief Lives*, Thomas Hobbes. The citation is from the edition by Oliver Lawson Dick, *Aubrey's Brief Lives* (London: Secker and Warburg, 1950), 150. Hobbes and Aubrey (who was nearly forty years younger) were friends, and Aubrey's "life" of the philosopher is the most extensive of those he wrote. For the friendship, see Dick, ibid.: xc–xci.

[8] The "Pythagorean Theorem."

A simple example of an "unbeautiful truth" is a list of positive integers. Mathematics is not frozen in time like a Grecian urn; solutions of old problems lead to new considerations. Though truth may not necessarily be beauty, beauty is truth, and for the mathematician impels to its own communications. As Helmut Hasse (who will be met again) remarks:[9]

> Sometimes it happens in physics again and again, that after the discovery of a new phenomenon, a theory fitted out with all the criteria of beauty must be replaced by a quite ugly one. Luckily, in most cases, the course of further development indeed reveals that this ugly theory was only provisional. . . .
>
> In mathematics this idea leads in many instances to the truth. One has an unsolved problem, and, at first, has no insight at all how the solution should go, even less, how one might find it. Then the thought comes to describe for oneself what the sought-for truth must look like were it beautiful. And see, first examples show that it really seems to look that way, and then one is successful in confirming the correctness of what was envisaged by a general proof. . . . In general we find a [mathematical] formulation all the more beautiful, the clearer, more lucid, and more precise it is.

As Hasse puts it elsewhere, truth is necessary, but not sufficient for real (*echt*) mathematics—what is also needed is beautiful form and organic harmony.

One result of this aesthetic is that the mathematician thinks of himself as an artist, as G. H. Hardy did:[10]

> The case for my life, then, or that of anyone else who has been a mathematician in the same sense in which I have been one, is this: that I have added something to knowledge, and helped others to add more; and that these somethings have a value which differs in degree only, and not in kind, from that of the creations of the great mathematicians, or of any of the other artists, great or small, who have left some kind of memorial behind them.

Or, as Hasse says even more forcefully, "The true mathematician who has found something beautiful, senses in it the irresistible pressure to communicate his discovery to others."[11]

Mathematics is the "basic science" sine qua non. At the same time, it is quite different from basic experimental science by being divorced from laboratory procedures. Even so-called applied mathematics only takes place on paper with pencil.[12] The hallmark of mathematics is logical rigor. However important or

[9] Hasse, *Mathematik als Wissenschaft, Kunst, und Macht* (1952), 18–20. Hasse also makes many analogies between mathematics and music in particular. This theme is also discussed in R. C. Archibald, "Mathematicians and Music," *American Mathematics Monthly* 31 (1924): 1–25. See also note 16 below.

[10] G. H. Hardy, *A Mathematician's Apology* (Cambridge: Cambridge University Press, 1969), 151 (original edition, 1940).

[11] Hasse 1952: 26.

[12] A laboratory may provide the idea for a piece of mathematics, but the actual doing of the mathematics is not with machinery. Computing machines are generally used to provide suggestive "experiments," a posteriori verifications, supplementary data extending a proof to a previously untreated finite range, or counterexamples by extensive numerical search. Recent "computer proofs" in mathematics, such as of the famous "four-color problem," or the existence of a projective plane of

suggestive or helpful heuristic or analogical arguments may be, it is only the mathematical proof according to accepted standards of logical rigor that establishes a mathematical result. Those logical standards may be and are disputed (and were in Nazi Germany), but given an accepted set of such standards, mathematical proofs according to them establish mathematical results that are true without qualification. On the one hand, a mathematical result is "sure"; on the other, however, all but the final results with proofs are, at best, incomplete mathematics: the mathematician's "experiments" are usually eminently unpublishable as such. This removal of mathematics from the concrete world contributes to the mathematical aesthetic. While there are notions of a "beautiful" experiment in the experimental sciences, in mathematics the aesthetic is purer for its removal from the natural irregularities of concrete life. "As for music, it is audible mathematics," writes the biologist Bentley Glass,[13] and perhaps the traditional[14] musical aesthetic is the one most closely resembling the mathematical; here, too, given the underlying assumptions, there is a purity of form that is part of the notion of beautiful. Deviations like Mozart's *Musikalischer Spass* or some of the less slapstick efforts of P.D.Q. Bach (Peter Schickele) are jokes because of their introduction of irregularities into a presumed form. Similarly, Littlewood presents as humorous an unnecessarily cumbrous presentation of a proof that can be expressed quite clearly and elegantly.[15] The papers of Hasse and Archibald cited earlier also stress the analogy between the musical and the mathematical aesthetic.[16]

In some sense, then, mathematics is an ideal subject matter; it is, however, made real by the actions of mathematicians. In Russell's well-known words:[17]

> Mathematics possesses not only truth, but supreme beauty—a beauty cold and austere, like that of sculpture, without appeal to any part of our weaker nature, sublimely pure, and capable of a stern perfection such as only the greatest art can show.

Nevertheless, mathematicians make tremendous emotional investments in the doing of mathematics. Mathematicians, despite their pure aesthetic, the divorce of their actual work from concrete reality, and the surety of their results, are not like petty gods in ivory towers playing at abstruse and difficult, but meaning-

---

order 10, do not alter this statement. In such problems, many sets of cases (to which mathematical theory had reduced the problem) needed to be verified, the number being well beyond human capacity. The theory of computing and computing machines, including the design of algorithms, is, in contrast, part of mathematics. The division into "pure" and "applied" mathematics, or into "theoretical" and "experimental" in physics, is a comparatively recent phenomenon.

[13] Bentley Glass, "Liberal Education in a Scientific Age," in Paul Obler and Herman Estrin, eds., *The New Scientist: Essays on the Methods and Values of Modern Science* (1962), 215–238, 233.

[14] Aleatoric music and computer-generated music are not considered in this sentence. On the other hand, systems like Schönberg's *Tonreihe* quite clearly are.

[15] Littlewood 1986: 49–52.

[16] However, the widespread belief in the frequent conjunction of mathematical and musical *ability* seems at best a dubious proposition. See G. Revesz, "Die Beziehung zwischen mathematischer und musikalischer Begabung," *Schweizerische Zeitschrift für Psychologie, und Ihre Anwendungen* 5 (1946): 269–281.

[17] Bertrand Russell, "The Study of Mathematics," *New Quarterly* (Nov. 1907), reprinted in Bertrand Russell, *Philosophical Essays* (London: Longmans Green, 1910), 71–86, quote from p. 73.

less, games. The final piece of mathematics is abstract, aesthetically beautiful, and certain; but it is not (nor could it be) an instantaneous or automatic creation. The *doing* of mathematics is as emotionally involved, often clumsy, and uncertain as any other work that has not been reduced to a purely automatic procedure.[18]

Thus, the nature of mathematical abstraction and mathematical causality, coupled with the popular ignorance of the nature of mathematical research and the removal of mathematics from the everyday world, seem to make mathematics one of the least likely subjects for the sort of investigation proposed. Yet some Nazi mathematicians and psychologists stood this reasoning on its head. At the same time, they emphasized with a peculiarly Nazi bias the often neglected roles of aesthetics and inspiration in creating mathematics. They argued that exactly the apparent culture-free nature of mathematical abstraction and mathematical causality makes mathematics the ideal testing-ground for theories about racially determined differences in intellectual attitudes. As E. R. Jaensch wrote in 1939:[19]

> Mathematics can simply have no other origin than rational thinking and mental activity (*Verstandestätigkeit*). "Irrational" mathematics would be a wooden iron, a self-contradiction. If, therefore, one discovers something worth exposure about the ways of thought (*Verstandeskräfte*) that still command the field on this area—and that happens in many respects with complete justice—so one can hereby only obtain help by bringing other forms of rational thought in more strongly—in no case however, through the conjuring up of irrationality. This way is simply excluded in mathematical thought. Even if in other scientific and educational disciplines it is possible artificially still to maintain the appearance that Reason (*Verstand*), as treated through that radical cure, still lives—in Mathematics it is impossible.
>
> Hereby, the question of mathematical thought attains the character of an especially instructive example—an "illuminating case" in Baco's [sic] sense—for the forms of logical thought and rationalism above all, but also in other areas of knowledge and in everyday life.

What is important to note here is the insistence that the supposed autonomy of mathematics from irrational influence makes it exactly appropriate for investigating various intellectual types. Just because of the rationality of its results, mathematics was deemed an excellent medium for perceiving the various important differences between different peoples' ways of thought. It did not prove difficult to discover, for example, a Nordic type, a Romance or Latin type, a Jewish type, and, in fact, several subvarieties of these. Jaensch's theory of types could be elaborated independent of or in conjunction with *Rassenseelenkunde*, or the theory of the "racial soul." This was done most prominently by the distinguished mathematician Ludwig Bieberbach, who will be discussed particularly in chapters 6 and 7. By delineating a "Nordic" mathematics distinct from

[18] Indeed, when automatic procedures or algorithms are well known in mathematics, they are simply cited.

[19] E. R. Jaensch and F. Althoff, *Mathematisches Denken und Seelenform* (1939), vii–ix.

French or Jewish mathematics, great emphasis could be placed (necessarily) on the mode of intellectual discovery as opposed to its fruits, and, therefore, on feeling and attitude toward the world.

However important this inversion of the usual attitude toward mathematics may be for investigation, there are at least two other reasons arguing for a study of mathematics in the Nazi period. The first is that among the substantial number of mathematicians who were sympathetic in varying degrees to the Nazi cause were several who attempted to associate the political argument with various philosophical differences within mathematics. This did not alter the truth of any mathematical fact, but it did declare that certain mathematical disciplines were "more equal" than other varieties. Nor was this simply a question of "pure" versus "applied," of theory versus immediately usable results. Both these beliefs and the ones about the salience of psycho-racial differences within mathematics also argued for the distinction in differences of pedagogical style. Put succinctly, a Nazi argument promoted by Bieberbach was that because Jews thought differently, and were "suited" to do mathematics in a different fashion, they could not be proper instructors of non-Jews. Indeed, their presence in the classroom caused a perversion of instruction. Thus an elaborate intellectual rationale for the dismissal of Jews was established, discussed, and defended.

In addition to these psychological, philosophical, and pedagogical arguments that, however seemingly perverse today, reveal that mathematics, at least in its "doing," if not perhaps in its "being," may be less culture-free than one thinks at first, there is yet another facet of mathematics in the Nazi period that deserves investigation. With the advent of Hitler, irrationalist themes in German thinking achieved political respectability—indeed, became the order of the day. Thus the historian Walter Frank could say:[20]

> Let us clearly understand one another, the intellectual is the exact opposite of the spiritually creative (*geistig Schaffenden*). The creator produces values. The intellectual defines the values produced by others. The intellectual is the clever man, the educated one, but also the one without character or personality. The greatest enemy of the creator (*Schöpfers*) is not the primitive man. For his instinct can now and then more easily comprehend great things than all the cleverness of the clever. The greatest enemy of creativity (*Schöpfung*) is always the clever man.

Similarly, Otto Dietrich in 1935 could remark:[21]

> National Socialism does not tend to dry abstract thinking. Its *Weltanschauung*, tied to the *Volk*, will open up once more learning (*Wissenschaft*) to flowing Life, and the infinite fullness of life to learning.

[20] Walter Frank, *Kämpfende Wissenschaft* (Hamburg: Hanseatische Verlagsanstalt, 1934), 30, as cited by Leon Poliakov and Josef Wulf, *Das Dritte Reich und Seine Denker* (1959), 51.

[21] Otto Dietrich was Hitler's press chief, latterly turned philosopher. The citation is from *Die philosophischen Grundlagen des Nationalsozialismus* (Breslau: Ferdinand Hirt, 1935), 38, as cited by Poliakov and Wulf 1959: 278. According to Poliakov and Wulf (276), Alfred Rosenberg complained about Dietrich and his ilk: "Where were the new philosophers ten and fifteen years ago?"

And Edgar von Schmidt-Pauli in 1932,[22] explaining the attraction as he saw it of the party he believed in:

> National Socialism corresponded to the spiritual (*seelischen*) position of the broadest layers of the German people, which, in the garden of errors [*Irrgarten*, usually translated "maze"] of the rationalism of the postwar years, yearned instinctively for powerful leadership.

In such an atmosphere, it was reasonable to fear that the common view of mathematics as the rational subject sine qua non might jeopardize its public standing, its role in the schools, the state funds it received. Not only irrationalists attacked mathematics in the Nazi period, but also some "rational" physicists of Nazi persuasion. Thus one finds Nazi sympathizers among mathematicians defending mathematics from the attitudes of some of their political brothers-in-arms, like Phillip Lenard. Lenard, a Nobel laureate physicist and early supporter of Hitler, argued in his *Deutsche Physik* (1936) that it was important for students to avoid studying too much mathematics in school! In his view, mathematics was the "most subordinate intellectual discipline" because, under Jewish influence, it had lost its "feeling for natural scientific research."[23]

Even conceding to these varied Nazi thinkers that the supposed autonomy of mathematics from the rest of culture is precisely what makes it the ideal medium for investigating the factual content of other a priori truths, it does not necessarily follow that an investigation of the sort proposed is called for. It is all too tempting to look back on the Nazi period in Germany and dismiss all the consequences of its point of view as detestable and patent falsehoods when considered rationally. The terrors of the Nazi regime make it even easier to reject all aspects of Nazi thought as self-serving propaganda, and any possible relationships to mathematics or mathematicians the result of compulsion or time-serving. Such a point of view overlooks the fact that the Nazi philosophy was an all-embracing worldview with its full complement of intellectual theorizing. The Nazi emphasis on rearmament would necessarily lead to an emphasis on the military uses of science, and a concomitant disparagement of science, whose immediate applications were not easily seen, might be expected. This point is only enhanced by a remark of Hans Heilbronn, "The application of mathematics to military problems was neglected in Hitler's Germany, certainly by comparison with England and the U.S. And in some cases the armament industry in the widest possible sense provided a refuge for anti-Nazi mathematicians who had been expelled from the universities, and could not emigrate."[24]

[22] Edgar von Schmidt-Pauli, *Die Männer um Hitler* (Berlin: Verlag für Kulturpolitik, 1932), 27.

[23] Philip Lenard, *Deutsche Physik* (1936), 1:6.

[24] Heilbronn, in a personal letter to Jonathan Liebenau (Oct. 10, 1974) in my possession. Similarly, Wilhelm Magnus (for whom there were no bars to a university position) told me of his own "retreat" into industrial work in 1936 in order to avoid the politics of the university (interview, 1982).) However, in 1940 Magnus became professor at the Technische Hochschule in Berlin, presumably to avoid the war industry. Perhaps the best-known case of such "industrial emigration" is Gustav Hertz. The nephew of Heinrich Hertz, his father was Jewish. Having shared the Nobel Prize

What is more interesting, as the journal *Deutsche Mathematik* (1936–1943) and articles by mathematicians of the stature of Ludwig Bieberbach and Wilhelm Blaschke make clear, is an attempt to distinguish within the mathematical community between mathematics that was "rooted in the national people" (*völkisch verwurzelt*) and that which was not. In other words, "Deutsche Mathematik" was not simply a question of expelling Jewish professors from the university or deemphasizing the roles of non-German nationalities in the creation of mathematics; it was the quite serious matter of discerning what was a typically "Nordic" mathematics suitable for and to the new German state and its aspirations. This task also involved the historical problem of finding common Nordic elements in the great German non-Jewish mathematicians from Kepler to Hilbert that could be shown as lacking among the Jews as well as the French.[25]

Given that something so seemingly value-free as mathematics can be imbued with ideological content, and that the Nazi period provides a prime example of such attempts, as well as the several different Nazi streams of thought affecting mathematics, it becomes interesting to see just what this "cultural conditioning" meant for mathematics, and what its effects were, if any. The positions taken up by propagandists for "Nordic Mathematics" cannot be dismissed as simple flattery of the powers regnant.[26] What, if anything, about Germany and mathematics provoked these German mathematicians to their opinions? Such problems have perhaps disturbing echoes today, when some academics talk about "the political structure of mathematics," when there is a confusion between the intrinsic nature of a discipline and the behavior of its practitioners toward outsiders, when some philosophers of science espouse a radical relativism that rejects all truth claims. Furthermore, as will be seen, reading the pedagogical concerns of German university mathematicians in the late 1920s and 1930s—concerns that antedated but were exacerbated by the Nazis—sometimes makes the similar concerns of some contemporary American mathematicians seem like "déjà vu all over again."

In addition, it is clear that the National Socialist typology cannot be dismissed out of hand as mere typology and so unworthy of further consideration. Psychological typology was popular in many circles between the wars, as well as earlier. A case in point is the American psychologist A. A. Roback, author of *The Psychology of Character* (1927) and many other works dealing with characterol-

---

in physics with his countryman James Franck (who emigrated in 1933), he lost his professorship at Berlin in 1935. Between then and 1945, Hertz was a technical director for Siemens, from 1945 to 1954 was in the Soviet Union, in 1954 became a professor in Leipzig, and received the National Prize (First Class) of the DDR (East Germany) in the following year.

[25] The English, of course, were Aryans. Hitler at first saw them as a natural ally, and in the prewar years Maxwell and Newton were included in the pantheon of Aryan science. For example, B. Thüring, in *Deutsche Mathematik* 1 (1936): 10, writes of the "North-German feeling for Nature at the basis of Kepler's and Newton's work" in contrast to Einstein's. Newton is also characterized as a "German researcher" by Thüring.

[26] To be a Nazi party member and a scientist did not make it necessary to propagandize a Nazified science. The famous physicist Pascual Jordan became a member of the Nazi party in 1933, but opposed Ludwig Bieberbach's "Deutsche Mathematik" ideas; see below, chapter 7.

ogy (including a 1927 bibliography of over 3,200 items). A story in *Time* magazine in 1935 reported on his attempts to distinguish Jewish students (positively in this case) by their writing style.[27] Roback clearly did not realize that his methodology, rather than the use others made of similar observations, might be at fault. Indeed, his report shows his obvious pro-Jewish bias. Ironically, the students examined by Roback were in a class taught by his colleague Gordon Allport (the future author in 1954 of *The Nature of Prejudice*). It is interesting that an anti-assimilationist Jewish researcher and scholar could be dedicated to typology that distinguished a Jewish type (defended by simplistic and erroneous statistics) as a mode of investigation without considering it insidious as late as 1935. *Time* certainly thought so.

From a methodological point of view, history that does not deal in biography must necessarily conceive of varieties of abstract individuals as its enactors. However much one talks about forces and movements, nevertheless, whatever motivates history, it is *made* by people.[28] It is nearly banal to observe that categories of individuals are constructed by abstraction of commonly held views; that a complex of views is associated with a category; and that no single individual placed in that category necessarily holds all the views associated with it, though such a person will presumably hold the defining ones. In addition to such attitudinal categories, there are all the natural categories of profession, nationality, socioeconomic status, race, religion, and so forth, into which people place themselves during their lives. If groups of people have a common heritage, or common activity, or common upbringing, one tends to look for other aspects of that commonality that may not be visible superficially. This makes it necessary and valid to sometimes speak of Germans, of scientists, and, as a subgroup of the latter, of mathematicians. It is not the intention to speak of a fixed national character. Henry V. Dicks, in a preliminary psychological study of Nazi ideology, remarks:[29]

> When speaking of "national character" we mean only the broad frequently recurring regularities of certain prominent behaviour traits and motivations of a given ethnic or cultural group. We do not assert that such traits are found in equal degree, or at all, in all members of that group, or that they are so conjoined that the extreme is also the norm. Neither do we assert that the traits are found singly or in combination in that group alone.

[27] *Time* magazine (Sept. 30, 1935): 35. Roback's original report is in *Character and Personality* 4 (1935–36): 53–60.

[28] So, for example, Eugen Weber remarks in an article on Romanian fascism: "All of which suggests that the major factors in a radical or revolutionary orientation are less strictly sociological than psychological: those cultural and, above all, chronological factors which make for greater availability, greater restlessness, greater receptivity, at least, to possibilities of change and of action to secure change." "The Men of the Archangel," *Journal of Contemporary History* 1 (1966): 101–126, p. 120.

[29] Dicks, "Personality Traits and National Socialist Ideology," *Human Relations* 3 (1950): 112. See also by the same author *Licensed Mass Murder* (Heinemann: Sussex University Press, and London: Educational Books, 1972).

Are there sociopsychological factors linking the professional activity of certain mathematicians, among them very fine ones, and their adherence to the National Socialist cause? Given the tremendous force and ubiquity of the Nazi *Weltanschauung*, there is the additional aspect of the potential relationship, if any, between the general forms and interests of mathematical scholarship during the Nazi period and this worldview. Indeed, one highly respected strain of the German intellectual tradition "conceived universal history as the progressive differentiation of the peculiar character of each nation."[30] When, under the Nazis, that "peculiar character" became the highest of values, the pressures for it to pervade all intellectual activity were monstrously increased.

One relevant question thus may be: how do mathematicians think? This question seems particularly interesting, since mathematicians are viewed by the lay public as an esoteric lot, the Brahmins of a largely unintelligible science. The curiosity about inhabitants of this purported *sanctum sanctorum* of mathematics is reflected popularly in such items as a rather negative article in *New Yorker* magazine replete with unsubstantiated assertions,[31] or the pejorative and banal view of the mathematician as a disembodied intellect who is "not of this world" until stirred to enter it. A far less trivial example within German culture of the mathematician as disembodied intellect, but construed positively this time, is in Robert Musil's monumental *The Man without Qualities*.[32] Here it is exactly the disembodiment that permits a dispassionate view of society. In any case, one way the question can be put is: how real is this disembodiment?

Musil, at one point in his career, was an applied mathematician; the cited pamphlet by Hasse, as well as books by mathematicians such as Henri Poincaré, G. H. Hardy, and Jacques Hadamard, attest to the mathematicians' own interest in "what makes a mathematician," and "how mathematicians think." Hadamard's book in particular, while addressing itself to the general problem of creativity, treats it by soliciting mathematicians' own introspective opinions of how they think.[33] Nevertheless, his book is not very illuminating on this issue, except to emphasize that mathematical thought does not take place like the step-by-step procedures of an automaton, that many mathematicians often think initially in vague images rather than in symbols, and that the thought processes are not clear to the mathematician doing the thinking. Many mathematicians have testified to this only vague knowledge of their thinking processes. Yet, one striking fact is that the *results*, when successful, are remembered until they can be written down *in extenso* with the usual mathematical formul-

[30] Hans Kohn, "Rethinking Recent German History," in idem, ed., *German History: Some New German Views* (London: Allen and Unwin, 1954), 29, in discussing Ranke.

[31] Alfred Adler, "Reflections: Mathematics and Creativity," *New Yorker* (Feb. 19, 1972): 39–45.

[32] Robert Musil, *The Man without Qualities*, trans. E. Wilkins and E. Kaiser (New York: Coward-McCann, 1953–60), orig. ed. *Der Mann Ohne Eigenschaften*, published posthumously (Hamburg: Rowohlt, 1952). Musil was an Austrian, and the novel is about Vienna, but it seems fair to adduce it as a representative of German (-speaking) culture.

[33] Hadamard 1949: 2.

ary, and that the occasion of the insightful event can be recalled after many years.

Thomas Kuhn's words;[34] "I hope to have made meaningful the view that the productive scientist must be a traditionalist who enjoys playing intricate games by pre-established rules in order to be a successful innovator who discovers new rules and new pieces with which to play them," would seem to apply with special force to mathematicians. Almost as an afterthought, Kuhn adds the caveat that perhaps what he has been saying only applied to "basic science," and[35] "the personality requisites of the pure scientist and of the inventor may be quite different with those of the applied scientist lying somewhat between."

Mathematicians as practitioners of "basic science" certainly use the metaphors of play as well as aesthetic ones in describing their work. Even if how mathematical innovation is accomplished is no clearer to mathematicians than to others, the occasion of that innovation is clearly remembered despite the vagueness of the intellectual processes. For one example, J. E. Littlewood writes (forty years after the event) of a famous result of his:[36]

> The problem seethed violently in my mind . . . and the "idea" was vague and elusive. Finally I stopped, in the rain, gazing blankly for minutes on end over a little bridge into a stream (near Kenwith Wood) and presently a flooding certainty came into my mind that the thing was done. The 40 minutes before I got back and could verify were none the less tense.

It seems, therefore, that attempts at delineating specific separate compartments in a mathematician's mind are hopeless.

Nevertheless, no discipline is ever entirely separate from its practitioners. Although mathematics was not an obvious academic "pressure point" for the Nazis, as disciplines as varied as biology, anthropology, history, German literature, and architecture were, a large number of prominent mathematicians were at least Nazi "fellow-travelers," or even open propagandists for the Nazi cause.

In an unpublished paper,[37] Thomas Reissinger has raised the question, "How is it that mathematicians [as a group] did not see through [the Nazi ideology and practice] at least a little bit better than their nonmathematically trained fellow men?" Though mathematicians are, as a group, as human and fallible as others when it comes to cultural, political, or ideological matters, this question does, nevertheless, have meaning. Reissinger draws an answer from Karl Popper's philosophy of science. In the physical sciences, observes Reissinger, one seeks "general laws," which then serve to explain particular occurrences by examination of "boundary conditions." In mathematics, the emphasis on general laws (perhaps subject to side conditions) is even more intense. The mathemati-

[34] Thomas S. Kuhn, *The Essential Tension* (1977), 237. This is a collection of essays. The cited essay is "The Essential Tension: Tradition and Innovation in Scientific Research," 225–239.

[35] Ibid.: 238–239.

[36] Littlewood 1986: 83.

[37] Thomas Reissinger, "Die Verführbarkeit der Mathematiker" (The seducibility of mathematicians), University of Mannheim, preprint no. 120 (n.d.).

cian, above all, seeks to prove conjectures through a deductive chain of reasoning, and logical and analytical training is directed to that end. History, on the other hand, is concerned with singular events, and concentrates on nontrivial boundary conditions. Thus, mathematical training, however it prepares the faculties for analysis, is not only of no aid in judging historical/political situations, it perhaps inclines toward misjudgment. Furthermore, intellect has no necessary connection with the ability to reason. This, certainly, is banal, but Reissinger argues further that the ability to reason about ideas depends upon free exchange with others leading to critical examination. The solipsistic aspect of mathematical training and practice does not, however, favor such uses of reason.

The reader may or may not agree with these ideas of Reissinger. What is interesting about them is their explanation of a lability of many mathematicians, opening them to uncritical acceptance of political slogans and ideological posturing, while eschewing any historicist notion of hidden social forces upon which historical events are merely epiphenomena. This is not to deny that, as in every discipline, one finds every sort of reaction to the Nazis among German mathematicians. Nor is it to deny that the majority of German mathematicians who remained in Germany were not active Nazi sympathizers, but simply attempted to do their best to uphold their discipline in a difficult time. But there were also many mathematicians, both young and old, who were opportunists and took advantage, either of necessity or shamelessly, of the situation, either to attack and destroy opponents, or just to advance themselves. Among mathematicians there were, as already mentioned, *völkisch* ideologists, as well as conservative nationalists who were too "unpolitical" to understand the difference between the Third Reich and a conservative or even monarchist government. There were prominent mathematicians who partook in the mystical appreciation of Hitler as German savior, and those who wrote letters of denunciation. All these many varied types deserve consideration because, beyond providing the trite proof that mathematicians are as human as anyone else, they reflect the state of mathematical activity at the time.

The first quarter of the twentieth century saw both Germany and mathematical science (in which Germans were prominent) confronted with multiple crises: in the former case, political and psychological; in the latter, technical and psychological. Simply to make some sort of equation here and so pass matters off is too pat and comfortable to be true; however, the nature of these crises and the temperaments engendering and confronting them deserve further examination.

CHAPTER TWO

The Crisis in Mathematics

THE spiritual crisis in German society prior to the First World War, which, subsequent to it, became translated into political terms, has been the subject of considerable study.[1] However, in mathematics as a subject matter, there was a contemporaneous crisis—or rather, crises—as alluded to earlier, in which Germans played a prominent role.

The term "crisis in mathematics" as used, for example, by Hermann Weyl[2] is usually taken to refer to the logical and foundational dispute aroused by the set-theoretic antinomies and Zermelo's pinpointing of the so-called Axiom of Choice, to be discussed below. However, while these questions still reverberate today, they reflect a changing view of mathematics and reaction to that change, which need to be elucidated first. This was the development of abstract conceptual notions and the increasing reliance on more general methods of proof and so-called existence theorems, in which the existence of a mathematical object with certain prescribed properties was logically demonstrated without that object ever being explicitly exhibited. As a setting of the "mathematical stage," this chapter is devoted to a description of these crises in a manner that I hope will be accessible to the reader with little or no mathematical background. In addition, this chapter provides more than just a setting. Issues involving the mode of intellectual discovery in mathematics became critical for a number of mathematicians aligned with the Nazis. For them, this was an issue of axiomatics versus valid intuition. In their terms, it was an issue of logic-chopping axiom-juggling against true insight into the mathematics naturally displayed. The set theory created by Georg Cantor and the axiomatization of it by Ernst Zermelo, together with his highly counterinituitive Well-Ordering Theorem, were particular bêtes noires. This may have partly been because Cantor was erroneously believed to have been Jewish.[3] Similarly, the dislike for abstract algebra may

[1] Among other works, for example, on the Youth Movement as a response to this crisis, see Walter Laqueur, *Young Germany* (1962); on the development of German ideology, George Mosse, *The Crisis in German Ideology* (1964), and Peter Gay, *Weimar Culture: The Outsider as Insider* (1968); on three "important outsiders" whose thought became "inside" as the result of the crisis, see Fritz Stern, *The Politics of Cultural Despair* (1961). A somewhat overrated but excellent study of a particular manipulation of German ideology in crisis is Siegfried Kracauer, *From Caligari to Hitler* (Princeton: Princeton University Press, 1947). Among the multitude of novels that reflect one aspect or another of the development of this crisis and its ultimate resolution in Naziism are Heinrich Mann's *Der Untertan* (Man of straw) and, as retrospect, Thomas Mann's *Dr. Faustus*. A reader who is a peruser of novels no doubt has his or her own favorites as well.

[2] Hermann Weyl, "Mathematics and Logic," *American Mathematical Monthly* 53 (1946): 2–13.

[3] See J. W. Dauben, *Georg Cantor, His Mathematics and Philosophy of the Infinite* (1979), esp. 271–299 and 140–148. Cantor's antecedents and religious education are discussed on 273–280. Can-

have been partly because one of its principal founders, Emmy Noether, was Jewish, female, and left-wing. Such reasons, however, are superficial. The mathematics of Cantor, Zermelo, and Noether would not have been favored by the "ideological Nazi" mathematicians in any case. At the same time, there were mathematicians who, while not Nazis in *Weltanschauung*, were spokespersons for the Nazi government, either by position or by choice. Even among these people, there were a variety of mathematical and political views.

Ludwig Bieberbach, as will be seen in chapter 7, became a leading protagonist of Nazi attitudes among mathematicians after 1933, and had, in the 1920s, converted from his earlier beliefs on one side of the crisis in the foundation of mathematics to the other. The leading exponent of the position to which Bieberbach converted was the famous Dutch mathematician L.E.J. Brouwer, who was a noncombatant in World War I and aggressively pro-German prior to 1933. Brouwer's ideas about mathematics being rooted in its human creators clearly would have been attractive in the Nazi period to someone like Bieberbach, who in 1940 would speak of "the rootedness of science in the people." Yet an affinity of ideas does not in itself indicate collaborationist mentality.

The story is much more complicated than this. People favoring Brouwer's ideas came from all facets of the political spectrum. To indicate how even more complicated the "crisis situation" was, French mathematicians who were in the forefront of criticism of Zermelo also, incidentally, were the founders of modern real analysis and measure theory—a subject Bieberbach once said was a playground for non-Aryans (but he would naturally condemn French mathematics anyway). This and other complications, such as the general attempt by some to discern the mathematics of Aryans from that of non-Aryans, or the psychological distinction made by some during the Nazi period between Brouwer's point of view and Hilbert's on the philosophy of mathematics, will be treated later in this book. Nevertheless, the general mathematical atmosphere needs to be understood in its outlines, so that the transporting of some of the controversies to the political realm will be clear.

One of the best-known instances of the development of increasing abstraction in mathematics is the story of invariant theory.[4] This theory has a long and interesting history, going back to the British mathematicians George Boole and Arthur Cayley. For our purposes, it is sufficient to note the following facts (the reader need not know the meaning of the mathematical terms briefly intro-

---

tor's nonmathematical writings include the mystical work *Ex Oriente Lux*. Also, to judge from a postcard in the possession of the Deutsches Museum in Munich, he and his family were at one time friendly with Julius Langbehn, the "Rembrandtdeutscher." The postcard confirms the breaking of this relationship; the reasons are not stated. For Langbehn, see F. Stern 1961.

[4] See particularly Charles Fisher, "The Death of a Mathematical Theory," *Archive for the History of the Exact Sciences* 3 (1966): 137–159; and idem, "The Last Invariant Theorists," *Archives Europèenes de Sociologie* 8 (1967): 216–244. See also Constance Reid, *Hilbert* (1970), and Bell 1940: chap. 20. Bell's works are written in a lively fashion, but are romanticized, biased, and unscholarly. Nevertheless, they remain the best-known and most widely read popular histories of mathematics. Their citation, however, is here marked *caveat emptor*. See also note 12 below.

duced). The main problem of invariant theory was the computation of all the "invariants" of what Cayley called "quantics,"[5] an "*m*-ary" quantic involving *m* independent variables. One of the questions this resolved itself into was whether there were perhaps, for each fixed *m*, a finite number of invariants from which all the others could be formed by certain simple arithmetic operations; such a set was called a finite basis for the set of invariants. Cayley in 1856 erroneously thought that for certain cases with m = 2, no finite basis existed. An orgy of computations of special cases was unleashed. In 1868 Paul Gordan showed that, in fact, for m = 2, there always was a finite basis; he did this by actually constructing an algorithm for computing the independent invariants that formed a finite basis. Gordan's own proof involved a symbolic representation for invariants that conceptually was a step in the direction of abstraction and generalization and away from the methods of Cayley and J. J. Sylvester, who went on with their students computing special cases.[6] The labor was immense, but of course the struggle now began (or was resumed) to find a finite basis for m > 2. Many special cases and classes were treated, but the general result proved elusive. An estimate of the labor involved in invariant-theoretic calculation may be indicated by the fact that Emmy Noether's 1907 dissertation (published in 1911), written under Gordan's supervision, involved an extension of some of his theorems to *m* variables; at the conclusion is a table of 331 forms in symbolic notation.[7] In 1888 David Hilbert heard about the problem, and proved very briefly that there was a finite basis in every case. However, he did not exhibit a procedure for finding a basis; he merely demonstrated its existence abstractly. Gordan is reported to have said of Hilbert's proof, "That is not mathematics, that is theology,"[8] a remark whose meaning is unclear. Gordan's friend and colleague Max Noether[9] took the comment as pejorative, but a student contemporary took it as saying that Hilbert's work was "god-given."[10] Perhaps each generation interpreted the remark, if it was made, to suit itself and its own conceptions. In any case, Gordan himself improved and simplified Hilbert's proof. Invariant theory of Gordan's sort went on, however, as witnessed by Emmy Noether's dissertation. In 1911, though, Emmy Noether became aware of Hilbert's methods, and in her hands, influenced by Hilbert, his theorem became one that had no mention of invariants and was just a theorem in the new axiomatic abstract algebra of which she was a founder. Much of the

[5] Quantics are homogeneous polynomials.

[6] Cayley and Sylvester, especially the latter, did try to prove Gordan's theorem also with their methods, but failed. The detailed story is complicated. See the literature cited below in note 12.

[7] B. L. van der Waerden, "Nachruf auf Emmy Noether," *Mathematische Annalen* 111 (1935): 469–476; and E. Noether, "Zur Invarianttheorie der Formen von *n* Variabeln," *Journal für die Reine und Angewandte Mathematik* 139 (1911): 118–154.

[8] Cf. Bell 1940: 403.

[9] An eminent mathematician of the period himself, a close friend of Felix Klein, and Emmy Noether's father.

[10] See Fisher 1966: 149. Cf. also Charles Fisher, "Paul Gordan," in *Dictionary of Scientific Biography*, ed. Charles C. Gillespie (1970), vol. 5, p. 472. The student contemporary was Gerhard Kowalewski.

rest of invariant theory was similarly subsumed into the "modern algebra" that received its most influential text in B. L. van der Waerden's *Moderne Algebra*, first published in 1931. In 1955, for the fourth edition, the title became *Algebra*, at the suggestion of H. L. Brandt (in 1951), to indicate that the mathematics it contained was not simply a fact "which was yesterday unknown and perhaps tomorrow will be forgotten." And in the introduction to the seventh edition (1966), van der Waerden remarks how his work, originally conceived as an introduction to the "newer abstract algebra," has become the "first introduction to algebra" for many students. As a result, material that once could be taken for granted, now needed to be introduced into the new text.[11] The old invariant theory became subsumed into the newer, more abstract subject matter, in which its identity was completely lost; even its theorems disappeared as separate entities.[12] This, of course, did not happen without a struggle; the very fact that Gordan's ambiguous and possibly apocryphal remark about Hilbert's proof is often cited to characterize him as someone trying to hold back the progress of the mathematics represented by such existential proofs indicates that not all mathematicians accepted the new way of looking at things. Indeed, Gordan, who did appear to have been open to novel ideas, as the leader of the pre-Hilbert school seems to have been saddled willy-nilly with opposition to the newer approaches. In any case, Hilbert's theorem marks a departure that culminates in the *Moderne Algebra* that reached a final form in the 1920s. There are, of course, other sources of contemporary axiomatic algebra besides invariant theory; for example, the subject matter that became group theory.[13] Invariant theory has been selected only as an illuminating and well-known case of the progress from early computation to mathematical abstraction: from Cayley to Gordan to Hilbert to axiomatic algebra, in which some of the important old theorems and problems completely lose their earlier identity. For algebra, the development spanned may perhaps be marked by Cayley's paper of 1856 at one end and van der Waerden's 1931 text at the other.

A similar development happens at roughly the same time in the concepts of function and continuous function.[14] Leonhard Euler's conception of a "contin-

[11] B. L. van der Waerden, *Algebra, Moderne Algebra* (1931–66).

[12] That is, insofar as algebraists were concerned, invariant theory was dead. The frequently made statement that it was killed by Hilbert is not, strictly speaking, true; but it is true that the ideas introduced by Hilbert led to a supersession of the old ideas and methods, and to a solution in principle, if not in explicit exhibition, of the theory's central problems. The old invariant theory is still pursued by some physicists and engineers, particularly those interested in continuum mechanics, and has had something of a revival recently. The above sketch is, of course, too brief to represent the whole truth. The interested reader is referred to Fisher 1966 and 1967; Karen Hunger Parshall, "Toward a History of Nineteenth-Century Invariant Theory," in David Rowe and John McCleary, eds., *The History of Mathematics* (1994), 1: 157–206; and Anthony Crilly, "The Decline of Cayley's Invariant Theory," *Historia Mathematica* 15 (1988): 332–347.

[13] For a history of early group theory, see H. Wüssing, *Die Genesis des Abstrakten Gruppenbegriffes* (1964). Interestingly enough, Cayley was prominent in the beginnings of this algebraic area as well.

[14] Two excellent sources for this history are A. F. Monna, "The Concepts of Function in the Nineteenth and Twentieth Centuries, in Particular with Regard to the Discussions between Baire,

uous function" was one that could be determined by a simple analytical expression for the whole range of the independent variables.[15] In 1821, Augustin Louis Cauchy gave his definition of continuity, which is essentially the modern one (and had already been stated independently by Bernhard Bolzano a few years earlier); nevertheless, the ideas of function and of continuity remained vague. In 1826, Nils Abel complained about the indistinctness of functions in analysis and blamed it for the large number of false proofs and uncertainty in the subject. Curiously enough, in 1837, Peter Gustav Lejeune-Dirichlet apparently had a quite general concept of function without mentioning Cauchy's notion of continuity, but while maintaining Eulerian ideas of it. Cauchy criticized the Eulerian ideas in 1844, but in 1851 Bernhard Riemann gave a completely Eulerian definition of continuity (or nondefinition, as his language involves defining a continuous function as one that continuously varies). By 1870, Hermann Hankel was saying that the only way to clarity was to free oneself from all residues of Eulerian ideas. It was Hankel who made explicit the idea present in Riemann's and Dirichlet's work that a function has no necessary general properties.[16] But then (almost contemporaneously with the developments in invariant theory leading to Hilbert's proof) a change that lasted finally did come about as a result of the work of Karl Weierstrass and Gaston Darboux.[17] The apparent reason was the acceptance of a modern concept of (a real-valued) function (of a real variable) as a result of the work of Dirichlet and Riemann. Functions were now simply correspondences between (or sets of "ordered pairs" of) real numbers. So, although an appropriate definition of continuity in the modern sense had been given as early as 1821 by Cauchy, no one had had any use for it, and it was not even clear what use it offered until the idea of function had been clarified.[18] Riemann, though he clarified the notion of function, mistakenly maintained the Eulerian notion of continuity. This was at least in part because he believed that every continuous function of a real variable can be represented everywhere by its Fourier series, which was not shown to be false until 1873 (by Paul du Bois-Reymond). Thus again, by the 1880s, the notion of function

---

Borel, and Lebesgue," *Archive for History of Exact Sciences* 9 (1972–73): 57–84; and T. Hawkins, *Lebesgue's Theory of Integration, Its Origin and Development* (1970). Original references for the ensuing sketch can be found in these sources if not otherwise footnoted.

[15] Monna 1972–73: 159; Hawkins 1970: 3–4.

[16] Monna 1972–73: 60; Hawkins 1970: 29ff. Hankel himself made errors of the sort only clarified when the concept of "nowhere dense set," stemming from Georg Cantor's work, was introduced and understood.

[17] Darboux's "Mémoire sur la théorie des functions discontinues" appeared in 1875. This contained an example of a continuous nondifferentiable function. Weierstrass's example, which had been constructed earlier, was published by du Bois-Reymond the same year. Earlier examples had been constructed but not published by Bernhard Bolzano and C. Cellérier and were without influence (see Hawkins 1970: 45). In 1878, Ulisse Dini introduced the "Dini derivates," and also published the first treatise on the theory of functions of a real variable. For the history of both integration and differentiation theory, see Hawkins, ibid.

[18] Cauchy's definition of continuity applies to the modern concept of function, but his entertaining this latter seems more apparent than real; see ibid.: 10–12.

and continuity seemed to have become finally clarified to the point where more problems would be and were being attacked. From the points of view of both integration and differentiation, these involved the consideration of various infinite sets of points. Riemann had considered in 1854 the question of when a function is integrable, and had formulated the "Riemann integral" independent of continuity considerations; thus, though "Riemann-integrable" functions have many properties closely related to continuity, Riemann allowed a class of explicitly discontinuous functions to enter his considerations. The study of various kinds of infinite sets to which the problems of integration and differentiation led were to end in a mathematical crisis still reverberating today. A basic question was, how general could one make one's reasoning without erring? If one took a "realist's view" and saw mathematical objects as quasi-physical entities, how should one treat infinite agglomerates? Indeed, could one treat them? This became especially important when the study of infinite sets led to certain antinomies, and when certain axiomatic procedures that had been used implicitly for some time with regard to infinite sets, on being made explicit, led to uncomfortable, seemingly paradoxical results, at the same time as they seemed essential for doing ordinary analysis.

Indeed, the reaction to even the ordinary distinction between general function, continuous function, and differentiable function, established by the work of Riemann, Weierstrass, and Darboux, was not easily accepted. In a famous letter, Charles Hermite wrote T. J. Stieltjes on May 20, 1893, "I turn away with fright and horror from this lamentable plague of continuous functions which have no derivatives."[19] This was the doyen of the Paris school, then seventy-one, writing to one of the most brilliant of his contemporaries, and eighteen years after Darboux's paper. And in 1905 Henri Lebesgue would write:[20]

> Although since Dirichlet and Riemann it is generally agreed that a function is a correspondence between a number y and numbers $x_1, \ldots x_n$, without preoccupying oneself with the procedure which serves to establish this correspondence, many mathematicians seem only to consider as genuine functions (*vraies fonctions*) those which are established by analytic correspondences.

Two aspects of the increasing concern with a careful handling of various kinds of infinite sets became especially critical and sources of continuing controversy. One was Georg Cantor's Theory of Sets, initiated in 1870. Cantor was the first person to carefully analyze the properties of certain infinite sets. His work (like Dirichlet's and Riemann's) was motivated by problems in Fourier series. The other (not unrelated) was Ernst Zermelo's elucidation of the Axiom of Choice in 1904.

The problematic infinite entered algebra as well, for once it was viewed as an axiomatic subject, not only was the structure of the most "naturally occurring"

[19] *Correspondance d'Hermite et de Stieltjes*, ed. B. Baillaud and H. Bourget, 2 vols. (Paris: Gauthier-Villars 1905), letter 374 (in vol. 2). Actually, in the context, the functions Hermite is talking about are defined by infinite series.

[20] Cited in Monna 1972–73: 71.

algebraic objects such as the real numbers or the integers infinite, but also there was no reason not to consider other infinite algebraic entities. Even Zermelo's Axiom of Choice enters algebra in a transmogrification known as Zorn's Lemma.

Although a detailed history of the dispute centering on Cantor's set theory and Zermelo's axiom would itself occupy a volume, a brief elucidation of the axiom may aid in understanding the controversy it stirred in western European mathematics and the attitudes that developed toward it. The controversy was not one that was long in coming, either. The volume of *Mathematische Annalen* following the publication of the extract from Zermelo's letter to Hilbert in which the "selection principle" (Axiom of Choice) is used to prove the Well-Ordering Theorem contained attacks by Emile Borel and Arthur Schoenflies, a priority claim asserting improved results from P.E.B. Jourdain, and a claim by Felix Bernstein that the Well-Ordering Theorem could be proved without using the choice axiom, as well as one of the enduring applications of Zermelo's theorem: Georg Hamel's construction of "Hamel bases" and "Hamel functions." In 1905 also appeared a critical discussion among French mathematicians, a criticism by Guiseppe Peano in *Rivista di matematica*, and sharp criticism from Poincaré in the *Revue métaphysique et moral*.[21]

Zermelo's Well-Ordering Theorem says simply, "Every set can be well ordered." "Well ordered" here is a technical term meaning that the set can be so ordered that every subset (in that order) has a least element. Zermelo proved this on the assumption of the axiom that (in the later formulation of 1908) "a set S that can be decomposed into a set of mutually exclusive parts, A, B, C . . ., each of which contains at least one element, possesses at least one subset S_1 that has with each of the considered parts A, B, C . . . exactly one element in common."[22] That is, one can construct a new subset by choice of elements from given subsets, irrespective of whether the set or subset in question is finite or infinite, irrespective of whether elements of the set can be placed in one-to-one correspondence with the integers (and so "enumerated") or not (as Cantor had shown was the case with all the points on a line segment).

Zermelo gave two proofs (1904 and 1908) of the Well-Ordering Theorem, both of which use the Axiom of Choice, and following the second answered his critics. The same 1908 volume of *Mathematische Annalen* saw Zermelo begin the attempt to put set-theoretic mathematics on an axiomatic basis. This had appar-

[21] See G. Peano, "Super theorema de Cantor-Bernstein et additione," *Rivista de matematica* 8 (1902–6): 136–157. It is only the "additione" that is relevant. An English translation appears in *Selected Works of Giuseppe Peano*, ed. H. Kennedy (Toronto: University of Toronto Press, 1973), 206–218. See also H. Poincaré, "Les mathématiques et la logique," *Revue de métaphysique et moral* 13 (1905): 815–835. Peano points out that in 1890 (that is, fourteen years before Zermelo), he had already declared any infinite Axiom of Choice illegitimate (see *Mathematische Annalen* 37 [1890]: 210). The critical discussion among French mathematicians is in "Cinq lettres sur la Théorie des Ensembles," *Bullétin des mathématiques* 33 (1905): 261–273.

[22] E. Zermelo, "Neuer Beweis für die Möglichkeit einer Wohlordnung," *Mathematische Annalen* 65 (1908): 107–128.

ently become necessary, as the Axiom of Choice was found to pervade much mathematics where its use had been previously unremarked. Indeed, one of the ironies of the detailed story is that much of the best work of some of Zermelo's opponents, such as René Baire, Emile Borel, and Henri Lebesgue, had implicitly used a choice principle.[23] In fact, ordinary mathematical analysis itself turned out to have the worm of Zermelo's axiom at the very heart of the subject, in what became known as the "principle of the least upper bound."

Zermelo divided his critics into four groups:[24] (1) those who opposed the Axiom of Choice, (2) those who opposed nonpredicative definitions, (3) those concerned with the possible relationship between the so-called Burali-Forti Paradox[25] in Cantorian set theory and Zermelo's proof, and (4) those who would attack "generating principles" and the theory of well-ordered sets.

As to the first, Zermelo reemphasized for the benefit of Borel and Peano that the Axiom of Choice is an *axiom*, not proved:[26]

> Insofar therefore as E. Borel and G. Peano in their criticism state the lack of a proof, they have merely adopted my own standpoint. They would have indeed compelled me to thank them if they had for their part proved my conjecture as to its unprovability, that is, the logical independence of this postulate, and thereby confirmed my conviction.[27]
>
> Now unprovability even in mathematics is, as is well known, in no way equivalent with invalidity, since indeed not everything can be proved, but every proof assumes again unproved principles.

Zermelo went on to examine the utility and previous unremarked use of the Axiom of Choice as evidence in its favor, at the same time disposing of critics who found it "not met with in their philosophy" (Peano) or who erroneously thought to substitute some other principle for it (Felix Bernstein).[28]

[23] Lebesgue had indeed recognized an earlier instance of an inexplicit use of the choice principle in his famous thesis (see "Cinq lettres" 1905: 267), and says of this instance somewhat ambivalently, "In my thesis I demonstrated the existence (in a non-kroneckerian sense and perhaps difficult to make precise . . .)." That is, he recognized that the principles he was expositing in the epistolary exchange should cause him to reject this portion of his own dissertation, but apparently could not quite bring himself to do so.

[24] Zermelo 1908.

[25] "So-called" because Burali-Forti did not believe there was a "paradox" in anything he published; the paradox was actually created by Bertrand Russell. See G. H. Moore and A. Garciadiego, "Burali-Forti's Paradox: A Reappraisal of Its Origins," *Historia Mathematica* 8 (1981): 319–350; and A. Garciadiego, "On Rewriting the History of the Foundations of Mathematics at the Turn of the Century," *Historia Mathematica* 13 (1986): 39–41.

[26] Zermelo 1908: 111.

[27] Following important work of Kurt Gödel in the 1930s, this in fact was done and published by Paul Cohen in *Proceedings of the National Academy of Sciences* 50 (1963) and 51 (1964).

[28] Zermelo 1908: 115. In his paper "Über die Reihe der transfiniten Ordnungszahlen," *Mathematische Annalen* 60 (1905): 187–193, Bernstein attempted to replace "Zermelo's principle" with one of his own concerning "multiple equivalence" (193). This is refuted by Zermelo (1908: 113, n. **).

Cantor's theory of infinites (*Mächtigkeiten*) therefore requires in any case our postulate, as does Dedekind's theory of finite sets which forms the basis of arithmetic . . . in the theory of completely discontinuous functions the principle is often unavoidable [as in the case of Hamel's application of it].

The question of nonpredicative definitions raised by Poincaré leads into the dark forest of linguistic philosophy. Suffice it to say that Zermelo pointed out that similar definitions had been current and accepted previously in other areas of mathematics without being viewed as "illogical" and that it would be difficult to avoid them. For Poincaré, however, the Axiom of Choice was unavoidable and unprovable. As Zermelo remarked, Poincaré's standpoint was in the last analysis that of a productive science based on intuition.

The last two groups of critics entered into technicalities of axiomatic set theory and the antinomies that had appeared there. This obviates their discussion here, as does the welter of confusion and misunderstanding, which shows how desperately set theory needed clarification.[29] One might note that, with the exception of Jourdain, all the mathematicians among Zermelo's opponents are honored today as among the founders of set-theoretic mathematics.

Zermelo's Axiom of Choice and Well-Ordering Theorem are only two issues arising from Cantor's set theory. The antinomies or paradoxes were another. At least one of these, Russell's Paradox[30] (to which the so-called Burali-Forti paradox[31] is a technical kinsman), is popularly known. This paradox concerns the set of all sets that do not contain themselves. Does it contain itself or not? If it does, it doesn't; if it doesn't, apparently it does.[32]

[29] For example, Bernstein and Jourdain in the same volume of the *Mathematische Annalen* claimed to have proved statements, each of which directly contradicts the other. This state of confusion also appears in the "semantical" paradoxes; see A. Garciadiego, "The Emergence of Some of the Nonlogical Paradoxes of the Theory of Sets, 1903–1908," *Historia Mathematica* 12 (1985): 337–351.

[30] In fact, "Russell's Paradox" was "discovered" considerably earlier by Zermelo, as is confirmed by a note in the *Nachlass* of Edmund Husserl, found at the University of Freiburg; see B. Rang and W. Thomas, "Zermelo's Discovery of the 'Russell Paradox,'" *Historia Mathematica* 8 (1981): 15–22. Cf. Zermelo 1908: 116–117. Zermelo apparently thought of the "paradox" as a problem to be handled within set theory.

[31] Actually, that Burali-Forti's argument is "paradoxical" is Russell's formulation, *not* Burali-Forti's. See note 25 above.

[32] What is often claimed to be a popular variant of this paradox, the so-called Barber Paradox of the barber who shaves all and only those who do not shave themselves, is really no paradox at all. The reasoning simply proves that no such barber can exist. The analogous deduction is that the set in question does not exist, but in "naive set theory," in which a set is any collection of objects, the nonexistence of a specified set is "inconceivable" and consequently "paradoxical." "Solutions" to the paradox consequently often involve a technical delimitation of the notion of "set." Another (true) variant of the paradox involves the adjectives "autological" and "heterological." An adjective is autological if it is self-descriptive (e.g., the adjective "English" is English, the adjective "many-lettered" is many-lettered), otherwise heterological. Clearly every adjective is either autological or heterological (and "most" are heterological). But "heterological" is itself an adjective; is it heterological or autological? The paradox, due to Kurt Grelling, is often attributed to Hermann Weyl because of a misattribution by Frank Ramsay. See Volker Peckhaus, "The Genesis of Grelling's Paradox," in *Logik und Mathematik* (Berlin: Walter de Gruyter, 1995). Grelling died in Auschwitz.

The object here, however, is to discuss the attitudes provoking and provoked by this crisis, and further substantive historical discussion is too tempting a digression to be indulged.

Zermelo's use of the Axiom of Choice in his sensational proof of the Well-Ordering Theorem involved operating on sets without actually constructing an execution of those operations. This led Henri Lebesgue to ask, "Can one demonstrate the existence of a mathematical entity without defining it?" The question came in an epistolary discussion between four prominent French mathematicians: René Baire, Emile Borel, Jacques Hadamard, and Lebesgue himself. Hadamard's answer to Lebesgue's question was "yes"; the others', "no."[33]

> Lebesgue, Baire, and you [Borel] adopt in this regard [to the distinction between what is determined and what may be described] Kronecker's point of view, which up until now I had believed peculiar to him. You answer negatively the question posed by Lebesgue . . . I answer in the affirmative, . . . the principal question [with respect to Zermelo's theorem], namely to know if a set may be ordered, evidently does not have the same meaning for Baire (no more than for Lebesgue and you) as for me. I would rather say: Is ordering possible? (And not even can *one* order, out of fear of having to think what is that *one*): Baire would say: Can *we* order? A completely subjective question in my opinion.
>
> There are thus two conceptions of mathematics, two mentalities that are present. I do not see in all that has been said up until now any reason for changing mine. I do not wish to impose it. All the more I would assert in its favor the arguments which I indicated in the *Revue Generale des Sciences* (30 March 1905), namely, . . . I believe the debate is at bottom the same that arose between Riemann and his predecessors, about the notion of function itself. The *law* that Lebesgue demands seems to me to resemble strongly the analytic expression that Riemann's adversaries strongly advocated. And indeed an analytic expression that was not too bizarre, . . . I do not see how we have the right to say "For each value of x there exists a number satisfying . . ., let y be that number . . ." while, because "the bride is too beautiful," we are not able to say, "For each value of x there exists an infinite number of numbers satisfying . . . let y be one of those numbers."

Hadamard makes the point here that there are two different views of mathematical activity in contention: one that maintains the implicitly subjective quality of all mathematics, as it is a uniquely human activity, and one that maintains a sort of ideal, objective existence for mathematics, the results obtained through human agency being, in fact, independent of any specific human creator, or any human being at all, except as to their perceptions. The tension between these views of mathematics, is, perhaps somewhat paradoxically, intrinsic to the subject itself. Exactly the distinction pointed out by Hadamard was to characterize attempts to resolve the logical difficulties that appeared in the foundations of mathematics. L.E.J. Brouwer and his school would maintain the inevitable subjectivity of all mathematics, while those gathered around David Hilbert came to

[33] See "Cinq lettres" 1905: 269–271.

describe mathematics as in themselves meaningless scratches on a piece of paper manipulated according to certain rules.

These two positions, as well as various others, are amusingly, fairly, and tellingly presented by A. Heyting, a disciple of Brouwer's, in the dialogue that opens his book *Intuitionism: An Introduction.*[34] Here a classicist, a formalist, an intuitionist, a literalist, a pragmatist, and a significist discuss the various positions on the nature of mathematics associated with their names. This is too lengthy to recapitulate here; however, Hilbert's position is put succinctly by Mr. Form as follows:[35]

> In daily speech no word has a perfectly fixed meaning; there is always some amount of free play, the greater, the more abstract the notion is. This makes people miss each other's point, also in non-formalized mathematical reasonings. The only way to achieve absolute rigour is to abstract all meaning from the mathematical statements and to consider them for their own sake, as sequences of signs, neglecting the sense they may convey. Then it is possible to formulate definite rules for deducing new statements from those already known and to avoid the uncertainty resulting from the ambiguity of language.

And Mr. Int expresses Brouwer's view:[36]

> Brouwer's program . . . consisted in the investigation of mental mathematical construction as such; without reference to the nature of the constructed objects, such as whether these objects exist independently of our knowledge of them. . . . In the study of mental mathematical constructions "to exist" must be synonymous with "to be constructed."

This returns us to Lebesgue's question: Can one demonstrate the existence of a mathematical object without defining it? René Baire went even further than Lebesgue and Borel in answering in the negative.[37]

> For me progress in this range of ideas consists in delimiting the domain of what is definable. And, in the final analysis, in describing its appearance, everything must lead back to the finite. That is, provisional rules for constructions that admit infinitely many steps are to be ruled out.

And Baire was quite explicit that given, say, the set of positive integers, "it is as far as I am concerned false to consider the parts of that set as given."[38] It would seem that even the set of positive integers itself is inconceivable as a whole from this point of view.

Mathematics either is or is not independent of its human creators. Zermelo, as well as Hadamard, recognized the distinction, as he wrote in the preface to his second proof that it "allows the purely formal character of well-ordering,

[34] A. Heyting, *Intuitionism: An Introduction* (1966).

[35] Ibid.: 3.

[36] Ibid.: 1 and 2.

[37] "Cinq lettres" 1905: 264.

[38] Ibid.

which has nothing whatsoever to do with spatial-temporal ordering, to come forward more clearly than in the first proof."[39]

Zermelo's three-page note of 1904 brought to explosion the crisis that had been seething in mathematics. In the nineteenth century, Kronecker had opposed nonconstructive existence proofs and Weierstrass's school, and had persecuted Cantor. What perhaps distinguishes Zermelo's existence proof is not only its generality, but the apparent hopelessness of ever constructing a well-ordering for an arbitrary set.

Once it was clear that Zermelo's Axiom of Choice was not newly discovered, but only newly enunciated and elucidated, and how much "standard mathematics" depended upon it, mathematicians fell into roughly two camps (or, more precisely, three). There were those who cried "backwards" and, like L.E.J. Brouwer, emphasizing the human agency in mathematics, rejected not only Cantorian set theory but pure existence proofs and the Aristotelian law of the excluded middle (so that a statement may be neither true nor false). Only constructive reasonings are allowed. One radical offshoot of this wing even rejected negation (G.F.C. Griss). Another form of constructive mathematics became prominent in the 1960s in the United States under the leadership of Errett Bishop. For all varieties of constructivists from Brouwer to Bishop, a large amount of mathematics has to be, if not jettisoned, then reformulated in a manner that most mathematicians find clumsy and unappealing.[40]

A second band, among whose early leaders were Zermelo and Hilbert, refused to leave (in Hilbert's words) the paradise into which Cantor had ushered them,[41] and set about trying to determine how to prune the thistles while maintaining the roses in it.

Most mathematicians, however, went the way they had been going, ignoring the "crisis in the foundations" with little more than an occasional sidelong glance. Nevertheless, it was apparent that the gradual expansion of mathematics and penetration of its logic had reached a point of scientific crisis. In T. S. Kuhn's terms, one might almost say that, roughly speaking, set theory prior to Zermelo's 1904 paper was in a preparadigmatic state, and Peano, Zermelo, and others were about to begin the attempt to order the chaos. Certainly the papers of P.E.B. Jourdain, Arthur Schoenflies, and Felix Bernstein in *Mathematische Annalen* 60, which concern themselves with Zermelo's theorem, have the flavor of preparadigmatic chaos.[42] The air of deep crisis that surrounded this situation

[39] Zermelo 1908: 107.

[40] This does not mean it is wrong, of course. The innovations of Augustin-Louis Cauchy and Karl Weierstrass were surely clumsy and unappealing to their contemporaries, though today they are de rigueur.

[41] "No one shall drive us out of this Paradise Cantor has created for us," D. Hilbert, "Über das Unendliche," *Mathematische Annalen* 95 (1926): 170. An abridged version appears in *JDMV* (1927): 201–215, where the quotation also appears.

[42] The history of set theory fits Kuhn's schema in another way as well: the paradigm initiated by Zermelo took such strong hold that alternatives were neglected and forgotten. In particular, there are no antinomies in Cantor's original set theory in which sets are constructed by the operation "set

and infected it had two sources. One was the position of set theory as the implicit foundation stone for all mathematics, and so a crisis here implicitly meant a crisis that permeated the subject. Zermelo had emphasized how his reasonings in part lay previously unrecognized behind the most basic mathematics. The other source of the air of crisis was that an influential group of mathematicians led by Brouwer decided not to make the transition to paradigmatic axiomatic formalism, but instead to return, far more sophisticated and much sobered, to a world before troubles had arisen. It was as though Rousseau's corrupted savage had forgotten the use of his club and returned to the state of nature, remembering only the unfortunate consequences of his having possessed it.

Hilbert's words about "Paradise," and Brouwer's insistence about the inevitable subjectivity of all mathematical activity, as well as his seeming rejection of his own applauded work,[43] raises the question of a psychology of mathematical thinking.

The mathematician's attitude toward the world of mathematics may perhaps be usefully compared to the infant's toward the world being newly discovered. The creative act of the mind alone, the projection of intuition onto reality, is reminiscent of the way the child learns about the world around him. When experience conflicts with intuition, intuition becomes modified, becomes adapted to reality. For the natural scientist *qua* scientist, this relationship is with a tangible reality of mental constructs. The mathematician, however, creates a reality that is then investigated; this reality takes on a life of its own, as it were, and offers yieldings and resistances to efforts at comprehension. The childlike wonder at and admiration of the universe by the pure scientist has often been remarked; similarly, the mathematician, when queried about mathematical activity, often refers to the metaphors of play. The technologist is concerned with the important immediate effects of manipulation of gross reality, the "pure scientist" with possible interpretations of experiments that in themselves are not important manipulations of reality, though their interpretations may lead to such. The mathematician, however, is at even one further remove, and is primarily concerned with symbolic manipulations of representations of a reality whose existence is largely mental. Mathematicians are players with their own creations. This allows for both a creative outlet of aggression, and a satisfaction

of." On this point, see Kurt Gödel, "What Is Cantor's Continuum Problem?" *American Mathematical Monthly* 54 (1947): 515–525.

[43] See L.E.J. Brouwer, "Intuitionistische Mengenlehre," *JDMV* 28 (1919): 203–208, p. 204, where Brouwer says: "I have also in my simultaneous [to promoting intuitionism] mathematical works free of all philosophy, regularly used the old methods, whereby never [the] less I strove (*bestrebt war*) only to derive such results, about which I could hope that, after finishing a systematic construction of intuitionistic set theory, they would find a place and be affirmed to have a value, perhaps in modified form, in the new educational construction (*Lehrgebäude*)." Brouwer did begin such a construction of an intuitionistic set theory, but, so far as I know, never achieved the aim here expressed.

in accomplishment.[44] The mathematician, even more than the physical scientist, has "the urgent need to control and make right what has been done, and what is being done," in Ella Sharpe's formulation.[45] The pleasure in knowing has a purely aesthetic component. The mathematical universe is at once the most demanding of universes, because nothing less than perfection is aspired to, and the one in which the mathematician—god—player is most in control. Physical reality intrudes neither to temper explorations nor to provide excuses for imperfections.

Both mathematician and infant feel paradoxically at once omnipotent and restricted. The infant's omnipotence is an illusion that suffers correction as wish fulfillment is thwarted, and with the discovery that abilities are in fact restricted. The mathematician's feeling of omnipotence arises because, no less than the infant, the mathematician is analogously the creator of the mathematical world, which is open to play. Again, analogously, the mathematician soon discovers that mathematics performed according to logical rules and adhering to standards of rigor has a "life of its own," as it were, and is not subject to completely arbitrary manipulation. Thus, the mathematician also is hemmed in by reality. The reactions of various mathematicians to the crises in their subject can also be seen in this light. The more abstract mathematics of Hilbert and Emmy Noether, Gordan's "theology," handles whole classes of problems, and provides existence proofs; but it cannot always treat definitively the specific problems giving rise to the questions examined. Sometimes, as the history of invariant theory itself shows, the explicit solution to a problem, and not just the existence of the mathematical entity, is necessary; on the other hand, often the existence is sufficient, and an explicit representation seems impossibly complex. One might view the more abstract mathematics as an advancement in maturity—as indeed its proponents did: a creating of new areas in which to exercise power, a more sophisticated view of the old frustrations. Or one might see it as a dismissal of true problems through only partially conquering them, and partially bypassing them. That is, the problem might "pass," or one might become "fixated" in it.

Furthermore, although mathematicians engaged in set-theoretic and foundational problems were of many nationalities, the concentration on Hilbert and Brouwer (who, though Dutch, was an ardent German nationalist) reflects the position at the time of German mathematics.

After the Franco-Prussian War and Bismarck's unification of the German states (omitting Austria) in 1871, Germany was the ascendant political power. It also was ascendant intellectually and mathematically. In 1822 Peter Gustav Lejeune-Dirichlet had gone to France in order to study mathematics because the

[44] Cf. Sándor Radó, "Paths of Natural Science in the Light of Psychoanalysis," *Psychoanalytic Quarterly* 1 (1932): 683–700, esp. 696, which speculates on the scientist's symbolic replacement of God's will with his own.

[45] Ella Freeman Sharpe, "Similar and Divergent Unconscious Determinants Underlying the Sublimations of Pure Art and Pure Science," *International Journal of Psychoanalysis* 16 (1935): 199.

general level of mathematics in Germany was low. (There was Karl Gauss, of course, but Gauss was sui generis, and perhaps not known as a good mentor.) After 1870, however, the European epicenter of mathematics shifted eastward to Germany. This certainly did not happen all at once, as names like Charles Hermite, Camille Jordan, Emile Picard, and Henri Poincaré, among others, testify. There were also equally prominent Italian and British mathematicians, but the central mathematical concerns gradually became those of German mathematics. As for American mathematics, the fact that American mathematicians often studied in Germany, frequently obtaining their doctoral degrees there, speaks for itself. Though Felix Klein recommended that the young David Hilbert and Eduard Study spend some time studying in Paris, which they did, the fact that he apparently always made such recommendations to gifted young German mathematicians seems, at least in part, based on positive memories of his own mathematical experiences in France in 1870 studying with Jordan.

By 1886, when Hilbert was a new doctor and made his Parisian voyage, the mathematical world was already frequently looking in a German direction; by 1900, when Hilbert gave his famous address (in Paris) on the mathematical problems of the future, Germany had succeeded France as the mathematical standard-bearer.[46]

German mathematicians were prominent in originating, developing, and treating set-theoretic questions, from Georg Cantor to Kurt Gödel and Gerhardt Gentzen.[47] Perhaps this partly explains, especially as the early development took place between 1870 and 1910, the number of mathematically prominent French critics of the development of set theory and the lack of prominent French participants. Zermelo and Hilbert represented one prominent position. On the other side were Brouwer and Weyl. Brouwer was not only a German nationalist but had published some of his best work in the *Mathematische Annalen*. Hermann Weyl, one of the most effective propagandists for Brouwer's intuitionism, was a brilliant student of Hilbert's who had somewhat uncomfortably and uneasily half-moved into the opposing camp. The crisis was also a crisis internal to German mathematics, and thus a crisis in mathematics.

The issue of whether mathematics was independent of its human creators was also a prominent one for Nazi mathematicians.[48] However, for them it took a slightly different form: was mathematics to be the product of an intuitive intercourse with the "real world" rigorized, or was it to be rigorous formalism

[46] For Hilbert's and Study's trips to Paris and Klein's attitude, see Reid 1970: 20–22. At the same time, Adolf Hurwitz also felt it necessary to master French mathematics so as to supersede it (ibid.: 21).

[47] Gödel, who made fundamental contributions to these problems in the 1930s, was of Austrian nationality, but German-trained. For Gentzen under the Nazis, see chapter 8.

[48] In discussing these philosophical and ideological issues, the phrase "Nazi mathematician" is here used as a shorthand for "mathematician committed to the positive ideological interaction between the discipline of mathematics and Nazi ideology." As will appear in later chapters, there were mathematicians who were politically attached to the Nazis without necessarily accepting the ideological consequences for their discipline discussed here, or even that such consequences existed.

constructed from clear definitions? In reality, both sorts of activity are necessary. However, it was the intuitive, the *anschaulich* to the exclusion of all else, that the creators of a "Deutsche Mathematik" placed on their standard. Thus Klein, that intuitive genius, suffered ancestor-worship at their hands. And so, the probabilist Erhard Tornier in the first issue of *Deutsche Mathematik* could raise the question, "Mathematiker oder Jongleur mit Definitionen?" (Mathematicians or definition-jugglers?) and say,[49]

> It is the typical Jewish-liberal thesis that the criterion of the right to existence of a mathematical theory is its "aesthetic beauty," whereby is meant a logically closed—the simpler the better—construction upon definitions. . . .
>
> Every theory of pure mathematics that actually has the ability to answer concrete questions that relate to real objects such as integers or geometric configurations, or at least to serve for the construction of theories so capable, has a right to life.

In addition, for the Nazis there was a great and naturally largely unremarked irony that Hilbert, the high priest of formalism, was an East Prussian Lutheran, while Kronecker, the forefather of antiformalism, was a Jew, and Cantor, while of partly Jewish origin, was reared a Lutheran and was something of a Christian mystic.[50] While the intuitionist and the Nazi might both be concerned with a subjective mathematics, there the resemblance ends. Intuitionism is far from "intuitive" in its elaborateness. Intuitionism means "constructively available to the mind" rather than "intuitively obvious." The subjectivism of intuitionism has to do with an estimate of the capacities of the human mind to make correspondences with reality, with a view of mathematics as a reality constructed by human intellects, and therefore never capable of being freed from them. As Arend Heyting, one of Brouwer's most prominent disciples, remarks:[51]

> The intuitionist mathematician proposes to do mathematics as a natural function of his intellect, as a free, vital activity of thought. For him mathematics is a production of the human mind . . . we do not attribute an existence independent of our thought, i.e., a transcendental existence, to the integer or to any other mathematical object. . . . Even if they should be independent of an individual act of thought, mathematical objects are by their very nature dependent on human thought.

The intuitionist distinguishes between "faith" in the existence of an object (which is rejected) and the actual "knowing" of it by having it to hand. For him there is no *credo* whatsoever, only *scio*.[52]

Brouwer rejected much of contemporary mathematics, including implicitly what many considered his greatest work. Brouwer's 1907 thesis (published in 1910), "Over de Grondslagen der Wiskunde" (On the foundations of mathemat-

[49] E. Tornier, "Mathematiker oder Jongleur mit Definitionen," *Deutsche Mathematik* 1 (1936): 8–9. For more about Tornier, see chapter 4, "Hasse's Appointment at Göttingen."

[50] See note 3 above.

[51] See P. Benaceraff and H. Putnam, *Philosophy of Mathematics* (Englewood Cliffs, NJ: Prentice-Hall, 1964), 42. This collection contains several selections from Brouwer and Heyting.

[52] Thus the famous credo of Kronecker: "God made the integers, all the rest is the work of man."

ics), and a 190[illegible] paper, "Over de Onbetrouwbaarheid der logische Principe" (On the untrustworthiness of logical principles), already contain the beginnings of his intuitionism. Nevertheless, in the years 1909–1913, Brouwer turned to topology, and proved several fundamental results, including a mathematical definition of dimension (1910–13), the principle of invariance of domain (1910–11), the Brouwer fixed-point theorem (1911), the n-dimensional Jordan curve theorem (1911), the existence of indecomposable continua (1909), and invented Brouwer degree (1911). In 1912 Brouwer became a professor at Amsterdam and a member of the Royal Netherlands Academy of Science. His inaugural address, however, was not on topology, but on intuitionism and formalism.[53] He thus returned after a spectacular five years in "conventional mathematics" to his original philosophical point of view, and was sufficiently self-aware to realize that this return involved turning his back on work that had won him fame.[54] From 1918 onward, intuitionism was the primary direction of his work, and much of his tremendous achievement of 1909–13 was not only left behind but necessarily viewed by him as faulty, though perhaps reparable. Indeed, Brouwer continued an active and inspirational interest in topology, but his point of view was highly idiosyncratic.[55]

This rejection of much of mathematics, this insistence on the nonexistence of mathematical entities separate from the mathematician conceiving them, represents an extreme version of the "fixated" point of view. Not only are the old problems unsolved, but their purported solution has led to the realization that much of what has come before must in fact be redone. There is an effort to claim that logical processes in mathematics are in fact dependent upon the individual performing them. In principle, this would seem to involve an almost compulsive repetition of mathematical acts, though of course in practice, intuitionists accept the precedent of previous work as well. Thus Brouwer writes:[56]

> And there can be no talk of the existence of a causal connection of the world independently of men. On the contrary, the so-called causal connection of the world is an outwardly working thought-force in the service of a dark function of the human will that thereby subjects the world more or less defenselessly, analogous to the way the serpent makes its prey defenseless through its hypnotizing glance and the cuttlefish through spraying its secretions. . . .
>
> Mathematical examination, respectively mathematical treatment, functioning at first in the service of the will of individual men, can now become represented, exactly like, at first, autonomous aggressive or defensive activity, as work in the service of a commanding will, be it the individual will of another man, be it the parallel will of a group of men, or of all humanity. . . .

[53] *Bulletin of the American Mathematical Society* (1913): 81–96.

[54] See note 43 above.

[55] E.g., Paul Alexandroff, "Die Topologie in und um Holland in der Jahren, 1920–1930," *Nieuw Archief voor Wiskunde* 17 (1969): 109–121.

[56] L.E.J. Brouwer, "Mathematik, Wissenschraft und Sprache," *Monatshefte für Mathematik und Physik* 36 (1929): 154, 155, 157.

> Now, however, for transference of the will, especially for the transference of the will mediated by speech, there is neither exactitude nor certainty. . . . Therefore, for pure mathematics, there is also no certain speech.

Similarly, Arend Heyting, Brouwer's disciple, remarks that as an intuitionist,[57]

> my mathematical thoughts belong to my individual intellectual life and are confined to my personal mind, as is the case for other thoughts as well. We are generally convinced that other people have thoughts analogous to our own and that they can understand us when we express our thoughts in words, but we also know that we are never quite sure of being faultlessly understood. In this respect, mathematics does not essentially differ from other subjects; if for this reason you consider mathematics to be dogmatic, you ought to call any human reasoning dogmatic.

Furthermore,[58]

> [Intuitionism's] chances of being useful for philosophy, history, and the social sciences are better [than for physics]. In fact mathematics from the intuitionistic point of view is a study of certain functions of the human mind, and as such is akin to these sciences.

In other words, one can never be exact enough, and yet just this, the mathematician must repeatedly attempt.

Hilbert's view cited earlier that Cantor's set theory was a "paradise" for mathematicians represents the view for advance to new maturity, to new omnipotence even, and holds out the promise of ever further advances. He sees mathematics as always making new discoveries, mathematicians surveying new vistas from some achieved "peak in Darien" and then going forth and conquering them. Thus his famous credo:[59]

> For the mathematician there is no *Ignorabimus* and in my opinion also for natural science, none whatsoever. . . . The real reason Comte did not succeed in finding an unsolvable problem lies, in my opinion, in that there is no such thing as an unsolvable problem. Instead of the foolish *Ignorabimus*, our solution in contrast is:
>
> We must know.
> We will know.

Thirty years earlier, in his famous address on mathematical problems at the international congress in Paris, his point of view was identical:[60]

> This conviction of the solvability of every mathematical problem is a mighty impetus during work; we hear within us the continual call: There is the problem, seek the solution. You can find it by pure thought; for in mathematics there is no *Ignorabimus*.

[57] Heyting 1966: 8.

[58] Ibid.: 10.

[59] D. Hilbert, "Naturerkennen und Logik," *Naturwissenschaften* (1930): 959–963, reprinted in D. Hilbert, *Gesammelte Abhandlungen* (1965), 3:378–387. The reference to Comte is to his failed attempt to cite the chemical constitution of the stars as an unsolvable problem. The last lines are often quoted in the original: "Wir müssen wissen. Wir werden wissen."

[60] *Göttinger Nachrichten* (1900): 253–297, reprinted in Hilbert 1965, 3:290–329, p. 298.

For Hilbert, the mathematician is a detective, probing the dim unknown and bringing it into clear light. For Brouwer, the mathematician is an uncertain master in the world of his creation. Hilbert's view is optimistic and believes in the reality of progress, Brouwer's is pessimistic and regressive—one must return and reconstruct. Both men devoted many years to the logical situation in mathematics, but their points of view and psychological positions were quite different, and, not surprisingly, they were personally antipathetic. According to Constance Reid, Brouwer regarded Hilbert as "my enemy" and once left a house in Amsterdam where he was a guest because B. L. van der Waerden, also a guest, referred to Hilbert and Courant as his friends.[61] According to A. A. Fraenkel, "A personal note was assumed above all in his [Brouwer's] discussions with Hilbert, they covered each other immediately with personal slanders (*Schimpfworten*)."[62] According to Paul Alexandroff, who studied with Brouwer, an attempt was made at a dinner in Emmy Noether's house in 1926 to heal the breach between the two men. Apparently this was accomplished through bringing out (at Alexandroff's initiative) their common opinion (presumably negative) of the famous complex analyst Paul Koebe. Unfortunately, the reconciliation did not last.[63]

For the Nazi mathematician, the issue was the ability of mathematics to "relate to real objects," where geometric configurations and integers were admitted as examples of "real objects." The question was one of restraining mathematical abstractions, of permitting only that which was related to a concrete physical world. If "Deutsche Physik" were to be predominantly experimental,[64] "Deutsche Mathematik" was to have some connection, however tenuous, with the mathematics of everyday: the integers, geometry, probability. The concept was vague (and needed to be so), and there is no use nor value in attempting to make it more precise. For Ludwig Bieberbach, the distinction was between defining π as the ratio of the circumference of a circle to the diameter, and defining it (as Landau did in his introductory book in analysis) as twice the smallest positive zero of the cosine (itself defined as a power series).[65] From a purely formal viewpoint, since both are characterizations of π, one could start with either as a definition and, with enough labor, proceed to the other. Landau's definition is not the common one or the naive one, and it may (or may not, since it serves the purpose of his task) be unwise on his part, but for Bieberbach it is typically Jewish. The Nazis made the distinction between the imposition of

[61] Reid 1970: 187.

[62] A. A. Fraenkel, *Lebenskreise* (1967), 16.

[63] Paul Alexandroff 1969: 122–123. Koebe in Alexandroff's memoir is somewhat coyly referred to solely as "den bekannten Luckenwalder Funktiontheoretiker" ("the well-known analyst from Luckenwald").

[64] See, for example, Lenard 1936: passim.

[65] Bieberbach, "Persönlichkeitsstruktur und Mathematisches Schaffen," *Unterrichtsblätter für Mathematik und Naturwissenschaften* 40 (1934): 236. See also the newspaper report on the lecture that is the origin of this article, in "Neue Mathematik," *Deutsche Zukunft* 14 (April 8, 1934), p. 15. Bieberbach's article is analyzed in chapter 7. This report was the source of an important conflict in the German Mathematical Society, as discussed in chapter 6.

mental constructions upon "natural" mathematical objects and the receptivity of a genius to what nature will show. In Bieberbach's words,[66] "With Jacobi it is the will of men to push through and his will against the given, with Euler, a playful at-oneness with things." In some sense, this ideological Nazi point of view is that Jews, for example, attempt to "create" mathematics, whereas the true German "discovers" mathematics. However, not only is this old distinction in philosophical points of view, according to them, racially determined, but also the "creation" point of view is perverse—at best, *undeutsch*, and so unsuited to the instruction of Germans. Since Jews a priori hold this *undeutsch* point of view, the educational detriment of Jewish instruction becomes apparent. It is along these lines that Bieberbach was thinking when he remarked on the independence of Erich Jaensch's theory of typological types from the theory of racial differences. Jaensch distinguished several different types of thought in mathematics and labeled these types with letters and subscripts; however, it turns out that[67]

> the S-structure is predominant and pure among Jews, with certain restrictions, also among the French. . . . Since the S-Type does not correspond to our prominent indigenous folk-type (*vorwiegenden eigenen Volksart*) in Germany, by the investigation of the form of mathematical talent (*Anlage*) tied up with the S-structure we obtain in any case a striking picture especially of Jews and also of French, not however of German mathematicians. In Germany mathematical talent (*Anlage*) is tied up with the J-types.

Thus, for the Nazis, what was subjective about mathematics was the various individual approaches to it; these could be categorized and grouped together, whereupon they fell, it so happened, into national groupings with variations. For the Nazis, the subjectivity was particularistic, while for the intuitionist it was universal. However, having once admitted that mathematics only contains subjective truths, if one is not careful, one begins comparing subjectivities. It is always easy to find those methods of approach to which one is sympathetic as being more valuable than unsympathetic methods. To make the results subjective can then lead, unless care is taken, to a demeaning of what is obtained by other methods, whose logical status may not be in question, but which are "unsympathetic." Instead of each person's method becoming merely his best way of coming to whatever result is obtained, there occurs a valuation of methods and persons according to a priori criteria.

It should be emphasized that intuitionism (although it finds nonintuitionist logic wanting) has no such intentional tendentious valuations as part of its formal program; nor is there any necessary logical connection between intuitionism and Nazi philosophy: the figure of Hermann Weyl alone is sufficient to dismiss such speculations. In fact, for someone thoroughly applying to mathematics the biologistic Nazi *Weltanschauung*, which referred all criteria of validity

[66] Bieberbach 1934: 239.

[67] Jaensch and Althoff 1939: 80–81.

to the *Volk* intuitionism would have been as reprehensible as any other form of "nonintuitive" logic. They would all be Tornier's "Jongleur mit Definitionen."

Yet, for all the differences between a scientific program and a political one, both intuitionism and the Nazi *Weltanschauung* saw science, particularly mathematics, as being subjectively determined. Furthermore, as discussed in chapter 7, among those German mathematicians who came to promote intuitionism was Ludwig Bieberbach. Brouwer's own behavior after 1933 also led to suspicions of such a relationship between Nazi sympathizing and intuitionism. In 1934, Brouwer negotiated with the Nazi government about a professorship in Göttingen.[68] On June 30, 1939 (signing himself Egbertus Brouwer), Brouwer wrote an encomium for the seventieth birthday of Theodor Vahlen, the aging, competent, but far from distinguished mathematician who, an *alter Kämpfer* (or "old fighter"), had joined Hitler in 1923 and had been an important bureaucrat in the Nazi educational apparatus, as well as *Herausgeber* (or "publisher") of *Deutsche Mathematik*. This gratuitous praise by Brouwer (as a scientist) of a mathematician whose professional work had not the least in common with his own, and whose primary activities for some time had been political, would seem to be an indication of his ideological sympathies on the eve of World War II.[69] After World War II, Brouwer was also temporarily suspended from his post by the Dutch government after the war in the *Zuivering* ("cleaning-up") of the universities, though he apparently was never penalized by a court.[70] The suspension, however, was *not* for pro-Nazi activities, and it would appear that, despite his friendship with Vahlen, Brouwer was *not* pro-Nazi. Two letters from Arend Heyting to Jonathan Liebenau, and letters from Hans Freudenthal, Walter van Stigt, and Dirk van Dalen to the author, all affirm that Brouwer was neither pro-Nazi nor anti-Semitic (though he did not offer any resistance to the German occupation of the Netherlands). Furthermore, these letters corroborate one another in some detail. It seems Brouwer was truly "unpolitical," maintaining friendships with Nazis and Marxists alike. Van Stigt says he has interviewed a Jewish woman who claims Brouwer hid and protected her during the war.

It is hard to be completely unsympathetic to either Brouwer's or Hilbert's philosophical point of view—each requires its own kind of courage. One must pay for set theory with the real and serious problems revealed by the antinomies, for the Axiom of Choice and an "easy" and intuitively[71] appealing the-

[68] The literature contains several undocumented references to such negotiations by Brouwer with varying dates and either Göttingen or Berlin as locations. However, in Brouwer's *Nachlass* in Amsterdam are copies of correspondence with Helmut Hasse, Erhard Tornier, and Theodor Vahlen relevant to a position at Göttingen. These show signs of having been torn and burned, and I do not know how they survived their attempted destruction; however, what remains makes the existence of such negotiations patent. How serious Brouwer was about actually eventually coming to Göttingen is unclear. I am indebted to Prof. Dirk van Dalen for copies of the surviving papers.

[69] See *Forschungen und Fortschritte* (15 Jahrgang #18), for June 1939. For more about Vahlen, see below, chapters 4–7.

[70] Personal communication from L. DeJong of the Rijksinstituut voor Oorloogsdocumentatie.

[71] Of course, intuition to a certain extent is based on experience and what one is used to. The existence and importance of homology groups was apparently "intuitive" to Emmy Noether (see

ory of real numbers with phenomena like the "Banach-Tarski Paradox."[72] To reject the dubious and attempt to rebuild on a firm foundation, while insisting on the existence of inevitable infirmities caused by the entrance of human subjectivity, is a path especially remarkable when it involves rejection of one's own previous work. To push forward with a heroic effort to conquer the difficulty rather than radically reject the past as too complex and uncertain to master seems no less courageous. Such decisions are even sharper today with the proof by Kurt Gödel and Paul Cohen that the Axiom of Choice is in fact independent of the usual axiom system on which mathematics is based.

Finally, there is the paradox that the actual completion of the Brouwerian program would of course lead to a new position of omnipotence for such mathematicians: all predecessors would have been replaced by this new revelation. Such "new revelations" have happened before in mathematics—for example, in the concept of function, and in the manipulation of infinite processes—and they may well happen again. As Abraham Robinson remarked in 1969, "There is every reason to believe that the codification of intuitive concepts and the reinterpretation of accepted principles will continue also in future and will bring new advances, into territory still uncharted."[73] It is my view, however, that by its own psychological nature, the Brouwerian program can never be completed, since it emphasizes and involves the continual, almost compulsive, reassessment and consideration of the inexactitudes of language, communication, and inevitable human subjectivity.[74]

The tensions revealed in mathematics by the "crisis at the foundations" are reflective of a general tension existing in mathematics and natural science, a tension that Thomas Kuhn has gone so far as to call "essential."[75] For, as Kuhn

---

Alexandroff 1969: 121), but hardly to a mathematician preceding her. Similarly, it may be natural for engineers and mathematicians today to have intuitions involving complex numbers, but this could hardly have been the case at that earlier time when even roots of negative quantities were considered illegitimate.

[72] The "Banach-Tarski Paradox" is briefly the following: It is possible to mathematically decompose the surface of a sphere into a finite number of different pieces and mathematically reassemble those pieces so as to form the surface of a sphere of twice the radius. Startling as this may sound, the key is the word "mathematical"; there is *no* claim that this can be done physically, as it, of course, cannot. For example, one of the mathematical pieces may perhaps be a point, which has no dimension and is physically unrealizable. In fact, it can be shown that some of the pieces in such a decomposition must be what mathematicians call "nonmeasurable." Not only are nonmeasurable sets not physically realizable, their mathematical existence depends upon the presumption of the Axiom of Choice. It has actually been shown that the minimum number of mathematical "pieces" in which the decomposition can be carried out is five, and one of these may be taken to be a point (see R. M. Robinson, *Fundamenta Mathematica* 34 (1947): 246–260).

[73] Abraham Robinson, "Some Thoughts on the History of Mathematics," *Compositio mathematica* 20 (1969): 188–193.

[74] See Heyting 1966: 1–12. For example, on page 9, Heyting writes, "If really the formalization of language is the trend of science, then intuitionistic mathematics does not belong to science in this sense of the word. It is rather a phenomenon of life, a natural activity of man, which itself is open to study by scientific methods."

[75] Kuhn 1977.

points out, scientific "education" is more a matter of indoctrination, at least in those areas in which "paradigms" have been established due to an initial systematization of the field.[76]

> First, education in the natural sciences . . . remains a dogmatic initiation in a pre-established tradition that the student is not equipped to evaluate. Second, at least in the period when it was followed by a term in an apprentice relationship, this technique of exclusive exposure to a rigid tradition has been immensely productive of the most consequential sorts of innovations.

This would seem to be especially true in mathematics. "Normal science," "normal mathematics," is the day-to-day elucidation and extension of the old results, an increasing broadening and refinement of them. In natural science, the distinction between such activities and the scientific "revolution" that causes the Kuhnian paradigm to change are signaled by the recognition of and concentration upon sources of dissonance.[77] Several examples of such "revolutions" in physical science are familiar to the general educated public. In mathematics, such "revolutions" are more esoteric; in fact, revolutions in the sense that some basic idea (like phlogiston, Ptolemaic cosmology, Aristotle's physics) is dethroned, never to rise again, do not take place in mathematics. Instead of revolutions in mathematics, one should probably speak of "fundamental innovations" that surpass their origins and usually incorporate them. Such innovations provide a new way of looking at the mathematical world that preceded them. As Donald Gillies has remarked, they may be "revolutions" in the political sense of the "Glorious Revolution" in Great Britian, a change that left the previous concepts untouched in fact but considerably restricted in importance or domain of validity. These are changes that also take some time for their full development (this is true in the political analogy as well, which may have reached its culmination with George I). As Joseph Dauben has remarked, whether there are revolutions in mathematics depends upon "what one means by the term 'revolution'." The debate started by the controversy between Dauben and Michael Crowe, and examined in Gillies' book *Revolutions in Mathematics*, seems to end in agreement that the problem is "definitional." A prime example is Herbert Mehrtens, who wishes to use the concepts developed by Kuhn while denying the existence of Kuhnian revolutions in mathematics. While Kuhnian revolutions do not take place in mathematics, revolutions in Joseph Dauben's sense of "discontinuities of such magnitude as to constitute definite breaks with the past" do. This leads to the perplexing philosophical question: "What is mathe-

[76] The quotation is from Kuhn, ibid.: 229. Compare also Michael Mulkay, "Some Aspects of Cultural Growth in the Natural Sciences," *Social Research* 36, no. 1 (1969) (an abridgment appears in Barry Barnes, ed., *Sociology of Science* [London: Penguin Books, 1972], 126–142), and Mulkay, *The Social Process of Innovation: A Study in the Sociology of Science* (London: MacMillan, 1972). Similarly Bentley Glass (1962: 220) has written, "As to the scientific spirit, there is little of that in either the conventional textbook or lecture. One meets it better in *Arrowsmith* or the *Life of Pasteur*."

[77] The use of Kuhnian constructs does not imply any necessary adherence to his (varying and much debated) interpretation of them, or that of others.

matics about?"—which fortunately is not germane for the considerations of this book.[78]

In any case, some (certainly not all) examples of such fundamental innovations are the initial development of "modern algebra" by Emmy Noether and Hilbert, the invention of set theory by Cantor, the recognition of the role and consequences of the Axiom of Choice by Zermelo, the independent introduction of a new species of integration by Henri Lebesgue and Grace and W. H. Young, the creation of metric spaces by Maurice Fréchet, and, to take a more recent example, the independent development of the dual theories of generalized functions and distributions by Jan Mikusinski and Laurent Schwartz. In addition to innovations such as these, there are innovations in mathematics that involve new ways to use old tools in old problems. The curious fact remains that a mathematical difficulty may be not only an anomaly (the "impulse functions" of physics), or a bewildering computational complexity (the finite basis problem of invariant theory), but also simply a lack of awareness of the appropriate uses for the tools at hand. Again, the introduction of a new mathematical tool may be involved in making precise the appropriate definition of an age-old concept (Lebesgue's integration theory and the meaning of function) or in abstracting the essence of certain procedures, this new clarity allowing for a new and fruitful development with unsuspected applications and implications (the theory of metric spaces). Yet all these recognized and conquered difficulties, all these new ideas and many others unmentioned, originate from practitioners trained in and mostly doing "normal" mathematics. As Kuhn remarks,[79]

> Yet—and this is the point—the ultimate effect of this tradition-bound work has invariably been to change the tradition. . . . At least for the scientific community as a whole, work within a well-defined and deeply in-grained tradition seems more productive of tradition-shattering novelties than work in which no similarly convergent standards are involved. How can this be so? I think it is because no other sort of work is nearly so well suited to isolate for continuing and concentrated attention those loci of trouble or causes of crisis upon whose recognition the most fundamental advances in basic science depend.

This certainly applies to mathematics.

Mathematical education with its tradition-bound work has a certain positive role to play in the creation of a mathematician. Not only is the mathematician thereby inaugurated into the processes of "normal mathematics," but the education itself tends to provide a definite way of looking at the subject matter that

[78] For the above, see Donald Gillies, ed., *Revolutions in Mathematics* (1992), particularly the introduction, and chaps. 1–5 and 15. On the philosophical question, one is tempted to agree with Michael Dummit (in Alexander George, ed., *Mathematics and Mind* [New York: Oxford University Press, 1994], 11–26) that mathematics is not about "things" at all but is the "systematic construction of complex deductive arguments" (p. 13). Yet Jeremy Gray (chap. 12 in Gillies 1992) convincingly shows that there are revolutions in mathematical ontology, in which the meanings of "things" change. Citations above are from Gillies 1992: 50 (Dauben), 316 (Crowe), 51 (Dauben).

[79] Kuhn 1977: 349.

reduces whatever dissonance may occur between perceptions of it as they are and as they might be wished. One of the great virtues of traditionalism in education (which is sometimes overlooked these days) is that it provides meaning. The investigator for whom all new perceptions are equally valid (for example, in one of Kuhn's preparadigmatic states of a science) is surely as much at a loss as the one who knows not how to conceive something new. Somehow good education must avoid stultifying while presenting a firm judgmental foundation. This is especially true in mathematics. In any case, the traditionalist aspect of mathematical education allows a mathematician to "get on with the job," but what happens when a "crisis" occurs? The pressures generated by the traditionalist aspect of scientific education tend to reduce the experience of cognitive dissonance.[80] Whether, for an individual, they are in fact sufficient to allow a traditionalist interpretation of the crisis, or perhaps just ignore it, or whether such an individual in fact becomes a "renegade" from tradition, depends on individual factors. There are, for example, the various "monster-barring" and "monster-adjusting" strategies cogently discussed by Lakatos,[81] which allow one to persevere in a given tradition when faced with "cognitive dissonance" in a mathematical or scientific enterprise.

Both Brouwer's and Hilbert's programs represent attempted resolutions of the tension brought to a pitch by the "crisis" in mathematics. In fact, their resolutions make antipodal emphases in weighing the seemingly paradoxical subjective-objective nature of mathematics discussed earlier. Most working mathematicians ignored the crisis as much as possible, as their counterparts in similar situations had always done. In A. F. Monna's words:[82]

> It is a common feature of the history of mathematics that progress was often obtained without mathematicians' troubling themselves much about the foundations; one observes this not only in set theory but also in analysis. I do not mean to say that a critical attitude was absent. . . . Perhaps mathematicians [of an earlier time] felt they had to seek definitions or descriptions of objects that already existed in a natural way and that all other objects were artificial or pathological. The deeper ground is then a problem of an ontological character. The idea that a mathematician creates his objects by means of definitions was alien to the mathematicians of earlier time [roughly before 1900].

Brouwer's approach has been considered "radical" because of his radical rejection of principles of classical logic as well as of many of the developments in

[80] The "theory of cognitive dissonance" arises from the work of Leon Festinger and his collaborators. See, e.g., his *A Theory of Cognitive Dissonance* (1957). A very readable presentation of an extended case study in the theory is L. Festinger, H. Riecken, and S. Schlachter, *When Prophecy Fails* (1956). The relevance of the ideas of "cognitive consensus" and "cognitive dissonance" for the study of the history of science was already noted by Mulkay (1969: n. 76). My own discovery and use of Festinger is independent of Mulkay, but as in his paper, it is true here that there is no necessary adherence to Festinger's theoretical interpretation of his constructs.

[81] Imre Lakatos, "Proofs and Refutations," *British Journal for the Philosophy of Science* 14 (1963–64), reprinted as chapter 1 of Imre Lakatos, *Proofs and Refutations* (1976).

[82] Monna 1972–73: 81.

analysis that were reaching their final form as he initiated his criticism. But in a way, Brouwer's program is, if anything, reactionary; he will hold only to certain constructive methods; only those objects that can be given in this way will be considered "real." And it is always remembered that these principles are the creations of human beings. Brouwer, by a "radical" restriction of what is logically allowable, constrains the mathematician to a narrow view of the subject matter, and to obtaining a smaller class of results, with more difficulty. The reward is in knowing that the Brouwerian standards of rigor have been adhered to. In some ways, in this program, the mathematician, who is forever involved with the objects manipulated, therefore seems their slave. At best, the mathematician is an uncertain master.

In contrast, Hilbert's program is one of ever further conquest, of pushing to the limits and then beyond to newly perceived limits. The investigator is the master of axiomatically defined quantities, and is free to investigate in the confidence that the work is objective. The potentiality of knowing whether a statement is true or false is tantamount to stating that it is either true or false.[83] Curiously enough, this is not at all obvious to intuitionists. The insistence on the inability to declare that every mathematical statement is either true or false, and so, if not false, must be true, seems contrary to "ordinary intuition" and "common sense."[84]

It is in this sense of mastery and confidence in objectivity that Hilbert's program breathes the optimistic spirit of free liberal inquiry and Brouwer's the pessimistic one of cramped and overdetermined investigation. However, from another point of view, Hilbert's program involves the total subjection of the mathematician's world to the mathematician's will. ("We must know. We will know.") It is reminiscent of Bacon's dictum about torturing nature on the rack to obtain her secrets. Brouwer's point of view, in contrast, involves the harmonious union of the mathematican with the world surveyed. The mathematician is a part of it, and only by explicitly recognizing that role in it can success be achieved, or at least as much success as possible. In other words, untrammelled inquiry means potentially conquering everywhere, but trammelled inquiry means something is preserved. Which program one prefers, and which one thinks perjoratively about, is a matter of taste.[85]

To briefly sum up, there are two extreme attempts at resolutions of the crisis in mathematics. One proceeds by allowing the problem "to pass," that is, by refusing to acknowledge that it cannot be overcome, and progressively seeking ever newer formulations and methods to do so. The other proceeds by becoming "fixated" in the problem, acknowledging its dominion, and retrenching to save whatever can be saved. In the first, the resolution involves the consump-

[83] Cf. ibid.: 76.

[84] The development of multivalued logics (where there are more choices than just "true" or "false") by Jan Lukasiewicz and others is independent of intuitionism.

[85] There are also other variant programs, and Hilbert's ideas in particular show a development. As already suggested, and as will be seen particularly in chapters 6 and 7 below, such programs became involved as well in the nationalist currents that swept Germany and ended in Adolf Hitler.

tion of aggression in ever newer conquests and detections. In the latter, aggression is (as Brouwer himself indicates)[86] consumed in the eternal attempt to be a harmonious portion of the mathematical world in which one dwells.

The crisis in mathematics would seem in considerable measure to revolve around the double question: "What is and what should be mathematics?" An attempt has been made to interpret the attitudes that seem to lie behind the two most important efforts to resolve the crisis that affected the German mathematical community—and so mathematics as a whole.

This crisis, however, was (and is) deeper than most scientific crises, for it involved not just a theory of combustion or of matter, but the nature of allowable thought-processes. Yet, precisely because it was so fundamental and so removed from everyday practice, many working mathematicians could simply "get on with the job" in the tradition in which they had been trained, without paying the crisis much heed.[87] Thus the strong traditionalist training had two effects: it allowed ongoing work to proceed under the tacit assumption that everything would come out satisfactorily in problems concerning logical fundamentals, so that little if any change would be required in the form, and none in the substance of the work; and it provoked responses like Hilbert's to save as much of the established order as possible—"save the appearances," as it were. For Brouwer, on the other hand, the pitch of dissonance was sufficient to evoke a total rejection of traditionalist teaching, and even of his own acclaimed work. Hermann Weyl attempted a working compromise between Brouwer and more traditionalist views, but never really satisfied himself, while it made him a heretic from both the formalist and the constructivist schools. As he eloquently put the situation:[88]

> From this history one thing should be clear: we are less certain than ever about the ultimate foundations of (logic and) mathematics. Like everybody and everything in the world today, we have our "crisis." We have had it for nearly fifty years. Outwardly it does not seem to hamper our daily work, and yet I for one confess that it has had a considerable practical influence on my mathematical life: it directed my interests to fields I considered relatively "safe," and has been a constant drain on the enthusiasm and determination with which I pursued my research work. This experience is probably shared by other mathematicians who are not indifferent to what their scientific endeavors mean in the context of man's whole caring and knowing, suffering and creative existence in the world.

In 1918 Weyl and George Pólya made a bet on whether in twenty years' time Brouwer's ideas would have come to dominate mathematics, Weyl taking the affirmative. According to Constance Reid, in 1938 Weyl admitted he had lost the bet, but asked Pólya not to make him concede publicly.[89]

Weyl noted in 1946 that the mathematical crisis had been going on for nearly

[86] Brouwer 1929: 155.
[87] Cf. A. F. Monna's remarks above, note 82.
[88] Weyl 1946: 13.
[89] Reid 1970: 211.

fifty years. Nevertheless, as the preceding discussion makes clear, it was after World War I that the sense of crisis among mathematicians really flourished.[90] At this time, it intersected with a crisis of confidence infecting the German professoriate.

[90] This is true of watersheds in other German-speaking cultural areas as well; for example, Expressionism in art, or "Bauhaus-style" architecture, or various post-Romantic musical systems, or quantum physics. For this last, see Forman 1971. World War I seems to have provided a cataclysmic political change, accelerating the development of a world already significantly changing both culturally and scientifically.

CHAPTER THREE

The German Academic Crisis

GERMANY in the Weimar period was notable for a remarkable efflorescence of arts, of letters, of the sciences. Peter Gay's small book on Weimar culture provides a survey of art and literature, while consciously and regrettably neglecting science.[1] Paul Forman has provided an insightful and suggestive essay concerning the possible relationship between the philosophical ideas current in Weimar and the physical and mathematical ones.[2] Kurt Mendelssohn's biography of Walther Nernst[3] describes the work of its hero and related science with passing mention of the social context. Mendelssohn does suggest[4] that times of great political and social turmoil are times of great scientific turmoil; but however well this superficially seems to fit the Weimar period in Germany, as a formula it is rather too glib to have much meaning without a deeper study of the social contextual relationships of science.

A main concern of this work is to study such relationships in the hothouse atmosphere of Nazi Germany with particular respect to mathematics. Believing that the *Third Reich*, while hardly an inevitable consequence of Weimar, was prepared there, the attitudes of academics toward society, toward politics, toward their subject matter during this period, become of interest.

The Nazi *Weltanschauung* itself had little to say about science or mathematics, and most of that was negative, but mathematicians who believed in Hitler's nationalist or cultural message made distinctions. For example, Erhard Tornier condemned axiomatics, Ludwig Bieberbach considered measure theory a subject fit only for non-Aryans, and Max Steck had no use for formalism. There certainly was no "party line" about such matters, and mathematicians who espoused similar cultural politics might disagree on how that politics affected mathematics. The very fact, though, that there were interactions of this sort suggests looking at the extra-mathematical social and cultural crises, and the professional reaction to them, at least briefly, even if this is a subject that has been often examined. Thus, this chapter looks at some of the features of Weimar academic society. When Weimar replaced Wilhelminian Germany, professors felt declassed; they resented the "unbelievable" German defeat in the war, and even more the conditions imposed at Versailles. They were "apolitical," but nevertheless carried on politics in an ideal atmosphere. In addition, anti-Semitism ran through many faculties. The German professor was generally a

[1] Gay 1968: 3.

[2] Forman 1971.

[3] Kurt Mendelssohn, *The World of Walther Nernst: The Rise and Fall of German Science, 1864–1941* (1973).

[4] Ibid.: 110, and chapter 7 below, passim.

conservative establishment figure who, under Weimar, had largely lost his establishment status.

Undeniably, the German academic community from 1918 to 1933 contained members of every political persuasion. Nevertheless, there was an academic culture of which the large majority of professors and students partook. An important part of this academic culture was of the notion of the professor as a prestigious state servant who had been declassed by the collapse of the empire and the establishment of the republic. "Academic freedom" was freedom in academic and personal matters, not the freedom of academics to speak out politically, and the life of the intellect remained confined to the academic realm without penetrating or affecting the community in any critical way.

Chapter 8 will look biographically at some particular mathematicians, but their attitudes need to be set in the matrix of attitudes held by that class so aptly termed "mandarin" by Fritz Ringer.[5] For although Ringer explicitly excludes natural scientists from the details of his study, by tradition, upbringing, and collegiality they were, as he says, "as much mandarin intellectuals as their colleagues."[6]

These "mandarin intellectuals" were also legally civil servants who had freedom in classroom instruction. In exchange for this freedom, though, they served the state. Leo Arons was a physics instructor who was, by chance, an active socialist. When the right to teach was withdrawn from him in 1898 by Prussian governmental fiat, overruling his own faculty, there was no great protest from physicists (or other academics for that matter).[7]

Less than forty years later, the attitude of professors of mathematics and natural science toward the national state had not changed much. Helmut Hasse, a distinguished mathematician, told Constance Reid in an interview around 1975:[8]

> My political feelings have never been National-Socialistic but rather "national" in the sense of the Deutschnationale Partei, which succeeded the Conservative Party of the Second Empire (under Wilhelm II). I had strong feelings for Germany as it was created by Bismarck in 1871. When this was heavily damaged by the Treaty of Versailles in 1919, I resented that very much. I approved with all my heart and soul Hitler's endeavors to remove the injustices done to Germany in that treaty. It was from this truly national standpoint that I reacted when the Faculty more or less suggested that such a view was not permissible in one of its members. It was also the background for my remarks to the Americans. They were talking about reeducating Germany, and I

[5] Fritz Ringer, *The Decline of the German Mandarins* (1969). This is an important thoroughgoing sociohistorical analysis of the development of social and political attitudes, 1890–1933, in the German professoriat.

[6] Ibid.: 6.

[7] See Hans Bleuel, *Deutschlands Bekenner* (Bern: Scherz, 1968), 50–53; Ringer 1969: 141–142; Edward Shils in M. Weber, *Max Weber on Universities*, trans. and ed. by Edward Shils (1974), 15 n. 16; and Dieter Fricke, "Zur Militarisierung des deutschen Geistesleben im wilhelminischen Kaiserreich: Der Fall Leo Arons," *Zeitschrift für Geschichtswissenschaft* (1960): 1069–1107.

[8] Constance Reid, *Courant in Göttingen and New York* (1976), 250.

said some strong things against this. It irked me that everything against Hitler was desirable, and everything that he had done was wrong. I continued to be a national German, and I resented Germany being trampled under the feet of foreign nations.

German professors of the time were always, first, "national." That inclination toward the state brought with it tremendous prestige, and that prestige could be a temptation that was ultimately deceitful and destructive. Furthermore, being "apolitical" produced a political naiveté enhancing such temptation. In 1936, Eberhard Hopf, another distinguished mathematician who was an assistant professor at M.I.T. and had become one of the best analysts of his generation, accepted a call to a professorship at Leipzig. On June 23, 1945, Hopf wrote Richard Courant, who was among the first professors dismissed by the Nazis and had emigrated to the United States:[9]

> Needless to say how deeply I have regretted my lack of political insight in 1936 when I decided to accept the call to Leipzig and to leave M.I.T., in spite of generous offers President Compton of M.I.T. made to me. When I fully realized what the men in power in Germany were heading for it was too late to return to the States. I and particularly my wife who was more reluctant about leaving the States have had to pay for my erroneous judgment of the situation. Within the last years, ill-willing people pursued us with a whole flood of denouncements and intrigues that caused additional trouble for us. (Our outspoken and strict avoiding any Nazi affiliations probably contributed to it.) My wife on whose shoulders lay, without outside help, the whole care for the four of us was forced to do half day work besides. Only by a serious breakdown she got rid of this burden. My attempt to fight against this kind of treatment only led to a threat of worse treatment. Since that time I had the distinct feeling of being watched and I was, therefore, more careful than before. But the constant swallowing of anger I found harder to stand than the many air raids in Munich.[10] Needless to say that the quick end of the last war phase came as a great relief to us and that we could not help looking upon the oncoming Americans as potential freers.

As early as 1934, in fact, Hopf had been considered as a potential faculty member for the "rebuilding" of the Göttingen mathematics department under the Nazis.[11] On September 30, 1946, Courant wrote Minna Rees at the Office of Naval Research in the same tones as Hopf had written him:[12]

[9] E. Hopf to Courant, June 23, 1945, p. 2, in Courant Papers, Courant Institute at New York University. This lengthy five-page letter was written in English.

[10] Hopf's return to Germany was to replace Leon Lichtenstein, who had died on August 21, 1933, of heart and kidney ailments. The position was left vacant for several years. Had he not died, Lichtenstein would certainly have been eventually dismissed as Jewish. In 1944 Hopf had become Carathéodory's successor in a mathematics chair at Munich.

[11] See exchange of letters, July 14–July 20, 1934, between his father Friederich and Helmut Hasse in the papers of the Göttingen Mathematical Institute. These papers (hereafter cited as MI) were discovered and organized by Norbert Schappacher and are available for inspection at Göttingen. Friederich Hopf was prepared to provide the Aryan proofs for his son and daughter-in-law as requested by Hasse.

[12] Courant to Minna Rees, Sept. 30, 1946, in Courant Papers (Courant Institute). An essentially similar view of Hopf appears in Norbert Wiener, *I Am a Mathematician* (1964), pp. 209–211.

> Hopf is perhaps the best representative of mathematical analysis in Germany. His field is close to applied mathematics; there is no doubt that from a scientific point of view he would be a very noticeable addition. Hopf is unassuming and a more or less scholarly type, whose interests are entirely centered around scientific matters. I have no misgivings about his political attitude, although he committed the major blunder of accepting a position as professor at Leipzig around 1936 although at the time he held an assistant professorship at M.I.T.; Hopf has repented this step. In view of his great scientific qualifications I feel that it should no longer be held against him. There is little doubt that Hopf would be welcome at various universities. The time is too short to make inquiries now, but I can say that we would be glad to have him work on our ONR [Office of Naval Research] contract.

The "Lex Arons" of 1898, which allowed the Prussian government to overrule the Berlin faculty in matters affecting *Privatdozenten* (beginning teachers not in the civil-service faculty), was, by and large, accepted by the German professors, despite the fact that it constituted "a militarization of the academic community."[13] From before that time to Hasse and Hopf in the 1930s, the German academic community could be characterized as "national" but otherwise apolitical. German academic freedom as defined by the government in 1898 did not include the right to be a politically active Social Democrat (even if there were no politics in the physics classroom); thirty-five years later, it did not include the right to be a Jew. The majority of the German academic community seems always to have acquiesced in the narrow definition of academic freedom as freedom for specialized investigation—*akademische Lehrfreiheit*—rather than "freedom of speech." In 1784, Kant made a famous distinction between the public and private uses of Reason (which somewhat inverts our contemporary understanding of public and private) in which he praised Frederick the Great as the one prince in the world [who] says, "Argue as much as you will and about what you will, but obey!"[14]

> The public use of one's reason must always be free, and it alone can bring about enlightenment among men. The private use of reason, on the other hand, may often be very narrowly restricted without particularly hindering the progress of enlightenment. By the public use of one's reason I understand the use which a person makes of it as a scholar before the reading public. Private use I call that which one may make of it in a particular civil post or office which is entrusted to him. Many affairs which are conducted in the interest of the community require a certain mechanism through which some members of the community must passively conduct themselves with an artificial unanimity, so that the government may direct them to public ends, or at least prevent them from destroying those ends. Here argument is certainly not allowed—one must obey.

[13] Fricke 1960.

[14] The following quotation is taken from Immanuel Kant, *What Is Enlightenment*, as translated by Lewis White Beck in Immanuel Kant, *Foundations of the Metaphysics of Morals and What Is Enlightenment* (1959), 87, and 91–92. It is true that Kant was primarily talking about freedom of religious opinion, but he clearly extends the idea to general affairs of state.

These were in fact the principles under which Weimar academics came to operate. Roughly contemporary with Kant's distinction is the creation by Wilhelm von Humboldt of the "modern German University." It may be, as Wolfgang Abendroth has argued, that Humboldt hoped and believed that every autonomous individual could be developed by the university through his critical self-consciousness into an individual capable of appropriate and responsible action. Such an individual could then help in academic or governmental affairs, and develop and realize a humanistic culture as a counterweight to the contradictions and dangers of ordinary bourgeois society. Thus Humboldt believed that the university would produce an autonomous individual who was *engagé*, but at the same time whose worldly actions were informed by learning and culture, and who thus could be meaningfully responsible for those actions.[15] It may be possible, but historical reality was otherwise.[16] Instead of political individualism, a *Kulturstaat* came into being: the state would support learning in a widely humanistic sense; in return, the educated would become the trained civil servants and defenders of the state. The state would be enlightened from the inside, as it were, and so its rule could be undisputed. The professor need not be political, only "national." The idea of the "apolitical" German academic who keeps the world at a distance is well known. Perhaps less well known is how the very idea of the "German University" determined the "apolitical" character of its faculty. This is not only relevant for the ensuing discussion of the effects of Nazi policy upon mathematics and mathematicians and the autonomy of scientific development, it is also contemporaneously relevant to the United States, since the Humboldtian German University was recommended as a model for English-speaking ones by personages no less than Matthew Arnold and Abraham Flexner, among others, and has influenced university education in both the United States and Great Britain—indeed, it could be argued that the very idea of a university as a place where research is done in tandem with teaching is a German one.[17] From Wilhelm von Humboldt's suggestions to the Prussian King Friedrich Wilhelm III concerning the principles on which the University of Berlin was founded in 1809 to his latter-day disciple Eduard Spranger in the early 1930s, this apolitical attitude was part of the idea of the German University.

Wihelm von Humboldt's famous essay "Ideas toward the Determination of the Boundaries of State Activity"[18] was written around 1792, but only published

[15] Wolfgang Abendroth, "Das Unpolitische als Wesensmerkmal der deutschen Universität," in *Universitätstage 1966, Nationalsozialismus und die Deutsche Universität* (1966), 193.

[16] E.g., Ringer 1969: passim; Abendroth, 1966; Fritz Stern, *The Political Consequences of the Unpolitical German* (1960), 104–134; Eric Voegelin, "Die deutsche Universität und die Ordnung der deutschen Gesellschaft," in *Die deutsche Universität im Dritten Reich* (1966), 241–282; Frederic Lilge, *The Abuse of Learning* (1948); S. D. Stirk, *German Universities through English Eyes* (1946).

[17] For Arnold, see Stirk 1946: 13–14; for Flexner, ibid.: 18–20.

[18] Wilhelm von Humboldt, *Ideen zu einen Versuch, die Graenzen* [sic] *der Wirksamkeit des Staats zu bestimmen*, [c. 1792], translated as "The Sphere and Duties of Government" by Joseph Coulthard, Jr. (1854).

posthumously by his brother Alexander in 1851, though parts had appeared previously. The principal bar to its publication seems to have been that it was considered subversive. (The epigraph to the essay is from Mirabeau, and it appears to have been an influence on John Stuart Mill's *On Liberty*.) Nevertheless, in the next generation after Kant, while von Humboldt advocated the development of the autonomous political individual for cultural reasons, he also preferred monarchical institutions to republican ones, and for reasons similar to Kant's. In a monarchy, all that was necessary was for the citizen to be law-abiding and not do anything that would threaten state interests; then the state would not care how he comported himself. It was, of course, desirable for the duties of being a citizen of a state, and of being a private person, to be as harmonious as possible, and that happens when the duties of a citizen are not particularly peculiar, so that no sacrifice (such as religious beliefs) is required of the private person. In fact, it is exactly such a harmony that von Humboldt is aiming at. It is easy to understand the position taken by Kant and von Humboldt. It is not just that they were living under the Prussian monarchy, but also that they had before them the examples of the decline of the Athenian city-state and the Roman Republic. From an eighteenth-century Prussian viewpoint, both succumbed to demagoguery—in the former case, ending in a disastrous war, in the latter, in bloody civil war and proscriptions, with stability only restored by a monarchy. Kant preferred the fairly enlightened monarchy of Frederick the Great. Von Humboldt may perhaps have desired a constitutional monarchy (as did Mirabeau, and as Friedrich Wilhelm III had avoided providing until his death in 1840). Although this analysis of the roles of citizen and state was written around 1792, von Humboldt lived to see not only the French Revolution dissolve into terror and war, but the aftermath of Napoleon I and the ensuing European conflict, Louis XVIII, and even Louis-Philippe (von Humboldt died in 1835). For von Humboldt, human self-consciousness should inspire (in the moral person) an empathy with others, rather than a cold, callous solipsism. This striving to be a perfect citizen who does not prejudice state interests brings happiness and feelings of fulfillment. The idea of perfection "may prove a warm and genial feeling of the heart and thus transport his existence into the existence of others."[19] In short, education was to be free so that the individual could best freely serve the state.

Eric Voegelin has argued that ideas like these, translated into an academic program, produced a university that, instead of transmitting the life of the spirit and intellect (*Leben des Geistes*), into the life of the community (*Leben der Gesellschafts*) stood as an "iron curtain" between them.[20] There is no question this was almost the opposite of von Humboldt's intention. His plan for the university *was* an attack on orthodoxy and state interference in education, it *did* produce the remarkable flowering of German scholarship in the nineteenth century, but it also fitted all too well the master-servant relationship demanded by

[19] Ibid.: 79.

[20] Voegelin 1966: 262.

a monarchy. The scholar was independent as a scholar and a state servant as a civil servant. Indeed, the law of April 7, 1933, under which many academics were dismissed by the Nazis, was a law for the "reform of the civil service."

Von Humboldt's ideas fairly soon produced the aforementioned "German professor," a memorable portrait of whom has been left by William James in describing the famous philosopher Wilhelm Dilthey in 1867:[21]

> He is the first man I have ever met of a class, which must be common here, of men to whom learning has become as natural as breathing. A learned man at home is in a measure isolated; his study is carried on in private, at reserved hours. To the public he appears as a citizen and neighbor, etc., and they know at most about him that he is addicted to this or that study; his intellectual occupation always has something of a put-on character, and remains external at least to some part of his being. Whereas this cuss seemed to me to be nothing if not a professor . . . as if he were able to stand towards the rest of society merely in the relation of a man learned in this or that branch—and never for a moment forget the interests or put off the instincts of his specialty. If he should meet people or circumstances that could in no measure be dealt with on that ground, he would pass on and ignore them, instead of being obliged, like an American, to sink for the time the specialty.

And Dilthey was a philosopher who stressed the importance of the flow of life!

This "mandarin tradition" maintained itself to the bitter end. The emphasis on freely given service to the state as the end product of German education appears in an essay "Hochschule und Staat" (The university and the state) written by von Humboldt's prominent latter-day disciple Eduard Spranger in 1930. Spranger writes:[22]

> The *student body* is the youth which has grown up in this epoch and must educate itself (*sich . . . bilden*) for such great responsibilities. The way thither is through learning and through service. *Through learning*: for the political world of today is the complicated historical product of numberless forces and factors which can only be directed if one has previously attempted to understand them. *Through service*: for the way to leadership has always gone only through the vestibule of obedience and morally and freely given (*sittlich-freien*) subordination. The highest thing which the German spirit brought forth was the ethic of freely given service. Those were the old thoughts about order of the German knights, it was the idea of the genuine monarchy and the sense of every noble succession thereto, it was the good core in old Prussia. If in the future there is any sort of nobility, it will be again a nobility of service.

Of course, students' inclinations are in part determined by their instructors, and Spranger closes this essay with a threefold warning to German postsecondary schools and their teachers, to students, and to the government, which he quotes from Nietzsche:

[21] As cited by Stern 1960: 112.

[22] Eduard Spranger, "Hochschule und Staat," in Eduard Spranger, *Gesammelte Schriften* (1973), 10:220.

All education (*Bildung*) begins with the opposite of all that which one now prizes as academic freedom, it begins with obedience, with subordination, with training, with service. And as the great leaders need those led, so do the followers need their leader: here rules in the ordered range of spiritual qualities (*Ordnung der Geister*) a mutual predisposition, indeed a sort of prestabilized harmony. This eternal order, to which things ever and again strive with appropriate weights, wishes to work against that culture which sits on the throne of the present, disturbing and destroying it. That culture wishes to demean the leaders to *its* service or bring them to humiliation; it flickers to those to be led when they seek their predestined leader and deafens with noisy means their seeking instinct. If, however, nevertheless the leader and led meant for each other have found themselves together, struggling and wounded, then there is a deep-seated wonderful feeling like the sound of an eternal lute.

Spranger concludes (in 1930) that if each of the groups he mentioned competitively strive to emulate this ideal for themselves, then there is enough work for the next decade and fruitless strife among them is unnecessary.[23]

Indeed, when in October 1932 the philosopher Theodor Litt wished to censure National Socialist student rowdies at a meeting of the corporation of German universities, Spranger dissented because he thought "the national movement among the students to be still genuine at the core, only undisciplined in its form."[24] Spranger's dissent is worth examining as the reaction and interaction to the Nazi regime of a conservative German professor who was imbued with the status of his profession. Among other items, Spranger warned Franz von Papen what dangers for the whole of Germany denunciation and lack of discipline at the universities must bring; he objected to the Nazi "Deification of the People" (*Vergottung des Volkes*) on religious grounds; he was offended when a professor (unnamed) was appointed to a newly created position of "political pedagogy" without his being consulted beforehand. He was easily outmaneuvered by the Nazi educational bureaucracy, and eventually withdrew his proferred resignation two months after he made it with a public declaration of his wish "to be able to devote his work as before to the German people (*Volk*) and State, in close connection with academic youth."[25] His influence in the university at an end, he retreated into a sort of "inner emigration" and continued teaching.

There *were* differences among the professors in the Weimar Republic; they included people of every variety of political persuasion ranging from Pan-Germanism to socialism, including "rational but not convinced" defenders of the republic like the famous historian Friederich Meinecke, radical pro-Nazi antirepublicans like the Nobel laureate physicists Philipp Lenard and Johannes Stark, as well as socialist defenders of the republic like the jurist Gustav Radbruch. However, as seen, the very conception of the German university could

[23] Ibid.: 223–224.

[24] Eduard Spranger, "Mein Konflikt mit der National-Sozialistischen Regierung 1933," *Universitas, Zeitschrift für Wissenschaft, Kunst, und Literatur* (1955), 457–473.

[25] Ibid.: 473.

lead to professors (of whatever political persuasion) who could also quite self-consciously see themselves as having nothing to do with the state, except to be a state civil servant. A consequence, at least in Wilhelminian Germany, was that the professional academic supported the state that gave him his livelihood. It is well known that this natural alliance between professors and their government ceased to exist for many during the Weimar period, but was reawakened in them by the promise of a nationalist government that stressed the dignity of being German, as the National Socialist one did.

A word more should be said, however, about the actual nature of that "academic freedom," which Nietzsche, quoted approvingly by Spranger, decried in favor of "obedience, subordination, training, service." For Spranger and Nietzsche are not speaking of students alone. What academic freedom meant (or did not mean) in late Wilhelminian Germany was castigated by Max Weber in a newspaper article of September 20, 1908 and a journal article of January 1909.[26]

Robert Michels, an open (though highly critical) Social Democrat, could not hope to "habilitate" (that is, qualify to be a teacher) in a Prussian university "as a result of the enforcement of the *lex Arons*," but also found it difficult to "habilitate" anywhere in Germany, apparently also because of his political beliefs. According to Max Weber, when this was mentioned by his brother Alfred at a teachers' congress, a Professor Theodor Fisher from Marburg said that Michels "could not, for quite different reasons, expect habilitation." Michels sought an explanation, and, says Weber:[27]

> He received a reply from Professor Fisher to the effect that the decisive reason was (1) "not just the fact of his social democratic beliefs but their public and exceptionally visible expression"; and (2) his family life: could Dr. Michels—who, lest we forget something "important," is an "aryan"—have even for a moment doubted that a man who would not allow his children to be baptised would be "impossible in any high ranking position"? The reply went on to say: "What a wonderful position you would have been able to obtain in Marburg where you were so well recommended and where many influential persons looked on you with the greatest favour! These persons have been very distressed and said it a great pity that you have wasted all this." The letter ends with the reproach that Dr. Michels used his house, of which Professor Fisher was the acting landlord, so badly that the house had still not been sold!
>
> The reproduction of these statements is not intended to put the writer of the letter in a personally poor light. On the contrary, I am, unfortunately, rather certain that—except for the last sentence which is irrelevant to this discussion, unless the landlord's "good conduct certificate" was to be taken into account in the habilitation proceeding—the content of this letter would be regarded in most academic circles as quite in order. It is characteristic of our public life in general and of the situation in our universities in particular. I cannot honestly hide the fact that it is my "personal" conviction that the existence and the influence of such views, because indeed of their very sincerity, are no honour for Germany and its culture, and that furthermore as long as

[26] Max Weber, in Weber 1974: 14–23.

[27] Ibid.: 17.

> such views prevail it will be impossible—as far as I am concerned—to act as if we possess an "academic freedom" which someone could infringe.
>
> I am convinced too—once again according to my own personal conviction—that religious communities, which knowingly and openly allow their sacraments to be used, in the same way as university fraternities and reserve officers' commissions are used—to make a career—richly deserve that contempt about which they frequently complain . . . it should be required in the interest of good taste and truthfulness that henceforward we ought not to speak of the existence of "the freedom of science and teaching" in Germany, as has always been done. The fact is that the alleged academic freedom is obviously bound up with the espousal of certain views which are politically acceptable in court circles and in salons, and furthermore with the manifestation of a certain minimum of conformity with ecclesiastical opinion or, at least, a facsimile thereof. The "freedom of science" exists in Germany within the limits of political and ecclesiastical acceptability. Outside these limits, there is none. Perhaps this is inseparably bound up with the dynastic character of our system of government. If it is, it should be honourably admitted but we should not delude ourselves that we in Germany possess the same freedom of scientific and scholarly teaching which is taken for granted in countries like Italy.

So in 1908, and in the following year, in the journal *Hochschul-Nachrichten*:[28]

> [There is] the assumption, made in all seriousness, that it is possible to separate the question as to whether a university teacher's expression of a particular belief, e.g., a politically or religiously "radical" belief, should prevent his retention of a professorial chair—to which the answer was naturally negative—from the other question as to whether the same sort of belief should stand in the way of appointment to a professorial chair.
>
> There is another equally widely shared view which asserts that the university teacher must, on the one side, "bear in mind" that he is an "official" when he acts publicly—as a citizen in elections, in statements in the press, etc.—but that, on the other side, he is entitled to claim the right that his statements in university classes are communicated no further. . . . If one links this latter viewpoint with the proposition that there is a significant difference between not permitting a professor to retain his chair and not allowing a person to be appointed to a chair when the disqualifying views are identical, one arrives at the following rather unusual conception of "academic freedom": (1) when an appointment is at issue, not only the scientific or scholarly qualifications of the candidate for an academic post may and should be examined, but also his submissiveness to the prevailing political authorities and ecclesiastical usages; (2) a public protest against the prevailing political system may justify the removal of the incumbent of a professorial chair from his post; and (3) in the lecture hall, where neither publicity nor criticism are allowed, the persons who have been appointed as university teachers may express themselves as they wish "independently of all authority."
>
> One sees that this conception of academic freedom would be ideal for one "whose

[28] Ibid.: 18.

wants are satiated" or for the "happy possessor of manifold goods" (*beati possidentes*), to whom neither the freedom of science and scholarship as such, nor the civil rights and duties of the university teacher have any significance; it is the ideal of those who wish to be at ease in the cultivation of the "station in life" in which they find themselves. And this "freedom" can naturally serve as a "fig-leaf" to cover up, to the greatest extent possible, the imparting of a certain political tone to university teaching in all those in which it is feasible.

Less than twenty-five years separate this characterization by Weber of his colleagues and Hitler's accession to power. As Hannah Arendt has remarked with equal acidity, "[German scholars] have proved more than once that hardly an ideology can be found to which they would not willingly submit if the only reality—which even a romantic can hardly afford to overlook is at stake, the reality of their position."[29] From Humboldt on, the tradition of "freedom in scholarship" and, concomitantly, a complacent acquiescence "to sing the tune of him whose bread I eat,"[30] was the accepted role of the majority of civil-servant university professors. In some sense the *Vernunftrepublikaner* (republicans by reason if not conviction) among the German professoriat after 1918 were more "traditional" than their conservative colleagues—their allegiance was to the state, such as it was. During the Weimar period, however, the majority of German academics seemed to reverse that traditional allegiance. Their alienation from political reality became an alienation directed against the state rather than acceptance of it. This presents at least two questions. First, if German academics truly were predisposed to be apolitical, why were they so much inclined against the Weimar government? Second, to what extent can the general German academic atmosphere as illustrated by "humanists" of every persuasion be held to apply to mathematicians? The plain fact is that there does not appear to be much in the way of explicit political statement by mathematicians.

The answer to the first question is complicated and consists of many interrelated factors. Clearly there is no room here to go into any detail; nevertheless, some of these may be briefly indicated.

One factor, without question, is that under Weimar, many academics felt their elite standing threatened by an officially more open society, which publicly advocated pluralism. This coincided with a "cultural crisis" that began in the 1890s and reached its height in the years after (and no doubt because of) World War I.[31]

A second related factor is a simple yearning for restoration of the standing of the imperial years, when life was simpler and more secure (at least for those of

[29] Hannah Arendt, *The Origins of Totalitarianism*, rev. ed. (1980), 168. Earlier (p. 146) she credits Hobbes with foreseeing the social creation of this general type: a "poor meek little fellow who has not even the right to rise against tyranny, and far from striving for power, submits to any existing government."

[30] Weber 1974: 20.

[31] Ringer 1969. Cf. F. Ringer, "The German Universities and the Crisis of Learning, 1918–1932" (Ph.D. diss., Harvard University, Cambridge, Mass., 1960).

some position). The pre-1918 orientation of most of Weimar society (including the Social Democrats) has been discussed by Heinrich Winkler:[32]

> The orientation towards the authoritarian system of the pre-1918 period . . . was in any case a mortgage on parliamentary democracy. This is true not only of those strata which turned towards National Socialism after 1929 or else contributed to its rise; it is also true to a certain extent of the representatives of Weimar democracy. The conception . . . that the chief task of parliament was to criticize the government, survived the November revolution of 1918. The most important characteristic of a parliamentary system, the confrontation between the governmental majority and the opposition, was constantly obscured by this anachronistic dualism. The tendency of the parties, not least the Social Democrats, to disclaim governmental responsibility in critical situations, can ultimately be traced back to an unconscious fixation vis-à-vis the political system of the Kaiserreich. This system had failed to motivate the parties to consistently fight for a majority of the voters; their exclusion from active participation in government had favoured instead the ideological orientation of political parties and their restriction within a particular social milieu. Initially, the Weimar party system was scarcely different from that of Bismarckian Germany.

Academics who saw themselves above the mundanities of party politics had even more reason to desire a return of a hierarchical system with their assured position in it; indeed, they had spent their youth in such a system and knew its benefits firsthand.

The defeat of Germany during the war made permanent a split in the German academic community that was already well developed prior to 1918. Friedrich Meinecke remarked in 1926 that the split between those academics who during the war years had adopted an extreme annexationist or ultranationalist point of view, not even wavering in 1916 in their war aims, and those who after 1915–16 had advocated peace on a rational basis, was the historical source of the split between the German academics who stood by the Weimar constitution and those who were its enemies.[33] Certainly the German academic community as a whole had in the first phase of the war issued a number of statements and petitions: the famous petition of the ninety-three signed by an additional 4,000; "Deutsche Reden in schwerer Zeit"; and the ultra-annexationist "Intellektuellen-Eingabe" organized by Reinhold Seeberg and Dietrich Schäfer, among others.[34] These early statements defended German war aims and looked forward to a triumphant Germany that had extended its territory. Indeed the group around Seeberg and Schäfer formed a committee "for a German peace" (i.e., a non-

[32] Heinrich Winkler, "German Society, Hitler and the Illusion of Restoration, 1930–33," *Journal of Contemporary History* 11, no. 4 (1976): 10–11.

[33] Friedrich Meinecke, address (pp. 17–31), in *Die Deutsche Universität und der Heutige Staat* (1926), 21–22 and passim.

[34] For a collection of such statements throughout the war, representing a variety of opinions, see *Aufrufe und Reden deutscher Professoren im Ersten Weltkrieg*, with an introduction by Klaus Böhme (Stuttgart: Reclam, 1975).

negotiated one).[35] As the fortunes of war shifted and a grim stalemate ensued, followed ultimately by German surrender and the declaration of the republic, the academic community split largely into the two groups mentioned. The distress of the largest group among the academics, which never ceased in its ambitions for a "victorious peace" or *Siegfrieden* as the final outcome, seems to have led to a sort of narcissistic rage at the republic and belief in the "stab-in-the-back"[36] legend, as well as a disgust with parliamentary democracy (which was the "system" of the victors).

Thus a third factor was the defeat of national aspirations in World War I—German scholarship was already predominant in several areas, including mathematics, so that the ultra-annexationists and their like among academics no doubt saw the extension of national hegemony as a fitting extension of an academic superiority already felt. The boycott of German science and scholarship after the war until the late 1920s by France and Britain certainly added to these feelings of disappointed rage and rejection of the political realities. This openly expressed itself on more than one occasion in a railing against the weakness of Germany, against a state "with neither defense nor honor nor governmental power,"[37] speaking of[38]

> the grotesque idea of healing Germany through the introduction of a "west-European" (read French) constitution—what is this other than an application of the plainly ineradicable conviction of 1789 that absolutely good constitutional forms exist, that everything depends on these forms, and that like Parisian hats on the little head (*Köpfchen*) of a German lady of fashion, one can modishly set (*draufstülpen*) Parisian laws on German development without thereby preparing the greatest disaster. . . . For us [the ideas of 1789] are foreign to our nature, they have injured us unspeakably (*unsagbar*).

Included in such complaints was the appeal for a new kind of leader. For example: "If the German way (*Art*) and Christian belief unite themselves, then we are saved. Then we will work with our hands and wait for the day when the German hero will come, whether he come as prophet or as king."[39]

If rage at unfulfilled expectations in the belief that Germany's defeat was a result of political double-dealing led to a large number of academic associations

[35] Ringer 1969: 190. See also Klaus Schwabe, *Wissenschaft und Kriegsmoral* (1969), 70–74, 95–97, and passim.

[36] According to Holger Herwig, the "stab-in-the-back" legend was the creation of the *Deutschnationale Volkspartei* politician Karl Helfferich and particularly Field Marshal von Hindenburg in testimony before a subcommittee of the committee of enquiry (*Untersuchungsausschuss*) on the origins of World War I. See Holger Herwig, "Clio Deceived," *International Security* 12, no. 2 (1987): 30–31.

[37] Kurt Sontheimer, "Die Haltung der Universitäten zur Weimarer Republik," in *Universitätstage* 1966: 30 (citing Gustav Roethe in 1924).

[38] Kurt Sontheimer, *Antidemokratisches Denken in der Weimarer Republic* (1962), 236, citing Adelbert Wahl in 1925.

[39] O. Procksch, *Schriften der Universität Greifswald*, no. 11, p. 23, as cited in Sontheimer 1962: 275. Also cited in Theodor Eschenburg, "Aus dem Universitätsleben vor 1933," in *Deutsches Geistesleben und Nationalsozialismus*, Tübingen (1965), 40.

with the right-wing opposition to the Weimar Republic, still, such highly charged political statements were in a certain sense "unpolitical." Indeed, they emphasized "national" interests over "state" interests, a theme also among the Pan-Germans, which the Nazis also appropriated.[40] German academics rarely spoke in practical political terms.[41] Their discourse almost always took place in terms of *Geist*, or, as Ringer puts it: "Their critiques of modern politics almost always ended in a resolution to increase the moral impact of learning upon public life."[42] As Kurt Sontheimer has pointed out,[43]

> The hypostatization of the political task of the university [to defend the rationally discovered true] to an abstract service to the state is a dangerous thing. For the Weimar patriot-professors who called their colleagues to service for the fatherland it was regularly (*in aller Regel*) an implicit request (*Aufforderung*) to deny service to the democratic republic and to serve another, presumably more German, idea of the State.

Thus German professors who made political statements were in a very real sense "unpolitical." They operated in an ideal world. Many of them encouraged opposition to the Weimar constitution and a parliamentary form of government—in the name of what? This presents a fourth theme, already alluded to, and perhaps the most important reason for the disaffection of the majority of the German academic community from Weimar. German academics and German intellectuals in general saw no necessary connection between "Western" and "German" spirit and civilization.[44] A famous exposition of this German need for a "nonpolitical" "cultural" state is Thomas Mann's *Betrachtungen eines Unpolitischen* (1918), the reflections of a self-defined unpolitical intellectual. That (practical) politics was not German had been stated explicitly by Mann: "The political spirit, anti-German as spirit, is with logical necessity inimical to anything German (*deutschfeindlich*) as politics."[45] As an unpolitical man, he exclaimed,[46]

[40] Arendt 1980: 236–243, esp. 237.

[41] This is one of the major themes of Ringer 1969.

[42] Ringer 1969: 252.

[43] Sontheimer 1966: 37.

[44] "Intellectual" is used here in the sense of Benda's *clerc*: as someone who speaks to the world in the tones of a spiritual guide; see Julien Benda, *La trahison des clercs* (The treason of the intellectuals), trans. R. Aldington (1969). On the German intellectual attitude, cf. among others, Stern 1961; Sontheimer 1962, 1966; Mosse 1964; Abendroth 1966; and Benda 1969: passim. Benda claimed (in 1928) that most of the moral and political attitudes adopted by European intellectuals since the Franco-Prussian War were of German origin (ibid.: 58), and Germany was in particular the origin of what he saw as a pernicious nationalist particularism in all European intellectual and political matters. Kurt Sontheimer ("Antidemokratisches Denken in der Weimarer Republik, *Viertel-jahresheft für Zeitgeschichte* 5 [1957]: 44) says that Weimar was called into being at a time when German intellectual life (*Geistesleben*) was rejecting the Western European Enlightenment tradition more decisively than heretofore. (This article should not be confused with the book of the same title already cited.)

[45] As cited by Abendroth 1966: 194; cf. Sontheimer 1957: 44.

[46] As cited by Abendroth 1966: 196.

> Away with the motto "democratic," foreign to our country. Never will the mechanical-democratic state of Western custom succeed with us. One only needs to Germanify the word, and say "inhering to the people" (*volkstümlich*) instead of "democratic"—and one names and comprehends the exact opposite: For "inhering to the German people" (*deutsch-volkstümlich*) means "free" inwardly and outwardly, but it does not mean "equal," neither inwardly nor outwardly.

Mann, of course, later changed his mind and became a defender of the republic,[47] nevertheless, he was the best known and at the time most gifted exponent of the need for a "German," "organic" form of government. Indeed, Mann's *The Magic Mountain* (1924) is perhaps the classic statement of Germany, intellectually between East and West, but a part of neither, seeking a "third way," and Hans Castorp ultimately descends to the reality of the trenches of World War I. Many German intellectuals besides Mann contrasted in 1918 and succeeding years the Western tradition, which had its roots in the French Revolution and the Enlightenment, with an organic, romantic, "German" tradition. The parliamentary democracy of Weimar was a foreign "Western" imposition on Germany. Its opponents sought an indigeneous German way of thought in the political realm. As Kurt Sontheimer remarks, it was this attempt to counter the Western European conception of parliamentary democracy and liberalism with a different German conception that deprived the Weimar government of a large measure of intellectual support.[48] This intellectual belief in a new and greater future for a peculiarly German *volkstümlich* form of government no doubt had some roots in the "unbelievable" defeat of the German armies during the war. The ensuing "cognitive dissonance" was perhaps yet another factor in the intellectual insistence that parliamentary democracy was "un-German" and had to be replaced by a more German form of government.[49] As already noted, academics and intellectuals participated in a whole spectrum of attitudes toward the Weimar government, and it may seem one-sided to emphasize these right-wing ones. However, there seems little doubt that the majority of German academics participated in this right-wing opposition, or, at best, supported Weimar because to do so was *vernunftig* (i.e., reasonable) rather than because of any dedication to republican ideals.[50] Academics certainly were not apolitical in the sense of not making political statements, and their antidemocratic thought did

[47] For a complete investigation of Thomas Mann's political journey, see Kurt Sontheimer, *Thomas Mann und die Deutschen* (Munich: Nymphenberger Verlag, 1961). Mann had come around to a defense of the Weimar Republic by 1922. What Kurt Sontheimer says of his attitude at this time—"The Republic is our fate and *amor fati* is the only correct relationship to it" (ibid.: 53)—is a succinct characterization of the attitude of those whose support for the republic was based on it being the rational thing to do.

[48] Sontheimer 1957: 44; many examples of these strivings can be found in Sontheimer 1962. In particular, "democratic-liberalism" and "parliamentarism" are often stigmatized as "French."

[49] For the use of the psychological idea of "cognitive dissonance," see chapter 2, note 80.

[50] Cf., e.g., Ringer 1969; Sontheimer 1957, 1962, 1966. The term *Vernunftrepublikaner*, or "republican by reason" (instead of by conviction), seems to have been coined by the historian Friedrich Meinecke (see Ringer 1969: 203).

not express a sudden revulsion against the Weimar Republic and a sea change in attitudes toward government. Rather, they wished to seek a peculiarly German and nonparliamentary form of government, and this, in part at least, because of a sense of cultural crisis that had its beginnings in the 1890s.[51] Academics *were* "unpolitical" in the sense that their political activity took place in a cloud-cuckoo-land filled with desires to protect art and humanity from "dirty politics," to seek refuge from the corruptions of mechanistic "civilization" in "culture,"[52] to carry on "unpolitical" discourse in a world of ideological purity, necessarily divorced from implementation, and not least to view parliamentary democracy, in the words of a popular book by a gifted right-wing publicist of the day, as the "Rule of the Less Valuable."[53] Such a quantity of "unpolitical" talk by the academic community had its undeniable political effect not only in reducing intellectual support for the Weimar government, but also in influencing the student body—it is well known that the German Student Association was already Nazi-dominated by the summer of 1931.[54] This is not to say that these professors necessarily were members of the Nazi movement—they certainly were not[55]—but that they prepared a soil in which it could flourish is undeniable.[56]

But what of mathematicians? Or natural scientists? The traditional academic gulf between "humanists" and "scientists" was no less deep in Weimar. Ernst Troeltsch wrote in 1924:[57]

[51] Ringer 1969, esp. chap. 5.

[52] The distinction between *Zivilisation* and *Kultur* appears prominently, for example, in Oswald Spengler's *Der Untergang des Abendlandes* (The Decline of the West), which appeared in the 1920s, but was scarcely original with him.

[53] Edgar Jung, *Die Herrschaft der Minderwertigen* (Berlin: Deutsche Rundschav, 1927). Jung, personally close to Franz von Papen, was among those assassinated on June 30, 1934, "The Night of the Long Knives."

[54] See, for example, Michael Steinberg, *Sabers and Brown Shirts* (1977); table 18, p. 92 there shows that at most universities by 1931, the Nazis obtained 30–40% of the vote in student elections. They were most popular in Jena, where they had 65% of the vote. However (ibid.: 91), these votes, even when not a majority, amounted to electoral control of student councils in eleven universities. In the summer of 1931, the Nazis took over the leadership of the *Deutsche Studentenschaft* at its meeting in Graz (ibid.: 111–112).

[55] On this point, see Helmut Seier, "Universität und Hochschulpolitik im Nationalsozialistischen Staat," in K. Malettke ed., *Der Nationalsozialismus an der Macht* (1984), 143–165. Here (p. 145) Seier cites, among others, Anselm Faust, "Professoren für die NSDAP," in M. Heinemann, ed., *Erziehung und Schulung in Dritten Reich* (Stuttgart: Klett-Cotta, 1980), 2:31–49, who estimates that only 1.2% of all university teachers were members of the Nazi party prior to 1933 (p. 42). See also note 95 below.

[56] See in addition to Ringer and Sontheimer, among many others, for example, Daniel Gasman, *The Scientific Origins of National Socialism* (1971), and Willy Hellpach, *Wirken im Wirren*, vol. 2, *Lebenserinnerungen, 1914–1925* (1949). Hellpach, an academic, was the candidate of the *Deutsche Demokratische Partei* for president of the republic in 1925. Many university professors seem to have been members of the party known as DNVP (*Deutschnationale Volkspartei*) (Ringer 1969: 201; cf. Hasse 1952). All writers on German academic matters in this period seem to comment on the strength of the DNVP among professors. For the role of this monarchist movement in the Weimar period, see Mosse 1964: chap. 13; Walter Kaufmann, *Monarchism in the Weimar Republic* (1953); and Lewis Hertzman, *DNVP: Right-Wing Opposition in the Weimar Republic, 1918–1924* (1963).

[57] As cited in Ringer 1969: 346.

It is the revulsion against drill and discipline, against the ideology of success and power, against the excess and the superficiality of the knowledge which is stuffed into us by the schools, against inellectualism and literary self-importance, against the big metropolis and the unnatural, against materialism and skepticism, against the rule of money and prestige, against specialization and bossism, against the suffocating mass of tradition and the evolutionary concept of historicism. . . . Furthermore, a profound intellectual revolution undoubtedly lies in the changes within scholarship which are today still little noticed. The need for synthesis, system, *Weltanschauung*, organization, and value judgment is extraordinary. The mathematization and mechanization of all European philosophy since Galileo and Descartes . . . is meeting with growing skepticism. . . . In the cultural and historical disciplines, people are defending themselves against the tyranny of the evolutionary concept, against mere summations and critical assertions.

Nevertheless, Klaus Schwabe remarks concerning the World War I propaganda of the professors that "as fellow-travelers (*Mitläufer*), the natural scientists should have reacted similarly to their colleagues in other disciplines," and cites the fact that of 352 professors signing a Pan-German-inspired statement in 1915, thirty-nine were natural scientists.[58] The total number of *Intellektuellen* signing this statement was 1,347.[59] Ringer remarks,

It is my impression that in their attitudes toward cultural and political problems, many German scientists followed the leads of their humanist colleagues. But I am unable fully to substantiate this conclusion, and it is certainly possible to imagine scientists taking a more favorable view than humanists of technological civilization.

Jeffrey Herf, meanwhile, has argued cogently that there was a movement, especially prominent among engineers (but among whose most prominent propagandists were "independent intellectuals" like Ernst Jünger and Oswald Spengler) to adapt technological society to reactionary uses.[60] Paul Forman has argued that German physicists and mathematicians during the Weimar period were infected by the irrationalistic stance, philosophically and intellectually popular at the time, that insisted on the cultural relativism and anthropomorphic character of *all* concepts. Forman demonstrates the probable influence of these ideas associated with Oswald Spengler and others on the development of such Weimar German scientific ideas as uncertainty in quantum physics and intuitionism in mathematics.[61]

[58] Schwabe 1969: 193 n. 38.

[59] Ibid.: 70. Among the prominent scientists who were signers were Gustav Mie, Richard Willstätter, and H. Struve. Of course, such a signature did not predict future politics. For example, Willstätter, a Nobel Prize-winning chemist and fully assimilated Jew, later resigned his university position in 1924 because of an anti-Semitic incident (see his *From My Life* [1965; original German edition, 1949], 361 ff.). A complete list of signatories exists in the *Nachlass* Schemann, available at the Universitätsbibliothek, Freiburg im Breisgau (*Handschriftenabteilung*).

[60] Ringer 1969: 6. Jeffrey Herf, *Reactionary Modernism* (1984).

[61] Forman 1971. For intuitionism, see also Walter van Stigt, "The Rejected Parts of Brouwer's Dissertation on the Foundations of Mathematics," *Historia Mathematica* 6 (1979): 385–404.

Yet, a check of the signers of the 1915 ultra-annexationist declaration mentioned above against the members of the German Mathematical Society in the same year reveals only five common names, none of them prominent as mathematicians.[62] Felix Klein did sign the well-known 1914 declaration of German intellectuals, "To the Civilized World," concerning the invasion of Belgium,[63] and it is often pointed out that Hilbert and Einstein did not,[64] but in fact Klein is the only prominent mathematician to appear among the original signatories. The argument presented earlier concerning the attitudes of German university professors during the Weimar period and to what extent those attitudes were based on the very ethos of the German university system depends (as do those of Ringer and Sontheimer) for its application to mathematicians and physical scientists on the assumption that they were no different by and large from their more outspoken "mandarin" colleagues in the humanistic disciplines.

What evidence is there of political interest on the part of the mathematicians of Weimar? Among the very few[65] academics of the far left in politics were at least four mathematicians: Emil Gumbel, Max Zorn, Fritz Noether, and Emmy Noether. On the other hand, with the exception of Gumbel, they do not seem to have been politically active.

Emmy Noether was a "*Salonkommunist*" and apparently had to move out of her lodging in a student boardinghouse in April 1933 because she was "a Marxist-leaning Jewess."[66] According to Paul Alexandroff, she was delighted, on her visit to the Soviet Union in the winter of 1928–29, by "Soviet scientific, and especially mathematical successes"; furthermore, her "sympathies were always unwaveringly with the Soviet Union; in which she saw the beginning of a new era in history and a firm support for everything progressive," despite the fact that "manifestation of these sympathies seemed both outrageous and in poor taste to most of those in European academic circles." Hermann Weyl says rather less strenuously that she "sided more or less with the Social Democrats; without being actually in party life she participated intensely in the discussion of the political and social problems of the day."[67] In any case, Emmy Noether was

[62] See above, note 59. H. Struve was a prominent mathematical astronomer, but not apparently a member of the Society.

[63] The ninety-three original signatories truly represented a Who's Who of German scholarship and science. An English text of the declaration along with a list of the signatories can be found in *Deutschland über Alles, or Germany Speaks*, compiled and analyzed by John Jay Chapman (1914), 37–42.

[64] E.g., Reid 1970: 137–138.

[65] Ringer 1969: 201.

[66] Alexandroff, from his eulogy of Emmy Noether (reprinted in Brewer and Smith 1981: 107; see next note). See also Emmy Noether's own, much milder description in letters to Heinrich Brandt (April 8, 1933, and April 26, 1933) as published by Werner Jentsch, *Historia Mathematica* 13 (1985): 5–12. Copies of the entire text of these letters are in my possession.

[67] Auguste Dick, *Emmy Noether, 1882–1935*, Beiheft to Elemente der Mathematik (1970). This edition reprints the informative obituary eulogies of Noether by Bartel van der Waerden (in German) and Hermann Weyl (in English). An English translation of Dick's book appeared in 1981 (from the same publisher); this volume contains van der Waerden's translated eulogy and Weyl's, as

sufficiently unpolitical to have been bemused and amused by the presence in her home of a student in SA uniform in 1933[68]—the SA (or *Sturmabteilung*) were the original "stormtroopers." As Hermann Weyl said, "Her heart knew no malice; she did not believe in evil."[69]

Fritz Noether was Emmy's brother and two years her junior. He became a well-known applied mathematician, and his case is interesting in part because it shows what sort of protest was still possible in 1933 for a "full Jewish" professor of mathematics.[70] At the time of Hitler's ascension to the chancellorship of Germany, Fritz Noether was teaching at the Technical University in Breslau. He was forty-eight years old, a wounded World War I veteran who had been awarded the Iron Cross. Thus he fell under the original exceptions clauses of the April 7, 1933 law. On April 26, 1933, a group of students complained to the Rektor that his presence on the faculty "in large measure contradicts the Aryan principle," and that there was little surety he would "work in the spirit (*Sinne*) of the national movement." Noether protested immediately and, after a very brief, self-imposed interruption "because my activity at the university appears not to be safe from disturbances," took up lecturing again. On August 25, the students complained again, speaking also of his leftist orientation. Among other items, they accused Noether of being active in the "league for Human Rights," of having signed petitions in favor of Theodor Lessing,[71] of having opposed the hanging of a portrait of von Hindenburg in university public space, and the like. The letter came before the final decision to remove Noether from the faculty, but it was hardly necessary for that decision, as the formulary thereto remarked that he was "100 percent Jew," "had signed the petition for Emil Gumbel"[72] (as, indeed, had Gustav Doetsch, whose political persuasion in 1933 was antipodal to Noether's[73]), and had a political position "against the national movement."

---

well as a translation of Paul Alexandroff's eulogy. Van der Waerden's and Alexandroff's statements are also reprinted in James Brewer and Martha Smith, eds., *Emmy Noether: A Tribute to Her Life and Work* (1981), and Weyl's appears also in his *Gesammelte Abhandlungen* (1968). The original appearances of these obituaries are: van der Waerden in *Mathematische Annalen* (1935): 469–476; Weyl in *Scripta Mathematicia* 3 (1935): 201–265. The translations of van der Waerden and Alexandroff referred to above differ. Citations here from Alexandroff are as in Brewer and Smith 1981; page numbers for citations from Weyl are as in Dick 1970. For above citations, see Alexandroff in Brewer and Smith 1981: 107; Weyl in ibid.: 59. Fraenkel (1967: 159) applies the word *Salonkommunist* to Emmy Noether. The location of Alexandroff's eulogy (Moscow, 1935) should probably be taken into account in evaluating its statements.

[68] The student in the SA uniform seems to have been Ernst Witt (see Clark Kimberling, "Emmy Noether and Her Influence," in Brewer and Smith 1981: 29 and 47 n. 13). Professor Kimberling has said (personal communication) that B. L. van der Waerden said Emmy Noether told him the student in question was Witt. The story of this student is well known and widespread, but he is usually anonymous.

[69] Weyl in Brewer and Smith 1981: 72.

[70] For all the information cited below about Fritz Noether and his fate, see the detailed article by Karl-Heinz Schlote, "Noether, F.—Opfer zweier Diktaturen," *NTM Schriftenreihe* 28 (1991): 33–41.

[71] The "Lessing case" is discussed later in this chapter.

[72] The "Gumbel case" is discussed later in this chapter.

[73] See below, chapter 4, "The Süss Book Project" and "Gustav Doetsch and the Philosophy of

The intention was to dismiss Noether according to section 4 of the April 7 law (he was opposed to the national state).[74] Incidentally, using this rubric meant Noether's pension would be reduced by 25 percent.

Noether appealed and categorically denied the nature of the charges against him—he had always been politically inactive—however, knowing there was no chance he could reverse his dismissal, he instead asked that section 5, which allowed movement to other positions, be used instead, after which he would petition to be emerited. The point of the difference was that in this way, he could not only retain a pension, but also his reputation as a loyal German civil servant. This was agreed to, and in fact Noether did receive such a pension. However, this was cancelled when he emigrated to the Soviet Union in 1933, where he became a professor at Kubischev University in Tomsk. He was present in Moscow in 1935 on the occasion of Alexandroff's memorial address for his sister. He attended the International Mathematical Congress in Oslo in 1936, where he gave a paper, and in 1939 was in a Soviet jail accused of treasonable activities.[75] The year 1936 marked the beginning of the "purge trials" in Moscow. Noether seems to have been the only mathematician to travel to Oslo from a Soviet location, presumably on his German passport, since Soviet citizens apparently were not allowed to travel to the congress. At the time of the "purge trials," the charge was made that Trotskyists had intrigued with the Germans, among others, to overthrow Stalin, and this may have had something to do with Noether's arrest, especially since Trotsky was living in Oslo in 1936. On the other hand, the last of the major trials was held in late 1938, and by the spring of 1939 the negotiations had begun that would lead to the Molotov-Ribbentrop pact of friendship and non-aggression between the Soviet Union and Nazi Germany in August of that year, eight days before the onset of World War II. Nothing was heard of Noether after 1939, and the mystery of his disappearance was only recently cleared up thanks largely to the determined efforts of his two sons, Hermann and Gottfried Noether (died 1992), and the *glasnost* of Mikhael Gorbatchev, which allowed the opening of previously secret files.

On November 22, 1937, Noether was arrested on charges of being a German spy who not only spied on the Russian armament industry but committed acts of sabotage against it. On October 13, 1938, he was sentenced to twenty-five years in prison and confiscation of all his belongings. It appears that all the evidence against Noether and the three Russians accused with him was falsified.

Mathematics." Both Noether and Doetsch were undoubtedly complaining about the procedural irregularities in Gumbel's case. There is no reason to think that Noether or Doetsch shared Gumbel's (or Lessing's) politics.

[74] Section 3 of the law called for the dismissal of Jews, but had the exceptional clauses about World War I veterans and old-time civil servants. Thus Noether had to be dismissed via a different rubric. For similar reasons, no doubt, Edmund Landau (see below, chapter 4, "Hasse's Appointment at Göttingen") was let go according to section 6 rather than section 3. Such Nazi punctiliousness about "legal niceties" at the time may strike us as curious, but were nevertheless the fact.

[75] See Alexandroff in Brewer and Smith 1981: 99; Maximilian Pinl, "Kollegen in einer dunklen Zeit I," *JDMV* 71 (1969): 203–204; and Dick 1970: 34.

It is possible but uncertain that Fritz Noether appeared on a Nazi list of people to be arrested with the German conquest of Soviet territory. In any case, by Nazi decree, he lost his German citizenship in 1938, and so was clearly not available in 1939 for the exchange of persons following the Molotov-Ribbentrop pact. Hitler's surprise attack on the Soviet Union began June 22, 1941, and the German army had tremendous early success. Fritz Noether, imprisoned at Orel, was charged with further acts against the Soviet Union, and on September 8, sentenced to death. On September 10, he was executed. Presumably the second sentence was to provide juridical justification for the execution of prisoners in Orel before its capture by the Germans. This latter happened on October 8, 1941. In late 1988, the Supreme Court of the Soviet Union decided that Noether had in fact not been guilty of any crime, and on May 12, 1989, Hermann Noether was officially informed of the "complete rehabilitation" of his father: "Please accept my deepest sympathy although I understand that no words can alleviate your pain."[76]

Emil Gumbel was a mathematical statistician of note who wrote books of political import as well as statistical articles. Among the former were *Four Years of Political Murder* and *From the* Feme-*Murders to the Reichschancellery*: books that dealt with right-wing assassination in the Weimar Republic. On July 27, 1924, Gumbel made the following statement about the dead German soldiers of World War I: "Now, I would not exactly say that they fell on the field of dishonor, however, nevertheless, they lost their lives in a detestable way." The uproar that followed had even supporters of the republic saying that Gumbel did not deserve to be on a university faculty like Heidelberg's, and everyone who had lost a son, brother, father, or husband felt wounded inwardly by such remarks. Gumbel at first apparently took a rather arrogant attitude toward the disturbance, but, when he was suspended from teaching, he ended by apologizing for the slip in his mode of expression and by saying that he did not mean to dishonor fallen German soldiers. Nevertheless, the controversy did not die. Gumbel, who seems to have been a communist already, spent the 1925–26 winter semester in Moscow at the Marx-Engels Institute; in 1926 he returned to Heidelberg. He was not given to mild expressions of opinion: he told a group of Marxist students with reference to the starvation in Germany at the war's end that an appropriate war memorial in Heidelberg was for him not a lightly clothed maiden carrying the palm of victory but rather "a single large turnip." Gumbel was promoted to *ausserordentlicher Professor* (roughly "associate professor") by the Baden ministry in 1930 over faculty protest, but continuing protests from faculty and students, including a threatened student boycott, eventually resulted in revocation of the promotion (under a new minister) and in Gumbel's losing the *venia legendi*, or right to teach in a German university, in 1932.[77] According to Willy Hellpach, the liberally inclined Baden minister, him-

[76] Schlote 1991: 40.

[77] All material about the "Gumbel case" is taken from the extensive collection of original sources in the Heidelberg University archive, unless otherwise indicated. As Dr. Hermann Weisert, the

self an academic, who was charged with handling the "Gumbel affair" in 1925, not only did rightist nationalists carry on year after year about the Gumbel affair as an example of the contemptibility of the Weimar Republic, but it still reverberated after World War II.

Max Zorn became a communist by way of the *Monistische Jugend* (or "Monist Youth") at age twenty-one in 1927. Although others have emphasized the right-wing associations of the monist movement founded by Ernst Haeckel,[78] its youth movement was as much an outlet for romantic rebellion as the larger German youth movement with which it was associated. For Zorn in 1927, this led to communism. A student at Hamburg, he wrote his dissertation, which established an important abstract algebraic result on "alternative division rings," under the direction of Emil Artin. Unable to acquire the *venia legendi*, or right to teach, because of his politics, he was forced to emigrate, and did so in 1934. Shortly after emigrating, he published a three-page set-theoretic paper (in English) that has associated his name forever with one of the most used of mathematical principles.[79] According to Zorn, after his compulsory dismissal in 1934 by the head of the Hamburg mathematics faculty, Wilhelm Blaschke, Blaschke bought him a steamer ticket to the United States. Max Zorn and another young mathematician, Günther Höwe, were both naval buffs and friends; however, Höwe told him after Zorn's dismissal that it pleased him, because otherwise his conscience would have compelled him to denounce Zorn—such was the temper of the times.[80]

Two other mathematicians of reputedly leftist persuasion were Kurt Reidemeister and Robert Remak. Robert Remak was a brilliant algebraist and number-theorist, though an apparently more than somewhat difficult colleague. He had the reputation of being a "communist," and there were also rumors in the 1920s that he was "not completely Aryan," but this seems to have been mostly a result of his sarcastic personality, eccentricities, and unkempt habits, rather than stemming from any real knowledge about him or any political activity on his part. In fact, he was Jewish. He was twice denied *Habilitation* and the right to join the teaching faculty in Berlin, in 1919 and 1923, again largely because of his demeanor and habits. However, finally, on January 11, 1929, Remak (who had received his doctorate in 1911) was accepted on his third attempt. In September 1933, he lost the right to teach, and after the *Kristallnacht* of November

Heidelberg archivist in 1988, remarked to me, the "Gumbel case" has been described "a hundred times," and almost no book discussing any aspect of universities during the Weimar period fails to mention it. Nevertheless, the first full-scale biography of Gumbel, by Christian Jansen, did not appear until 1991 (Heidelberg: Verlag Das Wunderhorn), and an extensive biographical notice of Gumbel and "the case" was written by Karen Buselmeier in 1979 and used as a foreword (pp. 7–31) to the reissue of Gumbel's book *Verschwörer* (1984). Gumbel died on September 10, 1966, in New York. See also Anselm Faust, *Der Nationalsozialistische Studentenbund* (1973), 2: 57–62.

[78] Gasman 1971: passim.

[79] *Bulletin of the American Mathematical Society* 41 (1935): 667–670.

[80] Above material on Max Zorn comes from Maximilian Pinl, "Kollegen in einer dunklen Zeit III," *JDMV* 73 (1972), under Hamburg, and an interview with Zorn in Bloomington, Indiana (Mar. 18, 1991).

8–9, 1938, spent two months in the Sachsenhausen concentration camp. Released, he managed to emigrate to Holland, only to be recaptured during the war and deported to Poland. Remak died in Auschwitz sometime during or after 1942.[81]

Unlike Max Zorn, Kurt Reidemeister[82] was no youthful leftist, but he was someone who believed strongly in the traditional notions of freedom of scientific inquiry and scientific universalism. His life intersected significantly with several people who appear in this book. Born in 1893, the man who was to become internationally famous as a geometer and topologist, one of the founders of knot theory, was at first more interested in philosophy than mathematics. As a nineteen-year-old, he heard Edmund Husserl's lectures in Freiburg, and took courses in Marburg and Göttingen (such travel was then the custom among German students). Four years of service in World War I (he advanced to lieutenant) interrupted his study, and after the war he went back to Göttingen, where he qualified as a secondary-school teacher simultaneously in mathematics, philosophy, physics, chemistry, and geology. Edmund Landau, not known as an easy examiner, was his mathematics examiner and dismissed him after only thirty minutes with the grade of "distinction." In 1920 Reidemeister followed Erich Hecke to Hamburg, completing a dissertation in algebraic number theory under Hecke's supervision in less than an additional year. In Hamburg, he met Wilhelm Blaschke, who turned him toward an interest in geometry, and Blaschke entrusted the brilliant student with cooperation on the second volume of his *Differential Geometry*. Indeed, just a few months after receiving his doctorate, Reidemeister gave a plenary lecture on this subject, quite different from that of his dissertation, at the annual meeting of the German Mathematical Society. However, his other cultural interests were not left by the wayside. He wrote stories and poems and lectured on Spengler's *Decline of the West* (which has a long section on varieties of mathematics). Although not yet "habilitated," he received a call to a professorship in Vienna (just two years after following Hecke to Hamburg). There he pursued philosophical as well as mathematical interests, studying the philosophy of Ludwig Wittgenstein with Moritz Schlick and being part of that famous philosophical circle, the *Wiener Kreis*. There he also married. In 1925, he went to Königsberg, where his well-known books on knot theory and combinatorial topology appeared. In January 1933, shortly before Hitler's accession to power, National Socialist students at Königsberg fomented a disturbance directed against the university Rektor. Reidemeister devoted a whole mathematics lecture to explaining why the behavior of these students was totally unsupportable and not compatible with rational thinking. As a consequence, he was dismissed shortly after January 30, at a time when three "non-Aryan" colleagues, Gabor Szegö, Richard Brauer, and Werner Rogo-

[81] Pinl "Kollegen in einer dunklen Zeit II," *JDMV* 72 (1971), 190–193; H. Behnke, *Semesterberichte* (1978), 39–41; Kurt R. Biermann, *Die Mathematik und ihre Dozenten an der Berliner Universität, 1810–1933* (1988), esp. 209–211. Numerous anecdotes are attached to Remak's name.

[82] Material on Reidemeister below is from Pinl 1972 and an obituary by Rafael Artzy, *JDMV* 74 (1973): 96–104.

sinski, all were left in office (until after the law of April 7). Perhaps this early dismissal was fortunate for Reidemeister. Wilhelm Blaschke, his erstwhile mentor, and a "realist"[83] who got on well with the powers that were, made efforts to reverse his dismissal or to find him another job,[84] and in autumn 1934, he was Helmut Hasse's successor in Marburg. The intervening time he spent in Rome. After World War II, when Blaschke was dismissed by the Allies as a Nazi collaborator, Reidemeister would be asked by Blaschke to return the favor and help him get reinstated.[85] While at Marburg, Reidmeister apparently attempted to find ways to publish the work of Jewish mathematicians. He published philosophical as well as mathematical work, the former almost, but not quite, his only production after World War II, and he took a strong interest in mathematical education. He died in 1972. That Reidemeister might be considered "leftist" because he in fact advocated the usual academic norms is a significant sign of the times he lived in.

Just how unpolitical the academic atmosphere was in mathematics departments might also be inferred from the memories of visitors to them at the time. Saunders MacLane (born 1909) and Edward McShane (born 1904), who would both become prominent American mathematicians, were also both students at Göttingen in the early 1930s. MacLane recorded (around 1975) his impressions of German politics in 1931:[86]

> Things were always in disorder, but they [the Germans] accepted that. Different people, of course, had different views. My impression that first year was that probably Hitler shouldn't be taken too seriously. Politics in Germany seemed a great big mess. I distinctly remember buying a pamphlet that was labeled "The 27 Parties of Germany." There were 27 of them, and the NSDAP—the Nazi party—was just one.

From 1933, MacLane remembered the Reichstag fire (February 27) and the March 5, 1933, elections, after which he recalled all sorts of regulations, talk, and some unpleasantness, but largely suppressed his memory of the time. As for McShane, he remembers discussing politics with MacLane and the Göttingen *Privatdozenten* during the autumn and winter of 1932–33, and claims that though he and his wife spent New Year's Day 1933 in Berlin, they heard nothing of Nazi riots taking place there until they received newspaper clippings from worried relatives.[87] The attitudes of American student visitors might well

[83] Max Zorn's description of Blaschke in an interview, Mar. 18, 1991.

[84] For example, he collected signatures on a petition for Reidemeister's retention. For a copy of Blaschke's petition, and his request for signatures, see the personal papers of Hellmuth Kneser in the private possession of his son Martin Kneser, Blaschke to Kneser, June 18, 1933. These are hereafter cited as HK. I am indebted to Prof. Martin Kneser for permission to see and to copy some of these papers.

[85] According to an interview with Werner Burau (Jan. 31, 1988), Reidemeister helped Blaschke get fully reinstated. In fact, though, he argued for Blaschke's being pensioned off with a research contract. See below, chapter 8.

[86] Reid 1976: 130. One should also note that MacLane's "Hitler shouldn't be taken too seriously" was less than a year after Hitler's surprising success in the 1930 elections.

[87] Reid 1976: 130.

be dismissed as political naiveté on the part of foreigners were it not that most of the native German mathematicians do not seem to have been much more politically active or interested. Exceptions were people like Gumbel, who was apparently a committed communist, and Reidemeister, who was personally interested in political philosophy (though apparently not politically active). This is in part an argument from the paucity of evidence of political activity, but as already delineated, it fits the standard mode of behavior that expressed the long-standing relationship between the German state and the professors in its universities. There were also some mathematicians initially convinced by Hitler. Günther Höwe, the quondam friend of Max Zorn, had translated one of Oswald Veblen's books from English into German and wrote him (in German) on April 14, 1933:[88]

> One is accustomed to calling what has occurred in Germany in the last weeks the "national Revolution." I would rather call it "Reform."
>
> To begin with, the number of people who, as a result of this occurrence, have suffered bodily or financial damages, is vanishingly small compared to the number that is customary in revolutions in France, and France is respected above all in England and the U.S.A., also even today among many Germans, as the most civilized nation in the world.
>
> Furthermore, all the essential (*wesentlichen*) changes were not carried out directly by the people (*Volk*) or even through "the barricades," but, as the German people has preferred (*liebt*) for centuries, by the government.
>
> Finally, everything happened without any damage to the Weimar Constitution, and the constitutional changes that were undertaken took place on the basis of possibilities for their change foreseen in the constitution itself. The large number of people (*volker* [sic]) voting in the election (90 percent) shows that the people (*Volk*) have placed themselves on the side of the government created by Hindenburg in a way democrats also find satisfactory.

Hitler's government had held elections on March 5, 1933, in which the *National Sozialistische Deutschen Arbeiter Partei* (NSDAP) obtained 43.9 percent of the vote. The *Ermächtigungsgesetz* or "enabling act" by which the dying parliament committed suicide and officially gave Hitler dictatorial powers had been passed on March 23, and the law of April 7 "for reform of the civil service" had just appeared. I do not know what Höwe's prior politics were, but his apologetics at this point would seem to represent either Nazi fellow-traveling or a singular naiveté. As has been seen, within a year, he would be struggling with his conscience about denouncing a former acquaintance—with his conscience impelling him to denunciation. Of course, it is true that many wealthy, aristocratic, and monarchically inclined right-wing conservatives supported Hitler at the time as a "wild man" whom they would soon tame, after he had served his purpose of bringing the plebs to their side as well as thoroughly destroying "parliamentarism."

[88] Veblen papers, Library of Congress (hereafter cited as VP), under H miscellaneous.

A striking example of such political naiveté can be seen in the distinguished German-Jewish mathematician Richard Courant. Immediately after World War I, Courant was a member of the Göttingen town council, toyed with running for the *Landestag* (or state parliament), and wrote a long anticommunist letter to a Göttingen newspaper. At the time of the Silesian plebiscite in 1921, Courant, as a born Silesian, agitated for the German cause. As late as March 30, 1933, Courant was blaming Einstein and other "agitators" for the anti-Semitic feeling in Germany and distinguishing between "good" and "bad" Jews:[89]

> Even though Einstein does not consider himself a German, . . . he has received so many benefits from Germany that it is no more than his duty to help dispel the disturbance he has caused. Unfortunately, as I see from the papers, a reaction to these events has set in. . . . I very much hope that it will be possible to deter the intended boycott [of the Jews] at the last moment. Otherwise I see the future very black.
>
> What hurts me particularly is that the renewed wave of antisemitism is . . . directed indiscriminately against every person of Jewish ancestry, no matter how truly German he may feel within himself, no matter how he and his family have bled during the war and how much he himself has contributed to the general community. I can't believe that such injustice can prevail much longer—in particular, since it depends so much on the leaders, especially Hitler, whose last speech made a quite positive impression on me.

Ironically, two years later, Courant, a "sadder but wiser" emigré in New York, would criticize similar naiveté on the part of the famous mathematician Carl Ludwig Siegel.

Werner Weber was a young *Privatdozent*, and a Nazi sympathizer prior to 1933, though he was a student of Emmy Noether and Edmund Landau (they called his dissertation "excellent").[90] In the summer of 1935, he was appointed professor at Frankfurt as a substitute for Siegel, who was then visiting at the Institute for Advanced Study in Princeton. Siegel determined to return to Frankfurt and attempt to drive out Weber.[91] While the fact that Siegel was a "pure Aryan" might have given him some private hope of success, by April 1935 so much had happened that one wonders at his expectations. It is true that Siegel was partly motivated by the fact that he wanted to protect his friends Max Dehn and Ernest Hellinger, both well-known mathematicians and Jewish. Both Dehn (born 1878) and Hellinger (born 1883) had served during World War I and had appointments dating prior to 1918. Thus they technically did not come under the law of April 7, 1933. Both were dismissed in 1935, but remained in Germany until early 1939 (Hellinger spent six weeks in Dachau, from mid-November 1938 until his release, apparently effected by his sister, already in

[89] Reid 1976: 139–140. The translation is hers.

[90] Wolfgang Kluge, "Edmund Landau, Schriftliche Hausarbeit vorgelegt im Rahmen der Ersten Staatsprüfung," thesis, University of Duisburg (1983), 128. Weber received his degree in 1929, and from then until Landau's forced retirement was his *Assistent*. For this last event, and more on Weber, see below; chapter 4, "Hasse's Appointment at Göttingen."

[91] VP, Courant to Veblen, Apr. 22, 1935, under Courant.

America). As for Siegel, he actually did not return to Germany until after Dehn's dismissal.[92] The motives he expressed in a letter to Oswald Veblen did not mention Weber, but rather fear (based on a telegram from the Frankfurt Rektor) that if he accepted an offer to teach at Princeton in 1935–36, he would lose his post in Frankfurt without getting any pension, and "as my health is not very strong and since I am obliged to give a part of my income to my father, it seems to be advisable for me to return to my place in Frankfurt and to try to live there for the next time. Moreover my friend Hellinger is still there and I will not leave him alone" (English original).[93] Siegel left Germany for Norway in March 1940, received another appointment at Princeton in June 1940, and left by one of the last boats for the United States.[94]

While Siegel may seem naive in hindsight, his "naiveté" was that of a good man truly concerned about his friends. The feelings of Courant or Eberhard Hopf cited earlier reflect what seem to have been much more common attitudes: either the initial feeling that as a "good German" one should not be personally affected (Courant), or a political naiveté whose depth is almost unfathomable fifty years later (Hopf). While the personal attitudes of mathematicians toward the Nazi regime will be discussed in more detail later, one important fact should be noted here: liberal-thinking democratically inclined German mathematicians, foreigners with a different point of view, anyone, like Reidemeister, who did not simply "go along" with the ruling dictates, are salient just because so many of the German-speaking academics, mathematicians or not, did "go along" more or less willingly.[95] There were also mathematicians, as well as other scientists, some quite prominent, who did not hesitate to advocate Nazi ideas. They were also exceptions. The majority of German academics simply went along: it was within the academic tradition, reversed their perceived loss of status under Weimar, and could be given academic justification. As Karl Dietrich Bracher has remarked, not only was politics a dirty business so far as the professoriat was concerned, but there was an almost schizophrenic split between classical-humanistic education and a *Realpolitik* concerned with power.

[92] Carl Ludwig Siegel, in his *Gesammelte Werke* (1966), vol. 3, pp. 468–471. (A brief history of mathematics at Frankfurt is given there as well, pp. 462–474.)

[93] VP, Siegel to Veblen, July 8, 1935, under Siegel.

[94] Norbert Schappacher, "Das Mathematische Institut der Universität Göttingen, 1929–1950" (unpublished), 49; a condensed version appears in H. Becker, H.-T. Dahms, and C. Wegeler, eds., *Die Universität Göttingen unter den Nationalsozialismus* (1987), 344–373, here 359–360. The German invasion of Norway began on April 9, 1940.

[95] On this point, see Faust 1973. While according to Faust (cf. above, note 55) only 1.2% of university professors were actually Nazi party members prior to 1933, only 10% of university professors were "actively political" in any way. Furthermore, there were a number of very active pro-Nazi professors who were not necessarily party members (like Phillip Lenard). Faust argues persuasively that pro-Nazi sentiments among the professoriat were far more widespread than is usually assumed. As discussed above, the general university atmosphere was certainly not inimical to Nazi rhetoric, despite the undeniable existence of prominent prorepublican professors like Gustav Radbruch and Willy Hellpach. For Tübingen as a case in point, see Uwe Adam, *Hochschule und Nationalsozialismus, Die Universität Tübingen im Dritten Reich* (1977), 31–32.

Moreover, this provided an opportunity readily seized by the "National-Socialistic pseudo-reformers" with a powerful slogan that overcame many private demurrers: "The hour of the synthesis between Spirit (*Geist*) and Power has now arrived."[96] The German university was a mixture of political unconcern and openly antidemocratic sentiment. As Bracher remarked, "The real fate [of the German university under Hitler] lay not in the crimes of a minority, but in the failure of the majority of the educated."[97]

The same was true elsewhere as well. For example, the well-known mathematician Karl Menger wrote Oswald Veblen on October 27, 1934:[98]

> What I could not write you from Vienna is a description of the situation there. . . .
> First of all the situation at the university is as unpleasant as possible. Whereas I still don't believe that Austria has more than 45% Nazis, the percentage at the universities is certainly 75% and among the mathematicians I have to do with, except, of course, some pupils of mine, not far from 100%.

It should be noted that this letter was written three months after Engelbert Dollfuss' assassination and the failure, partly through Mussolini's intervention, of the Nazi attempt at an Austrian coup d'état—*Anschluss* was still three-and-a-half years away.

There is evidence, moreover, of similar attitudes in Switzerland. The mathematician Henrich Behnke recalls in his memoirs that when he traveled to Switzerland in the summer of 1933, "Everywhere the children in the streets greeted the car with the raised-hand salute—even more than in Germany, German citizens cheered it with Nazi shouts, and Swiss hurried to express their respect for the new regime."[99] Though Behnke does say that more moderate attitudes prevailed among his Swiss colleagues—a general neutrality toward the Hitler government and a horror at the dismissals of professors—his experiences are echoed by letters from Heinz Hopf to George Pólya and Oswald Veblen. Both Pólya and Hopf had very famous mathematical careers. At the time, Hopf's was just beginning, and Pólya was already at mid-career. In May 1933, Hopf wrote Pólya,[100]

> It is quite unpleasant that now also here in Zurich nationalists and anti-Semites have become powerful. There are continual assemblies of "fronts," namely the "nationals," the "federals" and others among the students—(and even more among the schoolchildren)—these tendencies seem therefore to be rather strong.

In July of the same year, he wrote to Veblen,[101]

[96] Karl Dietrich Bracher, "Die Gleichschaltung der deutschen Universität," in *Universitätstage* 1966: 129.

[97] Ibid.: 142.

[98] VP, Menger to Veblen, Oct. 27, 1934, under Menger.

[99] Behnke 1978: 127. As early as 1923, Hitler obtained financial support from Switzerland. See Ernst Deuerlein, *Der Aufstieg der NSDAP in Augenzeugenberichten* (1974), 180.

[100] VP, under Pólya.

[101] VP, under Hopf.

> I am very dismayed that presently there is absolutely no prospect of a German Jew obtaining a position in Switzerland: professorships are not open, new ones will not be created since there is no money, a large number of good Swiss wait for positions as Assistent that are becoming open, and besides, the tenor among the students is very nationalistic–anti-Semitic.

While Heinz Hopf may have been one of the "rational" Swiss colleagues visited by Behnke when he crossed the border for a breath of fresh air, Hopf himself seems also to have been very aware of the pro-Nazi sentiment in Switzerland. Nor was he likely to have been "neutral," as his relative Ludwig Hopf, ten years his senior and Jewish, was one of the mathematicians dismissed by the Nazis from the *Technische Hochschule* in Aachen. Hopf's letter to Veblen was a vain attempt to find a place for this cousin who, after 1933, had visited him regularly.

Hopf's father, Wilhelm, had converted from his father's Judaism to his wife's Lutheranism in 1895. Heinz Hopf had been born in a village near Breslau (modern Wrocław) in 1894. For the Nazis, of course, Wilhelm Hopf was a Jew, but together with his wife he remained in Breslau in progressively worsening circumstances. Heinz Hopf visited his parents regularly until 1939, and attempted successfully to obtain a Swiss immigration permit for them. However, his father became seriously ill, and then World War II intervened. Wilhelm Hopf died in Breslau in 1942. Hopf's Jewish ancestry also caused him difficulties, and in 1943 he was threatened with having to return to Germany, but he managed to obtain Swiss citizenship guaranteeing his ability to stay in Zurich.[102] It is interesting to contrast the reactions of the "Jewish" foreign-born Swiss citizens, George Pólya and Heinz Hopf, to the menace across the border—Pólya chose to emigrate, Hopf to stay (despite the ability to leave).

The university professors, by their overwhelmingly antidemocratic "unpolitical" stance, their feeling of having been declassed by the Weimar Republic, and their long tradition of independence from and consequent obedience to the state, were for the most part readily able to accept the Hitler government (already on March 3, 1933, 300 postsecondary teachers had declared themselves for Hitler).[103] However, there was yet another traditional factor in German university life that made the Nazi success there easier: anti-Semitism.

While it is true that Jews were represented in German universities far out of proportion to their numbers in the population, especially as *Privatdozenten*, it is also true that academic advancement was extraordinarily difficult for them. Furthermore, a large number of these "Jews" were either baptized or at least *konfessionslos*, that is, officially without religion, and distant from any Jewish community.[104] Indeed, from the 1880s on, it was common in Berlin for the sons of Jewish families who became academics or professionals to be baptized, and for

[102] Günther Frei and Uri Stammbach, "Heinz Hopf," chap. 38 in I. M. James, ed., *History of Topology* (Amsterdam: Elsevier, 1999), 1002.

[103] Bracher 1966: 132. See also Faust 1973.

[104] Ringer 1969: 136.

the daughters to be baptized preparatory to marriage to a Christian, while those sons who went into business or industry would remain Jews.[105] Nevertheless, advancement was difficult, even for the baptized, especially prior to World War I.[106] Kurt Hensel (baptized) was promoted in 1902 at Marburg, at age forty-one, where he had been an "associate professor" (*Extraordinarius*), but without pay, since 1891—without pay because the Wilhelminian education minister told him, "You have anyway enough money without it."[107] Another example of "late promotion" was Paul Epstein, who was still a *Privatdozent* at Strassbourg in 1918 when the French dismissed him in the aftermath of World War I. He never was more than an "associate professor" at Frankfurt in the succeeding fourteen years before the Nazis came to power. There were many Jewish *Privatdozenten*, because being such an instructor was a "free" profession, outside an official career with payment from the state. Advancement to Extraordinarius and then Ordinarius, though, meant advancement through the agency of the already established professors and a status as a state civil servant. In 1909–10, over 93 percent of the *Ordinarien*, or full professors, in Germany were Protestants or Catholics, but less than 81 percent of the *Privatdozenten* were. Fritz Ringer cites this as a statistical indication of the bias against Jews complementing the anecdotal evidence that can be gathered for each academic discipline.[108] Under the Weimar government, the official state position on Jewish advancement in the universities may have improved somewhat, but advancement was still largely in the hands of the older academics, and so was a continuation of the prewar situation—although "corrections" like the eventual appointment of Friedrich Hartogs in Munich were made. Concerning anti-semitism in universities in the 1920s, Abraham Fraenkel describes his own "superficial experiences" as follows:[109]

> As concerns the universities, anti-Semitic tendencies were at that time [the 1920's] reversed from the situation before the Revolution [of 1918]—that is, in Bavaria they were much more pronounced than in North and West Germany. The naming, though not the preferment (*Beförderung*), of Jews to positions as full professors (*Ordinariate*) remained infrequent with the exception of the new city-universities of Frankfurt and Hamburg.

[105] Fraenkel 1967: 97.

[106] Examples among well-known mathematicians include Friedrich Hartogs (*konfessionslos*), who became a full professor at age fifty-three in 1927 in Munich (ibid.: 84); and Alfred Pringsheim (*konfessionslos*), a full professor at age fifty-one in 1901 in Munich (ibid.: 82). Max Noether (Jewish), age forty-four, became a full professor in 1888 in Erlangen (see his obituary by A. Brill in *JDMV* 32 [1923]: 211–233, pp. 212, 229). Noether never received a desired and deserved call to a larger university. The characterization of him as Jewish follows Auguste Dick 1970: 4–5. The name Noether, distinctly non-Jewish for centuries, was given his father at the time of the Baden edict of tolerance in 1809 (ibid.).

[107] Fraenkel 1967: 97.

[108] Ringer 1969: 136.

[109] Fraenkel 1967: 184–185.

Fraenkel was a religiously observant "Orthodox Jew," and though he himself had no complaint about anti-Semitism in either Marburg or Kiel, where he had posts prior to 1933,[110] as a young man of eighteen, because of some early success, he decided to study mathematics and become an academic "despite the restricted possibilities for advancement open to Jews"[111]—indeed, a prewar friend of the family who helped him get a start became a postwar anti-Semite.[112]

It is true that there were exceptions to the dismal promotion record of Jews at German universities—for mathematics, this was particularly true at Göttingen, where Karl Schwarzschild became professor and observatory director at age twenty-eight, in 1901, moving on to Potsdam eight years, later despite a refusal to be baptized.[113] Hermann Minkowski became an Ordinarius at Königsberg in 1895 at age thirty-one, but moved the following year to Zurich, and then, largely through the influence of his friend David Hilbert, to Göttingen in 1902. When Minkowski died in 1909, he was succeeded at Göttingen by Edmund Landau, who was thirty-two at the time.[114] But Schwarzschild, Minkowski, and Landau truly were exceptions, and their presence at Göttingen was aided by the fact that Felix Klein and David Hilbert were not themselves anti-Semitic,[115] in contrast to the academic profession as a whole, as brought out by the figures cited by Ringer.

German academic anti-Semitism of the 1920s was a continuation of attitudes already set by the 1890s. Peter Pulzer has documented the rise of political anti-Semitism in Germany and Austria in the 1870s and 1880s; in the forefront of the academic side of the movement was the famous historian Heinrich von Treitschke. Although Treitschke initially believed in the possibility of Jewish assimilation as a "solution," gradually he vacated this position for one of the Jew as the eternal foreigner within the people.[116] In 1893 the Austrian journalist Hermann Bahr interviewed intellectuals in various European countries, but primarily Germany and France, on the subject of anti-Semitism. Bahr himself was of the opinion that "anti-Semitism is the morphine addiction of small people" and concluded his preface with the statement:[117]

> He who is an anti-Semite is one out of the appetite for intoxication and the ecstasy of a passion (*Begierde nach dem Taumel und dem Rausche einer Leidenschaft*). He takes the nearest arguments. If one disproves them, he will seek others. . . . Therefore I wish in no way to "disprove" anti-Semitism, something that has been done a thousand times and is always in vain. I simply ask with what feelings and what answers the educated of different nations take a position toward this appearance in the people (*Volk*). Per-

[110] Ibid.: 185.

[111] Ibid.: 78.

[112] Ibid.: 27, 76.

[113] Ibid.: 86.

[114] Ibid.: 87.

[115] See David Rowe, "'Jewish Mathematics' at Göttingen in the Era of Felix Klein," *Isis* 77 (1986): 422–449, for a discussion of this fact.

[116] P. J. Pulzer, *The Rise of Political Antisemitism in Germany and Austria* (1964), 226, 298, for the first attitude; 249–250 for the second.

[117] Hermann Bahr, *Der Antisemitismus. Ein Internationales Interview* (1894), 2, 4.

haps this produces a very curious document for some future time on the state of mind (*Geist*) in 1893.

Not all Bahr's German interviewees were anti-Semites; indeed, one of the most distinguished, the historian Theodor Mommsen, was an ardent opponent of Treitschke and anti-Semitism. What is striking, however, is the sort of snobbish anti-Semitism that appears in the statements of such luminaries as the economists Gustav Schmoller and Adolf Wagner, the biologist Ernst Haeckel, the publicist Maximilian Harden (himself of Jewish ancestry, his real name being Felix Ernst Witkowski),[118] and the liberal (in the 1848 sense) writer Friedrich Spielhagen.[119] In brief, for such people, anti-Semitism was justified on a variety of grounds, provided only that it was not too plebeian. Thus they could condemn the rabble-rousing anti-Semitic agitator Hermann Ahlwardt, very much a presence in the 1890s, while "understanding" and condoning the sentiments to which he appealed.[120] This same sort of elitist anti-Semitism persisted in the universities through the 1920s despite the occasional Mommsen, Virchow, Hilbert, or Klein, and made the events of the 1930s easier for the professors to accept without a great deal of fuss. Indeed, anti-Semitism ran so deeply in elite German society, and the Nazi campaign evoked such feelings, that even resistance circles such as that around Carl Goerdeler felt that there would need to be some sort of "solution of the Jewish problem" even if Hitler should be overthrown.[121] Similar sentiments were echoed by such distinguished *Vernunftrepublikaner* as Friedrich Meinecke.[122]

Before leaving the subject of German academic anti-Semitism, it is well to point out that it was hardly just a German or German-speaking phenomenon in the 1920s or later, nor were mathematicians in other countries immune to it. Some examples from mathematics in the United States will suffice to indicate the problem even in such a purportedly "pure" subject. On March 30, 1927, C. C. MacDuffee, a student of Oswald Veblen, and then at Ohio State University, wrote to him:[123]

[118] Harden was a remarkable character of the late Wilhelminian period. An ardent and vocal opponent of aspects of Wilhelminian practice, he was also no friend of Weimar. Despite the latter, in 1918 he was attacked by anti-Semites, and in 1922 survived an assassination attempt. For the details of this last, see Gumbel 1984: 86–90.

[119] Bahr 1894. For Haeckel, see also Gasman 1971. Gasman argues that the wide distribution of Haeckel's Social Darwinist views provided a fertile medium for the growth of Nazi sentiments.

[120] For Ahlwardt, see Mosse 1964: 138–139.

[121] On this point, see Christof Dipper, "Der Deutsche Widerstand und die Juden," *Geschichte u. Gesellschaft* 9 (1983): 349–380. It should be remembered that Goerdeler had been a member of the DNVP, a conservative party closed to Jews after 1929. See note 56.

[122] Although no anti-Semite, and dismissed by the Nazis, Meinecke nevertheless managed in his 1946 book, *The German Catastrophe* (translated 1950 by Sydney B. Fay, [Cambridge, Mass.: Harvard University Press]) not only to criticize Prussian militarism and the German bourgeoisie, but to say about Germany after World War I: "Among those who drunk too hastily and greedily of the cup of power which had come to them were many Jews. They appeared in the eyes of persons with antisemitic feeling to be the beneficiaries of the German defeat and revolution" (32).

[123] VP, under MacDuffee. Henry Blumberg was a mathematician whose highly original work has only relatively recently received appropriate notice.

> While the atmosphere is still a little strained, I do not believe that [Henry] Blumberg and I are in a serious position yet. [H. W.] Kuhn was much agitated toward me about the time that he saw you, and for fear that he said or implied things about me, I determined to tell you the situation. . . .
>
> Although I think I can take care of myself here I am alarmed for Blumberg because of the degree of animosity toward Jews displayed by almost everyone in the University. The following is an incident, which I have never told to Blumberg, [and] is for you only at present. Blumberg was offered the position here by the late President Bohannen, and tendered his resignation to Illinois. Then a group from the department went to Dean Hitchcock (who is an absolute incompetent) and persuaded him to withhold his approval of the appointment. Kuhn tried to enlist my sanction for the move by saying that my advancement would be more rapid if Blumberg were not here. I told him that if such a thing were done Ohio [State] would be blacklisted by the [American Mathematical] Society, and moreover that I would stay just long enough to get another job. President Thompson confirmed Blumberg's appointment without the approval of Dean Hitchcock. However, I believe that Blumberg will never get a rise in salary here. He does not know anything of this.

Refugee scientists from Hitler's Germany were not always welcome in the United States; the reception of refugee mathematicians in America has been discussed in some detail by Nathan Reingold.[124] Part of this story is the "genteel anti-Semitism" that was widespread in American mathematics at the time. Several examples are cited by Reingold. Among them is George David Birkhoff, one of the leading American mathematicians of his day. In 1934, Solomon Lefschetz was the first Jew to be elected president of the American Mathematical Society. Birkhoff, only a few months older than Lefschetz, had this to say about the prospect (though in the end he was the reporter of Lefschetz's nomination for the presidency).[125]

> I have a feeling that Lefschetz will be likely to be less pleasant even than he had been, in that from now on he will try to work strongly and positively for his own race. They are exceedingly confident of their own power and influence in the good old USA. The real hope in our mathematical situation is that we will be able to be fair to our own kind. . . .
>
> He will get very cocky, very racial and use the Annals [Annals of Mathematics] as a good deal of racial perquisite. The racial interests will get deeper as Einstein's and all of them do.

As one writer of a Birkhoff obituary noted:[126]

> After that [the attempt to help save Göttingen as a mathematical center during the depression] he was instrumental in bringing to the U.S. the finest of exiled talent,

[124] Nathan Reingold, "Refugee Mathematicians in the United States of America, 1933–1934: Reception and Reaction," *Annals of Science* 38 (1981): 313–338.

[125] Ibid.: 324.

[126] D. D. Kosambi, "George David Birkhoff, 1884–1944," *Mathematics Student* 12 (1945): 116 and 119.

> though he could never have been mistaken for a pro-Semite by anyone who spoke with him for more than thirty seconds on the subject. . . .
>
> One wonders whether the war had changed Birkhoff's naive views about the importance of race and the essential glory of Nordics, which he held very strongly at least till 1934. Did he realize that "pure Anglo-Saxon" was even more meaningless than "American scholar" had been before him, and that "American" was not a race but a mentality? It shows the fundamental nobility of his character that he never allowed such views to interfere with his scientific judgment, nor to prejudice him in the slightest in the matter of adjudicating research fellowships and prizes. He had no hesitation in recommending for important posts people with whom he was not and did not want to be on visiting terms.

Birkhoff's actions may have seemed ambiguous—his views certainly were not. On the occasion of the semicentennial celebration of the American Mathematical Society in 1938 he gave a well-known speech, "Fifty Years of American Mathematics," containing the following passage, which must be read in the light of his publicly well-known views.[127]

> The second special group to which I wish to refer is made up of mathematicians who have come here from Europe in the last twenty years, largely on account of various adverse conditions. This influx has recently been large and we have gained very much by it. Nearly all of the newcomers have been men of high ability, and some of them would have been justly reckoned as among the greatest mathematicians of Europe. A partial list of such men is indeed impressive: Emil Artin, Solomon [sic] Bochner, Richard Courant, T. H. Gronwall, Einar Hille, E. R. van Kampen, Solomon Lefschetz, Hans Levy [sic], Karl Menger, John von Neumann, Oystein Øre, H. A. Rademacher, Tibor Radó, J. A. Shohat, D. J. Struik, Otto Szasz, Gabor Szegö, J. D. Tamarkin, J. V. Uspensky, Hermann Weyl, A. N. Whitehead, Aurel Wintner, Oscar Zariski.
>
> With this eminent group among us, there inevitably arises a sense of increased duty toward our own promising younger American mathematicians. In fact most of the newcomers hold research positions, sometimes with modest stipend, but nevertheless with ample opportunity for their own investigations, and not burdened with the usual heavy round of teaching duties. In this way the number of similar positions available for young American mathematicians is certain to be lessened, with the attendant probability that some of them will be forced to become "hewers of wood and drawers of water." I believe we have reached a point of saturation, where we must definitely avoid this danger.

A large number of the people in Birkhoff's list were not Jewish. They were certainly all distinguished mathematicians, and it is not clear how Birkhoff drew up his list. However, the inclusion of Lefschetz (who was in the chair for Birkhoff's address) must be counted a calculated insult, as Lefschetz came to the United States in 1905, lost both hands in an industrial accident in 1907, after which he earned an American Ph.D. in mathematics (at Clark University), and

[127] G. D. Birkhoff, "Fifty Years of American Mathematics," in *American Mathematical Society, Semicentennial Publications*, vol. 2 (addresses) (1938), 276–277.

had been a U.S. citizen since 1912.[128] Similarly, Birkhoff's use of the well-known Biblical phrase "hewers of wood and drawers of water" is an allusion, which his audience would not have missed in the context, to the enslavement of the Hittites by the Hebrews.[129]

Further down the Charles River, on May 13, 1935, the distinguished statistician Karl Compton, then president of M.I.T., prepared the following confidential memorandum for Vannevar Bush, Norbert Wiener, and Henry Bayard Phillips.[130]

> Professor Wiener raised the question as to the possibility of a future appointment at M.I.T. for Mr. Levinson who is rated as the outstanding mathematical product of M.I.T. according to present indications. Professor Wiener reported some inertia or undefined opposition to Levinson's appointment and asked whether there was a basis for this.
>
> I replied that there was general recognition of Levinson's ability but that in certain quarters there was a question as to whether Levinson is as outstanding a young man as Professor Wiener believes him to be. Professor Wiener believes, however, that this question will be settled one way or another by Levinson's further performance in the near future.
>
> I also pointed out the tactical danger of having too large a proportion of the mathematical staff from the Jewish race, emphasizing that this arises not from our own prejudice in the matter, but because of a recognized general situation which might react unfavorably against the staff and the Department unless properly handled. I also emphasized the fact that this attitude was in no sense prompted by any criticism of present members of our staff or of the present situation, but was prompted only by a desire to safeguard against a situation which might lead to criticism and unfortunate results.
>
> After discussing various aspects of the situation and Professor Wiener's own relation to it, we agreed that we could accept the following principles.
>
> 1. No man should ever fail to have fair consideration for a position on our staff because of his race or analogous characteristics.
>
> 2. Other things being approximately equal, it is legitimate to consider the matter of race in case the appointment of an additional member of the Jewish race would increase the proportion of such men in the Department far beyond the proportion of population.
>
> 3. The way is not barred to consideration of Mr. Levinson when the time of his possible appointment approaches and he should then be considered on his merits in light of the above principles.

Note that Compton adopts the notion of a "Jewish race" rather than just a religion.

In terms of general attitudes, one should also remember that Harvard began

[128] *American Mathematical Society Semicentennial Publications*, vol. 1 (New York, 1938), 236–240. That Lefschetz was in the chair appears in *Bulletin of the American Mathematical Society* 45 (1939): 5.

[129] Joshua 9:19–21, 25–27.

[130] Wiener Papers at M.I.T., under Levinson. See also Wiener 1964: 211–212.

to restrict Jewish admissions in 1926, and "Jewish quotas" at Harvard, Yale, and Princeton persisted until well after World War II.[131] Nor was the U.S. situation (any more than the German one) merely one of academic snobbery. In 1948, the *Atlantic Monthly* carried a debate about whether a Jew should change his name in order to obtain a job and live "more comfortably" (later condensed in *Reader's Digest*),[132] and the publication date of *Gentleman's Agreement* is 1947.[133]

The fact that snobbish anti-Semitism was commonplace in the mathematical (and academic) world of the United States, as well as in Germany, in the 1920s does not diminish the fact that in the German situation, it meant less concern for dismissed colleagues. It does, however, indicate that academic anti-Semitism among mathematicians, as well as others, was not a peculiarly German phenomenon. A similar point could be made about England. Nevertheless, in Germany, it seems to have substantially weakened whatever academic opposition there initially might have been to Hitler. Finally, the dismissal of Jews meant more posts available for those not so tainted, especially at a time of a considerable "academic proletariat."

The elitist attitude of German academics, however, did provide one small, though easily overcome, barrier to the ready acceptance of the Nazi regime by a large number of the professoriat. Hitler was simply too much of a plebeian demagogue (as Hermann Ahlwardt was for the anti-Semites of the 1890s). However, Hitler's first cabinet included people like Alfred Hugenberg, Konstantin von Neurath, and Franz von Papen, sturdy conservatives who, it was widely felt, would keep the "wild man" Hitler under control. Indeed, a Munich professor apparently wrote a horrified letter to Hindenburg's son about the formation of the Hitler government and received the reply "What do you want? After all we have the cabinet majority."[134] Within months, even formally, this "cabinet majority" was meaningless. As Helmut Kuhn has remarked, concerning the politics of the Harzburger Front (October 11, 1931),[135]

> The juxtaposition of the conservative honorable-bourgeois nationalism and the heightened nationalism of the brown stormtroopers led to bourgeois refusal to recognize the qualititative difference between them. . . .
>
> The classic or conservative nationalism of the educated bourgeois class, which included the professoriat, had itself, unnoticed, acquired a character (*Zug*) of resentful radicalism, which, while it indeed did not remove its distance from the manic-revolutionary nationalism of the Hitler movement, did, however, lessen it.

[131] Marcia Synnott, *The Half-Opened Door: Discrimination and Admissions at Harvard, Yale and Princeton, 1900–1970* (1979).

[132] *Atlantic Monthly* 181 (Feb. 1948): 72–74 and (Apr. 1948): 42–44. *Reader's Digest* 52 (June 1948): 13–18.

[133] Laura Hobson, *Gentleman's Agreement*, a novel (1947).

[134] Joseph Pascher, "Das Dritte Reich erlebt an drei deutschen Universitäten," in *Die Deutsche Universität im Dritten Reich* (1966), 49.

[135] Helmut Kuhn, "Die Universität vor der Machtergreifung," in *Die Deutsche Universität im Dritten Reich* (1966), 25, 26.

The Nazis understood this and, especially after Hitler's famous speech of January 27, 1932, to the Düsseldorf industrialists,[136] acquired a general respectability, even among the wealthy. Of course, this aim of the Nazi party to acquire "respectability" had been going on for some time—publicly, at least since Hitler's "oath of legality" at the "Reichswehr trial" in 1930, which was a sensation following closely after the stunning relative success of the NSDAP in the September 1930 elections. This "oath of legality," as well as the suppression of Nazi advocates of direct violence—the so-called *Stennesputsch* in 1931—added to this respectability. For many of the educated, including the professoriat, nationalism of any stripe became preferable to the greater sin of not being nationalist at all. After September 14, 1930, the Nazi party was the second largest in the Reichstag, and the one that seemed best able to fulfill nationalist aspirations. The stunning September success (previously the Nazis had had only twelve seats out of 475 in the Reichstag, now they had 107 out of 577[137]) showed it to be a dynamic party, as contrasted with the somewhat stodgy conservatives (DNVP) with their aristocratic and landed gentry connections. Hitler's message was a religious one; he came as a savior and messiah;[138] he moved the plebs and held out to the patricians the extermination of the parliamentarism they detested—for nationalist-minded conservative academics, he represented that mystical, impossible union of *Geist* and *Macht*, "spirit" and "power." He was truly a charismatic figure. The intrigues of von Schleicher, von Papen, and others may have helped bring Hitler to power, and Hitler was astute at the backroom politics that characterized the last months of Weimar;[139] nevertheless, he always projected the image of leading a movement for national renewal that was ultimately beyond politics. The mathematician Gerhard Kowalewski was not the only academic who felt about Hitler: "He has been sent to us by Providence."[140]

A final factor to be considered in the German academic crisis was the state of the universities. These were no longer placid places of learning, where acolytes received instruction from professors. The case of student disturbances provoked

[136] For a complete English translation of this speech, see Norman Baynes, *The Speeches of Adolf Hitler, April 1922–August 1939* (1942), 1:777–829. For an analysis of its importance, see Alan Bullock, *Hitler, A Study in Tyranny* (1962), 196–199.

[137] Their previous high point had been thirty-two out of 472 in May 1924.

[138] This has been frequently commented. One excellent book that explores this theme is J. P. Stern, *Hitler: The Führer and the People* (1975). The Nazis themselves were well aware of the religious quality of Hitler's message. For example, Joseph Goebbels noted in his diary in 1926: "Hitler speaks. About politics, idea, and organization. Deep and mystical. Almost like a gospel (*Evangelium*). Shuddering, with him one passes by the abyss of being" (as cited in Deuerlein 1974: 262), and Gregor Strasser in his letter of 1932 resigning all Nazi party offices says: "In my opinion the NSDAP is not only an ideological movement (*Weltanschauungsbewegung*) becoming a religion . . ." (original, and a translation were first published in an appendix to Peter D. Stachura "'Der Fall Strasser': Gregor Strasser, Hitler and National Socialism, 1930–1932," in Peter D. Stachura, ed., *The Shaping of the Nazi State* (1978), 88–130.

[139] E.g., among many other discussions, Eberhard Jäckel, "Der Machtantritt Hitlers—Versuch einer geschichtliche Erklärung," in *1933, Wie die Republik der Diktatur Erlag* (1983), 123–139.

[140] Victor Klemperer, *LTI*, Rotbuch Taschenbuch no. 35 (1987), 122.

by Emil Gumbel has already been mentioned; these occurred from 1924 to 1932. But while Gumbel's case was an academic *cause célèbre*, it was far from the only occasion for nationalist student disturbances during Weimar. According to Jewish student publications, as early as December 1919, "Hatred of Jews rules in academic life," and by the end of 1920, there had been violent confrontations between Jews and anti-Semites at all but nine (out of twenty-three) German universities.[141]

Walter Landauer, who had been an Assistent in Heidelberg in 1924, and then was employed in the United States at the Storrs Agricultural Experiment Station (Connecticut), took an ongoing interest in Gumbel's case. A letter to him on March 11, 1931, from Otto Pfeffer, editor of the *Heidelberger Tageblatt*, remarked:[142]

> We also hold the view that the sharp intrusion of party-political forces seems to have begun a very doubtful development of German university politics. However, it is unfortunately also the fact that a very noticeable national wave, often in nationalistic [NSDAP?] excess, flows through all of Germany. Only by looking from the direction of this movement is it possible, or so we believe, to find the correct basis for judgment. Germany is at this time like an overheated steam kettle.

He went on to speak of problems such as war guilt accusations and reparations and asks for an "understanding of the realities, under whose hard impress, the German people and not least the academic youth who are constricted to the point of tragedy in their ability to get ahead, live." As one sociologist later put it:[143]

> The general joblessness among the academically trained, especially the youthful academically trained [prior to 1933], favored a [political] radicalization. Precisely the bad economic situation during the crisis [of 1929] had caused many young people from petit bourgeois strata in Germany to study in order to beneficially use the time when they would otherwise be jobless. After completion of their studies, they often had to wait years to find a position suitable to their education. . . . The fact that in the professions toward which this academically trained youth strove, numerous Jews were in office, created in them resentment and the envy of competition. However, all the cited motives that could create an anti-Semitic mentality were no longer new. They have the same roots and same structure as the anti-Semitism of the nineteenth century already had.

[141] Hans Peter Beuel and Ernst Klinnert, *Deutsche Studenten auf dem Weg ins Dritte Reich* (Gütersloh: Siegbert Mohn, 1967), 13. Twenty-three is the author's count; Bleuel and Klinnert do not say. If Bleuel and Klinnert intend to include technical universities (like Hannover) in their count, then (by my count) this is all but nine out of thirty-four. In any case, the number is substantial. See also George Mosse, "Die Deutsche Rechte und die Juden," in *Entscheidungsjahr, 1932* (1966).

[142] Universitätsarchiv Heidelberg B-3075/13 (Gumbel file).

[143] Hans Paul Bahrdt, "Soziologische Reflexionen über die gesellschaftlichen Voraussetzungen des Antisemitismus in Deutschland," in *Entscheidungsjahr 1932* (1966), 153.

In any case, these students, "constricted to the point of tragedy," whose "national movement" was "genuine at the core, only undisciplined in its form,"[144] managed to vociferously complain about and disturb the lectures of faculty, whose sins included writing a satiric article that skewered the Nazi party (Gerhard Kessler, University of Leipzig, 1932); criticizing the ultra-nationalist stance of many university students and faculty in an article in a Swiss socialist newspaper (Bertold Maurenbrecher, University of Munich, 1931–32); believing only defensive wars were justified and opposing war memorials in churches (Gunther Dehn, University of Heidelberg and University of Halle, 1919–32); being suspected of being a leftist sympathetic to Leon Trotsky (Ernst Cohn, University of Breslau, 1932–33); comparing the Treaty of Versailles to those of Brest-Litovsk and Bucharest (Hans Nawiasky, University of Munich, 1931); and trying to mediate fairly between a Bulgarian student and some German ones—the Bulgarian, among other things, called a certain Nazi student a "swinish German blockhead (*Sauboche*)" (Karl Mühlenpfordt, T. H. Braunschweig, 1931–33).[145] Nor were all such disturbances only focused on "Jewish" or "liberal" professors; Mühlenpfordt, for example, was a thoroughgoing German nationalist. The general stance of a large number of the faculty in such instances seems to have been to reprove the students but distance themselves from their colleague (Mühlenpfordt is an exception here).[146]

One cannot leave this topic without at least a brief mention of one of the most famous as well as the earliest of these disturbances. On April 16, 1925, Theodore Lessing, a rather colorful philosophy instructor (*Privatdozent*) at the technical university in Hannover, wrote an article in a German-language Prague newspaper sharply attacking Field Marshal von Hindenburg, then a candidate for the Weimar presidency, saying, among other things, that were the war hero Hindenburg elected, "he would be only a representative symbol, a question mark, a zero. One can say: better a zero than a Nero. Unfortunately, history shows that behind a zero, there is always a future Nero hidden." The Hannoverian students were enraged. Lessing's lectures were disturbed, some rioting students were disciplined by the university, the public prosecutor initiated a preliminary investigation, and the students found no support at the education ministry, whereupon, spectacularly, 1,200 of the 1,500-strong student body abandoned Hannover and transferred to the technical university at Braunschweig (though apparently many soon returned). As to the faculty, of course, they

[144] Above, note 24.

[145] Faust 1973, 2:51–56. *Sauboche* is difficult to translate literally, being one of those slang slanders that carries much more connotatively than denotatively. *Sau* is "pig" or, more accurately, "sow," with connotations of filth, disgust, and sluttishness. *Boche* was French (and then adopted into English) slang for Germans in World War I and derives from a shortening of *caboche*, which meant (and means) "the back of the head," and so slang for "blockhead," but which also entered English as "cabbage." Thus *Sauboche* literally means a "cabbage-headed [German] pig," or perhaps, as translated, a "swinish [German] blockhead," and derives from vulgar French slang mixed with German. Comparable English slang for a German (though of different provenance and connotation) is "Kraut." In any case, it is clear why the Nazi student might be upset at the pejorative name.

[146] As analyzed by Anselm Faust (1973).

wished the unruly students to be disciplined, but they distanced themselves from Lessing, whose "unacademic behavior" made him "unworthy to be a member of the faculty." Most to the point, almost every other university student group in Germany issued declarations of solidarity with the Hannoverians. For example, in Munich, the students at the university and the technical university issued a joint declaration in which they "recognize (*bekennt*) unanimously the national character of the German university," and knew they were "one with all German students in emphasizing the necessity of taking a position against an un-German spirit and damaging influences on the national quality of the university under the cover of a wrongly understood academic freedom."[147]

As Michael Steinberg remarks in his study of the National Socialist student movement:[148]

> The Nazi students flourished in an environment that was critical of their style, not their message. They were accepted as part of the larger nationalist movement at the university. This situation not only encouraged the support of wavering students but reinforced the Nazi students' own sense of rectitude and association with the nationalist traditon.

From before the time Heidelberg students dipped their handkerchiefs in Karl Sand's blood[149] to the present day, there has been a radical nationalist *völkisch* segment of the German student body that has found all actions, including murder, appropriate to its cause of a greater German, but strictly *German*, national unity. Attempts at brutal suppression, as for example in 1819, only drove it temporarily underground, and in 1931, in the guise of National Socialist students who distinguished between "national politics" and the despised "partisan politics," it triumphed.[150] Even in 1820, it was more than tinged with anti-Semitism.[151] Hitler was surely accurate when he said in 1930, "Nothing gives more credence to the correctness of our idea than the triumph of National Socialism at the university."[152]

[147] A summary of the "scandal" can be found in Faust 1973, 1:50–52. Lessing was assassinated in 1933 in Marienbad. Among his other works were *Jewish Self-Hate*, which he applied to himself. The first edition was issued by Jüdischer Verlag, Berlin, in 1930, reissued by Matthes and Seitz, Munich, in 1984. For a run-in Lessing had with Sigmund Freud, see Kurt Hiller, *Köpfe und Tröpfe* (Hamburg: Rowohlt, 1950), 307–8.

[148] Steinberg 1977: 116.

[149] Karl Sand, a radical student activist, in 1819 assassinated the writer August Kotzebue as a traitor to the people (*Volk*), and was publicly beheaded in 1820. The book *Scheiterhaufen* by Christian Graf von Krockow (n.d., c. 1988), taking the book-burning of May 10, 1933 as its initial focus, attempts to trace the intellectual ideas among students and academics leading thereto, as well as the duration and consequences of these ideas beyond the Third Reich. Karl Sand and his execution are discussed there on pp. 31–33.

[150] The most complete description is in Faust 1973, 2: chap. 6.

[151] See Krockow n.d.: 33. Anti-Semitism came and went among various student groups until the 1880s, when a steady increase seemed to begin (see, for example, Pulzer 1964; Mosse 1964: chap. 10; and Steinberg 1977).

[152] "Adolf Hitler to German Students," *Die Bewegung* for July 8, 1930, as cited by Anselm Faust 1973, 1:9.

The universities during Weimar themselves were fertile fields for the growth of extreme nationalistic sentiments. For all the examined reasons, professors could believe in the separation of university and state, yet, as private citizens, rail against the republic. During the Weimar Republic, those who saw it as a new, different beginning for Germany, and a significant break from the past, like Emil Gumbel, suffered the consequences, which were slower in being made definitive than with Leo Arons decades earlier, but were nonetheless punitive or final. Furthermore, the dignity of many professors as Wilhelminian state servants had been placed in the service of the national aspirations for success in World War I. Not only did German professors feel declassed under the republic but, along with many of their countrymen, they felt deep injury to their sense of national pride, a pride that had only very recently been established (within less than fifty years)—nor did the Treaty of Versailles help in this regard. Students, perhaps even more than their professors, perceived an unfair debasement of Germanhood in which the republic had acquiesced, and so practiced "national politics." Furthermore, their professional outlook for employment was dim and growing dimmer. Thus both students and professors were filled with a narcissistic rage at unfulfilled expectations—expectations somehow "due," but not achieved. Both yearned for a national government that stressed the dignity of being German, and spurned "Western parliamentarism" as a corrupt, un-German system. It was the parliamentary states that had frustrated their expectations, unmanned the German soldier and the German state, so to speak; furthermore, many believed this had been done by unfair and deceitful means, since only such means could achieve such an end. The military began the "stab-in-the-back" legend,[153] but it was widely believed and used to good effect by Hitler. Thus, it was natural that the parliamentary Weimar Republic should be viewed as a corrupt and deceitful intrusion into the naturally Germanic order of things. As a consequence, both students and professors played at restoration dreams—restoration of a state with employment, and with Germanic dignity, Germanic order, and Germanic unity. Finally, anti-Semitism was an essential ingredient of this academic culture. While nothing new in German or other academic circles, the increase in German anti-Semitism following World War I has often been remarked. The Jew as quintessential "outsider" religiously and, especially as represented by the *Ostjuden* (Jews from the East), culturally became the antithesis of true Germanhood. The assimilated Jew became for many Germans a "master of deceit," hiding his true nature under artfully assumed disguises; for academics in particular, the Jew became an un-German intruder in the club, a pseudo-German who had had full civil rights (even in Prussia) for only about a century and who took needed posts away from true Germans, a disturbance in the desired German unity and dignity. Indeed, if one adopts the view of such tribal and "racial" nationalism,[154] then the assimilated Jew, viewed

[153] See above, note 36.

[154] For a discussion of anti-Semitism, totalitarianism, and tribal nationalism, see Arendt 1980: esp. 227–243.

as attempting to disguise his "race," is inevitably a social poison. In sum, a large part of the academic community was prepared emotionally and socially to respond hopefully and positively to Hitler's message, provided that it was given a sufficiently respectable guise. Of course, in this they were not any different from other German elites, like jurists or doctors, who by and large supported the Nazi regime.

Why should the "unpolitical" academic elite initially find the Nazis so attractive, although later many had second thoughts? Beyond the usual reasons already suggested, such as the need of an "academic proletariat" for jobs and the inspiration of nationalistic fervor in hard and unfair times, there were perhaps other psychophilosophical reasons. Just as science may be a defensive barrier between human beings and nature, so the scientist's work provides a potential protection from continually having to deal with the "real world." This may be particularly true of mathematicians, whose work basically involves self-generated mental creations. The mathematician, alone among scientists, has complete control of the objects of his manipulation. There are no inconveniences of apparatus to get in the way of the elaboration of an idea, only the mathematician's own knowledge and ability. For that very reason, of course, the requirements of mathematics are much more stringent logically. A mathematical proof has no room for experimental error; within the standards of the day, it should be logically rigorous. In some ways vis-à-vis his work, the mathematician recalls W. S. Gilbert's Lord Chancellor in *Iolanthe* (Act I):

> The Law's the very embodiment
> Of everything that's excellent
> It has no kind of fault nor flaw
> And I, my Lords, embody the Law.

This manipulation of self-generated concepts is even true for the applied mathematician (though these latter sorts of results may later require laboratory confirmation). Under Nazi rule, the mathematician could, perhaps more easily than colleagues in, say, history, biology, or architecture, separate "work" from "life," and, having made that compartmentalization, could retreat into that famous "inner emigration," if so desired.

Mathematicians were not much different from the rest of their colleagues—there are no signposts in Weimar academic culture saying either "But mathematicians were an exception," or "But mathematicians were more extreme."[155] Furthermore, as later chapters will show, during the Nazi regime, mathematicians not only exhibited the same range of behaviors as their colleagues, but shared the same fundamental attitudes and reactions. In particular, there are examples of the phenomenon of initial enthusiasm for Nazism followed by se-

[155] David Rowe (1986: 426) would argue that the Göttingen mathematicians give the lie to extending the notion of a "mandarin" German professoriat to mathematicians and natural scientists. What seems more likely is that the mathematicians at Göttingen prior to 1933 truly were exceptional among university faculties. Even so, figures like Richard Courant, a nonreligious Jew, certainly had "mandarin attitudes."

rious second thoughts, when it was "too late," among mathematicians. More importantly, the crisis in mathematics became intertwined with more political issues in Nazi Germany, especially because of the emphasis of some on Jews as unfit by reason of their presumably different mathematical and pedagogical style; however, it did so in ways that were never clear-cut. It is undeniable that in addition to the effect of the regime on mathematics, there were some mathematicians who attempted to foster it out of conviction, and with respect to the discipline itself in both organization and content.

CHAPTER FOUR

Three Mathematical Case Studies

As is well known, the Nazi regime did not at all fit the romantic notion of a monolithic totalitarian state in which orders are passed efficiently down a smoothly organized hierarchical system, to be carried out by successive layers of underlings. Despite the *Führerprinzip* (or "leadership principle") that articulated such a system, the Nazi government fostered competing bureaucracies that struggled with one another for control, often in the name of being the true ideological standard-bearer. The reason for and possible purposiveness of this situation is still being debated.[1] Needless to say, this conflict of bureaucracies also opened the way for much personal conflict and politicization of academic life.[2] This politicization, however, was not necessarily only in the form of directives issued from on high; it also was a struggle among individuals, each necessarily purporting that they represented the true aims of the Nazi state in their actions. This was true in mathematics as well, and not only in so-called applied mathematics. Especially after the war began, one might have expected national service to have taken precedence over any jockeying for personal position in militarily important applied mathematics. This was not the case. Although now it is possible to recognize how some attempted to act in the best interests of their discipline, whereas others had less noble motives, nevertheless, all necessarily had to maintain that they were acting politically—that is, in the best interests of the state. This chapter describes three instances of such politicization of mathematics under the Nazis. Two are from applied mathematics and one from more "pure mathematics." One is from mid-war, one largely from 1938–40, and one from the beginning of the Nazi regime. All demonstrate how, under the Nazis, academic infighting was, if anything, enhanced rather than reduced, and also provided with political weapons. Indeed, all weapons were political. The names that occur in each of these instances, two of them almost totally undescribed previously, will echo in others, and throughout this

[1] This discussion was begun by Franz Neumann's *Behemoth* (1942, 1944), a book that still demands consideration. As one example, for a detailed discussion of how these overlapping bureaucracies affected nuclear research, see Mark Walker, *German National Socialism and the Quest for Nuclear Power, 1939–1949* (1989).

[2] There were even two education ministry structures concerned with *Wissenschaft*, or learning, one run by the army and one by the ministry. See Helmut Heiber, *Walter Frank, u. sein Reichsinstitut für Geschichte des Neuen Deutschland, Stuttgart* (1966), 116, 645–646. A reader wishing to survey the party structure of the NSDAP can consult the fifty-three pages and charts of *Rang-und Organisationsliste [der NSDAP]*, published by W. Kohlhammer, Stuttgart, pursuant to the Allied denazification law of March 5, 1946.

book.[3] I have described these incidents in very full detail in order to bring out both the Byzantine aspects of professional mathematical life under the Nazis and the qualities of the different personalities interacting. Another reason for such a detailed description is that the Nazi competition among bureaucracies and among personalities, for prestige and for funds, in situations where there might well be unpredictable political issues, produces a somewhat fragmented picture. There is no smooth history here. Consequently, brief descriptions are necessarily somewhat misleading.[4] In addition, Nazi language and attitudes sometimes have to be seen to be believed. Though many of the people mentioned had no lasting mathematical impact, these sorts of detailed "case histories" seem more likely to give a feeling for the *Zeitgeist*—the atmosphere of the times—as it really was than brief scattershot mentions of many incidents.

The Süss Book Project

This first case study stems from mid-war. It deals with a project for the creation or reprinting of mathematical books considered militarily important. This turns out not to be some straightforward, unremarkable project (if somewhat late in conception), but instead to involve tremendous personal jockeying between the two principals: Wilhelm Süss and Gustav Doetsch. What makes this even more interesting are the personalities involved. Süss had been the leader of the German Mathematical Society since 1937 and was Rektor at Freiburg. Doetsch had been, before 1933, a somewhat left-of-center Roman Catholic; after 1933, he metamorphosed into a man whom one mathematical colleague described as "110% Nazi." Furthermore, both Doetsch and Süss were colleagues at Freiburg and had been jointly (and amicably) involved in 1934 in a right-wing effort within the German Mathematical Society. This will be treated in chapter 6, but because Doetsch's political transition is interesting, and perhaps not entirely untypical of other academics who adapted themselves similarly, it will also be discussed following the case study. The interest of this story is only enhanced by the fact that the "referee" between the mutually inimical principals was Ludwig Prandtl, certainly the foremost aerodynamicist in Germany, and perhaps in the world, at the time.

While this brouhaha over mathematical books to be produced at mid-war for their military utility may seem an unfortunate and accidental conflict of person-

[3] The first two instances are, in fact, mentioned briefly, but nothing more, in Herbert Mehrtens, "Angewandte Mathematik und Anwendungen der Mathematik in national-sozialistischen Deutschland," *Geschichte und Gesellschaft* 12 (1986): 339–340, and 322; and the first is briefly summarized in Herbert Mehrtens, "Mathematics and War: Germany, 1900–1945," in Paul Forman and José Sanchez-Ron, *National Military Establishments and the Advancement of Science and Technology* (1996); 115–116. This is a continuation of Mehrtens' 1986 article. The second instance is also mentioned in Renate Tobies, "Untersuchungen zur Rolle der Carl-Zeiss Stiftung für die Entwicklung der Mathematik an der Universität Jena," *NTM Schriftenreihe* 21 (1984): 39–40.

[4] This seems to be true of Mehrtens' description, mentioned in the previous note, where events seem conflated to fit Mehrtens's preconception of Wilhelm Süss.

alities, the fact is that such profound intermixtures of personalities and non-academic politics affected all aspects of mathematical life in Nazi Germany to a surprising extent. Once adherence to the state was established and its pariahs were eliminated, the system, with its overlapping bureaucracies, not only allowed conflict but might even be said to have encouraged it. Applied mathematics, even militarily valuable mathematics, was no exception to this general rule of academic (and other) life in Nazi Germany.[5]

On October 3, 1941, Johannes Rasch,[6] a physicist and research and development engineer at Siemens and Halske, wrote to the *Reichsforschungsrat*[7] in Berlin making suggestions about the need for new, updated, or German versions of works in applied mathematics. He remarked that in England there were detailed modern works on such subjects, whereas in Germany such books were hard to come by. Four areas—Special Functions (including orthogonal polynomials), Bessel functions, spherical functions, and tables of formulas—seemed to him especially important. As examples he appended a list of important mathematical works (from his point of view) that needed new German versions. Interestingly enough, among them were (for elliptic functions) the *Funktionentheorie* of Hurwitz and Courant (1929) and the *Fundamenta Nova* of C.G.J. Jakobi [sic] (1829), which had Jewish authors; Courant was a forced emigré in 1934. Six weeks later, on November 17, Rasch added an eleven-page letter on the same theme, in which he remarked that his suggestions for a start on the reorganization and establishment of the technically important mathematical literature were strengthened by an article by Thornton Fry in the *Bell System Technical Journal* for July 1941. Fry's lengthy article, written at the behest of the U.S. National Research Council, lamented the neglect of applied mathematics in the United States, especially at universities; commented on how much further ahead Europe was in this regard; and gave many examples of the importance of applied mathematics. These were repeated at length by Rasch, taking up three-fourths of his letter, which ends:

> In the U.S.A. the growing importance of mathematics for research and technology is recognized, and correspondingly it is proposed to undertake a plan that will further applied mathematics for the purposes of technology in greater measure than heretofore.

[5] This lack of coordination even applied to relations between the German army and air force during World War II. See, e.g., H. J. Fischer, *Erinnerungen*, vol. 2 (1985), 134 (where, for example, army munitions would not fit air force machine guns). Such instances of military lack of coordination may be familiar in democratic-republican governments at war, but seem startling in Hitler's Germany—because of our false view of its structure.

[6] The full text of Rasch's letters of October 3 and November 17, 1941 (see below) can be found in the Bundesarchiv Koblenz (BAK) in file R73 12976.

[7] The *Reichsforschungrat* (National Research Advisory Board), or RFR, was undergoing reorganization at the time. Presumably this reorganization of research led to some of Rasch's proposals being appropriately considered by the research division of the *Reichsluftfahrtministerium* (air ministry), or RLM; at least Alwin Walther thought so. See letter of Walther to Ludwig Prandtl, Aug. 20, 1942, in "Briefwechsel mit den Herren Professoren Seewald, Süss, Doetsch, Wegner, Blaschke," Mappe no. 1, in Prandtl *Nachlass*, in archives of the Max-Planck Institut für Strömungsforschung.

> The development of important branches of industry in Germany evoked by the war determines a similar task for German mathematics. In my opinion, besides pure mathematics, an applied mathematical science must be encouraged in the strongest way. There are many other indications for German mathematics that could be given; in the present letter I have purposely restricted myself to the article from the U.S.A.

Whether Rasch had any thought of eventual (as it turned out, imminent) American entry into the war is unclear from his text, as is whether Fry's text[8] (which appeared just after Hitler's invasion of the Soviet Union) tacitly had war work in view. There already had been a small move, shortly after the war's start, toward enhancing applied mathematics at German universities: in 1940, applied mathematics was allowed to be a primary subject of university study for future teachers (instead of just a "minor"). Nevertheless, in 1942, Ernst Mohr, while giving the curriculum for applied mathematics at Breslau, stated how important it is for applied mathematicians to study "pure mathematics" as well, and lamented the war-caused lack of adequate teaching personnel.[9] Similar remarks were made by Georg Feigl, also at Breslau, in the same journal.[10] Georg Hamel and Feigl had earlier collaborated on an apparently influential educational plan for training future mathematics teachers, whether in pure mathematics, applied mathematics, or physics.[11] In fact, *instruction* in applied mathematics suffered, because only after the basic study common to both "pure" and "applied" could one become specialized in applied mathematics, and Feigl's article complained as well about the poor training students were receiving in mathematics before coming to the university. In any case, there seems to be no evidence that Rasch's letters stirred much interest until mid-February 1942. Perhaps this was because the war was still going reasonably well for Germany, despite the frozen paralysis on the Eastern front. However, in mid-February, Wilhelm Fischer, the secretary of the RFR, wrote to Georg Hamel and Wilhelm Süss asking for a conference about Rasch's concerns. At the time, Hamel was head of the *Mathematische Reichsverband* (MR), the national organization concerned with mathematical pedagogy, and Süss was president of the German Mathematical Society (in which post Hamel had preceded him). The inspiration for Fischer concerning this meeting seems to have been Abraham Esau, the head of the physics division of the RFR, and it took place on February 25.[12] A little later, Süss spoke

[8] *Bell System Technical Journal* 20, no. 3 (July 1941): 255–272. Fry would later become director of Bell Labs.

[9] *Deutsche Mathematik* 6 (1941): 493–504. For more about Mohr, see "Hasse's Appointment at Göttingen" in this chapter.

[10] *Deutsche Mathematik* 6 (1941): 467–475. Although *Deutsche Mathematik* was founded with the express purpose of promoting a "German" (as opposed to "Jewish" or "French") mathematics, there is nothing of this flavor in these articles. This journal and the movement it represented are discussed in chapter 7.

[11] *Deutsche Mathematik* 4 (1939): 98–108; and Walter Lietzmann, *Mathematik in Erziehung und Unterricht* (1941), 128 n. 2.

[12] See BAK R73 12976, Fischer to Süss, Feb. 10, 1942; Hamel to Fischer, Feb. 16, 1942; Fischer to Esau, Feb. 17, 1942. Esau will be met again in his earlier role as Rektor at Jena (below, "The

about this conference with Erich Kamke, a well-known mathematician in Tübingen and author of a compendium on solutions and solution methods for differential equations. On July 19, Süss talked with Alwin Walther, head of a successful but partly secret "computing machine laboratory" in Darmstadt, and on July 22 a meeting to discuss Rasch's suggestions was held in Berlin, chaired by Friederich Seewald, an engineer and member of the research division directorate at the air ministry.[13] The other attendees were Süss; three other mathematicians—Erich Kamke, Ferdinand Lösch, and Curt Schmieden; Rasch; and an air force staff engineer named Weinberger. Lösch and Schmieden were active in aeronautical research (although Lösch had in the 1920s done very "pure" mathematics), and they agreed on the importance of the project. Süss was present *ex officio*. For efficiency's sake, he would in effect be the sole director of the project, though Lösch and Schmieden would be advisors who would see to it that the interests of aeronautical research were attended to. In fact, Süss's own mathematical interests were in geometry and rather far removed from those of the book project. Seewald promised support; however, he left a final decision to the world-famous Ludwig Prandtl, the head of the directorate, and certainly the leading aeronautical research scientist in Germany at that time. Seewald also asked Gustav Doetsch for his opinion. This was quite natural, as Doetsch was in the air force and was the author of a (still) classic book on the Laplace Transform (1937).[14] On July 27, Seewald wrote Prandtl approvingly about the plan, and on July 30, Süss did so as well. Seewald's letter contains the unexplained remark that those present at the meeting wished to keep the names of prospective authors of the book series confidential. This is, in a way, clarified in Süss's letter, which said not only that the usual confidentiality with respect to prospective books and manuscripts was in order, but that several of the prospective writers had insisted that if their names were to be mentioned, then Doetsch must under no circumstances discover them. Schmieden apparently explained the (unmentioned) reason for this to Seewald, who agreed. Süss then gave Prandtl, under the same assumption of confidentiality, the names of fifteen prospective works with authors.

Seewald did contact Doetsch (though apparently respecting the pledge of confidentiality as to prospective authors), but mentioned Süss as the leader of the project. On July 27, Doetsch replied. Doetsch remarked that the project was not only important but had been considered so for some time, and that a book

---

Winkelmann Succession"). As head of the RFR's physics division, he would convene the first German meeting concerned with nuclear power. For this and more about Esau, see Walker 1989.

[13] See Walther to Prandtl, Aug. 20, 1942, cited in note 7, for this chronology. All further citations concerning the situation under discussion come from the folder in the Prandtl *Nachlass* cited in note 7, unless otherwise indicated.

[14] This book first appeared in 1937. In 1943, under the Alien Property Act, Dover Publications reprinted (in German) this "world-renowned monograph" on an "operational method which has become increasingly important for wartime research." After the war, this one volume expanded into three (published by Birkhäuser, 1949, 1955, 1956), to which was added in 1958 an introductory monograph. This last was translated into English in 1970. All four volumes are still frequently cited in the mathematical literature.

on Fourier Integrals (in English) had been reproduced, and other such reproductions were in train. He added that the ZWB (*Zentrale für wissenschaftliches Berichtwesen*, roughly, Air Minister's Central Office for Scientific Reporting) should be in charge of making formal assignments of authors for such a project and dealing with all official aspects of it, whereas the scientific direction should be with the air ministry in Braunschweig, where there was an established "Institute for Higher Mathematics in Physics and Technology" (in fact, founded at Doetsch's initiative). The leader of this institute was Professor Wolfgang Gröbner, who, according to Doetsch, was extremely competent in such matters, certainly far more so than Süss, who was not an applied mathematician. Furthermore, said Doetsch, Süss was undoubtedly so laden with tasks (as well as holding the presidency of the German Mathematical Society, he was, among many other things, Rektor of the university at Freiburg at the time) that he could not give appropriate attention to the project. Doetsch also complained that Seewald had sent him no authors' names, and so he would have to ask "his authors" whether their names could be mentioned in connection with this project. Superficially, Doetsch's reply seems quite reasonable; in fact, as the sequel shows, he was trying to obtain a position of importance for himself, and to denigrate his quondam friend and present enemy, Süss. The institute "led by Gröbner" seems to have existed mostly on paper. Doetsch clearly was interested in combatting Süss, because he seems to have contacted two of his friends, Udo Wegner in Heidelberg and Wilhelm Blaschke in Hamburg, who immediately wrote Prandtl complaining about Süss leading such a project. Doetsch's own such letter, with a copy of his remarks to Seewald, followed on August 1. The tones of Wegner's three-page letter of July 31 and Doetsch's letter are revealing. Wegner's letter speaks repeatedly of "personal attacks" by Süss on Doetsch, which are for him the reason why Süss and "the men behind him" did not approach Doetsch, whose scientific qualifications were unquestioned. He claimed that Süss and "the men behind him" went behind Doetsch's back to gain personal advantage. He emphasized that Doetsch had "the unbounded trust of all mathematicians of rank and name" with respect to his personality as well as his science, and so on in similar vein.

Similarly, Doetsch told Prandtl that unfortunately, Süss had had certain personal differences with him for a year and a half that originated in Süss's behavior during Doetsch's absence from Freiburg because of the war. He further believed these were responsible for Süss' "going behind [his] back" in the matter of the book project, though he, Doetsch, was the expert for such matters. Doetsch went on to say that he had already been pursuing such plans for a substantial amount of time, and so the failure to notify him was especially deplorable, since it would have made a "wonderful collaboration" (*schöne Zusammenarbeit*) possible, in which naturally any personal differences he had with anyone would play no role. Another new point in Doetsch's letter was the importance of "anchoring" the project within the air force (i.e., with Doetsch and Gröbner) as opposed to contracting it outside (by the air force research directorate to Süss's group), so as to guarantee that "the matters do not lose themselves in pure mathematics."

What may seem surprising at this stage are not only the overlapping responsibilities, even within the air force, but their manipulation to serve personal grudges and sensitivities. This is surprising, however, only to the romanticized idea that the functioning of the Nazi apparatus, at least in matters of the war effort, would be rigidly hierarchical. In fact, even in wartime, personal bickering often took center stage. The present matter was settled reasonably quickly and satisfactorily largely through the offices of Ludwig Prandtl; in other situations, especially those involving mathematics or mathematicians and ideology or politics, such a fairly easy settlement did not necessarily occur, as will be seen. In any case, how this situation was resolved is informative, not least because, as will later appear, Wegner, Doetsch, and Blaschke seem to form a sort of constellation in their attitudes toward the regime, whereas Süss and Prandtl seem quite different.

On August 3, Prandtl, having already received Doetsch's letter, wrote both Süss and Doetsch. To Süss he suggested cooperation with Doetsch, as he was "the referee for mathematical affairs in the research directorate [of the air ministry]." To Doetsch, however, he wrote fairly sharply rejecting his suggestion of Gröbner as the scientific leader of the book enterprise. Doetsch had cited in Gröbner's favor that he had worked at a well-known and much envied applied mathematics institute in Italy founded in 1929 by M. Picone. However, Prandtl wrote Doetsch that the suggestion of Gröbner did not sit well with him, and

> Gröbner, who, as a South Tyrolean, aspires toward Germany, was supposed to be accommodated a few years ago somewhere here. . . . He is in his soul absolutely abstract-minded and had simply done work with Picone to earn his daily bread; what he in fact did there was rather complicated trash (*umständliches Zeug*), from which one can see that he in no way conceived of the applied mathematics problem, which could be solved much more simply as one of applied mathematics, but also again became bogged down in pure mathematics.

Not only was Gröbner unsuitable, but

> marginally, I note my impression of the papers of the whole Picone Institute, that one has no proper connection at all there with really practical tasks, but rather one only writes papers *ut aliquid fiat*, that is, to prove the Institute's right to exist.

Prandtl did add that he had not followed the later papers from Picone's institute. That Prandtl's very "applied" point of view was in the minority among mathematicians is perhaps illustrated by the "international mathematical congress" held among Axis and neutral nations in November 1942 in Rome. On this occasion there was an excursion to Picone's institute.[15]

[15] *JDMV* 53 (1943): Abteilung 2, 21–22. For more about Picone's institute and its attraction for German mathematicians, see Reinhard Siegmund-Schultze, "Faschistische Pläne zur 'Neuordnung'

Prandtl suggested some possibilities for leadership, including collaboration with Süss, or Doetsch leading the project himself ("perhaps the smartest thing"). On the same day, Wilhelm Blaschke wrote Prandtl briefly in support of Doetsch. Of the trio of Blaschke, Doetsch, and Wegner, Blaschke was certainly the most distinguished mathematician, as Wegner was the least. Indeed, Blaschke was the most distinguished mathematician of all those involved in this contretemps (Prandtl did not consider himself a "mathematician"—in his reply on August 4 to Blaschke, he said that he had to decide between Doetsch and Süss, though he could not feel himself sufficiently a mathematician to do so). Blaschke commented positively on Doetsch's personality ("extraordinarily friendly"), and said that the whole program with Süss must be "based, at least in part, on the fact that Süss has great personal ambition, which is connected with his success in being both Rektor [in Freiburg] and for many years president of the DMV [German Mathematical Society]."[16] These subjective remarks are followed by the factual statements that Doetsch has a higher mathematical standing than Süss, and the latter's mathematics, in any case, was far distant from applications.

Since only one of the two men could be funded, Prandtl's reply asked Blaschke to be a referee for their plans. He also wrote Doetsch telling him he had done this. However, his letter to Blaschke contained Süss's list, which he asked Blaschke to hold strictly confidential, citing Süss's earlier remarks about Doetsch in this connection and promising some kind of prospectus of Doetsch's plans shortly. This last he asked Doetsch to provide.

Doetsch's reply to Prandtl's first letter (of August 3) considered Prandtl's three suggestions for the scientific direction: collaboration with Süss; doing it alone; giving it to two other mathematicians, Walter Tollmien (Prandtl's student) and Friedrich Willers (both at Dresden). The third approach he certainly found good, and he suggested that some mechanism be found for Prandtl to keep his eye on the whole project ("an invaluable advantage"). He also reiterated that he knew how to put aside personal resentments in order to work substantively with Süss—"and I could, for example, also sit impartially at the same table with Herr Süss for the purpose of pertinent discussions."

Blaschke replied to Prandtl with generally positive remarks about Süss's list, with a few exceptions, and suggested as editor Doetsch, or alternatively Gröbner (unaware of Prandtl's opinion of Gröbner).

On August 11, Prandtl wrote Süss pressing him to collaboration with Doetsch, saying he knew that Doetsch was occasionally given to harsh remarks (which, however, were mostly well-founded), and noting that Doetsch knew how to separate the personal from the professional. Since he considered collaboration with Doetsch, "who supervises a perhaps smaller undertaking at present," "indispensable," then if Süss could not manage it, Prandtl said, Süss

---

der europäischen Wissenschaft, Das Beispiel Mathematik," *NTM Schriftenreihe* 23 (1986): 1–17, particularly n. 22.

[16] See chapter 6 for why this in particular may have irritated Blaschke.

should assign someone from his group for the task. Prandtl also communicated Blaschke's positive and negative remarks about Süss's list, promised continuing confidentiality for Süss's prospective authors so that Doetsch would not learn their names, and wished Süss's reply as soon as possible. On the same day, Prandtl also wrote Doetsch, reminding him that his list of undertakings had yet to be sent. However, the next day, Süss phoned Prandtl, in response to his letter of a week earlier, to explain that he had become involved because of his presidency of the DMV and was perfectly ready to confer with Doetsch. The next day Prandtl phoned the engineer Weinberger who had been an air ministry delegate at the Seewald meeting, confirmed the origins of the proposal and Süss's involvement, and talked about Doetsch. All this was communicated to Blaschke, to whom Prandtl also told Süss's statement that the whole business really lay far from his own interests, he had no ambitions about it, and he would gladly give it over as soon as he saw he was no longer necessary. Furthermore, Doetsch and Süss were both ready to talk out the matter between them, and (in an apparent change from Prandtl's earlier opinion, and contrary to Blaschke's) Prandtl remarked: "As to Doetsch, I have heard that he has made himself disliked through frequent sharp and partly unjustified ugly judgments." A copy of this letter was sent also to Wegner (as answer to his July 31 one), whom Prandtl clearly and rightly considered a much more minor figure.

On August 21, Süss penned a personal handwritten letter (from his vacation in the mountains) to Prandtl (also on holiday) about Prandtl's letter of August 11, which had pressed collaboration, and which had followed Süss on holiday, as well as about their phone conversation of August 12 and Doetsch. Süss, of course, was glad that Prandtl was prepared to see the project funded, was sympathetic to the air ministry research directorate's desire to influence the planning—it was why he initially chose Lösch and Schmieden—and understood the ability to fund only one project. However, if Doetsch was given influence corresponding to Lösch and Schmieden in the project, he saw its certain failure:

> I say this not out of my personal relationship with Doetsch, for I know that I am at least as free of personal resentments as he is. . . . However, I know him so well through observation for many years, that I am completely certain of what I say. As I feel, so do *all* colleagues without a single exception known to me in Freiburg and also in Stuttgart, where Doetsch was earlier, insofar as they have gotten to know him more closely. The verdict of Blaschke from afar cannot influence me in this. . . . I say all this not to demean a colleague, but *for the sake of the matter at hand*. I also . . . make no reproach against Doetsch of his sort (*aus seiner Art*). He cannot escape his background (*kann nicht aus seiner Haus* [sic] *heraus*) and is perhaps not to be held responsible for his peculiarities of character, which are the question here, and have caused the withdrawal of the trust of his colleagues.

However, Süss felt that the research directorate might perhaps have a better experience, "at least for a certain while."

Süss earlier in the letter said that he was prepared, as Prandtl wished, to meet Doetsch, that perhaps this could take place shortly, and that he had asked

Georg Feigl to prepare and lead such a meeting. It should be noted that Süss never dealt directly with Doetsch (though they were technically colleagues at Freiburg): for example, Weinberger had told him when Doetsch might be in Freiburg, and he delegated Feigl to be the motive force in the meeting. Son of a well-to-do importer, and professor (since 1935) at Breslau (modern Wrocław), Georg Feigl was well known, loved for his pedagogical ability, and respected for his mastery of the mathematical literature. A lover of chamber music and collector of books and stamps, he was, in the words of one colleague, "one of the most sympathetic human beings whom I have ever met."[17] Clearly he was the sort of person who might mediate, if mediation were possible, between Süss and Doetsch.

Süss also addressed the issue of Doetsch already having started a project of the sort envisioned. By checking with an engineer in the ZWB, Süss reported that the only project there was a "planned" reprinting of three English books, already considered in his plans, and so he did not understand how Doetsch could say "a similar undertaking is already being planned (*im Werden*) in the ZWB." Indeed, Doetsch's list of the works he was supervising seemed mathematically paltry, though, for discussion purposes, somewhat significant as to its content. It is unclear whether Doetsch sent this to Prandtl on August 13 (after his requests of August 4 and 11) or not until September 3. Doetsch's list contained one reproduced English language book and plans to reproduce three others (no doubt the books mentioned to Süss). He took credit for the publication of the first volume of Erich Kamke's compendium of solutions and solution methods for differential equations ("made possible solely through the financial and moral help of the ZWB") and said that talks for a second volume on partial differential equations, though presently at a standstill, might be revived. (Such a second book by Kamke also appeared on Süss's list. One should recall also that Kamke was one of the first people approached by Süss and was in Seewald's meeting that had so agitated Doetsch.) In addition, Doetsch mentioned two books by Wegner in manuscript, a work in progress (Laplace Transform tables) by two of his doctoral students, and a book being planned by a Dr. Droste, who was a teacher at Stuttgart. All of this (or this little, when compared to Süss's list and prospective authors) was dressed up in a way that emphasized Doetsch's importance. One interesting suggestion he did make, which reflected the wartime conditions, was for the utilization of the facilities and personnel of an institute for applied mathematics at a Hungarian university, which was directed by someone with a German doctorate, "because most German institutes are much weakened by the drafting of their personnel." In a September 3 letter to Prandtl, Doetsch also mentioned an approach from Süss through "an intermedi-

[17] See Kurt R. Biermann, *Die Mathematik und Ihre Dozenten an der Berliner Universität, 1810–1933* (1988), 204–207. The quotation (from Hans Freudenthal) appears on p. 205, and in the obituary of Feigl by Maximilian Pinl, *JDMV* 70 (1967–68): 53–60. Also see Hans Freudenthal, *Berlin, 1923–1930* (1987), and Alexander Dinghas, "Erinnerungen aus der letzten Jahren des Mathematischen Instituts der Universität Berlin," in H. Begehr, ed., *Mathematik in Berlin* (1998). (I worked from a preprint of an earlier version of this book, which Professor Begehr kindly gave me.)

ary," and with "good will on both sides" believed something good would emerge.

The meeting of Süss, Feigl, and Doetsch did finally take place on September 15 and 18 in Freiburg for a total of five hours. Emerging therefrom was a protocol signed by all three in which Süss and Doetsch resolved to work together, since there was little overlap in their plans. Furthermore, the work would proceed in "close and loyal cooperation" between the (old) National Council on Research (RFR) (now replaced by the Association for German Research [DFG]) and the air ministry research directorate. Süss would be director for the former's participation and Doetsch for the latter's. Together they would supervise a third mathematician to be named, who would run the institute where the books would be prepared under the aegis of the air force. The ZWB would also be worked into a role in the enterprise.

Both sides informed Prandtl: Doetsch immediately on September 18, but without sending a copy of the protocol until October 6; Süss on September 26. Prandtl was, needless to say, overjoyed. He arranged on September 22 to meet and talk with Doetsch in Berlin about furtherance of the plans when both would be there for a meeting, sometime around October 2. Some of the difference in character between Süss and Doetsch shows in their letters to Prandtl of September 26 and October 6, respectively. Süss's letter had an apologetic tone:

> If, a little while ago, I still thought that cooperation with Doetsch was unfeasible, so, after the resulting discussion, I am able to hope that it still can result in a successful piece of work with maintenance of all interests. . . .
>
> I am sorry that my arguments against the involvement of Doetsch in carrying out the plans caused you special concern. After I learned from you the position Doetsch held in the research directorate, and you expressed to me the desire that I should take note of it, I chose the method of holding a discussion with Doetsch and hope with your agreement and aid for good results.

However, he also told Prandtl that he had obtained money in the meantime from the DFG, so at present the funds of the research directorate were no longer needed, but presumably might be in the future for further undertakings. What the research directorate could possibly help with, though, were practical questions, like the acquisition of paper and leave or military draft exemption for authors or co-workers.

In contrast, Doetsch's letter spoke of planning a get-together with Süss and Gröbner in a week in order to plan, despite Doetsch's knowledge of Prandtl's negative opinion of Gröbner. He also gave his plans for books (for example, one further reproduction of an English text would appear in the next week, and discussions would be undertaken about the continuation of a book series for engineers).

One would have thought, like Süss, that progress would now go forward without further antagonism; however, one further flash of resentment appears in Prandtl's correspondence. On October 2, in Berlin, Doetsch and Prandtl spoke about the possibility of having an advisory group of mathematicians for the book project, and they agreed on Erhard Schmidt, a distinguished analyst in Berlin, and Blaschke. In his letter of October 7 to Süss congratulating him on the amicable outcome of his meeting with Doetsch, as well as his receipt of money from the DFG, Prandtl asked his opinion of this, since Süss and Doetsch were co-supervisors. He declined such a role for himself: "I myself am not really a mathematician . . ."; and Süss, of course, was primarily a geometer, when what were wanted were analysts. Furthermore, he advised Süss that Doetsch would proceed only in contact with him. Thus, Prandtl placed himself as intermediary, if necessary, between Doetsch and Süss.

Süss's reply pointed out that Schmidt was notorious for not writing things like letters of recommendation and needed to be approached in person or by telephone for advice, while Blaschke was as much a geometer as he himself. As analysts, for a mathematical advisory panel, Süss recommended Konrad Knopp or Hellmuth Kneser at Tübingen or Richard Grammel at Stuttgart, instead of Blaschke—any of them would work well with him and be ready to do the sort of paperwork Schmidt was unwilling to do. Another possibility was Constantin Carathéodory; however, it would be hard to involve him. A word is necessary about these suggestions for an advisory council. Richard Grammel at Stuttgart had had some kind of disagreement with Doetsch earlier (Süss had previously suggested that Prandtl ask Grammel about the negative aspects of Doetsch's character), but was clearly a mathematician active in technological applications. Konrad Knopp was, above all, a mathematician who tried to preserve his discipline from political influence in a difficult time.[18] Hellmuth Kneser was one of Süss's oldest and closest friends,[19] a highly respected mathematician (as was Knopp), from a distinguished mathematical family—in later years he would joke that he would only be remembered as either the son of Adolf or the father of Martin. Carathéodory and Schmidt (as Blaschke) were among the most internationally famous mathematicians of the time. Some of these people will be discussed in more detail later; however, it should be noted here that Constantin Carathéodory was a Greek national (though a German citizen) who essentially removed himself from political involvement during the Nazi years.[20] He also had written Prandtl on August 27, 1942 about the Süss-Doetsch dispute. In it, he spoke of a conversation between him and Doetsch a few days previously from which he gathered that for the present, Doetsch's plans with the ZWB were restricted to the reproduction of some English-language works. As to the support of a series of original (or partly original) mathematical books important for the war effort, Carathéodory said of Doetsch:

> From the insecurity shown by his demeanor it is very questionable whether such support would be lasting. After all, the whole affair can only function without friction

[18] Adam 1977: 242.

[19] Their close friendship dated back to 1928, when Süss became a faculty member at Greifswald, where Kneser had been since 1925. See HK passim.

[20] "Yet he knew exactly where he stood" and helped colleagues in difficulty. See the speech by Heinrich Behnke celebrating the centenary of Carathéodory's birth, *JDMV* 75 (1974): 151–165, p. 163.

if the man directing this task has the trust of the German mathematical community. For Doetsch, who for decades has been accustomed to quarrelling with every colleague with whom he comes in contact, this is not the case, while for Süss it is self-evident.

It is unclear from Carathéodory's letter, apparently written while he was on vacation in Freiburg, whether the stimulus to it was simply his conversation with Doetsch, or a request from Süss. What makes this more interesting is Süss's seemingly gratuitous mention of Carathéodory's name to Prandtl on October 14.

Prandtl, true to his supervisory-intermediary role, sent Doetsch Süss's letter, with a suggestion that he contact Erhard Schmidt personally, adding, "How would it be besides if we would also clean up your old dispute with Grammel on this occasion?"

Doetsch's reply to Prandtl on October 22 was furious, and the character of the man (and the times) can best be illustrated by citation:

> The fact that you wrote to Süss concerning an advisory committee for us and seemingly asked his opinion about it (for he gives such in his answer) has me somewhat bewildered. Hopefully you communicated to him that the idea came from me, because otherwise he will immediately conclude that you wish to give me, as it were, a council of observers on the side, which would make my position with respect to him more difficult. For I must unconditionally hold firm that I am the representative of the research directorate [of the air ministry] equipped with full powers, even as he is the representative of the research council. I had considered the men named by me to you on October 2 only for our internal use within the research directorate, as indeed also Süss will make use of the advice of his numerous friends without our wishing or being able to advise him there.
>
> The effect is indeed also as expected. Instead of Blaschke, who, so far as I know, is also much esteemed by you, Süss suggests his most intimate friend (*intimsten Duzfreund*) Kneser with whom he plays ball on all occasions (I could adduce numerous examples), and also his [Kneser's] Tübingen colleague Knopp, with whom I am also very friendly. For the goal I have in mind both come not at all into consideration. However, completely generally, and independently of these special persons: whom we adjoin to our group is no business at all of Süss, just exactly as we do not prescribe for him the selection of his advisors. . . .
>
> Stimulated by my contact with publishers like Dr. Springer . . . I wished as soon as possible to get together a group of people and have also thought of Grammel, whom you also mention in your letter. The expression "dispute" for my relation to Grammel is far from the mark and to be explained by the whispering campaign unleashed against me by Süss. If some things don't please either of us about the other, that is by far no dispute. To characterize how I stand with Grammel, I would like to tell you only that a short while ago we sat together alone at a meal in the *Haus der Flieger* for over an hour during which we chatted in the nicest fashion.

Doetsch also tells Prandtl of a planned discussion to move matters along with Süss, Kamke, and Gröbner in Berlin, in which he asks Prandtl "for a free hand" so that he can "oppose the necessary authority to Süss's expansive ways, which tend very much to taking over every area that comes into his view."

Prandtl's reply was all diplomatic apology—sorry he misunderstood, never made commitments, and the like. He looked forward to hearing from Doetsch orally about the planning meeting with Süss, Kamke, and Gröbner. By now, about a year had passed since Rasch's original letter. In fact, the upshot was that Süss's project went forward, whereas Doetsch's seems to have fallen by the wayside. A report of books in progress and their status that Süss sent to the RFR on March 25, 1944, contains eighteen items, on substantial topics by substantial authors, of which, far from being mere aspirations, a goodly number appeared either during or in the years immediately after the war.[21]

It almost seems as though, for Doetsch, prerogatives were more important than performance. In any case, this incident presents the spectacle of personal pique delaying and interfering with a mathematical project ostensibly important for the war effort. What mattered most to Doetsch, the air-force officer/mathematician, was his position. While personalities are what they are, there was no one willing or able to bring him in line. It was only possible to take the longer path of mollifying him until he sidetracked himself.

As to Süss's project, it continued throughout the war, providing employment for mathematicians in mathematically useful activity. His progress reports in March 1943 and March 1944 were written on stationery that denoted him as "Leader of the Mathematics Working Group" in the *Reichsforschungsrat*. Having received 30,000 RM on October 5, 1942, and 40,000 RM on February 28, 1944, for the book project, and used it up, on January 22, 1945, he asked for an additional 70,000 RM, which was granted on February 14. On March 7, Süss expressed his thanks but asked about 5,000 RM for running expenses. Also on March 15, Fischer made up 101.78 RM of Süss's cost overrun of the 1944 amount. As the Third Reich totally crumbled, Süss managed to extract money as usual. In fact, Süss's letter of January 22, 1945, to Fischer speaks of the need for the money "because all military research should be further strongly intensified and broadened."[22]

Gustav Doetsch and the Philosophy of Mathematics

In chapters 6 and 8, Süss will be discussed in more detail; however, it is worthwhile here to gain some perspective on Doetsch.[23] As already noted, his book on the Laplace Transform (1937) is a "mathematical classic." Born November 29, 1892, Doetsch enlisted on October 12, 1914, and served with distinction in the air force on both fronts in the First World War, receiving the Iron Cross, both first and second class, as well as several other medals. In 1920, Doetsch ob-

[21] BAK R73 12976, Süss to Fischer, Mar. 25, 1944; cf. ibid., "*Bericht*," sent by Süss to RFR, Mar. 27, 1943.

[22] See letters in BAK R73 12976.

[23] There is no published memoir of Doetsch, and all the material below on his life comes from Personalakten Gustav Doetsch, Universität Freiburg, unless otherwise indicated.

tained his doctorate at Göttingen with a dissertation in summability theory supervised by Edmund Landau.[24] After some time at Hannover, he went as *Privatdozent* to Halle, by which time he had begun to establish himself. To judge from his inaugural lecture at Halle, World War I and its aftermath left Doetsch a skeptic as to the value of rationalism. The lecture, entitled "The Meaning of Applied Mathematics,"[25] took up the problem that has sometimes been called "the unreasonable effectiveness of mathematics."[26] Noting that scientific theory of experiential events is only contingent on contemporary knowledge, and can never claim to have discovered the truth in an absolute sense, but can only approximate it, Doetsch saw a deep philosophical reason for this situation—namely, the absolute disjunction between experience and thought, which belong to "two fundamentally different worlds." "Every dying theory proves this," he stated. Whatever can be said about the *form* of empirical reality, "its *content* is simply *irrational*." This led him to the radical standpoint that "it makes *no sense* to call a theory right or wrong or to seek the 'true' theory of some area." Explicitly anti-Kantian, Doetsch maintained that "a mathematical theory can give no truth at all." For Doetsch, a "law of nature" was a description of a causal connection, not an explanation thereof. Nevertheless, he admitted that a Truth ("indescribable in words") exists, but can only be better and better approximated (by successive theories) and can never be apprehended. Doetsch contrasted himself with Felix Klein (at the time a "grand old man"), who, he said, saw mathematics as the real matter of substance (*das wirklich Wesentliche*) and our perceptions of the world as uncertain. For Doetsch, experience of reality and mathematical theory had *nothing* to do with one another; they do sometimes coincide, but he did not attempt to explain how this could be. That mathematical theory is only an approximation of (a somewhat mystically conceived) reality, and can only be such an approximation, even in the radical terms that Doetsch used, may not seem so striking to us, nor may his insistence that theories can never reveal the true essences of the objects to which they are applied. But Doetsch carries the matter one step further into a general cultural criticism:

> The characteristic of *the* spiritual epoch (*Geistesepoche*), which at the moment is in decline, expresses itself in such *rationalistic dogmatism* [the antecedent referent is Ostwald's theory of painting]. It is the spirit of—one might say—the national scientific era that in reality coincides with the ninetheenth century, and that in our time sinks into the grave with violent convulsions, in order to make way for a new spirit, a new feeling for life, just as, one epoch earlier, metaphysical speculation and idealism had lived out their existence, youthful natural science with its mighty successes, confident

[24] Kluge 1983: 126.

[25] While Doetsch later did work in subjects such as heat conduction, at this time most of his published work (naturally enough) was in the line of his summability theory dissertation, and unapplied. Nevertheless, he had no difficulty using this title.

[26] Doetsch's lecture is cited as reprinted in *JDMV* 31 (1922): 222–233. The many emphases indicated are in the original. "The Unreasonable Effectiveness of Mathematics in the Natural Sciences" is an essay by Eugene Wigner (1960).

of victory, stepped upon the stage and now also presumed to frame a *Weltanschauung* from its material. This brought into being . . . a *natural* scientific feeling for life. . . .

However, *the* epoch at whose beginning we no doubt find ourselves today is fed up with such rationalistic convictions. . . . Here I only wanted to indicate that, in the *natural-scientific domain itself*, which has served so many others as a model, the *mathematical* handling of experiential material is far from being able to communicate conclusions about the *essence* of the world, and therefore to provide real knowledge. The application of mathematics to experience has only the sense that it provides an ideal *tool* for the construction of an *edifice of thoughts based purely in intellect* (*Verstand*), which by reasonable (*vernünftiger*) arrangement can serve as an *approximate image* of reality in certain parts.

After such a peroration, one might expect something like the Goethean tag about all theory being gray, but Doetsch was more sophisticated, and could not drive home his point enough—he thus proceeded to quote instead Hegel's famous attack on mathematics,[27] saying it applies at least to applied mathematics. Finally, he sealed his argument by citing the 1921 edition of Hermann Weyl's famous book *Space, Time, Matter*, thus providing himself with the imprimatur of one of the most famous of his contemporary mathematicians: "Beyond all knowledge provided by individual sciences remains the task of *comprehending*,"[28] that "comprehending" being exactly what Hegel had said mathematics could not do. The printed lecture is studded with italics and exclamation points that lend it a passionate urgency of feeling. Nor was this some idea loosely packed together for the occasion. A footnote reveals that Doetsch had been working for a considerable time on a longer note dealing with such epistemological problems.

This talk has been discussed, not just because of its radical nature and somewhat daring phrasing (especially for someone already set upon a career as an academic mathematician), but because it seems to tell us a great deal about the man and the time. One senses Doetsch's passionate yearning for a new age to replace the old one now bloodily exhausted, and for a new sensibility. It is no surprise to learn that at the time Doetsch, believing that science was divorced from true reality and could never penetrate the "essence" of things, was apparently a religiously practicing Roman Catholic. As a young man he had been a member of a Catholic youth organization, and during 1926–28, while at Stuttgart, he was a member of the German Catholic League for Peace, serving as one of its directors in the second year. Doetsch is another example of that Weimar tendency to adapt to (and adopt) the acausal currents then prevalent in general educated society following the defeat in World War I.[29]

In 1931 Doetsch was one of the academics who signed an appeal in favor of

[27] See G.F.W. Hegel, *Phenomenology of the Mind*, trans. J. B. Baillie, 2d ed. (1931), reprinted New York (1977), 100–103. E.g., "[Mathematics is] a defective way of knowing . . . [whose fame] rests on the poverty of its purpose and the defectiveness of its material, . . . [it is] not a conceptual way of comprehending" (102–103).

[28] Hermann Weyl, *Raum, Zeit, Materie*, 4th ed. (1921), 9. In the Dover reprint (1959) of the English edition, *Space, Time, Matter* (1922), the citation appears on p. 10.

[29] See Paul Forman 1971, 1973.

Emil Gumbel, the leftist and pacifist statistician discussed earlier (Emmy and Fritz Noether were among the others).[30] In 1931 also, Doetsch moved to Freiburg as successor to Lothar Heffter, retiring as a director of the mathematics institute at age seventy. Heffter had worked hard to get Doetsch as his successor in a position that, while perhaps more mathematically prestigious than Stuttgart, was relatively ill-paid. Heffter (who lived to be nearly one hundred) is another witness to Doetsch's seemingly perpetual ill-temper. Heffter had continued giving some lectures until 1936, when, as he later wrote, Doetsch, "who up till then had always treated me respectfully" as his predecessor, arranged a farewell ovation for him in a lecture room covered with flowers. There follows the lapidary sentence: "Unfortunately, his attitude toward me later changed completely." In contrast, Heffter's relations with Süss were always amicable, and he seems to have been well liked by his colleagues.[31] In any case, naturally enough, the petition for Gumbel signed by Doetsch dealt with issues of Gumbel's academic freedom and the impropriety of the actions of the Heidelberg students, not with the validity of Gumbel's beliefs or statements.[32]

If, in 1933, Doetsch found his earlier yearning for a new age seemingly incorporated by the new government of Germany, he would not have been alone among his colleagues, either in the desire or in the perception of its fulfillment.

In 1934 Doetsch and Süss jointly played a conservative role in a crisis in the German mathematical community discussed in chapter 6.[33] In April 1935, Doetsch asked the Baden Ministry of Culture (as required) for permission to participate as invited in a seminar in Geneva to be led by "the leading French mathematician Hadamard." Whether or not Doetsch realized Hadamard was Jewish is unclear; in any case, in September 1935, he asked the same ministry whether he was allowed to belong to the editorial board of the international mathematical journal *Compositio Mathematica*, and was told: "There are no considerations against German mathematicians working on the international mathematical journal *Compositio Mathematica*. On the other hand, working on an editorial board on which Jews are found is not wished." He then asked on September 30, 1935, to transmit the ministerial decision as his reason for resignation from the board;[34] on October 21 he still had not heard an answer, and so still had not resigned, but eventually he did. In 1935, Süss was also a member of the *Compositio* board. Doetsch told him the ministerial decision that they must resign, and corresponding information had also come to Süss from the Rektor. However, the ministry orally told them that they were forbidden to give

[30] Nor was this Doetsch's only such signature that would raise eyebrows later among Nazi bureaucrats. See HK, Vahlen to Kneser, Jan. 18, 1934, and Kneser to Vahlen, Jan. 20, 1934.

[31] Lothar Heffter, *Beglückte Rückschau auf neun Jahrzehnte* (1952), 160–161, 165.

[32] For the petition and Gumbel, see chapter 3.

[33] For Süss's attitudes at this time, see also below and HK, Süss to Kneser, Dec. 7, 1934.

[34] *Compositio Mathematica* was founded by the famous Dutch mathematician L.E.J. Brouwer (who was a German nationalist; see chapters 2 and 6) in 1934 (at least, that is when "Fasciculus 1" appeared). There were originally and continued to be Jews on the large and international editorial board.

its decision as the reason for their resignation, and had to make up apparent grounds for resignation, such as overwork. Süss also resigned, but clearly thought the decision and how it was handled somewhat stupid.[35] Doetsch also asked permission to attend the international mathematical congress to be held in Oslo in July 1936. This was granted (though, in fact, for financial reasons, not all those granted permission were able to go).[36]

In October 1936, Doetsch applied to the local military commander in Freiburg to become a reserve officer in the air force (presumably on the basis of his earlier service). The then-Rektor of the university, a Dr. Metz (whose name had been given by Doetsch), was asked as usual to be as thorough as possible in providing information about Doetsch's "personality, conception of life, public appearance, position on the National Socialist state, economic conditions and perhaps military activity, as well as public activity in state, party, or other organizations." Rektor Metz replied that there was nothing negative to say "so far as I know" about Doetsch's political position, that his economic affairs were in order, and that nothing special was known about his public activity in any state, party, or other organization; but he called attention to Doetsch's old signature on the Gumbel petition as well as the fact that he was not a Nazi party member. From the documents, it is clear that the Rektor was not sure how this would affect Doetsch's standing, and wished to protect himself against a possible future discovery. In any case, despite his earlier "political error," Doetsch was accepted as a lieutenant, and went on maneuvers in July and again on October 4, 1938. While he had originally planned to make up missed lectures by adding extra hours, this proved unnecessary, for the "complete relaxation of the political situation" as a result of the Munich Conference allowed his return to the classroom. By March 1939 he had advanced to captain, and during August was on maneuvers. In December 1941 he became a major. His birth and baptismal certificates seem to have been lacking, and instead he used the list of ancestors used at marriage (this was a common pre-Nazi usage indicating the families joined in matrimony) to establish his racial credentials (necessary as both university teacher and officer). These were finally approved, but, perhaps surprisingly, not until January 25, 1944.

In 1937, Doetsch acknowledged on the required form his former membership in and resignation from the German Catholic League for Peace. After his rise to major, he quite naturally became assigned to the research directorate. In this context, inspired by his earlier visit to Picone's institute in Rome, he pur-

[35] HK, Süss to Kneser, Nov. 13, 1935. In addition to Doetsch and Süss, among the German mathematicians on the original editorial board were Emil Artin and Georg Feigl. All German board members except Artin had disappeared from it by 1936 (vol. 3). Artin, who would emigrate in 1937, continued as a member.

[36] A list of those originally permitted to attend appears in Personalakten Gustav Doetsch, Universität Freiburg, under March 7, 1936. However, not only were many on this list, including Doetsch, unable to attend, but comparison with the list of attendees in the Proceedings of the International Congress, Oslo, 1936, shows the presence of a considerable number of German mathematicians not on this particular list.

sued the establishment of his own air force research institute for mathematics. A previous attempt to found such an institute had been turned down by the education ministry, as it wished to establish such institutes through universities and technical universities in order to maintain academic contacts. Before Wolfgang Gröbner agreed to become the director of this institute *in posse*, sited in Munich, two other mathematicians had turned down the opportunity. One of the things that most impressed Doetsch about Picone's institute, and that he wished to emulate, was that it worked over problems arising in industrial practice in a purely scientific way, so that finally, the original practical question no longer was apparent. In this context, he told Alwin Walther that Picone's institute worked "highly scientifically."[37] As has been seen, not only was this consistent with Doetsch's belief that "reality" and mathematical theory had nothing to do with one another, but also it was precisely this aspect of this institute that caused Prandtl's disapproval. Prandtl, as an aerodynamicist, believed not only in the utility of mathematics, but also that that was the primary purpose of mathematics.

Shortly after Doetsch's incensed letter to Prandtl and his meeting in Berlin with Süss, Kamke, and Gröbner, Süss, acting as Rektor at Freiburg, wrote a letter to the air ministry research directorate stressing the shortage of mathematical personnel at Freiburg, and asking that Doetsch be furloughed so that he could take up his teaching duties again, as he was desperately needed. He informed Doetsch of this request as well, adding, "Your presence in Freiburg during the winter semester would also be very favorable for our mutual undertakings." Süss's request was denied.

Heinrich Behnke was a mathematician who came through the Nazi period without the slightest Nazi taint.[38] He said of Doetsch: "[During 1941–43] he was always Nazi in tone and aggressive against foreign colleagues. I could not have dared to ask Doetsch for help for endangered Germans and foreigners."[39] In the Allied denazification process after the war, Doetsch was apparently banned from the university. As Süss reported to Kamke:[40]

> In the meantime Herr D[oetsch] sought to become emerited in writing after a corresponding repeated interrogation before the investigating commission. The liaison officer to the military government asked me briefly (also earlier) about D[oetsch]. He called him an "activist Windmill" and was of the opinion that someone who denounced others did not belong in an academic chair, independent of political considerations. Besides, D[oetsch] is supposed to have tried recently to call the reproaches, to which he already once admitted and tried to trivialize, unjustified . . . the commission satisfied itself with presenting him from the documentary record with his own writings in 1934.

[37] See Walther to Prandtl, Aug. 20, 1942, as in note 7 (p. 4 of the letter).

[38] For Behnke, see chapters 5, 6, and 8.

[39] Letter from Behnke, Nov. 19, 1945, to authorities considering whether Süss should be readmitted to academic life ("denazification"). In Personalakten Wilhelm Süss, Universität Freiburg.

[40] HK, Süss to Kamke (copy), June 4, 1946.

In 1951 Doetsch was once more allowed to give lectures, and at the time Süss was a codirector of the Freiburg mathematical institute.[41] In his presence, the decision was taken that, while Doetsch would be allowed to give lectures, he would not necessarily receive any further faculty rights (not even library privileges), and, at the slightest attempt on Doetsch's part to cause any sort of disturbance, he would be dismissed. Furthermore, there was an explicit declaration that according to the documents in the Rektor's file, when Doetsch was called to Freiburg, he was not named a director of the mathematical institute. Assuming the completeness of these documents, this ended such a question. Nevertheless, should it happen that Doetsch were entitled to request the position of such a directorship, then "in order to maintain peace at the mathematical institute, a second institute would have to be established for Doetsch." Despite this decision of 1951, when Dr. Metz as Rektor wrote concerning Doetsch's attempt to enter the air force reserve, he said (November 4, 1936) that Doetsch was "professor of mathematics at the University of Freiburg and as such, at the same time, [a] director of the Mathematical Institute."

Whatever the legal situation, Doetsch seems to have accepted the restrictions placed on him. He lectured abroad in Madrid (1951) and Uruguay (1953), and at the Catholic University in Louvain, Belgium (1953). However, in 1954, he was denied funds to be a member of the official German delegation to the International Mathematical Congress in Amsterdam. In 1961 he was released from all teaching duties on grounds of age. Although Doetsch apparently received a large deputation on the occasion of his seventieth birthday the following year, he expressly forbade any press announcement by the university of this event. Similarly, he forbade such announcements of his seventy-fifth and eightieth birthdays, and died a few months before his eighty-fifth birthday, on June 9, 1977. Speculation on this reticence and retirement in old age is tempting but probably not worthwhile.

What are we to make of Doetsch? He seems a man of great passion and irascibility. Somewhat liberally inclined with a strong religious foundation as a young man, he also shared the pervasive Weimar mood of decay and decline, whose most elephantine expression was Oswald Spengler's *Decline of the West*.[42] Whether or not he saw the Nazi movement as the embodiment of the new age whose threshhold he had perceived in 1922, he may have already been moving rightward, or at least in a more nationalistic direction, when Hitler came to power. In any case, as many historians have testified, and as discussed in chapter 3, the Nazi movement had an evangelistic, often quasi-religious tone, and Doetsch seems to have moved from one mystically conceived orthodoxy to another. In part, no doubt, as an attempt to protect himself and his family from his earlier "mistakes," including one as recently as 1931, and in part out of genuine conviction, he joined up as a reserve officer at the age of forty-three, and became the "110% Nazi" described by Behnke. At the same time, he be-

[41] Süss's codirector was Gerrit Bol.

[42] The standard English translation is that of C. F. Atkinson (1929).

came *extremely* careful that any action he took had higher approval. How much of his exaggeration of his own importance was an intrinsic character flaw, and how much a self-protective device, is unclear. Absent his extreme irascibility, there must have been many Germans like him. In any case, his colleagues seem never to have forgiven him—an obituary is yet to appear. Nor, perhaps, did he forgive them.

Erich Kamke

Why should those in Süss's group not wish Doetsch to learn their names? Given Doetsch's cantankerousness, and his loud pro-Nazi sentiments, a case in point may be that of Erich Kamke. Kamke, it will be recalled, was involved with Süss's plans from the start. Already a distinguished mathematician by 1933, Kamke was one of the few faculty members with democratic beliefs at Tübingen in the early 1930s. Indeed, Hans Bethe, who had taken a position there in physics in 1932 (which he would soon lose because of Jewish grandparents), spoke of the general right-wing orientation of the faculty.[43] Nevertheless, Kamke was a "convinced democrat and ardent Nazi-hater."[44] In 1933 he had to relinquish his position on one of the university's governing bodies. Kamke's wife was a Jew who had converted to Lutheranism, and on April 19, 1937, Bernhard Rust's education ministry declared that all academics who were "Jewish-related" had to be retired. Thus Kamke was among those academics given the choice of divorcing his wife or losing his job. He chose the latter, and on November 1, 1937, was forcibly retired.[45] Already in 1936, the Tübingen *Dozentenführer* (that is, the local Nazi political leader of the faculty, in this case the mathematician Erich Schönhardt) had complained about mathematics in Tübingen:[46]

> Considering the fact that both the other chairs of mathematics are held by personalities, of which one is Jewish-related and makes no secret of his feeling against National Socialism [Kamke], while the other up until now has taken no opportunity to show a distinctively National Socialist attitude [Konrad Knopp], it is a pressing necessity that the third chair of mathematics be held by a man . . . who stands in the movement and knows how to make his attitudes felt.

Indeed, Kamke's daughter, Adelheid, was as strong-minded as her father. A "half-Jew," she was studying pharmacy in Tübingen in 1940 when she was disciplined for wearing trousers and smoking a pipe.[47] It was during the period

[43] Hans Bethe, as cited by Adam 1977: 29. For Tübingen in 1924–26, see Eschenburg 1965.

[44] Adam 1977: 28. The characterization is due to Theodor Eschenburg.

[45] Kamke was also retired on the basis of paragraph 6 (elimination of supernumerary positions) of the April 7 law, but realized that this was a makeshift and assumed his chair would be refilled. He had received notice of the retirement three months previously. See Kamke to Hellmuth Kneser, Aug. 1, 1937, in HK. It should be noted, however, that the final form of these laws was the "Deutsche Beamtengesetz" of January 26, 1937, further clarified on July 12, 1938. For this, see Martin Broszat, *Der Staat Hitlers*, English translation, *The Hitler State* (1981), 253–255.

[46] Adam 1977: 141.

[47] Ibid.: 114.

after he had lost his academic position that Kamke wrote his well-known magisterial compendium on ordinary differential equations and their methods of solution.[48] This made him almost everyone's natural choice for a book on partial differential equations (despite the great difference in topic) in Süss's project (only Blaschke demurred somewhat). However, that does not quite explain Süss's early turning to Kamke. That the two men became somewhat close may perhaps be indicated by a letter written by Kamke to Süss on February 24, 1943, which ends "Mit freundlichem Grüss—Ihr" (with friendly greeting, yours) instead of the mandated "German greeting," "Heil Hitler," as a close. This was an official business letter. Süss would, of course, use "Heil Hitler" in such official (though not personal) correspondence. Kamke apparently felt comfortable enough with Süss to write in this fashion. These sorts of remarks may seem to the reader to infer rather too much from trivialities, but as will appear below, the failure to say "Heil Hitler" contributed significantly to a mathematician's loss of employment, and Süss, who was always necessarily officially correct, was Kamke's "superior" both in the book enterprise and as president of the German Mathematical Society. Of course, Kamke was unlikely to use "Heil Hitler" in any case.

In autumn 1944, Kamke was threatened by the Gestapo (like all who were "Jewish-related") with being placed in a work camp; however, this was prevented by the efforts of Süss and the physicist Walter Gerlach on his behalf.[49] Given Kamke's position, it is clear that Süss's patronage provided a protection for him, and, from what happened later, an intentional protection. Why someone like Kamke might not wish to be involved with a man of Doetsch's reputation seems clear.

The Winkelmann Succession

Another case study from applied mathematics, this time dealing with academic promotion, can be found in the saga of the choice of an applied mathematics professor at Jena.[50] This is particularly informative as to how even international politics could affect a mathematical appointment, and how the various bureaucracies of party, state, and academe, as well as industry, could come into conflict. Jena is especially interesting on two grounds. One is the fact that the major industry in Jena was the *Zeisswerke*, and that that famous optical company tried

[48] According to H. J. Fischer, a mathematics Ph.D. who was a member of the SD (that is, the Nazi Security Service), Kamke's work on this was suspect because of his wife, but Fischer assured the relevant authorities that the compendium was truly "important for the war effort." See Fischer 1985, 2:69–70.

[49] BAK R73 12976, Süss to Fischer, Oct. 12, 1944. Kamke to the Rektor at Freiburg, Oct. 20, 1945, in Personalakten Wilhelm Süss. H. J. Fischer claims that, approached by Gerlach and the chemist Georg Graue, he helped save Kamke (1985, 2:129). Süss did contact Gerlach about Kamke.

[50] There are over 200 documents in the BAK file covering this case of choosing a successor for Prof. Winkelmann at Jena. All citations, unless otherwise noted, are from this file, but not necessarily individually cited.

to exercise considerable influence on appointments in engineering, physics, and mathematics, as well as on the university in general.[51] A second is that Jena is in Thuringia, the first German province to have Nazi participation in its government, in 1930.[52] In addition, one of the most famous of all Jena scientists was Ernst Haeckel (1834–1919), whose version of Darwinism has been seen by some as preparing an intellectual basis for Nazi ideology.[53]

On October 11, 1938, Professor Max Winkelmann, suffering from some sort of psychiatric depression, officially retired from the chair of applied mathematics at Jena. His assistant at the time was Ernst Weinel (a student of Prandtl, among others). The retirement was known beforehand, and planning had already begun by August to fill the chair, prospectively by the 1939 winter semester. However, Weinel, whose contract as *Assistent* ran out September 30, 1938, was considered a suitable temporary substitute "in the extreme case" that no chair holder was found by the beginning of the semester. The traditional faculty list of three ranked candidates had in first place Rudolf Weyrich, followed by Karl Marguerre and Karl Klotter. Weinel was mentioned following the list as someone who "deserved special mention" as a result of his "outstanding achievements and many-sided knowledge," though "promotion at the same university was to be avoided as much as possible." Of the three candidates, both Klotter and Marguerre were employed by the German experimental station for aviation, though after World War II Klotter would return to academe.[54] Weyrich was the eldest of the three and then at the (German-language) technical university in Brünn (Brno) in Czechoslovakia, where he had been for fourteen years.

Rudolf Weyrich was born in 1896 and had served in World War I, was wounded in battle at Brzeziny, and had consequently lost the sight in his left eye. He was an ardent German nationalist, and had retained his German citizenship as well as his standing as an officer in the military reserve. In September 1938 he found himself caught in the midst of the agitation concerning the Sudetenland part of Czechoslovakia (which would lead to the Munich Conference) and wished to return to Germany. As a German citizen, he could not vote in the upcoming plebiscite, and on the same ground his application to join the Sudeten German *Freikorps*[55] was denied. At the same time, he felt it impossible

[51] According to Professor K. H. Weise (interview, Mar. 16, 1988), this also served to insulate these departments somewhat from political activity. However, the effectiveness of this influence is uncertain, and clearly doubtful in mathematics. Cf. note 52 below.

[52] The only mathematician dismissed from Jena was Max Herzberger, an optical scientist who emigrated to the United States. Actually, already in 1930, the Nazi-influenced Thuringian ministry had, against the wishes of the faculty, declined to confirm a regular appointment for Herzberger. See Pinl 1972, under Jena.

[53] Gasman 1971. Haeckel was himself clearly an early "antidemocratic" and anti-Semitic (see Hermann Bahr 1894) thinker of the sort discussed in the previous chapter.

[54] Winfried Scharlau et al., *Mathematische Institute in Deutschland, 1800–1945* (1990), 181.

[55] *Freikorps*, in general, were unofficial military units. They originated after World War I, and these units continued to fight after the armistice. Within Weimar, their veterans formed a residue of extreme right-wing sentiment and were involved in incidents like the "Kapp *Putsch*" and the assassination of Walther Rathenau. The Sudeten *Freikorps* was formed on these models.

to return from a visit in Germany to Czechoslovakia, as the Czech government was requiring all civil servants, under threat of dismissal and punishment, to sign a statement that they did not agree with Konrad Henlein's proclamation demanding the return of the Sudetenland to Germany. Naturally, Weyrich would not sign, and so feared dismissal and imprisonment if he returned to Brünn. The Rektor at Jena was the physicist Abraham Esau, who was desirous of having an applied mathematician of Weyrich's repute as someone well acquainted with both pure and applied mathematics as well as mathematical physics. Apparently already on September 21, 1938, Dr. Wilhelm Dames of the education ministry had the idea of calling Weyrich to Jena, and asked Konrad Henlein, as leader of the pro-German agitation in the Sudetenland, whether in his view Weyrich could be spared in Czechoslovakia.[56] The Munich Agreement was signed the morning of September 30. Henlein seems never to have replied.[57] However, on November 16, the office of the "*Reichskommisar* for the Sudeten German areas" did reply, saying that Weyrich needed to remain where he was in Brünn, at least for a semester, until after a decision had been taken about the German universities in Brünn and Prague. It would appear, however, that by November 19 a decision had been taken that these German universities would remain. Weyrich was appointed to Winkelmann's chair, but for the present only as a "substitute" (*vertretungsweise*). Part of the reason for the appointment was Weyrich's importance (as a German citizen) "for the four-year-plan and defense of the country." Needless to say, Weyrich was confused as to his future—did he need to stay in Brünn[58] or go to Jena? He wrote Dames on November 21:

> I have always maintained that I would stay at the post that the Reich's officials indicate, and it goes without saying that as a German citizen of the Reich, and as an officer of the German Reich, will not leave my post if I were not expressly told to do so. However, as I have said, I would like to know soon where I am at so that I can direct myself in that direction.

However, although Weyrich was both politically and academically well-approved, and especially desired by the Rektor in Jena as well as apparently the faculty, on November 23, a letter went from the leadership of the Nazi university teachers organization (*NS-Dozentenbund*) in Munich to the education ministry

[56] Weyrich had spoken to Dames sometime before September 20, 1938, about wanting to return from Czechoslovakia to Germany. On September 21, Dames had indicated to Henlein his intention of calling Weyrich to Jena, provided he could be spared. However, the official faculty list of recommendations for the position was not prepared until October 21, with Weyrich heading the list, and on November 19 it still had not been transmitted to Berlin, since on that day Esau called Berlin wanting to be assured, before he transmitted the list, that Weyrich would come to Jena. This would seem to indicate that Weyrich was in first place on the faculty list at least partially as a result of pressure from the education ministry and the Rektor.

[57] Although Henlein was the leader of the Sudeten German Party, after March 15, 1939, when Hitler marched into Czechoslovakia, it was not he but his lieutenant, K. H. Frank, who was rewarded with a high post in the German "Protectorate of Bohemia and Moravia."

[58] Brünn was in Moravia, and in an area of large German population, but not in the territory ceded to Germany as a result of the Munich Agreement. See Keith Eubank, *Munich* (1963), 243.

declining under "present-day conditions" to approve Weyrich's appointment. The reason was "the fact that German professors at the German universities in Czechoslovakia had important political interests to represent."

At the same time, Weyrich, who in the meantime had also received an offer of appointment at Graz, was writing a friend in the education ministry, trying to get aid in leaving the "exposed place" where he had been for fourteen years, and going to Jena. He would of course stay at his post as a strong nationalist and so forth if that were the Berlin decision, and mentioned that his assistant could succeed him in Brünn.

On the twelfth of December, Dr. Huber,[59] Dames's superior in the education ministry, wrote Weyrich concerning Dames' desire for an expedited carrying-out of his appointment. This communication gives some insight into the bureaucratic politics of such academic appointments in Nazi Germany. Among the terms suggested by Huber as essential to the appointment was the agreement of the *Führer*'s deputy (Rudolf Hess). This was because Weyrich was in a foreign country, and if his appointment were to be turned down, say, for a political reason, he would be stuck without a position, which would make for "at least a moral claim" against the German bureaucracy. As Huber and Dames knew, there was danger of exactly that, though not because of any political unreliability on Weyrich's part. On the other hand (since Brünn was in Moravia), the *Reichskommisar* for the Sudetenland was no longer responsible for the German universities in Brünn and Prague after it had been decided to maintain these; therefore his agreement to Weyrich's leaving Brünn was no longer necessary. However, the agreement of the parliamentarian Ernst Kundt (a leader of the Sudeten Germans), who was now leader of the German group within Czechoslovakia, would be required, as would that of the German embassy (to be requested via the Foreign Office). It is clear that were these agreements acquired, Huber and Dames would have protected themselves from any complaint like those that had arisen from the *Dozentenbund*. As to that organization, Huber advised Dames that it was doubtful whether they should be asked to make recommendations about Weyrich's successor (in Brünn) since "such a procedure could have considerable retroactive effect on the usual practice." Five days later, Dames wrote Weyrich of his impending appointment at Jena with "as much speed as possible"; however, a statement was necessary from the Rektor at Brünn that he could be spared. On December 30, Dames wrote the necessary letters to Foreign Office officials. The intention was that Weyrich would finally take up duties in Jena beginning summer semester 1939. Weyrich, who did not receive Dames's letter until a month later (there does seem to have been some difficulty with mail from Germany to German nationals in Czechoslovakia), was eager to come. The Rektor at Brünn, the mathematician Lothar Koschmieder, provided the necessary assurance that Weyrich could be suitably replaced by

[59] Sometime on or before December 3, Esau had telephoned Huber to ascertain Weyrich's availability; receiving a positive answer, on December 3 he finally sent the faculty recommendations to the Thuringian education ministry. See above, note 56.

his assistant and obtained the necessary approval also of the local *Dozentenführer*, Armin Schoklitsch. All was going well until around March 10.

Since the end of World War I, there had been a movement for autonomy within Slovakia. These fascist autonomists had allied themselves with the Sudeten Germans and, shortly after the Munich Agreement ceding Czechoslovakian territory to Germany, had seized power in Slovakia. The situation reached a crisis on March 9, 1939, when negotiations broke down between the Slovakian autonomists (supported by the Germans) and the government in Prague. Seizing the pretext as always intended, on March 15, Hitler invaded Czechoslovakia, and the next day, in Hradcany Castle, the traditional seat of government, signed a proclamation ending an independent Czechoslovakia.

A few days before March 12, Schoklitsch told Weyrich that, at least temporarily, he could not go to Jena after all. It should be noted that Schoklitsch, an academic in a political position, had a say in the academic disposition of his nominal academic superiors—this seems to have been typical of the local *Dozentenführer*, generally a younger academic who obtained a relatively powerful position through political activity. In desperation Weyrich wrote Dames, wanting to know what his situation was, stressing as always that he had never "shied away from a task for the best interests of our people," and so on; however, he found the situation "*very* wearying." Schoklitsch notwithstanding, the matter was pursued in Berlin—on March 22, the German Foreign Office was informed from Prague that Weyrich had been well known to the Brünn consulate as a "reliable German" and "positively disposed toward the new Germany." Furthermore (contrary to Huber), Kundt's approval for Weyrich's leaving for Jena was not needed, since Weyrich was a German national; that of Schoklitsch (which had been officially given on January 19) was all that was necessary. On April 18, the Thuringian minister for education (*Volksbildung*) was empowered (by Dames) to hold negotiations with Weyrich as to his conditions for coming to Jena, and to reach an agreement with him conditional on Dames's final approval; at the same time, Weyrich was again appointed, "for the present as a substitute," to Winkelmann's chair. By May 12, the Thuringian ministry still had not heard from Weyrich and so inquired as to his intentions; Dames also inquired; on May 19 Weyrich wrote Dames a rather different letter than previously, citing difficulties with coming to Jena. Nevertheless, Dames, taking these as negotiating points, asked the Thuringian ministry for negotiations with the goal of Weyrich's appointment by October 1, or, at the latest, winter semester 1939–40. In the meantime, Weinel kept offering the lectures at Jena as the temporary replacement for Winkelmann.

What seems to have happened is that for Weyrich, the ardent German nationalist, Czechoslovakia as the German "Protectorate of Bohemia and Moravia" was rather more attractive than the Czechoslovakia that just recently had been considered inimical by the "new Germany." In his letter, he raised factitious points, like the fact that the semester in Brünn began in March; that Dr. Schoblik, his putative replacement (and a Sudeten German), had to give lectures in actuarial mathematics (as replacement for another faculty member gone to Germany) and so could not take over his complete teaching load; and that the Czech krone was worth more than the German mark in buying power, and he needed corresponding compensation. He asked the Thuringian minister to invite him as quickly as possible to negotiate for the Jena position. In the meantime, his attitude was:

> For the present I am still an employee of the Protectorate of Bohemia and Moravia and for the present not yet taken over into the service of the German Reich again, it is therefore impossible for me at the present moment to leave my post before I am finally appointed in Germany. . . . Were I professor at an "inner German" university, named by the Führer and thereby certain of my position, so I would *without further ado* follow the invitation of the education ministry, given, of course, that I could be spared. However now the situation is rather different.

It also turned out (though not mentioned) that Weyrich had received indications of an appointment in Prague, and had an inclination to accept this rather than Jena, presumably for all the reasons a metropolis is often preferred to a smaller industrial city.

Preliminary negotiations with Weyrich were finally carried out in Weimar on June 12, 1939, by a ministerial representative. Among other concessions preliminarily achieved by Weyrich were the elevation of the position from an associate professorship (*Extraordinariat*) to a professorship (*Ordinariat*), some compensation for the superior buying power of his savings in krone, preservation of the money paid in for a pension in Czechoslovakia, a substantial increase in the proffered salary, renovation of the office he would have as director of the applied mathematics department including the acquisition of an automatic calculator, and the right to give lectures in pure as well as applied mathematics. Weyrich promised a decision by July 31 as to whether to accept the offer, provided all matters left open, such as his pension, were settled at a sufficiently prior date. Unfortunately, in late June, the Thuringian finance ministry balked at the salary (Weyrich's demands from all academic sources amounted to 15,680 RM, to be increased by 1,000 RM after three years—the Thuringian ministry was only willing to offer 13,680, the "normal maximum," but accepted the potential increase).

Weyrich, who ten months previously had been desperate for any reasonable opportunity to leave Czechoslovakia, was furious, but did not reject Jena. Instead, he refused to fulfill his promise to decide by July 31 until such matters were settled. While such negotiations were going on, the Jena faculty was beginning to lose interest in Weyrich.

Dames was also angry; Weyrich had begged him in September 1938 to find him a position in Germany; now Weyrich was vacillating and demanding more than was usually offered. On August 8, Weyrich and Dames met in Berlin and agreed, according to Weyrich, on a total amounting to 15,000 RM (11,600 RM salary, 2,000 RM guarantee of lecture fees, and 1,400 RM as a "location supplement"). In the middle of September, the Rektor at Jena telegraphed Weyrich to assume the chair in applied mathematics, and Weyrich proceeded thither. On arrival in Jena, however, he discovered that he could not assume the chair because someone in Berlin had decided it was necessary for him to remain in Brünn.[60] Weyrich returned to Brünn, declaring on September 30: "It goes without saying that I have no intention of raising a storm of protest against the decision of the ministry nor wish to push through my transfer to Jena, under any circumstances"; however, under the circumstances, he would like the money promised him for Jena paid to him in Brünn. Weyrich also complained about German professorial carpetbaggers in Czechoslovakia, who retained their higher German salaries plus a considerable supplement from the protectorate—why should he, a man of reputation, who had spent fifteen years at "an exposed forward post" in Czechoslovakia, receive less? Since Weyrich was a bachelor,[61] he mentioned that he had to support his mother, who was over eighty, and "was planning to start his own household," implying he had not done this only because of a lack of the necessary funds.

Dames in the meantime had been called to military exercises and Weyrich's file passed to the physicist Wilhelm Führer. In early October Führer visited Weyrich in Brünn, and told him there was no altering the situation, he would have to stay in Brünn until he actually held the piece of paper naming him in his hands. Nevertheless, in late October, the national leader of the Nazi university teachers' organization (*Reichsdozentenführer*), G. Schulze, who was a professor at Munich and whom Dames had kept informed about Weyrich, wrote to Schoklitsch asking him to determine finally Weyrich's opinion about his call to Jena (as well as inquiring about Schoklitsch's own to Graz). It will be recalled that on November 23 of the preceding year, this same office had forbidden Weyrich to leave Brünn; in both instances we see the political apparatus that officially played a merely "consultative" role in university appointments actively retarding or promoting this one.

However, either by this time or shortly thereafter, Weyrich had several other fish to fry. Sometime after his conversation with Führer, Weyrich was apparently approached by a number of people, including K. H. Frank,[62] about becoming the Rektor in Brünn. In addition, by this time he had been approached not only for professorships at Jena and Graz, but for two in Prague (at both the University and the Technical University) and another in Breslau (modern

[60] On September 1, 1939, with the beginning of World War II, all German universities were temporarily shut. However, Jena apparently was allowed to reopen on September 5, and Brünn on September 8—it was December 8, though, before all universities were reopened. See Adam 1977: 188, and nn. 1–6 there. Perhaps confusion of university administrations stemming from these political developments and an after-the-fact decision to preserve the *status quo ante* until things were clearer has something to do with this curious episode.

[61] Robert Proctor (Racial Hygiene [1988], 121) claims that by 1938, marriage was essential to professorial appointment. This does not seem to have been true in Weyrich's case. In 1938, Weyrich was forty-two, and had never been married. It is true that at one point in the negotiations he vaguely spoke of possible future children, but this is in connection with potential income, and his lack thereof never seems to have concerned the officials dealing with his case in the slightest.

[62] At the time Frank was secretary of state to Constantin von Neurath, who had become the first "National Protector of Bohemia and Moravia" (Eubank 1963: 258).

Wrocław). Weyrich answered Schulze almost immediately saying that he was "fundamentally prepared to stay a while longer in Brünn if he should assume an academic office there," however, not as just a mere professor, since his call to Germany ought to result in some local academic preferment. To forego a return soon to the Reich was not easy for him; however, said Weyrich, he realized that "it is not permissible for us tested 'old fighters'[63] suddenly to leave the battlefield without having taken care that their replacements from Germany (*Altreich*) are trained for the successful continuation of the struggle." As soon as someone arrived from Germany and became sufficiently acclimatized (with Weyrich's help) to Brünn to lead the university, he would "gladly return to the homeland in order to spend the remainder of my life among my compatriots (*Landesleuten*)." In the meantime, he asked Schulze to see that, if possible, the post in Jena remain open for him, and that in Brünn he at least receive the monies promised to him for Jena. The letter is further filled with self-praise for his self-denial (which he noted would surely be recognized) at not returning immediately to Germany. Weyrich did not directly mention the possible assumption of the Brünn *Rektorat* to Schulze, but only hinted at it. Weyrich did, however, talk about it a great deal in finally replying to urgent requests from the ministry in Berlin to come to a decision. Though Dames and Weyrich had met on August 8 in Berlin, on the preceding day Dames had written him in Brünn asking for a decision within a week or two.[64] On November 7 the ministry, in the person of Wilhelm Führer, wrote again. On November 29, Führer wrote the Thuringian ministry requesting a report on the Weyrich matter, and on December 27 that ministry wrote Weyrich with a written financial offer and anticipation of his beginning in Jena in the first trimester of 1940. Finally, Weyrich answered them on January 4, but this does not seem to have been appropriately communicated to Berlin in a timely fashion, for, on January 20, Dames, returned from military exercises, wrote Weyrich a curt express letter asking for a reply within a week; on January 22, Weyrich telegraphed that he was replying, which he did nine days later. Weyrich remarked that it was not *his* fault he was not in Jena—had he had his way, he would have been there a year before, and when he finally did go in September, when a financial agreement had been reached, he was turned away. He also discussed the possibility of becoming Rektor in Brünn in a tone worth reproducing (emphasis in original):

> *Several* officials from the government and the party came to me and pressed me that I should at least for the present take over the *Rektorat* in Brünn. I could not decline, if only from *national grounds*. Namely, my standpoint is that I cannot answer to the *German people* (*Volk*), if I withdraw myself from a task here for which I have been

[63] *Alte Kämpfer* were members of the Nazi party prior to January 30, 1933. They considered themselves an elite of "true believers."

[64] The date August 8 for a meeting between Weyrich and Dames settling financial arrangements comes from a letter of September 30 from Weyrich to Dr. Nipper in the education ministry. Later, however, Weyrich would claim he had never received Dames's letter of August 7. It seems strange that if they met, Dames would not have mentioned he had written Weyrich the preceding day.

selected by profession. I also cannot believe that the education ministry could have any other viewpoint in this connection.

Weyrich, of course, agreed to be considered for Rektor conditional on not suffering financial loss through staying in Brünn, and that eventually he would have the possibility of "returning to my German homeland, say to Jena." A suggestion that was made (which he threw out as a trial balloon) was that he "should wait for his appointment to Jena, and then the request would be made that he should be placed on leave from Jena to Brünn." (The bureaucratic point apparently is that this would guarantee his Jena salary in Brünn.) He sent Dames's express letter to K. H. Frank and other Nazi officials and asked how he should answer. He was told that they prospectively intended to lay the whole matter before the "*Führer*'s deputy" (presumably Rudolf Hess's office, though not necessarily Hess himself). In any case, Weyrich said, "for further clarification," he wished to lay the whole matter before Minister of Education Bernhard Rust, so that someone authoritative might decide what he should do in the prevailing circumstances. He was hurt by the tone of Dames's letter, but was willing to understand that Dames during his leave for military exercises was not fully informed as to what was happening:

> The clear answer that you desire from me, under the prevailing circumstances, must be given not by me, but by others, namely the officialdom set over me . . . if you insist on the requested clear answer, so I hereby tell you expressly that I will give such a clarification only to the Minister of Education, Dr. Rust, and, indeed, after I have completely explained the whole matter to him.

Both Weyrich's answer to the Thuringian ministry in Weimar on January 4 and his letter to Dames complain about the niggardliness (in his view) of the Thuringian offer. He is "embittered" by this. The actual content of the two letters is similar: he hopes the post at Jena will remain open; he may have to stay in Brünn. However, the Thuringian one also contains sarcastic remarks about Weinel (who, after all, had been giving the lectures in Jena for three semesters now—in March 1940, when Weinel's "temporary" substitution had spanned almost two academic years, the Thuringian ministry finally bethought itself of a bureaucratic way to obtain more funds for him by naming him the temporary holder of the position).

As to Weyrich's putative appointment, discussions immediately moved to a higher political level, as his astonishing letter to Dames had threatened. On January 30, Frank sent a telegram marked secret to Dr. Werner Zschintsch, the executive secretary in the education ministry in Berlin, asking for a postponement of Weyrich's call to Jena until a decision could be reached on the Rektor at Brünn. It seems Weyrich was the candidate of Dr. Jury (the district leader [*Gauleiter*] of the lower Danube region), who did not care for the other proposed candidate, a certain Dr. Kriso.

On March 29, 1940, Weyrich and Dames formulated another agreement in Berlin delineating the financial and other conditions under which Weyrich would take over in Jena in a position elevated to a regular full professorship plus the directorship of the institute for applied mathematics, contingent on Weyrich's final agreement, which he indicated from Brünn on May 28 and gave explicitly in a further letter on May 30.

However, by now, nearly two years after the search for Winkelmann's successor had begun, things had changed in Jena. In the first place, the Rektor was no longer the physicist Esau, who had been in the forefront of pushing for Weyrich's appointment. By now he had gone to Berlin to a position in the DFG, where, as already observed, he played a role in the Süss-Doetsch controversy. His replacement, Dr. Karl Astel, a "racial scientist," had no particular interest in Weyrich. By April 8, the Thuringian education ministry as well as the university saw the prospective Weyrich appointment as having miscarried and suggested an exchange of positions between Weinel and Lothar Collatz, then in Karlsruhe. While the mathematics faculty did not wish to promote their own former *Assistent* Weinel permanently, their choice of Collatz (who died in 1990) was first-class, as he had already begun to show signs of what he would become, one of the leading applied mathematicians of his generation. Or, as they put it, "Dr. Weinel was better suited to a technical university [*viz.* Karlsruhe], and Dr. Collatz to Jena." Ten days later, Dames in Berlin was contemplating the possibility of calling Wolfgang Gröbner to Jena, should Weyrich not take up the post—it would seem that Gröbner had prominent mathematical friends working for him, among them Wolfgang Krull in Bonn and Helmut Hasse in Göttingen, though presumably not Prandtl.[65] On April 22, Dames told the Thuringian ministry that there was no place for Weinel in Karlsruhe, that Collatz was involved in "important war work" at the technical university in Darmstadt and would not be moved, and that they had to wait for Weyrich to decide between them and the technical university in Prague!.

On May 29, Astel called Rudolf Mentzel, then head of the scientific division of the education ministry, and said the Jena faculty no longer wanted Weyrich. Mentzel told Dames to stop the appointment until further notice. On May 30, the very date of Weyrich's acceptance letter to the ministry, the faculty put it in writing. They cited several reasons for their present rejection of Weyrich after two years of waiting: (1) that the chief promoter of Weyrich, Abraham Esau, was now in Berlin; (2) that even in October 1938, mathematicians had been opposed to Weyrich since he seemed more a "pure" than an "applied" mathe-

[65] Dames did not know Wolfgang Gröbner's address and so asked Krull to make him aware of his consideration for a position, and request that he send in the necessary documents (curriculum vitae, list of publications, ancestry and organizational membership forms for him and his wife). Gröbner apparently lost an assistant's position in Rome as a consequence of declaring himself a German citizen. Hasse had also been one of Weyrich's original referees. For Gröbner and Hasse, see in this chapter "The Süss Book Project" above and "Hasse's appointment at Göttingen" below. Gröbner's name is known to contemporary mathematicians as the originator of "Gröbner bases," an important technique for studying certain algebraic questions in mathematics, which is most efficient when implemented on a computer. Since Krull and Hasse were both algebraists, this may explain their interest in him (as well as Prandtl's noninterest).

matician, and indeed, on a visit to Jena he had said he wanted to hold lectures in pure mathematics;[66] (3) Weyrich's continual delays, and "grotesque demands and wishes," seemed unbelievable and unbearable—the all-too-conspicuous role played by his private interests was incomprehensible; (4) various events in the meantime made it impossible to value his character any longer; and (5) it would appear Weyrich placed little value on the position in Jena, and the hope for a successful collaboration of Weyrich with mathematics and physics now seemed hopeless. A new list of possible appointees would be forthcoming. On May 31, Astel forwarded this faculty statement to Berlin with a cover letter "agreeing with the faculty reasons in all points." By the fourth of June, he had forwarded such a list to Berlin.

On the other front, Weyrich's ambitions also plunged toward disaster. Also on June 4, Weyrich's sponsor for the *Rektorat* at Brünn, Dr. Jury, wrote directly to Bernhard Rust, the minister of education, saying that Dr. Kriso had been chosen for the post despite Jury's misgivings. He now lamented that "party-member Weyrich" had apparently been ill-served by the competition for the *Rektorat*:

> I can only emphasize that Weyrich had nothing to do with the whole question except as I, as *Gauleiter*, forbade the *party-member*[67] [emphasis in original] Weyrich to leave Brünn before I gave him permission to do so. It would have pleased me a great deal if one of the best *National Socialists* among the Brünn faculty would have undertaken the leadership of the university. That is the single and sole reason that Weyrich delayed his call to Jena.

Since Jury had now learned that Weyrich's call to Jena suddenly might no longer be in the offing, he put the greatest importance on Weyrich's receiving that call as soon as possible, so he could officially take over in Jena at the beginning of the next semester. He made this "personal request" to Rust: a loyal party member should not be disadvantaged by his faithfulness.

On July 10, Weyrich, apparently having heard nothing official since his May 30 letter to Dames, and since the Brünn trimester was drawing to a close, wrote him to discover where he stood. No answer. On July 23 he telegrammed. Apparently again, no answer, and so he sent a letter to Mentzel on August 1. Mentzel did answer on August 7, told him that Jena had changed its mind about calling him "eight weeks ago," and had pressed the ministry in this regard. Therefore Mentzel could do nothing "as I would wish in any case to avoid your entrance into a group of professors among whom you are inwardly unwelcome, which therefore would also have to make you personally uncomfortable." A copy went to Dames. That all forces were now lined up against Weyrich is indicated by a letter (no doubt requested) from the then-head of

[66] The original committee consisted of a mathematician, a physicist, and a representative of the *Zeisswerke*. No doubt Abraham Esau had strong influence on the committee.

[67] Jury's salutation to Rust addresses him not as education minister but as "party member."

Thuringia (*Reichstatthalter*) Fritz Sauckel[68] to Rust, stating that he "has convinced himself" that Weyrich, by the nature of his professional interests, was not the right man for the job and would not work well with the other scientists in the institute, and that calling him would be a mistake—he suggested following the faculty recommendations of June 4, 1940. As for *Gauleiter* Jury, he did not in fact hear officially about his plea until September 12, 1940, three months afterward, when Dames wrote him. He said that Weyrich had not behaved with respect to Jena in such a way that "one could conclude [he had] a special feeling of responsibility toward the large and important institute for applied mathematics, which especially now during wartime had important military tasks to take up and carry out." This letter also copies language taken from the faculty's letter (written by the Dekan) and Mentzel's. As to Jury's having said it was all his fault Weyrich couldn't leave Brünn, that played only a small role.[69] Dames would consider Weyrich for another post, if a university suggested him, but he would not coerce a faculty in any way on behalf of a specific person.[70] Unless Jury and Sauckel came to some other agreement, for now Weyrich had only possibilities in Brünn.

So Weyrich's attempted manipulations ended in disaster for him, and after two years there was still no official institute leader in Jena at whatever professorial level (Winkelmann had been an "associate professor"). The students, of course, were getting lectures from Weinel. But what about the institute's "important military tasks"?

The new faculty list of June 4 contained, as was traditional, three names. In first place was Walter Tollmien, then at Dresden, a highly respected student of Prandtl who would be his successor at Göttingen in 1946. For Tollmien, already one of the most well-known applied mathematicians in Germany, it was clear that the position would have to be upgraded to a full professorship, as Weyrich had obtained by way of negotiation. Tied for second were Lothar Collatz and Ernst Weinel. Of the three, Collatz was at thirty the youngest, but had already begun to show his great ability. To repeat the name of Collatz, already rejected earlier and informally by the ministry for the position because he was doing "important war work" elsewhere, seems somewhat unrealistic unless his rejection was anticipated. Weinel is described as basically an engineer whose research was called "many-sided," and whose lectures "always make a special impression because of their clear and masterly didactic construction."[71] Among

[68] Fritz Sauckel had been minister-president of Thuringia since 1932, after a long career in the Thuringian NSDAP, beginning with membership in 1923. In 1939, he was put in charge of labor deployment, and after 1942, of the "acquisition" of alien workers. His activities in these regards led to a death sentence at Nuremberg for crimes against humanity, and he was executed on October 16, 1945.

[69] The original draft of this letter had sharper language, telling Jury that he was partly at fault for Weyrich's sad situation. However, this was struck.

[70] Dames seems somewhat disingenuous here, as, in general, the education ministry had no necessary compunction about enforcing an appointment at a university.

[71] Cf. also the praise for Weinel in the original candidates list.

outside supporters were Prandtl, Richard Grammel, Georg Hamel, and Tollmien. It seems clear that the faculty, while it would have been overjoyed to get someone of the caliber of Tollmien or Collatz, and had the customary reluctance to advance one of its own *Assistenten*, was setting up a situation in which the authorities would pick Weinel. And so it proved. Rektor Karl Astel's own letter with the transmission of the faculty list opted for Weinel because he was the most easily available, and it was desirable to fill the position as soon as possible. In addition, Astel remarked that the leader of the *Dozentenbund* found Weinel "reliable," "conscientious," and "faultless in political ideology." This last was underscored by a letter of July 6[72] from the office of Rudolf Hess pressing for Weinel's appointment.

All would have seemed easy now, and indeed, Weinel became the official candidate in September 1940.[73] However, by February 1941, he still had not been officially appointed! The problem was that Weinel had been born in Strassbourg (in Alsace), and on October 1 he had been asked to provide the necessary proofs, such as birth and baptismal certificates, for himself and his family to prove that he was "of German blood." Weinel, however, was in Jena, about 250 miles away, and although France had fallen in June 1940 (and Strassbourg was in the occupied area), one can imagine Weinel might still have had some travel difficulties. In any case, by mid-February 1941, Weinel still had not provided the necessary proofs, despite several requests to do so. It should be noted that Weinel could continue giving lectures as Winkelmann's "substitute" (as he had for over two years) at the very low pay of an *Assistent* (only recently somewhat supplemented) without any such rigorous proofs; however, once he would officially succeed to the position (at a considerable increase in salary), he would become a government employee, and so the proofs were required. Toward the end of April, Weinel did bring some (though apparently not as required) documents together with copies thereof to the Thuringian minister for education in Weimar. Because his copies were incomplete, originals and copies were returned to him to make complete copies; nevertheless, superficial examination seemed to show that Weinel had in fact brought adequate proof of his "German descent." On April 28, 1941, Weinel was called up; about a month later, he seems to have been briefly in Jena, only to return to the army on June 5 in a "mountaineer troop." The mills of bureaucracy ground on, however slowly, and on September 13, from Weimar to Berlin went a request for Weinel's appointment, though clearly his present service on the Eastern front[74] precluded him from obtaining some missing supporting documents: his curriculum vitae, his health certificate, proof of his citizenship, and a statement of his means—however, all of these but the proof of citizenship (which included his qualifications as an "Aryan") were already present in Berlin. Even had they not

[72] Thus, by the time Weyrich wrote on July 10, the powers that were had already aligned against him.

[73] That is, appointment would follow if he satisfied the nonacademic Nazi desiderata.

[74] "Operation Barbarossa," Hitler's surprise attack on the Soviet Union, began June 22, 1941.

been, the Thuringian ministry remarked that "given [Weinel's] army service he could be recommended without any second thoughts." The Nazi party repeated its political approval on October 22. All was eventually ready for the final act of appointment, the certificate of appointment as a state employee signed by Adolf Hitler. This happened on December 9, 1941 (countersigned by Rust),[75] along with seventy-five other such certificates of appointment.

As soon as the signed certificate was in Weinel's hands, the appointment would be official. However, the struggle over the appointment was not over: a sort of grotesque coda was yet to ensue. In early June 1941, when Weinel was back in Jena prior to his final entrance into an army "mountaineer troop" on June 5, he apparently had a serious academic quarrel with a Professor König,[76] with whom he had had differences previously, and appeal was made to Rektor Astel to settle it. On June 3, 1941, Weinel wrote Astel a vituperative letter about König. Not until October 1, 1941, did Astel send a copy of this letter to the Berlin ministry as proving the impossibility of an appointment of Weinel at Jena, citing not only his own personal opinion, but also that of the academic senate, which had occupied itself with the matter. Karl Astel, incidentally, was no secret "bourgeois liberal"; quite the contrary, as a "racial scientist," he favored forcible sterilization for various population groups, and he was a member of the SS.[77] Prior to 1933, he had been a "sports doctor" in Munich. What can Weinel have written that so outraged academic sensibilities? Weinel's letter must be quoted to be believed. Whatever the content of the original quarrel was, Weinel begins by saying that König presented it to Astel in a "distorted and jesuitical" manner in order to raise opinion against him within and without the university. He continues:

> I have been silent about earlier stinking squabbles (*Stänkereien*) caused by Herr König, though I had the possibility of nailing down Herr König for the false and doubtlessly maliciously broadcast statements about me and to bring him to answer for them, just as I know that Herr König is definitely snooping around in my personal affairs and attempts in this fine way to assemble some sort of "material" against me. That in these machinations Herr König always hangs around him the cloak of true caring for the good of the university is even more pitiful since he plainly is too cowardly openly to show his hatred, which I have known about for a long time.
>
> I have faced these plots up until now with a deliberate calm, although I must acknowledge that the short and long of it is that some of the filth thrown around must stick to me. However, after Herr König has now tried during my absence to develop

[75] Transmission of the document was handled by Otto Meissner, who was state secretary under Ebert, von Hindenburg, and Hitler. He was acquitted in the "Wilhelmstrasse Trial" and later court proceedings, in part because of aid given secretly to opponents of the Nazi regime. He died in 1953.

[76] Presumably the mathematician Robert König, who had been a full professor at Jena since 1927. For Robert König, see also below, chapter 5.

[77] Max Weinreich, *Hitler's Professors* (New York: YIVO, 1946), 30 and 262. Astel committed suicide in early May 1945 (see Benno Müller-Hill, *Tödliche Wissenschaft* [1984], 82). For Astel's view of the "National Socialist university," see Poliakov and Wulf 1959: 279–280. For his position as "Sportarzt" in Munich, see documents in the Munich University archive.

> even stronger activity against me, I can no longer look on this poisonous potpourri without doing something.
>
> I ask you to undertake a thoroughgoing investigation of the words spread around against me by Herr König and other male washerwomen and to shut the foul mouths (*Mäuler zustopfen*) in question. In particular I ask you to investigate the role played by my director, F. K. Schmidt. Professor F. K. Schmidt had more than enough opportunity to detoxify and settle my bad relationship with Herr König, which originally was based only on Herr König's personal antipathy to me. Instead, Herr Schmidt has, at every opportunity which presented itself, urged both sides on: in any case, for my part I can attribute a part of my knowledge of Herr König's plots only to communications launched by Schmidt. I add to this that Herr Schmidt has lifted his mask a little bit for me, as he, in saying farewell, preached at me, culminating in the statement that my attitude and behavior in no way corresponded to National Socialist fundamentals. It is very difficult for me not to write a satire about this!
>
> I regret troubling you with such filth and having to ask you, during the time I need to be a soldier, to protect me from shots in the back.

The fact that this letter was written on June 3, but only used as a complaint against Weinel on October 1, may perhaps be partially explained by the intervening summer, a lack of meeting of the academic senate, need for investigation, or the like—nevertheless, it seems strange that Astel should wait four months before complaining to Berlin. As Rektor, and so *Führer* of the university, he had no need to depend officially on any other academic body. Perhaps a conflict between internal academic pressures and Astel's own active Naziism also played a role. In any case, citing Weinel's letter, Astel asked that Collatz be appointed at Jena, and suggested that Weinel go to Prague. He rejected Weinel's accusations against König and Schmidt, but "even were they justified, such a letter is an impossibility." It was clear to Astel that Weinel's appointment would not provide the necessary good relations between applied mathematics and neighboring disciplines, above all mathematics. Astel, however, never even complained to the Thuringian ministry, which, as late as September 3, knew of no facts that would speak against Weinel's being named "associate professor."

F. K. Schmidt, about whom Weinel complains, had been called to Jena from Göttingen[78] in 1934 to be Ordinarius and director of the mathematical institute. A practicing Roman Catholic, Schmidt insisted on maintaining contact with his Jewish colleagues. When Richard Courant was forced from the editorship of the famous "Yellow Series" of mathematical monographs in 1933, he was succeeded by Schmidt, who kept Courant's name on the title page, despite pressure to remove it. Even after 1938, on his own responsibility, Schmidt published Jewish authors. His continuing conflict with the Nazi authorities in Jena led to his gradually being sidelined as institute director, and even to his removal from the university's examination committee. By late 1941, the situation was such that

[78] According to a history of mathematics at Jena in the Jena University archive, Schmidt at Göttingen had been "active as a substitute for a full professor." For information about F. K. Schmidt at Göttingen, see below, under "Hasse's Appointment at Göttingen."

Schmidt gave up his professorship at Jena and, with the help of the applied mathematician Richard Grammel, found a position doing research on glider flight at Ainringen,[79] where he remained until war's end. Known for his politeness, as well as his ironic humor, it is not unlikely that he addressed Weinel exactly as reported.[80] The distinguished Danish mathematician Harald Bohr (Niels Bohr's brother), an early opponent of attempts to "Nazify" mathematics, characterized Schmidt as a "friend of Hasse and an extremely bold and honest man."[81]

If Astel took time complaining about Weinel, Berlin took time answering. Not until January 23, 1942, six weeks after Hitler's signature was affixed to the official appointment and nearly four months after Astel's letter, was an official reply received from Berlin. Written for Rust to Astel, it points out that on June 6, 1940, Weinel had been recommended by Astel, and that he had been officially in charge of the operational running of the institute since January 1, 1940:

> Aside from the fact that in the earlier papers there is no report that speaks against calling Dr. Weinel, . . . the *Führer* has signed the certificate naming Dr. Weinel. If there are not really serious objections against Dr. Weinel, therefore, I ask you to put aside your second thoughts. In the case of such objections, your report should be to me at the latest February 10.

On February 6, Dekan Scheffer in Jena, formerly Weinel's strong supporter, forwarded a faculty report to Astel, who on February 7 sent it to Berlin "with his full agreement." Here, in the situation, all possible stops against Weinel were played, given the earlier positive reports.

First, there had been several considerations against Weinel in the original committee. The faculty thought at that time to put aside these considerations since Weinel as instructor and scientist was judged qualified, and the appointment would smooth his professional path. These considerations had to do with whether Weinel was suitable to undertake the leadership of an institute in which several disciplines were bound closely together and there was interaction with the Zeiss factory. (To these remarks, someone in the Berlin ministry appended the note that no such objections appeared in the June 3, 1940, faculty report.)

Furthermore, said the faculty, Winkelmann's growing psychic depression,

[79] Eberhard Hopf (see chapter 3) also retreated into such work. E. Hopf to Richard Courant, June 23, 1945, in Courant Papers (NYU), under Hopf.

[80] For the above material, see E. Kunz and H.-J. Nastold, "In Memoriam Friedrich Karl Schmidt," *JDMV* 83 (1981): 169–181, as well as letters of E. Lehnartz (Dec. 1, 1947), Ferdinand Springer (Dec. 11, 1947), and Richard Grammel (Jan. 10, 1948) in the archive of the University of Münster (under F. K. Schmidt) describing aspects of Schmidt's career and protesting the injustice initially done him in the "denazification process" following World War II. Whether Weinel's complaints had anything to do with Schmidt's departure from Jena shortly thereafter is unknown to me.

[81] VP, Bohr folder, Bohr to Veblen, Aug. 11, 1934. For Bohr's role in the conflict within the German Mathematical Society that same year, see below, chapter 6, "The Bieberbach-Bohr Exchange and the 1934 Meeting of the DMV."

which led eventually to his retirement, had had an influence on his *Assistent* (Weinel), so that he felt constricted, and instead of love for the institute, felt an aversion to it—a different atmosphere was definitely in Weinel's own interest.

In addition, if it was generally understood that mathematics and applied mathematics should work together harmoniously, so it was essential in Jena that applied mathematics form a bridge between physics and pure mathematics. Similarly, interest in the work at Zeiss, and readiness for possible activity there, were required. Weinel lacked interest in such reciprocal activity with academic physics and industrial applications. The faculty had thought that Weinel was suffering in part from the unfortunate atmosphere toward the end of Winkelmann's tenure, and suppressed its second thoughts. They had the expectation that, once Weinel was on the faculty list of acceptable candidates, this would provide him with solid evidence of the trust the faculty had in him, and so he would devote himself zealously to the development of the institute and such concerns as its interaction with Zeiss.

However, since 1938, when Weinel had been entrusted with the care of the institute, he had not done so, despite ample opportunity and funds. He had not even attempted the slightest accommodation with physics, technical physics, or optics. On the contrary, ill-feeling and estrangement had grown, and so it was also not to be expected that his receiving the position would change his attitude.

Therefore the faculty could no longer hold to its recommendations of June 1940 and would not be responsible for the consequences if Weinel took over the applied mathematics position. He needed to be elsewhere with new people and new relationships. Furthermore, if the *Führer* had already signed the appointment form for Weinel (as Rust had indicated on January 23, 1942), so the faculty wished to point out that Astel had already written the ministry on the preceding first of October, four months earlier, asking that Weinel not be appointed.

Six weeks later, on March 20, the ministry answered. It found no serious objections to Weinel's taking the chair of applied mathematics in the Dekan's letter. All grounds now brought forward must have been known to the faculty in 1938, and yet Weinel had been warmly endorsed on June 5, 1940. Furthermore, said the ministry, if the faculty had complaints about collegiality, it was their job to alleviate them through their own efforts. It could help in the common work with Zeiss and give Weinel an opportunity he would certainly not throw away. Finally, to prevent Weinel's appointment now that he was a soldier on the Eastern front was completely impossible. The signed appointment form would be sent to Weinel, thus completing his appointment. And so it was.[82]

What are we to make of this tale? In a rapidly rearming and later at war

[82] Officially, Weinel's appointment was effective March 1, 1942. According to Winfried Scharlau et al. (1990), Weinel continued in a professorship at Jena until 1952, and he died in 1979. But this source is clearly wrong in maintaining that Winkelmann, whose psychological problems began this whole story, held his chair until his death in 1946.

Germany, the professorship of applied mathematics at a university with strong industrial connections would seem to be important—indeed, was acknowledged as such by bureaucrats in the education ministry. Yet the appointment was prey first to the violent currents of contemporary international politics, then to the manipulations of Weyrich, attempting to use these currents for his own benefit, then to those involving Weinel. As an old Nazi supporter and a wounded war veteran, Weyrich naturally expected to get the best deal he could from the educational bureaucracy; indeed, apparently he had approached Dames about his discomfort in Brünn even prior to knowing the potential availability of the position in Jena. At the same time, he realized that he needed to attempt to manipulate the rapidly changing political situation in order to maximize his potential benefit. Unfortunately, he miscalculated, and the last-gasp attempt of his patrons in Czechoslovakia to save him was already an admission of failure. As to Weinel, if the academic post were so important, the bizarrerie of its delay during wartime by Weinel's Alsatian birth, and hence immediate lack of the necessary documentation, is scarcely believable. Indeed, the Thuringian ministry later went around the red tape. However, friction between Weinel and some of the faculty had apparently been growing. The wife of Professor K. H. Weise recalled that Weinel quarreled with almost everyone at Jena.[83] In any case, he never had been really the sort of applied mathematician the faculty desired. Some of them (perhaps F. K. Schmidt) may even have seen him as something of a careerist Nazi *parvenu*.[84] In any case, while the education ministry saw Weinel's military service as making his appointment inevitable, the faculty may have seen it as an opportunity to be rid of him. Weinel's attempt to defame Schmidt and König as not true enough to the Nazi cause could thus be turned into a demonstration of highly uncollegial behavior.

What this story clearly shows, also, is the unimportance of the position in question. The faculty may have truly wanted Lothar Collatz (whom they repeatedly suggested, and who certainly filled the desiderata expressed by Dekan Scheffer). Instead, a great deal of time was spent negotiating with the much more minor figures of Weyrich and Weinel. Neither Weyrich nor Weinel was what the faculty said (in abstract) they wanted, and indeed, part of the difficulties that arose with both was their too-great interest in non-applied mathematics. Yet both were offered the appointment. In Weyrich's case, this was presumably out of a mixture of mathematical adequacy, obvious high political reliability, and the personal interest of the physicist Abraham Esau (apparently, as Rektor, going somewhat counter to the mathematicians). In Weinel's case, the offer was presumably a combination of availability *faute de mieux*, and an evidenced political reliability. Weinel's apparent aggressive support for Nazi ideas (as demonstrated by his remarks about F. K. Schmidt) may have been partly occasioned by his need to find a political substitute for his missing papers

[83] Dr. Hans-Heinrich Weise to the author, letter dated Jan. 27, 1992.

[84] As will be seen below, Schmidt already had had unfortunate experiences with two such individuals at Göttingen, namely, Werner Weber and (at the time) Ernst Mohr.

in order to obtain the position, but it may also have been sincere. One could be a sincere Nazi, and also (or especially) not unwilling to take political advantage of one's "less progressive" academic elders. This very activity may have frightened more moderate faculty, who simply wanted to get on with being faculty and avoid political intrusions. On the other hand, such agitation was no doubt calculated to fall in with the political beliefs of Astel. The ministry eventually became exasperated with Weyrich and, for different reasons, the faculty with Weinel. The ministry, however, never felt it sufficiently important to impose a candidate for the job. Despite Dames's desire not to "coerce" a faculty, he plainly could have done so—as, for example, in the infamous imposition of the nonphysicist Wilhelm Müller into the theoretical physics chair abdicated by Arnold Sommerfeld at Munich.[85] Sommerfeld, one of the leading theoretical physicists of the early twentieth century, was replaced by Müller, an applied mathematician whose expertise was classical mechanics. The Munich appointment, however, had great *symbolic* importance, signifying the coming of Nazi doctrine to pure physics; the Jena appointment had *no* symbolic importance, only a practical one. As will be seen in the next section concerning the issues surrounding the leadership of the Göttingen mathematical institute, such symbolic issues were presumably quite important. But absent these, once the regime's basic demand—elimination of all Jews and clear political undesirables from academic posts—was satisfied, no one seems to have cared much about academic mathematics and the role it played, including even the use of applied mathematics for military purposes in wartime.

Hasse's Appointment at Göttingen

Before examining further the consequences for mathematics of this official indifference at high levels coupled with a rabid adherence to Nazi beliefs and doctrine among some (but a minority of) practicing mathematicians, it will be instructive to look at one more example of mathematical decision-making under the Nazis. The Süss-Doetsch quarrel and the succession to Winkelmann at Jena seem to show academic life at its worst (made even more dismal by the continual involvement of political considerations, where political feeling is masked by issues of academic "turf" and collegiality). Nevertheless, these incidents do not reveal the fabled (and real) dynamism of the Nazi movement as it affected mathematics. However, that is displayed in a third incident, involving Helmut Hasse's appointment to Göttingen, which, unlike the other two cases, has had

[85] See Alan Beyerchen, *Scientists under Hitler* (1977), 165–166. A comprehensive biography of Müller has recently been published by Freddy Litten: *Mechanik und Antisemitismus, Wilhelm Müller, 1880–1968* (2000).

more than brief mention in the literature before. Also, unlike the previous two instances, there were no military or industrial considerations.[86]

On April 7, 1933, the famous Nazi "Law for the Reorganization of the Civil Service" was promulgated. German academics were civil servants, and under this law, Jews (except those who had been appointed prior to 1914, or had served in the front lines during World War I), political unreliables "who did not at all times unreservedly promote the national state," and those whose positions were judged supernumerary were to be dismissed.[87] On April 25, 1933, the education ministry sent a telegram[88] to the Göttingen curator relieving six faculty members of their duties, placing them on leave but with full pay. Three of these were mathematicians: Richard Courant, Emmy Noether, and Felix Bernstein. None of them was directly dismissed, as none of them fell at the time under the April 7 law: Bernstein had been a civil-service appointee since 1911,[89] Courant had been a front-line soldier in the war, and Emmy Noether was a *nichtbeamtete* faculty member (i.e., without civil-service status). On May 6, however, such personnel were also included. On April 28, 1933, the Dekan of the Faculty of Mathematics and Science (the engineer Max Reich), apparently on his own initiative, advised Edmund Landau to "temporarily put off his lectures"; by agreement, these were given in summer semester 1933 after a week's delay, by his long-time *Assistent* Werner Weber.[90] At the same time, Reich advised the mathematicians Paul Bernays, Paul Hertz, Kurt Hohenemser, Hans Lewy, and Otto Neugebauer that "until the final decision as to their rights, they should not make use of their right to teach (*venia legendi*)." These were all *Privatdozenten* at the time. Bernays, Hertz, Lewy, and Hohenemser were Jewish; Neugebauer was (falsely) suspected of being a communist. The unexpectedness of the law of April 7 is shown by the fact that Hohenemser had taken the civil-service oath in order to be a qualified teacher at the Institute for Applied Mechanics in Göttingen just six days before.[91] The warning to Neugebauer again indicates Nazi bureaucratic confusion. Shortly before Reich's advisory of April 28, the university Kurator, Justus Valentiner, had asked Neugebauer to become the oper-

[86] For previously published discussions of aspects of this complicated incident, see Reid 1976; S. L. Segal, "Helmut Hasse in 1934," *Historia Mathematica* 7 (1980): 46–56; Schappacher n.d. All material about the incident not otherwise footnoted can be found in the published version of Schappacher's article, Becker et al. 1987: 344–373.

[87] For example, see the "Einleitung" by Hans-Joachim Dahms, in Becker et al. 1987: 26.

[88] A copy is reproduced in ibid., between pp. 28 and 29.

[89] Felix Bernstein was appointed to the Göttingen faculty in 1907 and since 1911 had a regular *Extraordinariat*. In 1919, he was denied promotion to full professor, but finally in 1921 was given a "personal full professorship" *against* the will of the faculty (only Courant, Hilbert, and Runge spoke for him). See Schappacher n.d., and M. Frewer, "Das Wissenschaftliche Werk Felix Bernsteins," thesis, Göttingen, 1979. A published paper drawn from this last appeared in *JDMV* 83 (1985): 84–94.

[90] See Schappacher n.d., and Kluge 1983: 94.

[91] H.-J. Dahms, "Einleitung," in Becker et al. 1987: 25; also, see an article by Cordula Tollmien on the Kaiser-Wilhelm-Institut für Strömungsforschung in ibid.: 472.

ational head of the Göttingen Mathematical Institute,[92] since Courant was on leave. On April 29, Neugebauer wrote to Valentiner, resigning his new office as institute director because of the letter from Reich, as well as to Reich, in both cases with biting sarcasm.[93] Hermann Weyl was named acting institute director. However, he lasted only until October 9.[94]

But it is not the oft-described dismissals at Göttingen that are of concern here, though there will be occasion to treat aspects thereof at a later point, but rather what happened afterward. As prelude occurred the infamous boycott of Landau's classes. Landau had taken Reich's advice in April: on May 5, 1933, he wrote Weyl (addressing him as the "acting operational director of the Mathematical Institute" instead of by name) that his lectures would begin May 8 (the semester had already begun late on May 1), but "in case I do not learn soon enough that the situation is clarified, I will ask our colleague Herr Weber as my *Assistent* . . . to begin in my place."[95]

Indeed, Weber gave Landau's summer-semester lectures. On May 9, Landau had been told by the Dekan that "clarification of his legal situation" could not occur before a few weeks had passed, and nothing happened during the summer semester. The new catalog of lectures for the winter semester contained Landau's announcement of what he intended to lecture on. No word from either the Kurator or the Dekan had come to him. What happened next is best described in Landau's own sober words.[96]

> I had therefore to assume as self-evident that from the point of view of those in charge (*von zuständiger Seite*), even if actually a little care was still necessary, all precautions had been taken in order for me to take up again my usual lecturing as was my duty.
>
> On Nov. 2, about 11:15, as I wished to leave my office and go to the large auditorium to begin my lecture, the entry hall was filled with about 80 to 100 students who let me through unhindered. In the lecture hall was *one* person. Plainly therefore a boycott in that sentries at the doors had prevented (without force) the students willing to work from setting foot in the lecture hall.
>
> What happened—and it happened with the collaboration of many who should be my pupils—leads me to the opinion that the sole consequence is my application for emeritization or pensioning.

[92] MI, Kurator to Neugebauer, Apr. 25, 1933. Neugebauer in fact generally substituted for Courant as operational director in the latter's absence: e.g., MI, Kurator to Courant, Mar. 6, 1933. Reich's letter must have been received on April 28 or 29, to judge from Neugebauer to Valentiner on Saturday, April 29 (where April 27 is mistakenly indicated for April 25).

[93] MI, Neugebauer to Valentiner, Apr. 29, 1933; Neugebauer to Reich, Apr. 29, 1933.

[94] On that date, Weyl wrote the Göttingen Kurator asking for his release as of January 1, 1934. He wrote from Zürich and did not return to Göttingen. See his letter in Archiv der Universität Göttingen (hereafter UAG) 3206b.

[95] Handwritten note in MI.

[96] UAG 3206b, Landau to Valentiner, Nov. 5, 1933. The whole letter is reproduced in Kluge 1983: 95a–95c, and the cited portion also published in Schappacher 1987.

Landau goes on to ask for this change in status and, at the same time, leave, and adds a postscript:[97]

> I only send this letter today because soon after this occurrence a student appeared in my office (by chance Professor Prandtl and Dr. Heesch were present at the time) in order to substantiate this happening, which according to him was the work of no organized group. I advised him to give me this substantiation in writing by Nov. 4, and in a form suitable for transmission to the Kurator and the ministry (which he did), so that I had a foundation which was based not only on memory for the preparation of my further steps. I enclose (without comment, however, with the deletion of his signature) a copy of his letter,[98] which naturally only strengthens my wish [for retirement].

According to Heinrich Kleinsorge, at that time the head of the organization of mathematics students, the boycott was actually put together through student channels by those who were members of the SA (*Sturmabteilung*, the original "stormtroopers," significantly less important after June 30, 1934, though still an avenue for demonstrating Nazi sympathies).[99] Kleinsorge himself was in hospital on the day of the boycott,[100] and the student who appeared in Landau's office was his deputy at the time, the brilliant young mathematician Oswald Teichmüller, who was also a convinced Nazi.[101] The Kurator replied in agreement with Landau's request on November 19; Landau received leave to work at Groningen (in the Netherlands) that was later extended until the end of the winter semester. On February 7, he was retired according to paragraph 6 of the April 7 law (which provided for the elimination of supernumerary professors). Landau received his regular pay until July 1, 1934, and thereafter the appropriate retirement pay until his death of a heart attack in 1938—this sort of Nazi legalism was characteristic not only of Landau's, but of other similar cases as well at this time.[102]

Since Weyl (who had been in Switzerland since mid-August) had resigned on October 9, with the enforced retirement of Landau, Gustav Herglotz was the only regular unretired professor left. Göttingen had been the leading light of German mathematics; now there was danger that it might become a mathematical backwater.[103] Since July 29, 1933, Weyl had been negotiating with F. K.

[97] Ibid.

[98] A copy of this letter has recently been discovered in the *Nachlass* of Erich Kamke and published by N. Schappacher and E. Scholz (with assistance of others): "Oswald Teichmüller—Leben und Werk," *JDMV* 94 (1992): 28–30. For a partial translation, see chapter 7.

[99] Ibid.: 5, citing an interview with Kleinsorge on Mar. 2, 1985.

[100] According to the manuscript version of Schappacher and Scholz 1992.

[101] Schappacher and Scholz 1992: n. 14. However, Constance Reid (1976: 156), apparently following a suggestion of Kurt Friedrichs, while agreeing that Teichmüller was the student in Landau's office, states that Werner Weber was the "commander" of the boycott. This seems dubious.

[102] In March 1939, the decision was taken that if Landau's widow decided to emigrate to the United States, then she would no longer be paid the appropriate pension money. See UAG 3206b letter of Mar. 11, 1939, from the education ministry to the Göttingen Kurator.

[103] The mathematics faculty, from *Assistenten* to *Ordinarien*, lost eleven members in 1933–34; cf. also MI, Weyl to Artin, June 9, 1933.

Schmidt that he might visit Göttingen and give algebra lectures in place of Emmy Noether (and perhaps others).[104] After Weyl's resignation, his former *Assistent*, Franz Rellich, became institute director,[105] and renewed the invitation.[106] Schmidt came. Rellich also managed to obtain the appointment of Helmut Ulm and Arnold Schmidt to *Assistent* positions.[107] However, Rellich had only a low-level appointment and, like Ulm and Arnold Schmidt, was not yet an established figure, while F. K. Schmidt had only a visiting appointment.[108] There was need for a prominent mathematical figure who was suitable politically to take over leadership in Göttingen. Furthermore, in mid-December, Rellich was ordered to report on January 7 for ten weeks to a field-sports camp near Berlin. This was, in fact, a mistake, since Rellich, as an Austrian citizen, was not subject to such forced training regimens. When he arrived at the camp, he was not admitted on these grounds.[109] However, on December 27, the Kurator had, after some hesitation, replaced Rellich with Werner Weber as acting director of the mathematical institute.[110] Rellich himself would lose his position at Göttingen six months later, on June 18.[111]

On February 13, 1934, the Dekan asked Weber, as acting director of the institute, for recommendations as to who should replace Weyl, and implicitly become the new operational director.[112] Three days later, Weber recommended Helmut Hasse, then at Marburg, with Udo Wegner in second place.

Werner Weber (1906–75) was at the time twenty-eight years old. He had studied with Emmy Noether, but as she was not authorized to supervise dissertations on her own, his dissertation had been approved for the Ph.D. in 1929 with Landau as "co-approver." It was marked "excellent." Weber had been Landau's *Assistent* since December 1, 1928, in which capacity he often corrected proofs of Landau's work. Landau was a stickler for absolute accuracy in such matters, and on occasion, when the printed work did not meet his specifications, he would send in a list of typographical corrections to be printed, including such matters as misplaced commas.[113] Despite the evaluation of his thesis,[114] Weber's abilities as a mathematician were apparently modest, and his editorial

[104] MI, Weyl to Schmidt, July 29, 1933, and Aug. 9, 1933; Weyl to Kurator, Aug. 9, 1933.

[105] Herglotz actually served in the post for two weeks, but he had neither interest in nor gifts for administration.

[106] MI, Rellich to Schmidt, Sept. 15, 1933.

[107] Both would shortly lose their appointments. For Ulm, see chapter 8.

[108] For F. K. Schmidt's personality, see "The Winkelmann Succession" above. He was certainly "politically unreliable," to use Nazi jargon.

[109] Rellich, curriculum vitae prepared Dec. 6, 1932, in UAG 3206b; MI, Rellich to Kurator, Dec. 13, 1933, Valentiner to Weber(?), Mar. 17, 1934, Tornier to Valentiner, Mar. 20, 1934.

[110] Apparently Weber did not have quite all the prerogatives of Rellich, but what he did not have is unclear. See Schappacher 1987: 353, and Schappacher n.d., citing UAG.

[111] MI, Tornier to Rellich, June 18, 1934.

[112] See Weber memoir, note 120 below, Anlage 2.

[113] E.g., Edmund Landau, *Collected Works* (1984), vol. 7, 454–456, which are the typographical corrections to a paper published in Spanish.

[114] An examination of "dissertation grades" would seem to show Emmy Noether as generous, and Landau as a somewhat harder judge.

work for Landau led to the widespread joke that his greatest ability was "being able to tell the dot on an italic *i* from the dot on a standard *i*."[115] The story that Weber disliked Landau because the latter had presented a paper of his to the Göttingen Academy only to follow it immediately with an improvement[116] seems to be false. In 1929, Landau did present for Weber a paper drawn from Weber's dissertation that was entirely in the spirit of Emmy Noether ("Remarks on the Arithmetic Theory of Binary Quadratic Forms"), but neither the publications of the Göttingen Academy[117] nor the comprehensive list of Landau's published works show any such follow-up note. Indeed, at the time Landau was working on quite different subjects.[118] Whatever his feelings about Landau, Weber apparently had also been a Nazi sympathizer and member of the SA for some time, though he did not join the NSDAP until May 1, 1933.[119] In fact, he was an example of the young dynamic convinced National Socialist who saw Hitler's rise as restoring German national prestige and honor, and who would complain about other party members in whom the pure flame did not burn sufficiently brightly. This is certainly the role he played in the turbulent events that followed, ending with Hasse's final assumption of the operational directorship of the institute in July 1934.

On January 18, 1940, Weber handed in to the office of Rudolf Mentzel in the education ministry a 227-page memoir concerning these events, together with appendices comprising copies of seventy documents for another 166 pages.[120] A copy was also sent to Theodor Vahlen, who was a mathematician and who had been a predecessor of Mentzel as ministerial director for science in the Nazi education ministry. Though in January 1940 the war was going very well indeed for Germany, Mentzel was astonished to receive such a massive report on events of six years previous that in no way related to the war effort. Furthermore, the scientific division of the ministry had had half its personnel called up, and he told Weber he should have waited until after the war to present such a memoir, since to do so now "ran the danger of being seen in Germany's most difficult time"[121] as a complainer. In a handwritten reply, Weber said the memoir was intended only for a later time as presenting examples of problems that might arise in the future, as indeed he had mentioned in his original cover

[115] This story was told to me independently by several people, including the late Prof. Hans Zassenhaus (Ohio State University), who was Weber's contemporary.

[116] Schappacher n.d.

[117] The paper appeared in the *Göttingen Nachrichten* for 1929, pp. 116–130.

[118] Landau 1984: vol. 9.

[119] E.g., BDC under Weber, Schering to Landt, Dec. 13, 1937. Weber's party number was 3,118,177. Party memberships were backdated to specific dates, so this does not mean "on May 1" but sometime thereafter.

[120] Now in the BAK. This will be referred to in subsequent notes as Weber.

[121] Weber: Mentzel to Weber, Jan. 23, 1940. Mentzel was at the time also president of the national German research foundation (DFG). For more on Mentzel, a typical Nazi careerist, who also appears in "The Winkelmann Succession" above, see Kurt Zierold, *Forschungsförderung in drei Epochen* (1968).

letter. Weber saw the Göttingen problems as a conflict between party and state.[122] Mentzel (who had been addressed in the cover letter as an old "Göttinger"), in his answer, withdrew the accusation of "complainer" and considered the matter closed temporarily. It would seem that one of Weber's motivations for filing these documents in 1940 was a fear that he might not survive his army service, and a wish that his side of the story be known. It is unclear when Weber wrote this *apologia pro vita sua*, but the many appended documents would speak to his having planned it in 1934, before he was forced to leave Göttingen. If it was written much later, the fact that temporal events are described in it sometimes to the quarter-hour is remarkable; if it was written around 1934, it seems strange to wait over five years to deposit it with the ministry. In any case, while a major source for the events to be related, it needs to be used with care, not just because of its author's evident bias.

A word is in order here also about Helmut Hasse (1898–1979).[123] Hasse was born in Kassel; his mother had been born in Milwaukee, but was raised in Kassel from the age of five. In 1913, his family moved to Berlin, and in 1915, at age seventeen, he took the "emergency leaving examination" (*Notabitur*) at his Berlin gymnasium in order to join the Imperial Navy. Stationed in the Baltic (where he also did cryptographic work), he studied the Dirichlet-Dedekind lectures in number theory, the famous book that was an inspiration to many algebraists and number theorists of his generation. During his last year in the navy, he took advantage of a stationing in Kiel to hear mathematics lectures by Otto Toeplitz. Demobbed in December 1918, he immediately went to Göttingen to study mathematics. There he fell under the influence of Erich Hecke. However, in 1919, Hecke left Göttingen to become one of the first *Ordinarien* at the newly founded University of Hamburg. Hasse, in the meantime, had come across Kurt Hensel's number theory book in a Göttingen bookstore, part of which discussed Hensel's new theory of "p-adic numbers." Fascinated, Hasse left Göttingen to study in Marburg with Hensel. Within a little over a year, he had obtained his doctorate, and within another his *Habilitation*, a remarkable record for only three years of formal study. Both dissertations involved the famous "local-global principle," discovered by and still associated with Hasse.

Hasse's first teaching position was at Kiel, where Toeplitz was still Ordinarius. Politically, Hasse was a very conservative and ardently patriotic man, and he became a member of the DNVP (*Deutschnationale Volkspartei*),[124] a right-wing somewhat aristocratic party originally led by Kuno Graf von Westarp, but after 1928 by the industrialist and publishing magnate Alfred Hugenberg, who would enter Hitler's first cabinet. Anti-Semitism was an increasing feature of the

[122] Weber, cover letter to Mentzel.

[123] The following material on Hasse draws on Harold Edwards, "Helmut Hasse," in *Dictionary of Scientific Biography* (1970); Günther Frei, "Helmut Hasse (1898–1979)," *Expositiones Mathematicae* 3 (1985): 55–69; a conversation with Prof. Manfred Knebusch (Regensburg, Feb. 25, 1988), who had been a postwar student in Hamburg; and conversations with Jutta Kneser and Prof. Martin Kneser, Hasse's daughter and son-in-law (Göttingen, Feb. 18, 1988).

[124] For the DNVP, see chapter 3, note 56.

DNVP over the years, and in 1929 the party was closed to Jews. Nevertheless, the party continued to have Jewish and "half-Jewish" adherents until 1933.[125] Despite his DNVP membership, Hasse was no anti-Semite, and, for example, remained friendly with Hensel until his death in 1941, although Hensel was certainly a Jew by Nazi standards—a thoroughly assimilated and baptized one. Though Hensel had been emerited in 1930, nevertheless the Nazis removed from him the *venia legendi* (right to teach) that emeriti retained, because of his Jewish grandmother.[126] Also, in 1933, after the announcement of the first dismissals from Göttingen, Hasse had organized from Marburg a collection of fourteen letters from colleagues about Emmy Noether's scientific importance requesting that she at least be enabled to have a modest existence with a small circle of advanced students. Toeplitz, Hasse's instructor at Kiel, was also Jewish. Nevertheless, Hasse's somewhat rigid principles, patriotism, and nationalism—Manfred Knebusch describes Hasse in the postwar years as having "all the strengths and weaknesses of a Prussian officer"[127]—caused him to view Hitler as a national hero and to apply for membership in the Nazi party in 1937.[128] This rigidity seems to have been maintained to the end of his life. Arno Cronheim, who also studied with Hasse in 1948–50, describes him as a man with whom it was at that time impossible to talk politics, and later on, in the 1960s, when Hasse visited Ohio State University where Cronheim was teaching, as someone with the same inflexible political ideas: for example, the Oder-Neisse line needed revision, and slavery in America had been a good institution for blacks.[129]

From Kiel, Hasse went to Halle as Ordinarius in 1925 at age twenty-seven, and then became Hensel's successor in Marburg. At the time of his going to Halle, Hilbert had already said, "There is no doubt that Hasse now ranks among the younger mathematicians who are in the forefront of those eligible for a professorship of mathematics."[130] Small wonder, then, that not only Weber, but also Hermann Weyl, saw Hasse, a very distinguished mathematician whose political sympathies were strongly right-wing, as the suitable future operational director of mathematics at Göttingen and as the man who might rebuild it after all the dismissals.[131]

However, Weber's recommendation of Hasse was far from wholehearted. It begins:[132] "For the new holder of Weyl's chair I recommend in first place Prof. Helmut Hasse in Marburg. I have decided on this, although real second thoughts of a political sort are against him." Weber goes on to say that Weyl's

[125] Mosse 1966: 243–244.

[126] For Hensel, see Pinl 1972: 199; and Fraenkel 1967: passim.

[127] Manfred Knebusch, interview (Feb. 25, 1988).

[128] Cf. above, chapter 3, and Segal 1980. Hasse's application is treated in detail below.

[129] In a letter to the author, July 23, 1984. In this letter Cronheim (who was an Auschwitz survivor) emphasizes Hasse's brilliance as a mathematical pedagogue. Cronheim was also a member of a group of students who petitioned the Soviet authorities to let Hasse lecture in Berlin.

[130] Frei 1985: 58.

[131] Schappacher n.d.; Reid 1976: 160. Cf. Weber, Anlage 54.

[132] Weber, Anlage 3.

chair[133] was the most important one in Germany, so ought to be filled by a world-renowned mathematician. Of the few who came in question, "all" except Hasse fail to meet "formal requirements" (such as being German citizens, or having clear racial credentials). However, Weber realized that the real lack (from his point of view) of not having an enthusiastic National Socialist in the job would not be filled by Hasse. He calls this lack a "tormenting mosquito" in the constitution of the faculty.

However, says Weber, should Hasse be turned down for, or himself turn down, the Göttingen job, then only one person came into question: Udo Wegner. Wegner was a valuable scientist (though no Hasse), a fine teacher ("almost unique"), and someone whose political activity for National Socialism deserved the highest praise and extended to far before Hitler's accession to power. He also had the complete trust of the students. For any professorship other than Weyl's, Weber would recommend Wegner "in *first* place" (emphasis in original).

In fact, already on January 6, that is, a month earlier, Weber had recommended to the Kurator that Wegner be called to Göttingen, calling it an "urgent" (*akut*) question and reporting on Wegner's work. However, Wegner's name had certainly been bruited about even earlier. In a letter to Ludwig Bieberbach on July 18, 1933, Hermann Weyl discussed "student agitation" for Wegner as Courant's replacement—a development whose prospect he called "ruinous for Göttingen in my opinion"; in a longer letter to the Dekan on August 10, a similar sentiment appears.[134]

Who was this Udo Wegner, who, as has been seen, would later be a supporter of Doetsch in his controversy with Süss? Born in 1902, Wegner took his doctorate at Berlin in 1928 with a dissertation in algebraic number theory supervised by Issai Schur, with Bieberbach as second referee. In 1929, he completed his *Habilitation* at Göttingen, where he became Courant's assistant, leaving there for a full professorship at the technical university in Darmstadt, which was his position at this time. In 1937, he went to Heidelberg to replace the expelled Jewish professor, Arthur Rosenthal, where he became Dekan (for mathematical and natural sciences) and would remain until 1945. According to his chemist colleague, Karl Freudenberg:[135]

> His appointment [to Rosenthal's position] went through without the knowledge of a portion, and indeed a non-National Socialist portion, of the faculty. . . . In a normal appointment process, as was customary prior to National Socialism, Wegner would certainly not have been appointed at Heidelberg.

At Heidelberg, Wegner would attempt in a comprehensive manner to improve the status and professoriat of the areas that were his responsibility, partly arguing that institutions as rich in tradition as Heidelberg (it was the oldest

[133] The operational directorship was not tied to a particular chair; cf. Weber, p. 3.

[134] Ibid.: Anlagen 53 and 54.

[135] As cited by Birgit Vezina, *Die Gleichschaltung der Universität Heidelberg* (1982), 155.

German university) should not lag behind others like Frankfurt and Tübingen. However, he also behaved in the expected Nazi fashion.[136]

> He pursued the Habilitation of the mathematician Dr. Hans Joachim Fischer, a keen National Socialist, and indeed in opposition to the recommendation of his mathematical colleague Seifert [who had succeeded to the chair of the other mathematician forcibly retired as a Jew, Heinrich Liebmann]. Wegner declared that a decree existed according to which inadequate academic performance could be compensated for by political service. . . . In the faculty he built his position of power ever more recklessly according to the "*Führerprinzip*." The operation of the *Führerprinzip* went so far that he, as it later appeared, misused the authority of his office in that he undertook . . . the granting of doctorates which in every way went against the proper procedures for doctorates and were kept secret from the faculty.

He also attempted to create an institute for aeronautical research at Heidelberg, with efforts apparently beginning in early 1943. By July 13, 1944, this was still not achieved. On August 19, 1944, however, Wegner sent a secret report to the Baden education ministry outlining the working groups for such an institute, which would involve mathematicians, chemists, biologists, physicists, metallurgists, economists, and philologists. The institute was (despite this late phase of the war) to be primarily concerned with civilian air travel. Nothing seems to have come of it, though once more the failed efforts at such an institute seem to reflect both bureaucratic inertia and the "business as usual" attitude of bureaucrats, even as the war wound to its disastrous close for Germany, something that the payments to Süss in 1944 and 1945 discussed earlier also reflect. It also may reflect the simple disregard even for possibly war-relevant "pure research" (airplane improvement) until very late in the war. There seems to have been no conflict with Prandtl's Göttingen institute (perhaps the reason why the Heidelberg project was to be for "aeronautical research" rather than aerodynamical research).[137] It may be in this context that Doetsch and Wegner became friends.

At the end of the war, Wegner not only lost his position in the denazification proceedings, but also was arrested and charged with criminal acts. It would appear that after settlement of these charges, Wegner went to France for a while. In 1949, he wrote Courant two ingratiating letters ("I allowed myself to ask this question, because I know that you, dear Herr Courant, always have good advice, for you have helped me in all life situations in the most touching way"). These had to do with obtaining funds for the expansion of a school whose aim was "not only to impart information, but also to stimulate creative thinking and to educate cosmopolitan, well-reared, self-secure, and freely thinking people . . . [to aim at] the overcoming of nationalistic narrowness through education to European thinking." These words were written in 1949 by a man who fifteen years previously had been esteemed as having served the Nazi cause long before January 30, 1933, and were addressed to a forced emigré! Courant

[136] Ibid. For more about H. J. Fischer, see below in chapter 6.

[137] See documents in BDC files, under Wegner. There was a concern about opposing U.S. or British postwar domination of air travel.

replied (in English) that he could not help, adding acidly: "I was interested in learning that you had left your position in France. As a matter of fact, I heard some rumors here that you were considering going even farther away." In 1957, an aeronautical institute in Brazil that had many professional recommendations as to Wegner's ability noted his former position as Courant's assistant on his curriculum vitae and wrote for a recommendation. Courant gave Wegner a high professional rating for the job envisioned, and then added the comment:

> Undoubtedly you know that he belonged to the over-naive and maybe slightly opportunistic people who joined the Nazis. As a consequence my contacts with him have been discontinued. But I feel that he has amply suffered for his mistake and that his talents and other qualities should be used as much as possible.

Wegner did not have another academic position in Germany until 1956, briefly at Saarbrücken, after which he also taught off and on at various universities until he retired in 1972 and returned to Heidelberg. Repeated attempts by this author to interview him in spring 1988 were in vain. He died in 1989.[138]

However naive or opportunistic Wegner may have been, in 1934, at age thirty-two, he stood before a tremendous opportunity—a convinced National Socialist and competent mathematician, he might soon earn his political reward by being appointed to one of the most significant mathematical positions in Germany. Already on May 24, 1933, shortly after Courant was placed on leave and before the boycott of Landau's classes, Weber and the leader of the National Socialist student group met with Wegner about the posibility of his succession to Courant's chair. Wegner, however, demurred, saying that Hasse was better suited to a Göttingen chair than he, and in any case, he was unsuited to being operational director of the Göttingen institute.[139] It was also clear to Weber that other chairs would become vacant at Göttingen and that Hasse was not only a world-famous mathematician, but also "from a political point of view was not a conscious exponent of the opposition [to National Socialism],"[140] an opposition that Weber thought was true of all other mathematicians of Hasse's caliber.[141] Thus, Hasse was acceptable, and if he came, he would soon (Weber thought) have a thoroughly politically reliable figure beside him as another of the *Ordinarien*, one who would eventually take over the directorship. Also recommending Hasse was animosity between him and Courant—and, who knew, "[Hasse] might eventually become a better National Socialist than many who presently acted radically."[142] The faculty recommended Hasse or Blaschke on

[138] The material on Wegner's career is drawn from Biermann 1988: 361; Vezina 1982: 155; letter from E. Hecke to E. Wolff, Nov. 25, 1945, in Personalakten Blaschke, Universität Hamburg, BAK R21 10261; and letters from Wegner to Courant, Nov. 28, 1949, Courant to Wegner, Dec. 8, 1949, and Courant to S. S. Steinder, Feb. 18, 1957, in Courant correspondence (Wegner file) at Courant Institute archive in New York.

[139] Weber, pp. 1–2.

[140] Ibid.: 4.

[141] Ibid.: Anlage 22, particularly pp. 283–284.

[142] Ibid.: 5. Cf. below for Hasse's "development" in this direction in Weber's eyes.

February 20, 1934.[143] Sometime in March, apparently, Hasse visited Wegner and said that he did not intend to take up the Göttingen appointment unless, simultaneously, someone else was also named who could master the currents in the student body.[144]

Hasse was appointed, and on April 13, he met Weber in Göttingen in order to find out what the situation of the institute was so that he could appropriately negotiate his acceptance with the ministry in Berlin. For example, he was worried that he might have trouble insisting on the replacement of Landau's chair.[145] Indeed, not only did Hasse not want to lose any positions, but he wished to use the occasion of his appointment to increase the number of positions.[146] The ministry agreed to let Hasse fill not only Landau's chair but also Felix Bernstein's, and consequently he was inclined to accept the appointment. With this in mind, he visited Göttingen again on April 23 to speak with Weber and Rellich. While in Berlin, he apparently also suggested Wegner for one of the empty chairs, for already on May 3, Wegner heard that he was on such a list of candidates.[147]

As a result of this second meeting, Weber began to have second thoughts about Hasse's allegiance to National Socialist principles and discipline. The immediate source of these was Hasse's desire to appoint assistants in the department without the involvement of the Rektor and his insistence on having a woman for secretary instead of some needy *alter Kämpfer* or wounded war veteran. Weber saw the former desire as "in reality a step against any political control of the assistants by the *State* [emphasis in original]." Hasse did not have "the necessary discipline with respect to official regulations that, without reservation, were a part of a true '*Gleichschaltung*,'" and "in the future it would be necessary to watch Hasse closely."[148] While these are Weber's after-the-fact recollections, by April 27, he was writing Udo Wegner that Hasse was a "reactionary of the worst sort."[149] This four-page letter reveals not only Weber's attitudes and rhetoric as a dynamic believer in the "new Germany," and the temptations offered his close friend and fellow-believer Wegner, but also the position of the Göttingen mathematics students, as the following extracts show.

> If this man [Hasse] comes to Göttingen he will seek to institute a system that reverses *everything* thus far achieved by the National Socialist state and without exception replace it with what existed in red Prussia. . . . Hasse is plainly of the opinion that now, after the National Socialist revolution has finished raging, the fighters are tired and will gladly allow him a free hand for the reestablishment of all the arrangements cast

[143] Ibid.: 19.

[144] Ibid.: 18. Weber took this as indicating Hasse's reservation about the National Socialist revolution at Göttingen, especially with regard to the Landau boycott.

[145] Landau's forced retirement (above) was technically on grounds of his being supernumerary.

[146] Weber, pp. 19–20.

[147] Ibid.: 24–25.

[148] Ibid.: 28–29. *Gleichschalten* was Nazi jargon for being harmonized to the Nazi revolution.

[149] Ibid.: Anlage 14 (Weber to Wegner, Apr. 27, 1934).

away as scrap iron. Only a completely blind person or a first-class enemy of the state can speculate on this happening.

It would be a double misfortune if this man received the leadership of the institute. . . . In case Hasse comes to Göttingen, the struggle against him will be carried out especially by the students in a large way, under the single precondition that we can have a hope of confirming the correctness of our action through positive achievements in a National Socialist sense. The pivot of this hope is you. Only if we can throw you on the scale as a counterweight to Hasse, does our action have a purpose. . . . Are you in a position, say in a month, to set up a mathematics program carried by the spirit of the new times we live in, which, should you be nominated to a Göttingen professorship, you intend to carry out . . . on account of lack of a leader you remain our only hope [to carry out such a program]. Your answer therefore will reveal whether now you can already realize something that without you would be an issue for the distant future. If your answer is positive all our equipment is in place. . . . I will withdraw my previous viewpoint in the matter of Hasse. . . . Simultaneously I will inform Hasse and warn him in a way he cannot misunderstand not to accept the call to Göttingen. Should he, however, come here . . . a protest action of the German students will arise, *etc*. . . . [News from a camp in Kiel is] that single student organizations will, without exception, be supported in all battles against university professors by the general student body. If you answer no, then we will renounce any battle.

Wegner answered four days later, on May 1. Before following this story further, it is necessary to try to understand exactly Weber's complaint against Hasse. Superficially, it seems hardly credible, however dedicated and dynamic a Nazi Weber was, that merely Hasse's insistence on control of *Assistent* and secretarial positions could have led to his being viewed as a "reactionary of the worst sort" when, a scant two months earlier, Weber had been recommending Hasse as someone who was at least a Nazi fellow-traveler if not yet a true believer. Weber would put down precisely the sources of his change of heart toward Hasse.[150] They seem to have been the following:

1. When Hasse gave a lecture in Göttingen on December 12, 1933 (a lecture that Weber did not attend), he began it with recognizable barbs against the National Socialist idea of making all sciences, especially mathematics, subject to political considerations.

2. Weyl had appointed two politically inappropriate assistants on October 1 and November 1, 1933.[151] It was this which prompted the Rektor[152] to decree that henceforth he had to approve all nominations for *Assistent*. Hasse's desire to reverse this decree showed his lack of sympathy with National Socialist ideas.

[150] Ibid.: Anlage 22. These are *not* there set out in the form below, which I have adopted for clarity.

[151] Arnold Schmidt and Helmut Ulm. However, Weyl sent his letter of resignation on October 9, and, though it was effective January 1, did not return to Germany. As mentioned earlier, the appointments were actually obtained by Fritz Rellich.

[152] Friedrich Neumann, apparently a good friend of Erhard Tornier (for whom, see below).

3. Hasse's statement that were he to undertake a position in Göttingen, he would want a man beside him who could "master the currents in the student body," meant someone with the spine to defy the students' National Socialist sympathies.

4. As someone who was well known as a strong German nationalist, Hasse had been assumed to have been honestly assimilated (*gleichgeschaltet*) into the National Socialist revolution. One evidence that this was not true was that on the occasion of his talk in December 1933, the "friends of Jews" in the mathematics institute made propaganda for him, placarding a room "in a style reminiscent of the announcement of dance entertainments."

5. Hasse was the candidate of all those who secretly wanted Courant reinstalled in Göttingen.

6. While Hasse was certainly a famous mathematician (and so technically suitable for the leading mathematical position in Göttingen), in fact his disciplinary viewpoint could not be harmonized with the National Socialist revolution.

7. Göttingen was a special place as the leading mathematics institute in Germany. The right of the National Socialist revolution was to fill such important places as the directorship of the mathematics institute with its spirit—Hasse was not the man to do that. In fact, the only man to do that and also be important scientifically, in Weber's opinion, was Udo Wegner.

Actually, Hasse's view of some of these matters is known, since he corresponded with the British mathematician Harold Davenport about them. Davenport and Hasse had common mathematical interests and were brought together by the well-known British mathematician (and expatriate American) Louis Joel Mordell. One of the original motivations for their acquaintanceship, which blossomed into a real friendship, was Hasse's desire to improve his English, and, as will be seen, Davenport posted to him English novels and newspapers.[153]

Far from being an enemy of Courant, as Weber had earlier surmised, Hasse was (in accord with [5] above) actually quite friendly with him. Indeed, when negotiating with the ministry in April about acceptance of the Göttingen position, Hasse raised the "intricate question of Courant's position."[154] As to the issue of appointment of *Assistenten*, it appears that formally and formerly the mathematics institute was not under the university but was directly under the ministry, which appointed *Assistenten* on the nomination of the institute director. Now, however, the *Assistenten* were part of the *Dozentenschaft*, and so under the *Dozentenschaftsführer*, who was subordinate to the Rektor (as *Führer* of the university). The Rektor thus demanded a role in their appointment. Hasse fully realized the political implications of the situation, no less than Weber supposed. As he wrote Davenport, were he to subscribe to this demand,

[153] Hasse's letters to Davenport are available in the Davenport folder in the archives at Trinity College, Cambridge. This material is cited hereafter as HDT. Davenport's letters to Hasse are (presumably) under seal with Hasse's papers at the Niedersächsische Staats und Universitäts Bibliothek. Hasse wrote to Davenport sometimes in English, sometimes in German. Letters cited that were originally in German are indicated below as translated.

[154] HDT, Hasse to Davenport, Apr. 24, 1934. Cf. ibid., Hasse to Davenport, May 2, 1934, Nov. 1, 1933, and Jan. 2, 1934.

> the appointment of assistants would be controlled by the *Dozentenschaft*, i.e., by a body whose main task is political. You will understand that I am not willing to submit myself to this regulation . . . I have allowed no mistake about my opinion about this point in Berlin.[155]

As to Hasse's December 12, 1933 lecture, *his* description of his nonmathematical comments is interesting in showing what Weber (at second-hand) would think inimical to the National Socialist revolution in mathematics. According to Hasse, the Göttingen Rektor (the Germanist Friedrich Neumann) thought the climax of science had occurred in the eighteenth century, and that it had been declining ever since. Hasse contradicted this and

> also said that now that so many renowned scholars had had to leave Germany, a failure in the development of our German scientific research would heavily support all those people who hold that National Socialism has impaired culture and civilisation and brought about barbarism. Finally I said that from my point of view there is no contradiction between being a good nationalist and being a good scholar.

His mathematics lecture followed.[156]

As is clear, Hasse did not believe in the politicization of mathematics (or any academic activity), but as will be seen below, Weber was wrong to think him only a tepid supporter of Hitler at this time. What Hasse did not understand was that National Socialism meant the politicization of all life according to its *Weltanschauung*. This effectively was his irremediable and unredeemable flaw in Weber's eyes.[157]

Weber's turning to Wegner as a counter-candidate to Hasse had not been his first attempt to prevent Hasse's assumption of the directorship at Göttingen. He had originally tried to approach the issue via the National Socialist teachers' organization at the university, whose treasurer he was. However, he was apparently not on very good terms with the leader of this organization, Hermann Vogel, and when he mentioned at an evening meeting his worries about Hasse being an arch-reactionary, Vogel broke into a fury, saying that he had expressly asked Weber earlier whether Hasse was acceptable and had been told yes. Weber's protests that his meeting just that afternoon with Hasse had changed his mind were to no avail, and he felt himself accused of having knowledgeably and carelessly brought an enemy of the state into a leading position in the university.[158] Weber's reaction is interesting. He might have carried on in what he calls "a bourgeois fashion" by taking his argument with Vogel to higher authority, and explaining his changed attitude toward Hasse, thereby defending himself from attack. But, says he: "As an SA-man, I could indeed not think this way . . . my honor was in danger." He had learned to take responsibility, and so

[155] HDT, Hasse to Davenport, May 2, 1934.
[156] HDT, Hasse to Davenport, Dec. 19, 1933.
[157] See also Segal 1980.
[158] Weber, p. 32.

resigned his post as treasurer.[159] He believed he had given a blank check to a professor insufficiently checked out beforehand, and so had "plunged German mathematics into the abyss."[160] But Vogel did not want to accept Weber's resignation; indeed, he seemed to have forgotten their quarrel completely. However, a few days later, he did remember it, and seemed as well to become part of the Hasse support, unwilling to do the slightest thing against him. According to Weber, Vogel behaved this way because Hasse was "convenient to him." Since the National Socialist teachers' group Vogel led had in fact evaluated Hasse positively initially (on the grounds, presumably, of Weber's earlier recommendation), Vogel could support Hasse easily if that were convenient, and also distance himself from Hasse, blaming Weber for his appointment, if *that* were convenient. The convolutions of Weber's opinion of Vogel's behavior toward him and his motivations should not cause further delay. However, the details suppressed here are an example of the sort of infighting that went on between shades of Nazi, each striving to prove oneself the true participant in standard-bearing for the "national will."[161]

In any case, Weber felt that his honor was threatened, and that if Hasse were "victorious," it would be stained forever, at least if he did not somehow make his changed attitude felt. And, says Weber (writing for his superiors in the education ministry):[162] "For the German whose honor is in danger, ordinary laws no longer apply. He partakes in the greatest fate that a German can have . . . to be personally responsible only to the *Führer*." Thus Weber resolved to fight the Hasse appointment on his own. If one examines the items in his bill against Hasse as described above, it is clear that, astonishing as it may seem, Weber's opposition really was occasioned primarily by Hasse's insistence on controlling *Assistent* appointments, with this apparently lending credence to an already festering suspicion.

And so, Weber wrote Wegner on April 27 the letter cited above. The suggestion to do so was Oswald Teichmüller's.[163] Were the plan requested of Wegner formulated, then on its basis the students would be assembled to support Wegner rather than Hasse. However, Wegner was something of a reluctant dragon. While ready to formulate the requested "mathematical-political" program, he did not want Weber to do anything precipitous against Hasse, but rather to seek political counterweights to him in other mathematics professors to be appointed at Göttingen.[164] Wegner paid a visit on May 3 to the astronomer Heinrich Vogt in Heidelberg, supposedly influential in university matters. He learned that a professorship for him at Göttingen was indeed under consideration, and received assurance that Weber's post as *Assistent* was not in danger.

[159] Ibid.: 33–39, Anlagen 12 and 13.
[160] Ibid.: 35.
[161] Ibid.: 39.
[162] Ibid.: 39.
[163] Ibid.: 43. For Oswald Teichmüller, see below and chapter 8.
[164] Weber, Anlage 17.

Indeed, should Hasse take over the institute, Weber was likely to become his plenipotentiary.[165]

Weber, spurred on by Teichmüller, decided to visit Wegner for further discussion, and did so on May 8 when they both traveled to see Vogt. While the latter did not think there was any possibility of preventing Hasse's appointment, he promised to do his best to alleviate Hasse's potentially pernicious influence through helping with a counterweight appointment. Wegner's reluctance to step forward is again revealed by his making Weber promise not to mention him as such a counterweight.[166] Why should Wegner, apparently an ardent National Socialist, be so chary of reaping the fruits of the revolution, when like-minded colleagues were not? Indeed, as already noted, Wegner had also earlier demurred in favor of Hasse. Part of the explanation may be that Wegner had no qualms about profiting from the revolution, but, as an "old Göttinger," thought it impolitic to appear as though he were pushing himself forward. In addition, Wegner had recently had a somewhat calamitous mathematical difference with Hasse and another famous algebraist, B. L. van der Waerden. In 1931, while still Courant's *Assistent* at Göttingen, Wegner published a theorem, which later that same year Hasse showed to be incorrect (citing an example noted privately some years earlier by Emil Artin). Hasse also proved a more general correct theorem. Wegner admitted his error, suggesting an additional hypothesis that would still leave his theorem different from Hasse's. However, in August 1933, B. L. van der Waerden demonstrated by a simple example that this, too, was incorrect. Wegner slightly generalized van der Waerden's counterexample and promised a further investigation. This, however, never seems to have appeared.[167]

In any case, on this occasion, Wegner and Weber had a ruminative stroll by the Neckar in which Wegner explained his view of the difference between Jewish and Aryan mathematicians. This is of some import, because of the "Deutsche Mathematik" movement that was devoted to such differences.[168] Wegner seemed to lay this difference to the mastery by Aryans of applied branches of mathematics (in its earlier unjudaized conceptions),[169] and a total rejection of Landau's

[165] Ibid.: 48 and Anlage 18. Vogt had been a member of the NSDAP as of January 1, 1931, as well as an SA leader. He had come to Heidelberg from Jena at the explicit recommendation of his teacher Phillip Lenard because he was an "old National Socialist." See Vezina 1982: 85 n. 303, and pp. 146–147. Also Beyerchen 1977: 142 and 250 n. 4; and Poliakov and Wulf 1959: 295–296, where Lenard recommends Vogt as a science referee for the *Völkische Beobachter*, the official Nazi party newspaper.

[166] Weber, p. 49.

[167] See *Mathematische Annalen* 105 (1931): 628–631; 106 (1932): 455–456; 109 (1934): 679–680. Weber was suggesting Wegner for Göttingen at the same time that Wegner was corresponding with van der Waerden on this matter. I wish to thank Stephen Weintraub, who first pointed out these papers to me.

[168] Weber, pp. 50–51. For "Deutsche Mathematik," see below, chapter 7.

[169] Wegner said this despite (because of?) having been Courant's *Assistent*. Courant was an applied mathematician. Similar ideas appear in the book *Deutsche Physik* (1936) by the Nazi party member and Nobel Prize–winner Phillip Lenard. For Lenard, see Beyerchen 1977.

"methods." Though Wegner never did write up his "program,"[170] nevertheless Weber pursued the matter with Vogt, naming Wegner now openly as his choice. Vogt, after some delay, apparently attempted to influence the ministry against Hasse and, presumably, for Wegner; however, this came to naught.

This academic and more or less official route, however, was not the only direction in which Weber attempted to forestall Hasse. On his own, he decided to involve the SS. In the Nazi bureaucracy, if an initial approach somewhere had failed, there were always other competing agencies to which one could turn. Whether this was a purposeful device or an accidental one may be debated—what is certain is that it led to much infighting and jockeying for position among Nazi officialdom. Some sense of this has already been presented; here it will be seen in full flower.

Johann Nikuradse[171] was an applied mathematician of Georgian (Soviet Union) background, but apparently a German citizen whose activities on behalf of the Nazi party dated back to 1923. He was working in Prandtl's aerodynamical institute when seven of his colleagues there accused him of espionage and the theft of books. The accusation was delivered into the hands of an erstwhile Captain Johann Weniger, who was the leader of the SS information service in Göttingen at the time—the pretext was that Weniger had already checked out the political status of the aerodynamical institute's members. Nikuradse seems to have been actively disliked by his colleagues, in part because he tried to use his good job with Prandtl to denigrate them. However, Weniger was also carrying on an intrigue against Prandtl, whose loyalty he suspected. Consequently, he showed Nikuradse the accusation, and Nikuradse then demanded a university judicial hearing. As a result of this hearing, held under SS pressure, all seven of Nikuradse's accusers were fired. Prandtl (who had innocently supported Nikuradse at the hearing) protested that his institute was being destroyed by the mass expulsion, and some of the seven were reinstated. However, internal dissension in the institute did not stop, and two months later, Prandtl felt himself compelled to fire Nikuradse for "pressing official reasons." He thus sealed Weniger's enmity toward him.

Accidentally, it seems, Weber made Nikuradse's acquaintance following his dismissal by Prandtl, and through Nikuradse came to know Weniger. Having been told that Weniger had direct contact with Theodor Vahlen, the mathematician who ran the scientific division of the education ministry, Weber decided to lay the Hasse matter before him.[172] Indeed, it seems that Weber met Nikuradse on the morning of May 11, and the same afternoon was in conversation with Weniger. Over the next little while, it appeared that Weniger believed there was a brilliantly organized band of conspirators in the Göttingen mathematics department who had obtained Hasse as a coworker; and that Hasse had traveled

[170] But see *Deutsche Mathematik* 2 (1937): 6–10, for an adumbration of what it was.

[171] Material on Nikuradse is taken from Cordula Tollmien, "Das Kaiser Wilhelm Institut für Strömungsforschung," in Becker et al. 1987: 464–488, p. 474.

[172] Weber, pp. 52–54.

to England to have secret conversations with Courant and plan his return. Because of these beliefs, Weniger wished to turn the mathematics institute upside-down, and even wanted to bring down the Rektor and the leader of the Nazi teachers' organization. He openly declared that Göttingen would one day thank *him* as its savior. Such sentiments were too much even for Weber, who says, "This megalomania can probably be excused by the head wound he received in the war."[173] Nevertheless, Hasse would later have to answer the absurd charge that he had been in England. Actually, Hasse had apparently planned a trip to England in spring 1934, a trip that he said "will be greatly favored by all my Nazi colleagues here. They consider it almost as a national duty that every German who has friends in England should go there and make 'Kulturpropaganda' for Germany."[174] Presumably the Göttingen negotiations prevented this trip, but Weniger somehow knew of Hasse's plans. Hasse did make a trip to England in early 1935,[175] and also apparently again (with official approval) to lecture at Manchester in March 1936.[176]

Indeed, though Weber had doubts about Weniger's mental stability, he was glad to know someone whom he believed had Vahlen's ear. Consequently, he prepared the long report for him on the succession to the Göttingen directorship that has been cited earlier.[177] In addition to attacking Hasse, this ten-page report was replete with praise of Udo Wegner as the only suitable person for the directorship, and was accompanied by a six-page scientific *vita* detailing Wegner's mathematical qualifications.[178] He also presented Weniger with a fifteen-page report on "The Mathematical Institute at Göttingen since the Revolution."[179] This, in addition to further complaints against Hasse and praise of Wegner as the only person suited to the Göttingen position on both mathematical and political grounds, also contains a description of the reactions to Courant's "forced leave," and of Weyl's attempt to preserve what he could of traditional Göttingen mathematics under Nazi pressure and the decimating dismissals. A particularly striking part of Weber's pejorative view of the traditional Göttingen mathematics institute is his view of the Landau boycott and retirement. Landau, after all, had been one of the signatories on his dissertation, and he had been Landau's *Assistent* for several years. Yet he called the action against Landau and its aftermath "the first really decisive success of the revolution [within the mathematics institute]."[180] Similarly, he praised the driving out

[173] Ibid.: 54–55.

[174] HDT, Hasse to Davenport, Oct. 28, 1933.

[175] HDT, Hasse to Davenport, Apr. 9, 1935.

[176] HDT, Hasse to Davenport, Dec. 22, 1935 and Feb. 3, 1936.

[177] Weber, Anlage 22.

[178] Strikingly, for someone of Weber's beliefs, referees suggested by him included Hans Rademacher, who, unbeknownst to Weber, was of left-liberal persuasion and had emigrated in 1933, and Otto Haupt, whose wife was "half-Jewish." (For Haupt, see *JDMV* 92 [1990]: 169–181, pp. 179–180; for Rademacher, see Max Pinl and Lux Furtmüller, "Mathematicians under Hitler," *Leo Baeck Yearbook* 18 [1973]: 130–181, p. 151.)

[179] Weber, Anlage 25.

[180] Ibid.: 309.

of Emmy Noether (who had been his actual dissertation supervisor). The seemingly trivial matter of female secretaries was also raised here.

Given the positions to fill in Göttingen, especially the institute directorship, which was arguably the most important mathematical position in the country, the political demands of the Nazi regime could be met in two ways. One might seek the very best mathematicians who were politically acceptable to fill the empty positions. An alternative, however, was to find the most dynamic true believer in the National Socialist cause who was mathematically acceptable, with the idea that if the most important post in German mathematics were filled by such a person, then mathematics in Germany would become more quickly *gleichgeschaltet*. Weber seems to have oscillated between these two antipodal positions, leaning, however, toward the second, and finally adopting it.

Weniger forwarded the Weber materials to Berlin; however, Vahlen replied that there was already a commitment to Hasse.[181] Nevertheless, he apparently left the impression with Weniger of appointments for *both* Wegner and Hasse, perhaps even with the former taking over the institute directorship. This may have all been in Weniger's imagination, but in any case, on May 18, he traveled to Berlin for a tête-a-tête with Vahlen about Hasse and Wegner.[182] Weniger seems to have achieved at least a delay in Hasse's formal appointment, but was much less anxious to facilitate Wegner's. Indeed, he suggested to Weber that perhaps Wilhelm Blaschke or Paul Koebe would be more suitable than Wegner.[183] He solicited further information about Blaschke from Weber (who wrote Wegner for his opinion of Blaschke), but decided he would promote Wegner after all, provided he were willing to fall in with Weniger's plans to purge the mathematics institute still further. Wegner declared himself willing to fall in with these plans; also, he wished to try to help Weber's career.[184]

At this point, Weber thought all was as well as could be expected: Vahlen had promised to reexamine Hasse's appointment, despite the ministry's preliminary offer to him in April; Wegner was a possible appointment, if not instead of, then in addition to Hasse. Hasse, believing that only the usual bureaucratic complications were responsible for the delay of his official appointment, visited Göttingen on May 25, failed to meet with Weber, arranged for temporary living quarters, and was astonished to find that Weber had not yet posted his course of lectures for the next term. Returning the same day to Marburg, he wrote on May 26 to Weber to say that he had had a telephone call from Berlin that

[181] From ibid.: Anlage 53; much earlier, Weyl wrote Bieberbach for a referee's report on Wegner, and one can infer from Weber, Anlage 54, that the report was negative and that Weyl had forwarded it to the appropriate authorities. To what extent Bieberbach's pro-Nazi sympathies were known at the time of Weyl's letter (see below, chapter 7) is unclear. In any case, Bieberbach was already clearly a good nationalist—someone who would count with the ministerial authorities beyond being a Berlin professor—and Weyl obviously anticipated a negative report.

[182] Weber, p. 58.

[183] Ibid.: 61. Both Blaschke and Koebe were far more distinguished as mathematicians than Wegner, and the faculty had already considered Blaschke as suitable.

[184] Ibid.: Anlage 27.

morning saying that his appointment as institute director was signed and on its way to the Göttingen Kurator, that he insisted that an announcement of his lectures to begin on May 31 be posted, and that he would arrive on May 29 to take over and wished to have a meeting on May 30 with all students and *Assistenten.*[185]

On May 25, Weber had promised Weniger to write Wegner about his willingness to go along with the plans for further purging of the institute. On May 26, Berlin had appointed Hasse, whose appointment, only a week earlier, had supposedly been subject to further investigation. Two things apparently had happened. Probably, Vahlen, though an *alter Kämpfer,*[186] did not take the complaints against Hasse as seriously as Weniger imagined he did. More importantly, Vahlen apparently left his deputy in the Prussian cultural ministry, G. A. Achelis, in full charge for two days. It would appear that Achelis was not as radical a Nazi as Vahlen,[187] and hastened to appoint Hasse—whether Vahlen actually approved such a step prior to his temporary departure is unclear. The document sent from Berlin to Hasse appointed him "subject to further decision." Weber thought, naturally, that Achelis had made the appointment on his own to create a fait accompli, and that Valentiner was in league with him to appoint Hasse. He claimed that Achelis's responsibility was much later confirmed.[188]

In any case, Weber was full of anger, and convinced in a tragic sense of his own responsibility for Hasse's coming to Göttingen because of his original recommendation. He betook himself to Weniger with his copy of Hasse's appointment notice (which he received on the same Saturday, May 26, that Hasse was writing to him about the confirmation of his appointment and announcing his arrival). Weber was resolved not to relinquish the operational control of the institute to Hasse, whatever it cost him.

At this point, another character enters the drama, one who would here play something of the role of first buffoon, and then scapegoat—though later, for unconnected reasons, he would be in a concentration camp under a death sentence. Ernst Mohr was an impecunious mathematics student who lived two doors from Weniger.[189] At F. K. Schmidt's request, Mohr had been appointed by Weber as Schmidt's paper-grader. On the same May 25, Mohr's girlfriend went to Schmidt and asked whether there was a chance Mohr might get a regular (and higher-paying) post as an *Assistent* in the institute. Somehow or other, she

[185] Ibid.: Anlage 28.

[186] Theodor Vahlen joined the Nazi party in late 1923, apparently in admiration of Hitler's attempted *putsch* of that year (*Deutsche Mathematik* 1 [1936]: 389). More accurately, since the NSDAP was declared illegal, Vahlen joined a substitute organization, the *Grossdeutsche Volkspartei*, and was named "Gauleiter for Pomerania"; with the relegalization of the NSDAP in 1925, he held the same office in it. He also served as an NSDAP member of the German parliament.

[187] Weber, p. 79. Cf. Ulf Rosenow, "Die Göttinger Physik unter dem Nationalsozialismus," in Becker et al. 1987: 374–409, p. 381, where Achelis speaks of retaining the famous Jewish physicist Max Born at Göttingen.

[188] Weber, pp. 69–80, 85–88. This is full of talk about hypothesized forgeries and nonexistent letters.

[189] Ibid.: 175–176.

and Mohr received the impression that Hasse, together with Schmidt, had long ago planned who the *Assistenten* in the institute would be, and that Weber and some other *Assistenten* would be let go.[190] Weniger considered Mohr as one of his "trusties" and, in Weber's view, attempted gradually to replace him with Mohr in his plans to "take over" Göttingen. Mohr would then end as Weniger's delegated satrap in Göttingen. In Weber's eyes, not only Weniger and Mohr, but also Hasse, were primarily concerned with political power in Göttingen (which seems an almost classic instance of projection).[191]

Weber met Weniger around 8 P.M. on May 26, and they spoke together with Mohr. Weniger was as startled as Weber at Hasse's sudden appointment and resolved to go to Berlin on Monday, May 28, to see Vahlen. Weber agreed to pay his travel expenses and sent him an express letter to that effect. Weniger also asked Weber to assemble individual student letters against Hasse's assumption of the directorship. Weber did not like this "demagogic action," but went along—at least until he received Hasse's letter the following morning (a Sunday).[192] While the student letters were in fact not sent, their content (dictated by Weniger) seems to have been betrayed by some students (fearing Hasse's possible successful vengeance?) to F. K. Schmidt.[193]

As to Mohr, at Weniger's direction, he maintained a constant agitation against Hasse, citing repeatedly his purported information from Schmidt. On May 28, Hasse sent Weber a letter saying he would take over the institute next day at 5:15 P.M.,[194] citing Sunday, May 27, as the official date of his appointment.[195]

Apparently on May 28, Weniger spoke with Vahlen. He sent Weber a letter that Hasse's professorial appointment at Göttingen was regarded by Vahlen as a fait accompli; furthermore, its consideration antedated his ministry. However, Vahlen promised Weniger that he would withdraw Hasse's appointment as institute director. Weniger would push for Wegner for this post as well as for a new personnel program, and, in this regard, he desired that a second student letter be sent, which he likewise dictated. This not only demanded that Wegner be institute director, but that Rellich, Ulm, Arnold Schmidt, and F. K. Schmidt be replaced, since as Hasse's "followers" they were "insufferable" (*ganz untragbar*). It began with a complaint that "the completely judaized mathematics institute at Göttingen had been waiting for over two semesters for National Socialist leadership."[196] To a flyer requesting these letters, Weber had no trouble signing Heinz Kleinsorge's name, since he "erroneously" believed that Kleinsorge (the

[190] Ibid.: 82, 91.

[191] Ibid.: 77.

[192] Ibid.: 84–85.

[193] Ibid.: 91, 98, 99.

[194] German academic time was always given as "c.t." (*cum tempore*) or "s.t." (*sine tempore*). "c.t." meant an actual meeting fifteen minutes thereafter; "s.t." meant "on the dot." Hasse announced his assumption of authority as at 5 P.M. c.t.

[195] Weber, Anlage 33. When following such correspondence, one can not help but remark on the great efficiency of the German postal system at the time.

[196] Ibid.: 93–94 and Anlage 34.

leader of the mathematics students' organization) would see it as an organizational affair.[197] Apparently, Mohr betrayed the flyer to F. K. Schmidt, or at least so Weber deduced. Kleinsorge complained about the use of his name, and Weber, who was far from excited by letter campaigns, handed all copies, except for the few already delivered, as well as the plates, over to him for burning. The "first letter" had not yet been sent and never was.[198]

Hasse must also have complained to the Dekan, because the latter went to see Weber to ask why Hasse's prospective lectures had not yet been posted. Weber answered, "I do not wish that these lectures should take place," and expatiated that Hasse was the worst sort of reactionary and in all good conscience he could not hand over the institute to him. The Dekan said that Weber should have informed him and other authorities, and they parted as Weber was on the way to his lecture. In the meantime, Kleinsorge and Heinz Wolff, the leader of the National Socialist student organization, agreed that *officially* neither of their groupings had any interest in Weber's business. Weber was, after all, acting counter to an official appointment by a Nazi ministry. Following his mathematics lecture, Weber, accompanied by Kleinsorge, Teichmüller, and Mohr, went to the Rektor's office and saw Hasse leaving. As Weber went in to see the Rektor and the Dekan, who were together, Kleinsorge and Teichmüller prepared to wait outside; however, Mohr pushed his way in, and the Rektor declined to expel him. Weber repeated his political objections to Hasse. Then Mohr began, rather histrionically, to try to persuade the Rektor that Hasse was completely impossible as a personality. According to Weber, to the Rektor's astonishment, Mohr shouted, "We don't want anyone who is only a scientist, we need someone with character!"[199]

When Mohr was done, it was suggested that Weber confront Hasse with his objections. While this went against the grain for Weber—"the activist doesn't desire to convince the opposition, he wishes to compel them"—he nevertheless accepted the suggestion. By now, only Hasse's departure would satisfy him, and no accommodation was possible. In Weber's eyes, Hasse had shown himself politically unreliable, and nothing would change that. Thus, when he and Mohr met Hasse together with the Dekan around five o'clock that afternoon, he explained that he was against Hasse's being named director, primarily because Hasse wished complete and sole control of the appointment of assistants in the institute, and secondarily because he interpreted Hasse's remark to Wegner about needing a strong man beside him as wishing to resist the National Socialist currents in the student body.[200]

Hasse explained that the second was a misunderstanding—he only wanted a scientifically proficient person as his aide in academic matters. As to the first, he had already recognized officially the right of the Rektor to control the place-

[197] Ibid.: 96 and Anlage 36.
[198] Ibid.: 100–102.
[199] Ibid.: 105.
[200] Ibid.: 106–109.

ment of *Assistenten*. Weber would not be placated; what mattered for him were Hasse's attitudes: he was and always would be a German Nationalist (i.e., a member of the DNVP,[201] and thus no true National Socialist). Hasse asked where Weber got the idea he was a member of the DNVP, and pointed out that the National Socialist government had appointed him. Weber replied he believed they were changing their mind thanks to an agent acting on his behalf, and pulled out Weniger's letter with its closing sentence, "In any case, V. promised me to withdraw the directorship of the institute from H."[202] Hasse was shocked, especially since he had had no inkling of Weber's attitude in April. Then Mohr broke into accusations against Hasse stemming from his understanding of what F. K. Schmidt had said. Hasse said he had planned to urge Udo Wegner's appointment to a chair at Göttingen as well. Clearly, he could not lecture in an institute led by Weber; however, he had been appointed director, and he asked Weber to turn over the directorship and the institute keys to him. Weber refused.[203] He and Mohr left; but despite Teichmüller's urging, they did not send the "first" anti-Hasse letter. Indeed, Kleinsorge (perhaps worried about the consequences when Hasse did take charge) opposed such a move.

That evening, Hasse telephoned the ministry. Weber says that Hasse attempted to do what "all non-National Socialists" prefer to do: when in political conflict with an "old National Socialist," to cast doubt on the latter's own National Socialist convictions.[204] In any case, Hasse had heard that Weber had made disparaging remarks about Bernhard Rust, the education minister, and this became one of the charges laid against Weber in the ensuing disciplinary proceedings. Hasse also withdrew his acceptance of the Göttingen position unless the ministry reconfirmed him in it. As the foregoing shows, Hasse clearly would accommodate himself to obtain the position, despite his firmly telling Harold Davenport in February, "It goes without saying I shall not take the position there [Göttingen] unless I can get absolute and full power to do what *I* [emphasis in original] like in every respect."[205] As to his opponents, Weber, supported by Teichmüller, decided that he would change his position only if so commanded by Adolf Hitler(!) or another "leading" National Socialist "whom he trusted" (like his stormgroup leader). Weber regretted that the actions of Weniger's man, Mohr, had muddied for officialdom the clarity of his own grounds for the action he took.

The ministry decided rapidly. Vahlen determined that Hasse should stay in the Göttingen position, but that a "strong Nazi" would also be given a chair at Göttingen. The disciplinary procedure against Weber began immediately. He had hearings on May 31 and June 4. Weber insisted that his action had been according to a "categorical imperative," and he had aimed only at success, not at correct academic etiquette. As to the charge about Rust, Weber distinguished

[201] See chapter 3, note 56, and the sketch of Hasse's biography above.

[202] Weber, pp. 109–110, 93.

[203] Ibid.: 112–113.

[204] Ibid.: 113; cf. Weinel *contra* Schmidt in "The Winkelmann Succession," this chapter.

[205] HDT, Hasse to Davenport, Feb. 17, 1934.

between mildly pejorative remarks about the ministry (which he had made) and about the minister (which he had not).[206] Hasse wrote on June 5 from Marburg saying he had no fixed plans for who should be *Assistenten* in Göttingen,[207] and on June 7, the Kurator apparently sent a report on Weber to Vahlen, who decided to quash any disciplinary action. Weber was suspended from his post as of December 31, but in fact this amounted to a one-month extension of his appointment, which had been due to run out on November 30.[208] Indeed, it was extended again until March 31, 1935.[209] An attempt, which failed, was made to find him a post in Greifswald.[210] Weber would have liked to have stayed in Göttingen; however, the ministry told him this was impossible, and so he spent a semester as a substitute for Carl Ludwig Siegel in Frankfurt,[211] and then went to Berlin at Ludwig Bieberbach's instigation.[212]

In fact, the day after he refused Hasse the keys to the institute, Weber gave them back to the Dekan. While Weber had conceived of his act as a "purely political" one based on his April 23 conversation with Hasse, in fact, for most of the other anti-Hasse agitators (except Teichmüller), Mohr's denunciations were also important. It now appeared that Mohr had in fact applied for an assistantship under F. K. Schmidt on May 25; thus his actions against Hasse could be easily interpreted as attempts at self-advancement. Weber, seeing an act he conceived of as purely political tainted by his association with Mohr, whose motives were not so "pure," returned the keys to the Dekan at Kleinsorge's suggestion. This happened less than two hours after Weber received official notice of the opening of disciplinary procedures against him; so he stressed that his return of the keys was motivated solely by his changed estimate of Mohr.[213]

Until Hasse and the "strong Nazi" professor to be put beside him arrived in Göttingen to take official charge, F. K. Schmidt was appointed as temporary institute director.[214] The two letters Weniger had ordered up (in vain) had fallen, as has been seen, into Schmidt's hands. The second (which had been destroyed except for a few copies) requested the dismissal of Schmidt, among others, and Weber had signed Kleinsorge's name to it.[215] As a consequence,

[206] Weber, pp. 142–150, Anlage 38.

[207] Schappacher n.d., citing UAG.

[208] UAG 32066, Vahlen to Kurator [Valentiner] (copy), July 10, 1934. Weber, Anlagen 58, 59, 63, and pp. 206–207.

[209] Ibid.: Anlagen 66 and 67.

[210] HK, Tornier to Kneser, Dec. 9, 1934; Kneser to Tornier, Dec. 10, 1934. Cf. Weber, p. 220.

[211] Carl Ludwig Siegel was one of the most distinguished mathematicians of his generation. Mathematicians (including Siegel, himself; see above, chapter 3) are usually incensed that someone as mathematically insignificant as Weber should have been Siegel's replacement. However, the appointment was only for the summer semester. More to the point, questions of academic value only rarely played a significant role in academic placement under the Nazis. The symbolic importance of the Göttingen position may have made Hasse a partial exception, but even here the issue of his political acceptability was paramount.

[212] Weber, pp. 221–222; for Weber and Bieberbach, see below, chapter 7.

[213] Weber, pp. 131–134.

[214] Ibid.: 122.

[215] Ibid.: 96, 100–102, and Anlage 36.

Schmidt achieved (through the Nazi student leader Heinz Wolff) Kleinsorge's resignation as leader of the mathematics students. Automatically his replacement was his deputy, Teichmüller.[216] Ironically, Kleinsorge seems to have acted as a moderating influence on Weber, while Teichmüller was just the opposite.

The "strong Nazi" who would come to Göttingen together with Hasse was not, however, Wegner. In fact, Vahlen apparently did not think Wegner's mathematical ability suitable to a chair in Göttingen.[217] The man chosen as a "strong Nazi" mathematically to come to Göttingen with Hasse was Erhard Tornier, one of "Wegner's worst enemies."[218]

Tornier, like Wegner and Weber, was an ardent Nazi who, as a mathematician, had ironically spent his formative years working primarily with Jews or "half-Jews." This was also true of other "dynamic Nazis" in this tale: in Wegner's case, Issai Schur and Courant; in Weber's, Emmy Noether and Landau; in Tornier's case, Richard von Mises and Willy Feller. These are not the only mathematicians who are examples of this curious phenomenon, which commands notice (and will surface repeatedly), but for which there seems no overarching explanation. Nor are the examples among Nazi academics only among mathematicians: one need only think, for example, about Martin Heidegger and Edmund Husserl in philosophy.

Erhard Tornier was eight years older than Wegner and six years older than Hasse, but, although he received his doctorate at Marburg in 1922, he did not become a member of the German Mathematical Society until 1930, when he was thirty-six. At that time, he was teaching in Kiel.[219] The brilliant young Willy Feller, who would become one of the most distinguished probabilists of his generation, had, after his Göttingen Ph.D., obtained a position in Kiel, and Tornier was working with him. Earlier, Tornier and Hasse had been students together in Marburg, and during Hasse's time in Halle, Tornier had habilitated there.[220] In 1931, Tornier asked the then-Ordinarius at Kiel, A. A. Fraenkel, an algebraist and logician who, incidentally, was an orthodox Jew and Zionist, for a position at Kiel "so he could be in [Fraenkel's] vicinity and might learn much [from him]."[221] Fraenkel, with Feller in mind, acquiesced. In 1932, Tornier

[216] Weber, pp. 125, 151–152.

[217] Ibid.: 138. As noted earlier, Wegner himself was unsure of his suitability.

[218] Ibid.: 121.

[219] Tornier was born in 1894, not, as sometimes reported, 1897. See membership list of the DMV in "Verzeichnis der Mitglieder," printed at the end of volume 40 (1931) of *JDMV*.

[220] In 1930, according to Scharlau et al. 1990: 142. However Scharlau is wrong about Tornier's going to Berlin; he went to Kiel in 1929 (cf. ibid.: 187). According to Schappacher n.d., Tornier and Hasse wrote a small joint paper. However, no such paper appears in the comprehensive list of either Hasse's or Tornier's papers in J. C. Poggendorf (*Biographisch-Literarisches Handwörterbuch der Exakten Naturwissenschaften*). However, Tornier and Feller *did* write a joint paper that appeared in 1932 in *Mathematische Annalen* 107: 165–187.

[221] Fraenkel 1967: 154. Fraenkel also knew Hasse slightly (ibid.: 153). The information he provides on Hasse's ancestry, however, is erroneous. Also, according to Scharlau et al. (1990: 142), Tornier had been Fraenkel's temporary replacement in Kiel while Fraenkel spent 1929–31 as a visiting professor in Jerusalem.

joined the Nazi party.[222] Tornier's own work in the foundations of probability was related to the offbeat (by later mathematical judgment) attempt by the famous applied mathematician Richard von Mises to establish axiomatic foundations for probability theory. Von Mises's work in this area is now disregarded, and Tornier's is forgotten by most mathematicians, the approach of the Russian mathematician A. A. Kolmogoroff having superseded theirs. However, Tornier's work seems to have been solid and well-thought-out.[223] With Hitler's accession to power, Tornier revealed his co-worker Feller as a non-Aryan and hoped to inherit Fraenkel's professorship.[224]

Thus the "strong Nazi" to be placed beside Hasse was an old acquaintance. By Sunday, June 3, Tornier was in Göttingen, and on the next day, an official announcement that he would lead the institute until Hasse's arrival appeared.[225] Tornier actually was sympathetic to Weber's "political action," and also worked for a while on an attempt to reverse Hasse's appointment. At the same time, he apparently now encouraged Hasse to give up Göttingen voluntarily.[226] As to Weber, while it was clear that he would have to leave Göttingen, Tornier tried to obtain another, even better, position for him.[227] He did this in the face of Hasse's insistence that suitable punishment for Weber's "lack of discipline" was a precondition for his assuming the directorship to which he had been named. Hasse also insisted that, as institute director, he had to be consulted about anything pertaining to its rebuilding.[228]

With Tornier in Göttingen and Hasse back in Marburg awaiting ministerial reconfirmation, the story was not quite over: like the coda in a Beethoven symphony, there would be yet one more iteration. For in this situation, despite Tornier's obvious distrust of Udo Wegner,[229] Weber decided to promote his friend again. However, all was in vain: Tornier wrote Wegner terming him suspect in fomenting the Göttingen trouble; Wegner met with Tornier to disabuse him, and traveled to Berlin to defend himself before Vahlen. Vahlen withheld approval or disapproval, wanted to hear no more about Wegner, and eventually named Theodor Kaluza as Courant's successor.[230] (Hasse had succeeded to Weyl's chair, and Tornier temporarily to Landau's.)

Throughout June, Hasse, on the one hand, considered who should be appointed to Göttingen, and on the other, resisted Tornier's efforts to become de facto operational director. Indeed, he suggested to Tornier, as though it were a threat, that any premature action on his part would further shake his already

[222] BDC, under Tornier. His party number was 1,114,305.

[223] See literature cited in note 86, especially Schappacher 1987: 366 n. 111, and Schappacher n.d.

[224] Fraenkel 1967: 155. Fraenkel left Germany, never to return, in February 1933. Fortunately, he had a job in Jerusalem waiting for him (ibid.: 195).

[225] Weber, p. 176 and Anlage 45.

[226] Ibid.: 173–174, 192, 180.

[227] According to Weber (ibid.: 198).

[228] MI, Tornier to Vahlen, June 25, 1934, and Tornier to Hasse, June 25, 1934.

[229] Weber, pp. 121, 174, 194–195.

[230] Ibid.: 193–200.

wavering decision to take over the Göttingen post, and he emphasized, apparently with Achelis's support, the appropriate role of the faculty in such decisions.[231] Since February, Hasse had also carried on a friendly correspondence with Courant, principally about Courant's status (future emeritization) and Rellich's position at Göttingen.[232] It is this correspondence, as well as the association with Harold Davenport, that may have led to the rumors of Hasse conferring with Courant in England.

Toward the end of June, Tornier was informed by the ministry that Hasse would arrive at the beginning of July.[233] However, on the thirtieth of June, according to a story later told by F. K. Schmidt, Tornier wrote Hasse an express letter telling him his life was in danger, and that he should come to Göttingen, where Tornier would protect him.[234] Whether Hasse actually did so is unclear. Saturday, June 30, 1934, was the infamous "Night of the Long Knives," or *Röhmputsch*, during which many supposed threats to the Nazi state from within or without the party were murdered.[235] On that very Saturday, Tornier posted a notice that[236]

> by order of the Ministerial director Prof. Dr. Vahlen, Professor Hasse will begin his work here at the beginning of July. Since all old difficulties have been removed, both Vahlen and I expect that no new artificial ones will be created.

Hasse arrived on July 2 and immediately posted a notice:[237]

> Today I have taken over the leadership of the institute together with Dr. Tornier.

On the sixth of July, Berlin officially named Tornier a director of the mathematical institute, an appointment forwarded the following Monday to him, which post he naturally accepted.[238] Until Weber found another place, Tornier attempted to smooth matters between him and Hasse.

What of Weniger and Mohr between Weber's refusal to turn over the keys to Hasse on May 29 and Hasse's final arrival to take over in Göttingen on July 2?

[231] Hasse to Herglotz, June 11, 1934; Hasse to Tornier, June 23, 1934; Hasse to Herglotz, June 23, 1934; in Herglotz correspondence in Niedersächsische Staats- und Universitätsbibliothek (Handschriftenabteilung). It would seem F. K. Schmidt was kept informed by Herglotz at Hasse's request.

[232] Courant to Hasse, Feb. 7, 1934; Hasse to Courant, Mar. 2, 1934, and Apr. 24, 1934; Courant to Hasse, Apr. 28, 1934; Hasse to Courant, late May 1934; Courant to Hasse, June 2, 1934, and June 21, 1934. In Courant papers retained by the Courant family.

[233] Weber, p. 205.

[234] Segal 1980, citing Harald Bohr to Oswald Veblen, Aug. 11, 1934, in VP, Bohr folder.

[235] Including Röhm and the SA leadership, and, among many others, General Kurt von Schleicher and his wife, Gregor Strasser, Gustav von Kahr, and Edgar Jung. For a study, see Hermann Mau, "The Second Revolution—June 30, 1934," in H. Holborn, ed., *Republic to Reich*, trans. Ralph Manheim (1973), 223–250.

[236] Copy in MI.

[237] Copy in MI.

[238] Copy in MI, Valentiner to Tornier, July 9, 1934. The version in Weber, p. 209, is apparently wrong.

Weniger was elderly,[239] and given to megalomaniacal and conspiratorial imaginings,[240] but, as an *alter Kämpfer*, installed as a local SS functionary. In his memoir, Weber repeatedly despairs of the character of this man he felt forced to work through. Weniger also seemed to have been treated in a somewhat bemused and long-suffering fashion by the ministry, and never taken seriously by them.[241] Promoting Wegner in Berlin at the end of May, Weniger announced to Vahlen the forthcoming anti-Hasse student letters whose writing he had ordered. But, as already seen, these letters were never sent; indeed, what did arrive in Berlin was a letter from one student who wished to withdraw the letter he believed had been sent.[242] On May 30, the Rektor arrived in Berlin to discuss Weber's refusal to give institute keys to Hasse, complained about Mohr's activity, and described Weber's later return of the keys because of his sense of Mohr's untrustworthiness.[243] At this juncture, Weniger attempted in vain to destroy Weber's character to the assembled officials, since Weber had blamed "his man," Mohr. Failing all comprehension of the situation, he actually continued to attempt to promote Wegner with the ministry until about June 10.[244] Weniger and Weber had a last meeting in Göttingen on June 4. This was as to be expected: Weber defended the honorableness and political purity of his behavior; Weniger viewed himself as the leader of the action, and Weber as his insubordinate underling.[245] As previously observed, Weniger's fantasies extended to an imagined complot of Hasse and Courant in England; the rumors of this were so widespread that Tornier on June 21 felt compelled to post a notice denying them.[246] In order to be rid of this difficult man who, nevertheless, had the authority of being an early Nazi, Tornier recommended him for a post with the *Deutsche Forschungsgemeinschaft* (DFG) in Berlin,[247] and by early July Weniger was there. Here Weniger behaved as at Göttingen: he was the true radical Nazi needed to strengthen the DFG's National Socialist feeling. Indeed, here too, Weniger's opinions were attributed to his head wound. He was fired.[218] However, this was not without negative consequences for the head of the DFG, Johannes Stark.[249] Weniger did not quite disappear from the world of mathematics on departure from Göttingen, for Professor Hans Zassenhaus has described his appearance at the annual meeting of the German Mathematical Soci-

[239] Weber, pp. 203–204.

[240] Ibid.: 135–136, 186, 201.

[241] Weber, pp. 139, 158; but see below, note 248.

[242] Ibid.: 137. The scene seems almost something out of Gilbert and Sullivan.

[243] Ibid.: 138.

[244] Ibid.: 139, 141, 186.

[245] Ibid.: 185–187.

[246] Ibid.: 201 and Anlage 55; a copy of the notice is in MI. Cf. Weber, p. 188.

[247] Weber, p. 204.

[248] Helmut Heiber 1966: 827–829. According to Heiber, Weniger's position in Göttingen had been head of an SS political information service, and the ministry praised his work in Göttingen to the DFG.

[249] Zierold 1968: 207. For more on Stark, see Zierold, ibid.: 173–212, and Beyerchen 1977: passim.

ety that fall in Bad Pyrmont, which will be discussed in chapter 6. According to Zassenhaus,[250] Weniger made an appearance in Nazi uniform and gave the (consistent) impression of a somewhat rabid radical Nazi. It is unclear why Weniger attempted to be at this meeting, though perhaps it was as Vahlen's representative. In any case, he was barred as a nonmathematician.

If Tornier and Vahlen decided Weber deserved only a wrist-slap and lecture—indeed, that as a true National Socialist, he had acted politically in a forgivable way—and if Weniger was something of a buffoon who had to be (barely) tolerated as an early Nazi, someone had to be punished for the insubordinacy at Göttingen. That person was Ernst Mohr. As Weniger's agent, Mohr had inserted himself into the anti-Hasse movement, much to Weber's disgust, even though Mohr "spoke splendidly."[251] Weber distanced himself from Mohr, whose motives seemed to him far from "pure," and whose accusations against Hasse were "unproved." Mohr was a competent mathematician: Hermann Weyl, his dissertation supervisor, had judged his 1933 doctoral dissertation as good, even if it showed little truly new insightful originality.[252] It seems apparent that Mohr desperately wanted a position as *Assistent* under F. K. Schmidt,[253] and was afraid that if Hasse brought in his own people, he would lose out. Tornier held a hearing for Mohr. At this hearing, Ernst Witt[254] testified that the previous Easter, Mohr had said that the establishment of a secret police was un-German, and caused people to fear for their personal safety. Four students testified that on Hitler's forty-fifth birthday (April 20, 1934), which date also happened to be Mohr's birthday, Mohr said to one of them something like, "It would be capital if the government, which is going to fall soon anyway, should already stumble over the price of eggs." The same students also reported that Mohr had spread as fact the widespread rumors of Ernst Röhm's homosexuality.[255] F. K. Schmidt and two of the students testified that Mohr had told Schmidt sometime in the preceding winter, that he thought the methods used by the government were "Slavic" and "un-German."[256] On the basis of these accusations, Tornier forbade Mohr from entering the mathematical institute.[257] On June 8, Tornier met with Wolff, Kleinsorge, Teichmüller, and Weber and revealed that in addition, he had discovered that (against all decency) Mohr was carrying on simultaneously

[250] Private conversation of Zassenhaus with the author, Nov. 17, 1987. Zassenhaus was not present at Bad Pyrmont, so his information is secondhand.

[251] Weber, p. 203.

[252] MI. Weyl's report is dated July 6, 1933. Weyl compares Mohr's work unfavorably with that of Richard Brauer and Issai Schur, but these were truly distinguished mathematicians.

[253] The logical choice for him in Göttingen at the time, given Schmidt's field.

[254] For Witt, see also above, chapter 3, note 69, and below, chapter 8. Witt was opposed on mathematical grounds to Wegner's receiving a professorship at Göttingen; see Weber, p. 196.

[255] Röhm, who would be murdered on June 30, was in fact a known homosexual. After the murders of the *Röhmputsch* (above, note 235), many scurrilous rumors were spread about his private life.

[256] Weber, p. 190.

[257] Ibid.: 191. MI, June 5, 1934. Tornier informed Weniger of his action as well.

with two women, and that he had revealed Kleinsorge's role in the anti-Hasse action.[258] Thus, the reasons for Mohr's punishment had nothing to do with the action against Hasse. At the same time, the *fact* of his punishment, together with his involvement in the affair, allowed him to serve as scapegoat for it.

Mohr did leave Göttingen and went to the technical university in Breslau (modern Wrocław) with that other acquaintance of Weniger, and Göttingen expellee, Johann Nikuradse. In Breslau, he completed his *Habilitation* on an applied mathematics subject in 1938.[259] In May 1935, Nikuradse wished to visit Göttingen for the summer semester for the sake of his research and to bring Mohr along. Hasse welcomed Nikuradse but refused Mohr entry, even as Nikuradse's vital assistant. He also referred the matter to the ministry, stating explicitly that the basis for Mohr's being banned from the mathematics institute was "his behavior in the occurrences on the occasion of my coming here," and that the action had been his jointly with Tornier.[260] There the matter presumably died. Mohr taught applied mathematics at Breslau both at the university and technical university, and on May 29, 1942, he became Hans Petersson's replacement in a chair of applied mathematics in Prague.[261] In March 1943, on Nikuradse's application, Mohr was also declared key to "the carrying out of work important for the war." Sometime after that September, Mohr apparently told a female friend he thought Hitler was a megalomaniac and the German military situation hopeless. He was led to this by the overthrow of Mussolini (July 24, 1943) and the illegal listening to English radio broadcasts. She denounced him. (It was not the first time Mohr's female acquaintances had gotten him in trouble by repeating what he had said.) On May 15, 1944, the Gestapo arrested him. On October 24, 1944, the court sentenced him to death as one of "our enemies' propagandists for [German] dissolution (*Zersetzungspropagandist*)." A month later, the sentence was confirmed. Quite exceptionally, the sentence was not carried out, and in April 1945, it would appear that Mohr was rescued by Soviet soldiers.[262] After the war, Mohr became professor at the technical university in Berlin, where his seventy-fifth birthday was celebrated in 1985. He died in 1989.

[258] Weber, p. 192.

[259] Mohr's earlier work had been quite unapplied. See *Festschrift Ernst Mohr zum 75 Geburtstag arn 20. April 1985*, ed. K. H. Förster (1985), 9, and Weyl's report (note 252 above).

[260] MI, Nikuradse to Hasse, May 24 and June 11, 1935; Hasse to Nikuradse, June 3 and June 15, 1935; Hasse to Vahlen, June 15, 1935.

[261] For Petersson, see chapter 8.

[262] Hans Ebert, "Mathematiker in KZ 1944/45," 24 pages. The author is grateful to Herbert Mehrtens for a copy of this document, a copy unfortunately lacking its references. Also, BDC entries in Mohr's education ministry file from May 15, 1944 forward. Cf. *Festschrift* 1985. The citation is from Ebert, p. 12. According to a letter from Fischer to Süss, July 28, 1944, in BAK R73 12976, Mohr had listened to foreign radio broadcasts "under the influence of his wife, who is Russian." The mathematician Hans Rohrbach, then Ordinarius at Prague, had brought Mohr's case to the attention of Fischer (an official in the education ministry) when it seemed that only imprisonment was in the offing for Mohr. A detailed description of the "Mohr case," both at Göttingen and with respect to his death sentence, that differs somewhat in nuance from that given here has been published by F. Litten: "Ernst Mohr—Das Schicksal eines Mathematikers," *JDMV* (1996): 192–212.

One can only surmise why F. K. Schmidt, a devout Roman Catholic[263] and no Nazi sympathizer, testified against Mohr. However, Schmidt seems to have had an ironic temperament, and he may have obtained some satisfaction from turning the Nazis' weapons dialectically against them when he had the opportunity. A view of his putative *Assistent*, Mohr, as one more young man attempting reprehensibly to obtain academic advantage politically in the Nazi atmosphere would have been sufficient to account for his evidence.[264] Kleinsorge (no doubt in revenge for his demotion), together with Weber, composed a letter to Vahlen about F. K. Schmidt's "traffic with Jews."[265] The fate of this letter is unknown, but Tornier, who seems to have viewed Schmidt somewhat equivocally, was not upset to see him leave Göttingen in the autumn.

The further history of the Göttingen mathematics institute has been well described by Schappacher.[266] As to Hasse and Tornier, they would be on opposite sides in the autumn of 1934 in an important incident involving the German Mathematical Society that will be discussed in chapter 6. As a consequence of this incident, Tornier and Hasse had a placard war on the public notice boards for the mathematics and statistics institutes in 1936. On April 4, 1936, Hasse wrote Vahlen that a new posting by Tornier "personally attacks my qualifications and I now cannot but be compelled to request my immediate leave until I am replaced." However, on April 9, not Hasse but Tornier went on leave (on grounds of illness), and on April 15, Hasse announced he would begin his lectures shortly.[267] As of October 1, 1936, Tornier was called to Berlin, taking his *Assistent* with him.[268]

Much earlier, Hasse had already managed to sideline Tornier in the operation of the mathematics institute. Indeed, just one month after taking charge in Göttingen, Hasse wrote the Kurator:[269]

> Ministerial Director Vahlen has told me orally that Dr. Tornier's codirectorship of the mathematics institute has merely the same meaning as usual for all full professors at the institute up until now; however, the operational business lies in my hands alone. A decree to this effect will arrive shortly.

The institute for mathematical statistics (which had been led by Felix Bernstein before 1933) moved into the bottom floor of the mathematics building in September 1934, and Tornier became Bernstein's successor as its director. This was even somewhat appropriate mathematically, as Tornier's specialty was probability theory. As director of the mathematical statistics institute, Tornier would, of course, deal directly with the ministry, but the importance of that

[263] Prof. K. H. Weise, in an interview with the author (Mar. 16, 1988), characterized Schmidt as an "old Jesuit." Cf. Weber, Anlage 27 (letter from Wegner to Weber, May 28, 1934).

[264] Compare F. K. Schmidt's behavior later with Weinel at Jena, and Weinel's complaints.

[265] Weber, p. 212.

[266] In Schappacher 1987. Schappacher n.d., contains this and a more extensive version.

[267] Schappacher n.d.; cf. Reid 1976: 178.

[268] MI, Ministry to Hasse, Aug. 25, 1936; Schappacher n.d.

[269] MI, Hasse to Valentiner, Aug. 1, 1934.

position was far inferior to Hasse's.[270] When Tornier left Göttingen for Berlin, his position in statistics was filled by Hans Münzner.[271] Though Münzner had been Felix Bernstein's assistant, he apparently became one of those young radical Nazi academics (again with a Jewish mentor) who sought to benefit from the Jewish exodus. According to both Bernstein's wife and his daughter, Münzner wanted Bernstein's books and tried to prevent them from being packed up for their owner. (Bernstein was fortuitously already in the United States when Hitler came to power.) Apparently, Münzner ordered various young Nazis to run every few minutes through the room where the books were being packed and shout "Heil Hitler."[272]

After Tornier went to Berlin, he apparently was divorced and led a rather excessive life—he was pictured walking with a notorious prostitute leading a tortoise on a leash.[273] On December 21, 1938, he requested emeritization because "weakness of his intellectual powers" caused a permanent disability, and this was done on January 28, 1939, effective at the end of April.[274] Actually, he was forced to this action because he had seduced an underage girl.[275] The grounds were stated by the ministry in the following way:[276]

> According to the carefully founded report of the director of the Psychiatric and Neurological Clinic at the University of Breslau, the personality of Professor Tornier varies substantially from the norm. He suffers from a constitutionally conditioned weakness of character which simultaneously represents a weakness of his intellectual powers and makes it impossible for him to fulfill the duties of his office.

By initiating his own retirement, Tornier managed to avoid official punishment, the initiation of a disciplinary procedure against him having been started in June. The local party court also decided against any action.[277] Among the accusations against Tornier was an inability to pay his debts.[278] Nevertheless, in

[270] MI, Hasse to Valentiner, Aug. 8, 1934. While this letter is clearly from Hasse, somewhat peculiarly the secretarial code marked on it is T/Tr rather than H/Tr or To/Tr, as was usual for Hasse or Tornier respectively (Tr is for Traphagen, the then-departmental secretary). Most likely this is a typographical error.

[271] MI, Valentiner to Hasse, May 11, 1936, citing ministry to Valentiner, Apr. 30, 1936. But Münzner's official appointment apparently did not come until about two years later: ministry to Kurator, Göttingen, July 25, 1938, in MI, under copy made Aug. 23, 1938 (presumably for Hasse).

[272] Marianne Bernstein-Wiener, private conversation. See Edith Magnus Bernstein's memoirs (unpublished, but in the possession of Elizabeth Smith of the Department of History of Science, Harvard University), supplement, p. 6.

[273] Reid 1976: 178.

[274] BDC, Tornier folder.

[275] HK, Süss to Kneser, Apr. 8, 1939. Süss speaks of "some legal offenses," specifically mentioning this offense, "running up debts, and the like." The context of these remarks was that both Bieberbach and Süss, though Süss very ambivalently, thought of Kneser as a possible replacement in Berlin.

[276] BDC, Tornier folder, Jan. 18, 1939.

[277] BDC, Tornier folder, Koch-Schweisfurth to *Gaugericht* Berlin, Sept. 22, 1938, and Franke to *Oberste Parteigericht*, Mar. 8, 1938.

[278] BDC, Tornier folder, Franke to *Oberste Parteigericht*, Mar. 8, 1938.

1941, an effort was made to expel Tornier from the Nazi party "because as a consequence of a nervous breakdown, he had not been able to care any more about the party since 1937."[279] At the time Tornier was living in Warsaw, and in February 1944, at his own request, he became a schoolteacher in the *Generalgouvernement*[280] and was temporarily assigned to a high school (*Oberschule*) in Radom.[281] Tornier survived the war, and afterward returned to the sort of mathematics involving von Mises's probabilistic ideas that had interested him much earlier. He also became interested in parapsychology and the statistical justification of parapsychological experiments. Tornier published and carried on an extensive mathematical correspondence with Hilda Geiringer, who was von Mises' widow and also a forced emigré, as she was preparing for publication the work left behind by her husband.[282] As the correspondence reveals, she also gave Tornier money. She was apparently aware of Tornier's Nazi past, but was willing to let political bygones be bygones.[283] Throughout the correspondence appears evidence of Tornier having psychological difficulties. He died in 1982.

Thus Hasse, from August 1934, had the mathematics institute under his sole control (though formally Gustav Herglotz and later Theodor Kaluza, as full professors, were also "directors"). Weber evaluated Hasse after the fact in his memoir as well, and decided:[284]

> From a political point of view, Hasse proved to be thoroughly trustworthy. He was one of those men who first acknowledged the new State when they had to represent it in a prominent position. After that was achieved, the Third Reich found approval in his eyes, and he represented it in the future out of complete conviction and without question and also vis-à-vis foreign nations. These motives for his National Socialism might appear to some particularly distinguished and give Hasse the air of a "Superman"; however, they will remain foreign to the old activists of the movement.

Weber even calls this development of Hasse's personality "natural" and "organic," and speaks of it as justification for his original favorable judgment of Hasse. His anti-Hasse action, however, had saved his honor. Nevertheless, according to Weber, Hasse, because he was appointed deceitfully, should have

[279] BDC, Tornier folder, De Mars to *Reichsleitung der NSDAP* (*Reichsschatzmeister*), Feb. 20, 1941. Presumably Tornier had not been paying his party dues.

[280] This was the Nazi term for a part of occupied Poland.

[281] BDC, Tornier folder, *Regierung des Generalgouvernements* to education ministry, Feb. 4, 1944.

[282] Richard von Mises, *Mathematical Theory of Probability and Statistics*, edited and complemented by Hilda Geiringer, appeared in 1964. Tornier is cited a number of times, and pp. 98–112 are devoted to "Tornier's frequency theory." Footnote 2, p. 60, begins: "Our presentation will borrow decisive ideas from E. Tornier. . . ." Footnote 9, p. 103, reads: "Many of the proofs and comments of Chapter 11 to which we here refer are actually from Tornier's work."

[283] For examples, see Tornier-Geiringer correspondence, Geiringer to Tornier, Jan. 24, 1966, June 25, 1966, Aug. 6, 1966; Dec. 18, 1967. She also addressed him as "Dear Friend" (Apr. 9, 1967). The Tornier-Geiringer correspondence is now housed at Harvard University. I am indebted to Magda Tisza, Hilda Geiringer's daughter, for the opportunity to see it, and for the information about her mother's attitudes toward Tornier.

[284] Weber, p. 223.

been removed; the Göttingen mathematical tradition had declined; the only person who would have been capable of rescuing it was, of course, Wegner.[285]

Similarly, Teichmüller, at the time of Hasse's arrival in Göttingen, had formulated a political-pedagogical survey of the faculty in which Hasse was termed "a great algebraist," "a German nationalist," "about forty-five,"[286] and "apparently still incapable of fitting in with the regime (*sich gleichzuschalten*). At present small and hateful. A call for him to come here is in the offing; we cannot approve of it."[287] Nevertheless, Teichmüller chose Hasse as his doctoral supervisor. Hasse thought quite highly of him.[288]

On May 2, 1935, Tornier, on behalf of the Göttingen branch of the Nazi teachers' organization, prepared a formal report on Hasse (for what purpose is unclear) in which he argued that Hasse was inappropriate as operational director of the mathematics institute. He demeaned Hasse's abilities as a teacher and in recognizing mathematical talent, and, while admitting Hasse was an important mathematician, attributed his success "more to a completely unusual working power and self-discipline . . . than direct ingenuity." He called Hasse's mathematical area (algebra and number theory) "essentially Jewish,"[289] and his accomplishments "one-sided."

However, wrote Tornier, "Prof. Hasse is completely unobjectionable with respect to his character, though it is determined by two tendencies that lead to the most surprising consequences." One of these tendencies was "the dash of Jew" (*jüdischer Einschlag*) in Hasse. (Hasse, in fact, had a great-great-grandmother who was a baptized Jew.[290]) The other was his coming from "an old family of jurists" (Hasse's father had been a prominent judge), which led him to see all of life (including mathematics) from only a formal point of view. Such tendencies led him to telegraph a wreath signed "The Göttingen mathematicians" to Emmy Noether's funeral in Pennsylvania.[291] They also "prevent him from understanding contemporary times; despite this, in any case, from time to time, he makes every effort in this direction, which, however, is in vain and only contributes to making him laughable in the eyes of the students." Nevertheless, he was "never Marxist, rather actually a German nationalist."[292]

In fact, as Hasse's correspondence with Davenport reveals, he was from the outset a political supporter of Hitler, though, as Tornier so rightly says, from a nationalist perspective. In 1933, Davenport had sent him a clipping from the London *Times* concerning a recent speech of Franz von Papen that seemed full

[285] Ibid.: 224–226.

[286] He was actually thirty-six years old.

[287] Weber, Anlage 64.

[288] Reid 1976: 179, citing Harold Davenport.

[289] Weber's earlier comment on Hasse's work.

[290] BDC file on Hasse. See also below. Tornier's different version has Hasse's ancestry slightly wrong (as does Fraenkel 1967). The BDC file in this regard is corroborated by the personal papers of Martin and Jutta Kneser (Jutta Kneser is Hasse's daughter).

[291] For Emmy Noether, see chapter 3.

[292] Tornier's *Gutachten* is in BAK R21 838.

of threats.[293] Hasse's reply represents above all pride in resurgent nationalism, regretting events like the initial expulsion of Jewish academics as an unfortunate "sacrifice" that he hopes will be rectified. In May 1933 (and even much later), many conservative nationalist academics no doubt thought like him—among other mathematicians, Erhard Schmidt.[294]

> Many thanks for the *Times* cutting. Though I can understand the mentality out of which it was written, I daresay it is rather saucy, in particular the point about our going to war in 1914 having been an "ambitious folly." To-day, one knows only too well that Germany was far more threatened by the encircling policy of your King Edward and all its consequences than England was by our going through Belgium. Although it is certainly bad that Germany is losing all sympathy from the other countries, she had to show some grit and backbone at last after all those years of willingly bending down under the dictate of Versailles. That is what I liked in v. Papen's speech. We have now the great answer from Hitler to all the stir round our borders and in particular to the Geneva proceedings. I quite agree with every single word of it. This speech, and the foreign policy which I hope will follow it accordingly, is exactly what I hoped for together with so many of my countrymen when I gave my vote to the Nazis. It is somehow tragical that this sort of foreign policy could not be brought about without the personal drawbacks for learned men in Germany. As it is, one has to take them as a sacrifice and to hope that reason will come back in due course. As to v. Papen's speech, I must say that the *Times* is certainly wrong by taking it as any sort of a new-war-fanfare.

In October and November, amid mathematics, Hasse sent news of dismissals of mathematicians without comment. To judge from a letter of December 9, Davenport had let his side of the correspondence lapse somewhat, though he did send Hasse a subscription to the *Manchester Guardian Weekly* and a *Times* calendar for the new year.[295] Perhaps Davenport wished to expose Hasse to very different opinions about Germany and world affairs. In any case, for a number of years, thanks to Davenport, Hasse received either the *Guardian Weekly* or the *New Statesman*, both left-of-center British publications. They do not seem to have altered his opinions much.

Thus, he derives from this reading (emphasis in original)[296]

> that English public opinion seems to be badly prejudiced against Germany. The papers pick up every bit of *bad* news they can get hold of about Germany, but they hardly think it worthwhile to write about *good* things brought about by National Socialism. One would not expect enthusiasm, since the whole movement has a certain

[293] *The Times* (London) for May 15, 1933, reports on the speech; a leader (i.e., editorial) for May 16 comments on it.

[294] See below, chapter 7. The text that follows is from HDT, Hasse to Davenport, May 18, 1933. It was written in English. "Edward" in this text should presumably be "George."

[295] HDT, Hasse to Davenport, Jan. 2, 1934.

[296] HDT, Hasse to Davenport, Jan. 29, 1934. Written in English. Though the journals Davenport sent Hasse could not circulate openly in Germany, there seems at this time to have been no difficulty in receiving them by mail.

> and very outspoken tendency against the Allies because of the Versailles Treaty. But one *would* expect a certain detachment and aloofness, otherwise so strong in the English. It is not *their* business after all.

So it was with Hasse after precisely one year of Nazi hegemony. And so it went. He thought Kurt Mahler "a very clever mathematician,"[297] but never said a word deploring his enforced emigration. By 1938, with the Austrian *Anschluss*, Hasse's nationalism (like that of many of his conservative colleagues) reached triumphal proportions:

> We are still under the overwhelming impression of yesterday's great events in Austria. A political dream has been realized that has moved for over 120 years the best of our nation. We listened on the wireless to the enthusiastic welcome given to Hitler in Linz. . . . You will readily imagine the great admiration that everybody here has for Hitler's wise policy which made this possible in spite of France and others. He has achieved within 5 years what the Burschenschaft of 1820, the Revolutionaries of 1848, what Bismarck, and Wilhelm II, and the Republic could not bring about.

Hasse also recalled Austrian agitation for *Anschluss* when the German "Society of Natural Scientists and Doctors" met in Innsbruck in 1924.[298] By the time of this letter, Hasse had already applied for membership in the Nazi party.

Through it all, Hasse and Davenport carried on a mathematical correspondence, and again Hasse received the *New Statesman* and an annual *Times* calendar for the new year. In late 1938, however, the Italian Jewish mathematician Tullio Levi-Civita was dismissed from the editorial board of the reviewing journal *Zentralblatt der Mathematik* as a result of political pressure placed on the publisher, Ferdinand Springer. Otto Neugebauer, who had continued as the journal's chief editor after emigrating to Copenhagen, took this occasion to ask other editors and reviewers to join him in a protest resignation. Davenport was one of these reviewers, though not an editor, and he complied. The editors who resigned were Harald Bohr, Richard Courant, G. H. Hardy, J. D. Tamarkin, and Oswald Veblen. Courant also took the occasion to withdraw his name from the famous "Yellow Series" of mathematical books published by Springer. This, while an act of conscience on Courant's part, was also desirable from the point of view of the Nazi education ministry.[299] At this same time, Hasse *joined* the *Zentralblatt* editorial board.[300]

A consequence of Davenport's resignation was Hasse's decision to end their relationship, which, while very productive mathematically, had become increasingly strained politically. His letter to Davenport, written in German, unlike

[297] HDT, Hasse to Davenport, May 4, 1934.

[298] HDT, Hasse to Davenport, Mar. 13, 1938. Hasse neglects the fact that Bismarck rejected an attempt to incorporate Austria, rather than trying to do so.

[299] See, e.g., VP, Bohr folder, Bohr to Veblen, Nov. 3, 1938; Veblen to Bohr, Nov. 22, 1938; telegrams Bohr to Veblen (received Dec. 1, 1938) and Veblen to Bohr, Dec. 1, 1938; VP, Courant folder, Courant to G. D. Birkhoff, Nov. 19, 1938, and Birkhoff to Courant, n.d. HDT, Neugebauer to Davenport, Nov. 25, 1938. Birkhoff demurred about sending in a resignation.

[300] See title pages for vol. 17 (1938), vol. 18 (1938), and vol. 19 (1939) of the *Zentralblatt*.

most of his preceding letters, reveals just how *gleichgeschaltet* Hasse had become. The friendship had been sufficiently close on Hasse's part to take him two months to decide to break it off. Throughout the letter, the familiar *Du* for close friends is used (translated as "you").[301]

> Dear Harold,
> As I learned in November, you have laid down your cooperative work on the *Zentralblatt*. You have troubled us deeply and offended us by this step. With it you have placed yourself formally in a front which is directed against a German scientific undertaking. This and the realization coming repeatedly to our view that you also besides stand in a front which wishes ill to National Socialist Germany for ideological reasons, is the reason why joy in the continuation of our up until now friendly exchange of thoughts is taken away from Clärle [Hasse's wife] and me. And therefore, I also ask you not to renew the subscription to the *New Statesman* for me. I thank you, that you have let it be sent to me for so long a time, and also for the *Times* calendar, that at Christmas I again received from you.
>
> Clärle and I feel melancholy that the many weeks and months that you have experienced in Germany have led to no other result than your present thoroughly negative position.
>
> Yours,
> Helmut

In the aftermath of World War II, Davenport sent Hasse food and cigarette packages, but was apparently not receptive to a renewal of the former friendship.[302] They did both attend the Boulder (Colorado) Number Theory Conference in August 1963, and Hasse looked forward to their meeting there.[303] From my own memory of that event, Hasse and Davenport were certainly not inimical to one another, and were formally friendly—but one would never have guessed that they had once been warm friends. In July 1966, Hasse's seventieth birthday and retirement was celebrated at Hamburg. Davenport, who spent the summer semester 1966 in Göttingen with C. L. Siegel, was invited to give an address and refused.[304]

Thus, Weber, Teichmüller, and Tornier, the radical Nazis at Göttingen, all of whom initially were prominently anti-Hasse, came correctly to view him as thoroughly assimilated to the National Socialist state. Indeed, Hasse seems to have been the sort of man whose first academic loyalty was certainly to mathematics, but who had no objections to operating within the Nazi state. He would fight stupidity and chicanery where he found it, but was not opposed to the Nazi government in any fundamental way. Weber's view that Hasse was "not a conscious exponent of the opposition [to National Socialism]" was an understatement valid only from his own radical viewpoint. One reason to attempt to

[301] HDT, Hasse to Davenport, Jan. 25, 1939.

[302] HDT, Hasse to Davenport, Nov. 22, 1946, Jan. 24, 1947, and Dec. 2, 1952.

[303] HDT, Hasse to Davenport, Apr. 26, 1963. This is a postcard in German. Hasse does not seem to have written in English after 1939.

[304] HDT, Siegel to Davenport, Oct. 26, 1966. Siegel speaks of Hasse's "Nazi friends" at Hamburg.

understand Hasse is that, despite unavoidable idiosyncratic differences, he seems representative in these prewar years of his "unpolitical" mathematical (and academic) colleagues. Men like Erhard Schmidt, Wilhelm Süss, and Hellmuth Kneser, equally conservative at the outset, might gradually have fallen out of sympathy with the regime, but Hasse was earlier on in a more prominent mathematical position, and, while never sympathizing with the politicization of academic life, also never seemed to understand this was an inevitable result of Nazi policy and attitudes.

On October 29, 1937, Hasse applied for membership in the Nazi party. There have been several partly inaccurate indications in the literature about this application, and so it seems appropriate to tell the story as revealed by the primary documents.[305] In addition, these documents reveal a tale of bureaucratic confusion that would be funny and worthy of a satire by James Thurber were it not so serious—the Nazi bureaucracy was nothing if not concerned with small details. To understand what follows, it is necessary to realize that while a condition to be a civil-service employee (which included academics with regular appointments) under the law of April 7, 1933, was the absence of any "full-Jewish" grandparents, the analogous condition to be a member of the NSDAP was much stricter. For this, one could not have had a "full-Jewish" ancestor living after 1800. Hasse's application went for comment to the local *Gauleitung* (South Hannover-Braunschweig). The leader of the teachers' organization, Dr. Artur Schürmann,[306] found no objection, and the local Göttingen party leader reported that everyone

> agrees that Professor Hasse has succeeded in his work [as leader of the mathematical institute] toward the conscious goal of raising the institute, almost completely denuded by the furloughing of Jews, to new heights and scientific importance. . . .
>
> He does not lack in comradeliness and readiness for action. Without doubt, he belongs to the most politically reliable German university teachers and could be a model for a number of party members among them. His *Weltanschauung* stands, in every respect, unconditionally with the ideas of National Socialism. . . . What must especially be recognized is his constant concern for further understanding for the new Germany among his many foreign acquaintances, not only through words, but also through selfless deeds.

Here we have the "organically developed" Hasse in Weber's phrase. However, the next paragraph adds:

[305] BDC file on Hasse. The date October 29, 1937, appears on the party membership questionnaire filled out by Hasse. If accepted, his membership would be backdated to May 1, 1937. Unless indicated otherwise, all documentation of this story is from this BDC file.

[306] Artur Schürmann, a member of the institute for agricultural economics at Göttingen, was extremely powerful in Göttingen affairs under the Nazis. Heinrich Becker calls him the "political controller of the university." See H. Becker, "Von der Nahrungssicherung zu Kolonialträumen: Die Landwirtschaftlichen Institute im Dritten Reich," in Becker et al. 1987: 410–436, esp. pp. 423–429, as well as other articles in that volume.

> It is now still my duty to indicate that Hasse has among his forebears a non-Aryan great-great-grandmother (born 1775). Although this fact is known to all competent authorities, this application is supported by them without reservation, and I request therefore that you also decide in this way.

People who had applied for membership in the NSDAP received a yellow card indicating them as "party applicants" until a decision was reached on their membership.[307] On February 14, 1938, the leader of the smaller district office of the party, a certain Achilles, wrote the *Gauleitung* in Hannover and the group leader in Göttingen excitedly saying that if the yellow card had been handed out to Hasse, it must be returned by him, and otherwise it should not be given him, since Hasse did not fulfill the preconditions for membership ("see German-blooded, Aryan, etc."). Hannover, however, stood firm; by March 8, the application was already with the national leadership—they suggested withholding the yellow card, and if a red membership card were issued, then it would be sent directly to Achilles, who would have to initiate a procedure for refusal in the regional party court. Two years after his application, on October 24, 1939, Hasse was admitted for membership in the party, backdated to May 1, 1937, and assigned membership number 5,642,208. Two weeks later, however, the return of the card was requested, and another one sent in its place: an error had been made; Hasse's number should be 5,639,530, not 5,642,208. Though it was two years later, Achilles had not forgotten, and on December 7, 1939, questioned the local Göttingen leader (who by now had changed) as to Hasse's satisfaction of the "German-blooded, respectively Aryan, etc." condition for membership. "Settlement of this matter in the fastest possible way is necessary and I request an immediate settlement of it." Two weeks later (to Achilles' frustration), the present Göttingen leader replied that Hasse was "1/16 Jew as a consequence of a baptized great-great-grandmother," the approval of the membership application being laid to the previous Göttingen leader. He ends, "I myself had occasion to talk with Hasse. He made a very good impression on me, even by the strictest standards." Nevertheless, by January 4, the new local *Gauleiter* had decided that either Hasse had to resign from the party, or the local party court would be asked to expel him. Hasse should be asked privately to declare his voluntary resignation. However, if Hasse had performed special service to the party, "he could submit an appeal for mercy (*Gnadengesuch*) via the bureaucracy to the *Führer*. . . . Such an appeal has some hope of success, however, only in case there actually are special services." There was some bureaucratic confusion about which membership card was which, but by mid-January this was settled.

At the end of January, Hasse, rather than resigning, decided to submit an

[307] The institution of "party applicant" was introduced by Rudolf Hess in 1937. "Party applicants" (*Parteianwärter*) paid party dues and were allowed to wear the party badge. See copy of the Military Government Information Bulletin no. 37, for April 15, 1946, p. 2, reproduced as part of Report no. 182 (1949) of OMGUS (Bad Nauheim) on German views of denazification, in the National Archives (U.S.). This was not declassified until 1993.

appeal to be allowed into the party, mentioning that he had a brother, Albrecht Hasse in Berlin, who was, he believed, a party member. A letter of inquiry went off to Berlin with party officials mentioning the "non-Aryan portion" found in the forebears. Until a decision was reached on Hasse's appeal, the party court could do nothing, and so Hasse's red party membership card was returned to him at least for the nonce. Schürmann, at the university teachers' organization, again spoke up for Hasse as a "willing and useful" member of his organization. In March, Berlin mentioned that Albrecht Hasse had only completely established his ancestry back through his grandparents, and was given to April 1 to complete the forms, and in April, Berlin told Göttingen about the Jewish forebear named Itzig, and that they were checking the validity of Albrecht Hasse's party membership since there had been no appeal for mercy in his case. On April 20, 1940, the appropriate official of the *Gau* wrote Göttingen: "After checking the content [of the appeal] I believe it scarcely possible that the *Führer* will decide positively in H[asse's] favor." However, according to regulations, he could not process the appeal further until Hasse had left the party; if he would not do this voluntarily, he would have to be expelled by a party court. Only then could the appeal be forwarded to the chancellery in Berlin with a recommendation. Furthermore, Hasse needed two photographs with front and side views, and a list of forebears in which the non-Aryan position was clearly marked. No proofs of this list, however, were necessary.

Göttingen did not answer in May, nor in June, despite repeated requests. In July, the district leader (no longer Achilles) wrote Göttingen repeating what had been heard from the *Gau* official in charge of appeals; at the same time, he wrote this official saying that his demand for Hasse's prior departure from the party had never been requested in similar appeals, and asking whether some special exception was being made in Hasse's case or the rules had been changed. On July 22, Göttingen told the district that Hasse was presently in the navy, and the whole question should be postponed to war's end, and this message was transferred to the *Gauleitung*. However, Hasse had been posted to Berlin, where he led a group working on mathematical problems involving curves of pursuit,[308] though he frequently returned to Göttingen to deal with business there. He inquired at the chancellery as to the whereabouts of his appeal—they couldn't find it (presumably because the *Gau* had not forwarded it), and so on January 10, 1941, asked Göttingen to research where it was. Göttingen wrote the *Gau* and finally, three months later, on April 16, 1941, they replied that the chancellery had decided that a decision could be postponed until Hasse left the armed forces.

As to Hasse's brother Albrecht, in July 1940, he was asked by his local group to leave the party voluntarily; otherwise he would be expelled.

This tale indicates one side of Hasse's personality: he seems to have been truly the traditional "unpolitical" German nationalist professor, who, in Max Weber's words, "sings the song of him whose bread I eat." Quite another side is revealed in his dealing with Tornier, or in his position at the Bad Pyrmont

[308] MI, letter to Hasse, dated Nov. 4, 1944.

meeting of the German Mathematical Society in September 1934.[309] He was a proud man with strong opinions and a strong spine and was not afraid to tell Constance Reid after the war of the distinction between the National Socialist and the conservative nationalist.[310] Although initially removed by the Allies from his university post after the war, after a research position at the Berlin Academy, in May 1949, he became professor at the venerable Humboldt University in (East) Berlin. In 1950, he went to Hamburg, from which he retired in 1966, but remained in a nearby suburb for the rest of his life.[311]

Hasse seems to have pursued whatever political position was necessary for the furtherance of mathematics and his career as a mathematician, and at the same time to have strongly resisted political incursions into mathematics, insofar as that was possible. He seems to have always believed in the attempt to rationalize the actions of the regime in a sane and German nationalistic direction. This was his fatal flaw in the eyes of the likes of Tornier and Weber, and ironically also in the view of the Allied victors after the war, as well as that of many American mathematicians.[312] On March 22, 1939, Carl Ludwig Siegel, having returned to Germany from Princeton,[313] wrote Oswald Veblen:[314]

> After the November pogrom, when I returned from a trip to Frankfurt, full of nausea and anger at the bestialities in the name of the higher honor of Germany, I saw Hasse for the first time wearing Nazi-party insignia! It is incomprehensible to me how an intelligent and conscientious man can do such a thing. I then learned that the foreign-policy occurrences of recent years had made Hasse into a convinced follower of Hitler. He really believes that these acts of violence will result in a blessing for the German people.

Siegel's view of the importance of Hitler's foreign-policy successes in positively influencing "intelligent" and "conscientious" people should not be underestimated, and not only in Hasse's case.[315] On the other hand, the insignia seen by Siegel was probably the sportsman's badge that Hasse had been awarded.[316]

[309] See below, chapter 6.

[310] Above, chapter 3; cf. Segal 1980.

[311] Frei 1985, and Harold Edwards, "Helmut Hasse," in *Dictionary of Scientific Biography* 1970.

[312] Hasse was a very controversial figure in American mathematical circles in the 1950s and early 1960s, and some refused to be in the same room with him or to have him invited to meetings because of his Nazi past.

[313] See chapter 3.

[314] VP, Siegel folder.

[315] Compare Erhard Schmidt in chapter 7. Many writers have commented how Hitler's foreign-policy triumphs (as well as his domestic economic ones) had made him extremely popular. The height of this popularity was in spring 1939. Hitler was even known as "General Unbloody" (Norbert Frei, *Der Führerstaat* [1987], 132). Ludwig Prandtl, despite early resistance to the anti-Semitic policy, also seems to have been positively influenced by these domestic and foreign policy successes. See Cordula Tollmien, "Das Kaiser-Wilhelm-Institut," in Becker et al. 1987: 476–477.

[316] See Hasse's response to a 1939 census of Nazi party members in the BDC file on Hasse. Hasse's daughter (Frau Jutta Kneser) has told me that he always wore this decoration (interview, Feb. 18, 1988).

In the light of the circumstances just detailed, Hasse's postwar explanation to Constance Reid seems somewhat disingenuous:[317]

> My endeavor at that time was to keep up Göttingen's mathematical glory. For this I needed the consensus of party functionaries at the university whenever I wanted to get some distinguished mathematician to fill a vacancy in Göttingen. Among these functionaries I had one close friend and one who was leaning towards helping me. They asked me to join the party so that they could help me better. It is true that I gave in and applied for membership. But on my application I put that there was a Jewish branch in my father's family. I was almost sure that this would lead to my application being declined. And so it was. The answer, which I received only after the outbreak of the war, was that the application was not going to be acted upon until the war was over. In the meantime, however, I had been able to help several mathematicians who were having political difficulties at other universities, by offering them positions in Göttingen.

Yet Hasse's pursuit of party membership may actually have been motivated by his concern for Göttingen mathematics. Konrad Knopp is an example of a mathematician who was operational director of his institute, though never a Nazi party member, or even fellow-traveler; however, Knopp was at Tübingen, which did not have the international mathematical prominence of Göttingen. One can be sympathetic to Richard Courant's postwar diary entry: "Met Hasse. Mixed feelings."[318]

. . .

The three case histories discussed have shown us a variety of mathematical personalities, and a variety of attitudes toward the regime. If Hasse is best characterized, though somewhat inadequately, as a traditional German-nationalist "good Prussian," as his daughter and son-in-law, his post war student Manfred Knebusch, Weber, Tornier, and he himself do, others were political manipulators, like Weyrich, and youthful parvenus, like Weber, Wegner, and Weinel. Men like Süss, Prandtl, and Hasse did business with the Nazis in the interests of mathematics and mathematicians. In the circumstances, such attempts at maintenance of professional integrity were faced with assault from those trying to take personal advantage of the political situation, like Tornier and Doetsch. Weyrich and Weinel may have suffered academic insult, but their main objective seems to have been moving political levers toward academic success. Though Doetsch may have had problems explaining away his political transition from left to right, that hardly rationalizes his insistence on personal prerogative above all, and his attempts to use politics to achieve it. The dynamic belief in Hitler's "new Germany" expressed by the likes of Weber and Wegner was genuine. Both show inklings of disciplinary considerations, which, however, are suppressed in favor of their political beliefs. In Wegner's case, this was not un-

[317] Reid 1976: 203.
[318] Ibid.: 263.

mixed with ambition. Hasse also ardently believed in the "new Germany," but not in a politicized academy. On the other hand, Mohr seems to have been a student desperately needing an appointment and trying to ensure one by political action. *Alte Kämpfer* might be canny like Weyrich (though ultimately failing), or deluded in a sense of self-importance like Weniger, but they could not be easily overlooked. Despite all these various differences and those of other personalities and situations discussed below, one theme seems to unite these case histories: the Nazi regime simply was not concerned about practical academic matters, at least insofar as mathematics was concerned. It did not care about applied mathematics in the midst of the war, and it excused Weber's academic insubordination in the name of ideology. Indeed, ideological correctness seems to have been the one thing that was important at the level of the education ministry. Academic mathematical disputes were not so much settled from above as allowed to wear themselves out. All competitors fought at least ostensibly in the name of the same ideology. It is possible that the sorts of bureaucratic conflict discussed here, which were not untypical throughout the "Behemoth" of the Nazi bureaucracy, were not the result of any well-thought-out plan to keep all such power at the top of the hierarchy, but were simply the result of the fact that the *only* thing that mattered in the Nazi state was power achieved in the name of ideology. Though parts of Franz Neumann's famous analysis of National Socialism may be open to question, his statement that in National Socialism, "nothing remains but profits, power, prestige, and, above all, fear,"[319] seems true enough, and our case histories exemplify this in academic mathematics. The only condition was that all activity, all struggle, was carried on in the name of ideology. It is almost as though some subterranean Social Darwinist principle were being allowed to operate that encouraged such personal disputes, in the tacit knowledge that the victor was right by right of victory.

From these three case histories, one can see that while conventional academic mathematics had its defenders, like (for all their differences) Süss, Hasse, and F. K. Schmidt, whatever success they had depended upon the fact that no one sufficiently high in the hierarchy was concerned that any particular dispute be settled one way or another, since formal ideological correctness was usually already a given. Thus, for example, Weber and Teichmüller could fight Hasse in the name of ideology, and only by being a good National Socialist was Hasse able to sideline and "defeat" Tornier. We will see this sort of phenomenon in other aspects of mathematics during the Nazi period, such as the attempt to introduce the *Führerprinzip* into the German Mathematical Society, but first we will consider the daily conditions in which mathematicians pursued their discipline.

[319] Neumann 1966: 397.

CHAPTER FIVE

Academic Mathematical Life

ALTHOUGH the Nazi regime was oppressive politically and ideologically, insisting on strict control of public expression relating in any way to governance or policy, it might be thought that work activity, so long as it was consonant with state objectives, would not be particularly affected. This might seem especially true for academics, who, as discussed in chapter 3, had a long tradition of adherence to whatever state might be, as an exchange for *Lehr-und Lernfreiheit* (freedom in teaching and learning) within academic and professional subject matters. Of course, some subject matters, such as history, anthropology, biology, and German letters and literature, would obviously be affected by Nazi politics and ideology; but many others, especially mathematics, with its aspect of *de novo* creativity, and even applied mathematics, would seem immune to such pressures. That is, the academic mathematician, both as mathematician and as academic, would not seem to be necessarily much affected by the regime, but could presumably retreat from political expression (if one had ever even been so interested) and simply continue mathematical work, while striving to keep the discipline as free of politics as possible. And there were some mathematicians who seem to have achieved this—Konrad Knopp at Tübingen is one notable example.

In fact, however, Nazi ideology and politics entered the academic mathematician's life in both the role of professional academic and the role of creator and teacher of mathematics in particular. In this chapter, concentration will be on the former role. There has been some inkling of this in the previous chapter, for example in some of the material dealing with Gustav Doetsch or Helmut Hasse. The distinction made by Hannah Arendt between dictatorial and totalitarian regimes seems particularly appropriate here.[1] The totalitarian Nazi regime, not content with assuring obedience to its wishes by the general populace through the usual dictatorial methods, wanted to isolate each individual so that no bonds other than those forged by the state connected members of the "community." This notion appears already in Hitler's *Mein Kampf*; its most lapidary expression is perhaps, "If the German people in its historic development had

[1] Arendt 1980. Totalitarianism was a "revolt of the masses" that ended in all individuals being portions of the state machinery, in which each individual's connection to others was mediated by the state. Dictatorial regimes were those that merely insisted on enforced obedience to the dictator or monarch. This distinction is still controversial; partly because Arendt's book came out at the height of the "cold war," and the distinction has since frequently been used by right-of-center elements in the United States to defend their actions. However, that an intellectual idea becomes adapted to political purposes that may be disapproved does not necessarily invalidate the idea. For a highly critical review of Arendt's book that nevertheless sees it as breaking new paths, see Pierre Ayçoberry, *The Nazi Question*, trans. Robert Hurley (1981), 130–133.

possessed that hard unity which other peoples enjoyed . . . ," and it was this that National Socialism strove to achieve.[2]

While National Socialism spoke of a *Volk*, each "atomized" member of it was defined as a member of the community by being a cog in the state machinery. Thus the life of the academic mathematician was replete with purposeful intrusions of a nonacademic nature intent on making each individual a subservient member of the state, without connections to others except as mediated by the state. Since academics had always legally been civil servants, such intrusions were easy, and perhaps even easily acceptable. Gustav Doetsch's resignation from the board of *Compositio Mathematica*, and the seeming gradual acclimatization of Helmut Hasse after the conflictful beginning of his term in Göttingen, are some small particular examples already seen.

Under the Nazis, academic life became a progression of increasing restrictions and continual evaluations. These latter had no respect for academic rank (in fact, were intentionally subversive of it). They also were sometimes very different for the same individual. What was important was that they were made continually. These reports were a mixture of political and academic opinions. Occasionally they were straightforward; more often they seemed to veer between the poles of outright condemnation and fulsome praise.

An extreme case of this destruction of university hierarchy, already seen, was the confrontation between Werner Weber and Hasse (not to mention the associated report on F. K. Schmidt's "associating" with Jews), in which Weber was hardly rebuked at all. Whether or not these reports were effective in a particular case is almost irrelevant. They *could* have an effect and sometimes did—the appearance of political correctness on the part of university faculty was enforced from its lower ranks, who would be eager to fill a vacant position. Reports of the *Studentenschaft*, whether unofficial, like Teichmüller's, or official, were, if anything, even more subversive in nature *whatever their content*. In this atmosphere, it is easy to understand how the young Ernst Mohr would attempt to obtain academic status politically. Many of his colleagues, students and young professors, were doing so.

Not only were academics evaluated on a political scale, they were also increasingly restricted in foreign contacts, travel, and time for study. For example, a requirement for admission to a regular faculty appointment was a field sports camp, an instance of which was mentioned in the previous chapter. In general, while there were dynamic Nazi "true believers" and propagandists among mathematicians, many of the political judges thought that the general attitude among mathematicians was wanting. One dynamic young Nazi mathematician, however, perhaps inspired by these field sports camps, created mathematics camps that proved to be pedagogically valuable.

Nazi policies meant that university students declined in both number and quality. This was true even before World War II, but the war naturally acceler-

[2] Adolf Hitler, *Mein Kampf*, trans. Ralph Manheim (1943), 396. Cf. also 435, 572, 602 (on National Socialist unions), and numerous other passages.

ated this phenomenon. The war also reduced the number of available faculty, but, of course, they had fewer students to teach. One consequence for mathematics was that the number of female students, originally discouraged by the Nazi government, increased. There were even a few female faculty (like Helene Braun at Göttingen). The decline in technical (including mathematical) faculty and research personnel as a consequence of military service also led, especially after the war had turned for the worse for the Germans, to an attempt to recall some for civilian war service. This effort, which had to battle various bureaucratic difficulties, some motivated by political jealousy, was led by the engineer Werner Osenberg. Since Hitler had little concern for even technical education, this devolved into a competition of bureaucrats. How successful Osenberg's effort was is unclear and probably always will remain so.

The decline in student quality was largely due to a decline in effective secondary-school instruction. The Nazis believed in a sort of scientific reductionism, but for them the basic science was biology. Even in pre-Nazi times, German lay opinion was not entirely favorable to mathematics. This led to the spectacle of a respected mathematician (Robert Koenig), not a Nazi sympathizer, giving quasi-biological justifications of mathematics, just as the war began. Even earlier, in late 1933, a young and gifted mathematician (Gerhard Thomsen), sensing what the Nazi atmosphere meant for the exact sciences, pleaded for more consideration of them, but using perforce Nazi ideological language (see below on the value of mathematics in the Nazi state). A Nazi psychologist who was something of a mathematician manqué (Erich Rudolf Jaensch) argued (as did others) for a new kind of rationality (whatever that meant) that would be more salubrious.

Secondary mathematical education declined precipitously under the Nazis, but its infestation with propaganda was already present during the Weimar Republic. Under the Nazis, this propaganda had more deadly meaning. At all levels, from elementary school to university, the importance of biology, physical well-being, and "spiritual character-building" was stressed. Some even thought mathematics could be used for this latter purpose, thus justifying it.

Erich Bessel-Hagen and the General Atmosphere

One source for the nature of academic mathematical life under the Nazis is the *Nachlass* (material left at death) of Erich Bessel-Hagen. Erich Bessel-Hagen was born in Berlin in 1898, the son of a surgeon. He attended the University of Berlin, receiving his doctoral degree under Constantin Carathéodory with a dissertation in the calculus of variations. Carathéodory thought Bessel-Hagen's dissertation the first important advance in the theory of discontinuous solutions for problems in the calculus of variations since his own work of 1905. While at Berlin, however, Bessel-Hagen had also heard lectures from the famous classicist, Ulrich von Wilamowitz-Möllendorf, and was inspired by him to a life-long devotion to classical antiquity as well. He not only had a wide-ranging knowledge of various areas of mathematics, but also was well versed in classical Greek, Greek philosophy, and the history of Greek mathematics. At the University of Bonn, where he arrived in 1928, he had a position in the mathematics department with a "special concern for history and pedagogy," and while he advanced through the academic ranks, he never had a regular "slotted" position.[3] Actually, his given family name was Hagen, but the family was directly related to the famous nineteenth-century mathematician and astronomer Friedrich Wilhelm Bessel, and legal permission was obtained to alter the name. Bessel-Hagen seems to have been a somewhat retiring personality who suffered from a severe constitutional illness; photographs show a face accentuated by large glasses. Certainly he was the butt of jokes among his colleagues; although he died in 1946, this author can recall, as a graduate student in the 1960s, hearing his name used as the object of mathematical jokes. The topologist Bela von Kerékjártó made one such joke publicly—in his 1923 book *Vorlesungen über Topologie*, the name Bessel-Hagen appears in the index with a reference to p. 151. If one turns to p. 151, Bessel-Hagen's name does not appear; however, the diagram representing a sphere with two handles and four holes—that is, a caricature of a face—does. In fact, Kerékjártó's book stemmed from some Göttingen lectures, and Bessel-Hagen was teaching in Göttingen at the time. Bessel-Hagen, who was a devout Roman Catholic, in fact seems to have been one of the most honorable and honest of men. He seems to have represented the best sort of "inner emigration" during the Nazi period. After the great German and Jewish (but unreligious) mathematician Felix Hausdorff[4] was retired, ostensibly because of age, in 1935, Bessel-Hagen seems to have been the only one of his former colleagues to visit him.[5] In the Bonn catalog of professors issued in 1968, he is described with unusual feeling as having been a "model of the noblest humanity. Unselfish goodness and purity of desire paired in him with an incorruptible love of truth and faithful friendship."[6]

Perhaps because of Bessel-Hagen's historical interests, perhaps because of his "love of truth," he apparently kept a collection of university edicts and related academic correspondence during the Nazi period, many of which found their way into his papers at the University of Bonn. They are uncommented, and are simply witnesses to the reality of those times. Because he was exempt from military service on account of ill health, he remained in the Bonn department of mathematics, at times essentially the only regularly appointed mathematician there.[7]

[3] Biermann 1988: 185. Most of these facts come from a catalog of deceased professors published by the University of Bonn in 1968 and available in its archive, p. 65.

[4] See chapter 8.

[5] Herbert Mehrtens, "Felix Hausdorff, Ein Mathematiker seiner Zeit," a brochure printed by the University of Bonn reproducing a speech on the occasion of unveiling a memorial tablet to Hausdorff in 1980, p. 31.

[6] Catalog of deceased professors, University of Bonn (note 3 above).

[7] Bessel-Hagen papers, University of Bonn. All material on academic life cited below in various sections and not otherwise referenced comes from this source.

The temper of the times showed itself early. Although he was only thirty-two years old, and had been at the University of Bonn for only two semesters, a certain Professor Eckhardt was suggested in early April 1933 as the new university Rektor by the National Socialists. Eckhardt remarked that this was because he was the only Ordinarius at Bonn who had been a member of the NSDAP back in the days when it was forbidden.[8] Though young, Professor Eckhardt had some sense of what German academic decorum had been prior to 1933 and declined the honor on grounds of inexperience. Immediately a rumor began that the declination was because his wife was partly Jewish (her mother's maiden name was Hamburger), and consequently as Rektor he would have had to fear conflicts with the students. On May 6, Eckhardt wrote an open letter to his colleagues complaining about the "slander"; his wife had "no drop of Jewish blood"; before the marriage, he had established an "exact genealogy"; Hamburger was a "purely Aryan Hessian peasant and soldier family," presumably an elision of Heimburger. These and some of his additional comments are an excellent example of the rhetoric of ordinary academic Nazis. For example, he says that the rumor "contains the infamous imputation that I would have as a man allowed a half-Jewess to be admitted into a party in which the Aryan principle is of decisive importance."

Professor Eckhardt may well have been a participant in the Bonn local book-burning that took place at universities across Germany on May 10. The Bonn announcement had a well-organized line of march (National Socialist students and female students were among other separate groupings); two faculty spoke for twenty minutes each before the midnight burning (in Berlin, Joseph Goebbels was among the speakers).

Open disapproval of unwelcome ideas and the encouragement of suspicion were atmospheric. They were accompanied by regulations. In July 1933, all civil servants, and appointees or workers in official agencies (therefore academics), were ordered while on the job or inside their place of work to greet one another with the Hitler salute. It was "expected of civil servants [including academics] that they would also greet one another outside of the workplace similarly." This regulation may at first have had some difficulty, because six months later, the regulation was expanded to state that while the salute could be given silently, the only words allowed to accompany it were "Heil Hitler," or just "Heil." Incidentally, anyone physically unable to raise his or her right arm was instructed to use the left if possible.

As servants of the state, all faculty had to take a "civil-service" oath. Under the Nazis, this was transformed into a personal oath to Adolf Hitler. In autumn 1933, the *Dozentenschaft*, a National Socialist "union" of all faculty without civil-service appointments (and others), was formed. Faculty (civil-service) dismissals

[8] However, Eckhardt does not appear in the list of early Nazis at Bonn in the memoir by the orientalist Paul Kahle: *Bonn University in Pre-Nazi and Nazi Times, 1923–1939* (1945 [prepared 1942]).

had begun the preceding spring, but the law requiring faculty to be resworn to Adolf Hitler was not promulgated until August 20, 1934.[9] The new oath read:

> I swear: I will be loyal and obedient to the *Führer* of the German Reich and people, Adolf Hitler, observe the laws, and conscientiously fulfill the duties of my office. So help me God.

The oath was taken in the following prescribed manner: with the left hand on the heart, in the presence of two witnesses, and the right hand raised, and spoken out loud.[10] Heinrich Behnke has recalled a theological colleague who comforted him by saying, "The oath can never include anything that goes against your conscience, for the oath is sworn to God."[11]

Strikingly, even those Jews who were still faculty members because they fell under one or another of the exceptions of the April 7 law, and had not yet been otherwise forced out of office, took this oath, and were allowed to do so. Thus, for example, among mathematicians, Arthur Rosenthal took the oath on November 6, 1934,[12] Felix Hausdorff on August 7, 1934,[13] and Otto Toeplitz on August 28, 1934. This oath-taking applied even to professors who had been forcibly furloughed in 1933, as was the case with Toeplitz.[14] Apparently there was a delay of some sort in the new oath-taking, however, for a year later, in July 1935, the education minister, Bernhard Rust, sent a sharply worded memorandum stating that taking the oath was an "obligation" of a civil servant, and that anyone who refused to take it or took it with reservations would lose his position: "The oath of loyalty by its very nature allowed no reservations." Aside from the much-discussed embodiment of the Nazi state in the "will" of Adolf Hitler,[15] by oath-taking, academics, including Jews, legally accepted Hitler as their lawgiver.

But what of the actual elimination of Jews and other "undesirables"? Outside of Göttingen and a few other exceptions, the law of April 7, 1933, seems to

[9] As an example of prior oaths under the Weimar government, at Hamburg one had previously sworn "obedience to the Hamburg constitution." See Personalakten Hecke, Hamburg. Hecke himself took the "Hitler oath" on November 27, 1934.

[10] This is the formulary procedure followed for Arthur Rosenthal at Heidelberg; see Personalakten Rosenthal, Universität Heidelberg. The oath is cited from this source (among many) also. The formulary procedure varied from place to place; the oath did not.

[11] Behnke wrote two descriptions of his academic life under the Nazis. One is contained in his autobiography *Semesterberichte* (1978); the other is an unpublished manuscript, "An einer deutschen Universität im braunen Sturm," a copy of which was kindly given to me by Prof. Karl Stein in Munich. Comparison of the two—sentences and paragraphs are sometimes taken over wholly from the second to the first—reveals that the second was probably a draft version for this material in the published book. The incident described here is from the unpublished manuscript (Behnke n.d.: 21).

[12] Personalakten Rosenthal, Heidelberg.

[13] Personalakten Hausdorff, Bonn. Thus Hausdorff took the oath before it became an official requirement.

[14] *Nachruf* Otto Toeplitz, in Personalakten Toeplitz, Bonn.

[15] E.g., Arendt 1980: 373–375; Broszat 1981: x; Stern 1975: chaps. 7 and 8.

have been enforced also with respect to its exceptions. Thus Hausdorff did not lose his position at Bonn until he was retired at age sixty-seven in 1935. Similar remarks apply to several other cases, though "encouragement" to retirement might come in the form of student boycotts, as in Edmund Landau's case. In April 1933, all university employees (including faculty) had to fill out a questionnaire corresponding to the April 7 law. The answers to these questions were clearly researched as to exact accuracy with respect to "race" by Nazi officials. For example, the now-forgotten mathematician Heinrich Liebmann, author of 106 papers,[16] had on April 13, 1933, no doubt accurately listed the religion of his maternal grandfather, one Karl Neumann, as *evangelisch* (i.e., Lutheran). Liebmann retired at age sixty in 1935 (after student disturbance of his lectures). A few months later, on February 21, 1936, the local (Baden) education ministry was informed that Liebmann's maternal grandfather was "born an Israelite." Liebmann died a few years later; however, his daughter (whose grandparents were all fully Aryan) could still receive the assistance coming to her as a professor's daughter at the University of Munich.[17]

Political dissidents were another category to be expelled, but not until July 24, 1933, were university personnel ordered to sever all connections, if there ever had been any, with socialist organizations, as well as to indicate whether such had existed. Refusal meant dismissal, and signing a formal declaration form was required.

Dozentenschaft Reports

While such measures aimed at eliminating individuals who were not welcome to the Nazis, others aimed at making professional ties Nazi ones. In October 1933, the *Dozentenschaft* was formed as the single and obligatory Nazi organization for those academic personnel in a university who were not yet civil servants (that is, had not yet been appointed to a "regular" teaching position). This was done on a statewide basis, and thus the decree for Bonn came from the Prussian ministry in Berlin. Professors of civil-service rank could connect themselves with the local *Dozentenschaft* "provided they were prepared to work together on the *Dozentenschaft*'s tasks." The story of Helmut Hasse at Göttingen in 1934 has already demonstrated what some of the *Dozentenschaft*'s tasks might be. All other academic organizations were absolved into the *Dozentenschaft*; no other grouping was allowed. Among its other activities were an office for labor service, as well as one for "defensive sports" (*Wehrsport*). The usual dictatorial imposition of leaders and subleaders (*Führerprinzip*) would apply.

Also in October, the use of the Hitler salute (*deutscher Gruss*, or "German greeting") to open and close classes was "requested" at Bonn, and presumably around this time in other universities as well. At Bonn, this request was number

[16] Pinl 1972: 162–167.

[17] See Personalakten Liebmann, University of Heidelberg.

five among a list asking for, among other things "active" participation at a lecture on racial science, support of concerts and theater in Bonn, and collection of material concerning the university's "defensive activity" during the French occupation of the Rhineland (1921–23).

The *Dozentenschaft*, usually led by a junior faculty member, was also responsible for giving reports on faculty as to their suitability for academic positions. These reports varied greatly in style and content. While political reliability was usually the primary criterion discussed, research ability and teaching ability were also mentioned. Frequently, the *Dozentenführer* had no idea about an individual and had to ask a politically trusted member of his discipline for a report, which would then be passed on verbatim.

A sampling of some of these reports on mathematicians gathered by or reported by these "teachers' leaders" gives the flavor of the times.[18] For example, on May 31, 1934, Professor Ernst Pohlhausen in Danzig, a former Rektor there, wrote a positive recommendation for the mathematician Wolfgang Haack, of the sort that he might have written two years previously, mentioning Haack's many excellences as a teacher and reputed fine scholarship (he was not in Pohlhausen's area of mathematics). However, the day previous, the *Dozentenführer* had already sent off a positive recommendation to Berlin that dealt primarily with Haack's character, about which he could only say "the best." Mainly this was because, in straitened economic circumstances for three years, Haack had nevertheless been a successful teacher and productive scholar, and

> Haack has a clear-thinking critical mind, whose very frank judgment and unconditional maintenance of what he has once recognized as correct have caused him multiple difficulties. On the other hand, exactly these qualities, together with his amiable nature, have won him attention among his colleagues.

Appended was the report of the *Unterführer*, which repeated the economic circumstances, the devotion to teaching, and the research excellences, and ended:

> Although during the time of struggle (*Kampfzeit*) of the National Socialist movement, Herr Haack had been uncomprehendingly opposed to it, it is to be assumed that his entering the motorized SA, to which he has belonged for a long time, accompanies an inner conversion.

Such mild political justifications of a faculty member probably occurred reasonably frequently,[19] and were themselves reasonably accurate. Lady Macbeth's advice, "To beguile the time, look like the time," was obvious to any young academic seeking a position in a time of severe ideological pressure. Furthermore, because all organizations were replaced by National Socialist ones, one could join such "innocent" Nazi groups as a motorcycle corps or a glider corps where the politics were less salient, as well as the association of *all* teachers, the NSLB, thereby establishing a "connection" with the movement. A similar option in the early days was the *Stahlhelm*, a German-nationalist organization of World

[18] The following citations of reports come from BAK R21 838 unless otherwise indicated.

[19] In fact, they do in my sample.

War I veterans.[20] Though "old Nazi fighters" might be suspicious of people with such limited political activities, such sufficed, at least at first, especially since the party membership rolls were only opened at infrequent intervals after 1933. Indeed, a non-Nazi like the mathematician Oskar Perron could give a recommendation that never mentioned politics, as he did for Fritz Lettenmeyer in December 1934.

For all that, many such reports were also far from mild. The reports of Teichmüller and Tornier on Hasse have already been examined. Though Teichmüller was a student, and so his report not an officially requested one from the *Dozentenschaft*, his summary comments on other Göttingen faculty are also of interest. For example, Wilhelm Cauer was judged "on average a bad teacher," "a liberal who has only contested his Jewish ancestry since 1933," a member of the NSKK (*Nationalsozialistisches Kraftfahrerkorps*)[21] only since 1933, and a seeker of positions. Heinrich Heesch was "unimportant," a "poor teacher," "no National Socialist," "a characterless and odd fellow." F. K. Schmidt was "able in his tiny special area, and pedagogically not without ability," "does not fit in with the new political way (*nicht gleichgeschaltet*)," "a philosemite" who "tries in a crabbed way to make himself loved by the students. The mathematics students' organization boycotted his last excursion trip for his students." Ulm and Rellich were characterized as "Courantclique," the former "long rejected by the *Dozentenschaft*,"[22] and the latter a "liberal philosemite." Hilbert's assistant, Dr. Arnold Schmidt, was accused of promoting "refusal of military service[23] and the lies about war guilt." Werner Weber received unstinted praise, while Tornier was noted as an old National Socialist whose scholarly and pedagogical abilities were unknown. As to Herglotz, Teichmüller simply said he was a "big shot" as a scholar and "good teacher" but "colorless"—nothing more.[24]

Judgments of Teichmüller's negative sort were not infrequent from a local *Dozentenschaft* as well. Sometimes, though, they were honest and straightforward. For example, the report of a certain Schönhardt on Erich Kamke, on November 25, 1934, after praising his mathematics, his teaching, and his interest in students, said:

> With respect to politics, Kamke takes now as previously an oppositional attitude. In particular he does not agree with the elimination of Jewish influence. This attitude must be connected with the fact that his wife is a half-Jew. His character is honorable and upright.[25]

[20] The *Stahlhelm* ("Steel Helmet") was a right-wing World War I veteran's organization founded by Franz Seldte, which even ran a presidential candidate in 1932. After the Nazi rise to power in 1933, joining the *Stahlhelm* was briefly a way to display the "right" attitudes without joining a National Socialist organization. Seldte himself joined the Nazi party in 1933, and the absorption of his organization into the SA and other Nazi organizations was total by 1935.

[21] This is the paramilitary motorcycle corps referred to above.

[22] Though Tornier later saw to it that he became a member of it.

[23] See Teichmüller, below in chapter 8.

[24] Weber, Anlage 64

[25] Similarly, from the opposite point of view, Paul Kahle (1945: 20) could speak of two Nazis whom he detested as Nazis as "honest."

These evaluation letters could also, however, obviously be used for defamatory purposes. Such a case is E. Schlotter on the mathematician Heinrich Kapferer (in late 1934?), a former teacher of his. After saying he had known Kapferer since beginning as a student in 1928, and commenting on the narrowness of his research area, Schlotter told a story about a test he took with Kapferer, which he twisted to comment on Kapferer's inability to judge "very good" performances.[26] He then went on to say: "Unfortunately with respect to scientific research he always seems to make a similar impression, as a standard authority (Prof. Doetsch)[27] has repeatedly said to me." Pedagogically, said Schlotter, Kapferer tries hard, but is unsuccessful, his lectures often only being understandable when one knows beforehand their content. As to politics, Schlotter pointed out that Kapferer was a follower of the former Center party[28] and:

> After the national arousal he became a member of the *Stahlhelm* (though he were never a front-line soldier). If one enters his office, he is gladly accustomed to be the first to offer a greeting, so he can say "Good Morning." One gets the German greeting [*viz. Heil Hitler*] only if one says even more quickly "Heil Hitler" and so he cannot do otherwise. Nevertheless he should be harmless in a political sense as in every other.

This recent student of Kapferer's concludes that "scientifically and pedagogically" considered, he should never receive a full professorship. The vision of Kapferer and Schlotter hastening to be the first to speak seems almost to have comic-opera qualities, were it not for its seriousness for Kapferer's career.

If Schlotter's nonpolitical remarks might even have had some substance, some of these *Dozentenschaft* reports were vilifications or even denunciations. One such example is the report on Hellmuth Kneser in January 1935. At the time, Kneser was a distinguished mathematician and one of the *Ordinarien* at Greifswald. He will be discussed further in chapter 6. The *Dozentenschaft* report on him reads as follows:

> Scientifically, Professor Kneser is, as one often hears, definitely at the top. As to his pedagogical ability, so let me say at once on this occasion that his lectures certainly have a respectable level though often lacking in pedagogical depth and clarity.
>
> Earlier on the left politically,[29] since November 1933 he has been a member of the motorized SA; however, not much seems to have changed in his whole attitude. Character: Valued in the first years of his stay in Greifswald, in the succeeding time he has been rejected more and more. His fatality (*Verhängnis*) seems to be his wife, who is possessed of an unbelievable urge to show off, mixes in everywhere, would gladly make "politics and diplomacy," and has brought her basically weak husband completely under her fateful influence. The consequences are plotting quarrels with colleagues, hateful discussions, etc., which in a small city like Greifswald naturally make

[26] Following an oral examination, Kapferer asked Schlotter what grade he expected, telling him that both the examination and his exercises during the semester warranted a grade of "very good." However, said Kapferer, he did not know whether he was permitted to give Schlotter the grade "very good." "Very good" was in fact a moderate grade (like a B).

[27] Doetsch, in fact, was not in Kapferer's field.

[28] The Center party was Roman Catholic.

[29] This is surely false. As will be seen, Kneser initially had conservative right-wing political views.

an unwelcome impression, and have put Prof. Kneser more and more in the role of a comic potentate. Perhaps all this would change in a new environment and larger city.

To show how varied in substance such reports might be, some eighteen months later, the Rektor at Tübingen wrote concerning the selection of a successor to Karl Kommerell, retiring at age sixty-five:[30]

> I urgently wish to obtain exactly for the professorship in question a personality who towers above the political average of "nothing negative known" and is to be valued *positively* [emphasis in original] from a National Socialist standpoint, and is suited to university political engagement (*Einsatz*). That the number of candidates coming into question is further restricted by this condition is not unknown to me.

In this connection he finds himself "satisfied," "as is the leader of the *Dozentenschaft* Prof. Schönherdt [*sic*]," with the suggestion of Kneser (whose name is consistently misspelled with two *e*'s).[31] Schönherdt's replacement, Walter Schwenk, who had complained about the political reliability of the other two mathematics *Ordinarien* at Tübingen, Knopp and Kamke, said,[32]

> Though Herr Kneser had, as was reported to us, not involved himself politically earlier, since the seizure of power he has engaged himself a great deal for the movement. He is in the SA and has already stood up a great deal for National Socialist goals in a university political connection.

As another example of vilification, the Marburg *Dozentenschaft* had this to say *in toto* about the complex analyst, Egon Ullrich, who followed Hasse to Göttingen:

> Dr. Ullrich was *Privatdozent* for mathematics in Marburg until autumn 1934, when he went to Göttingen as *Assistent* and *Privatdozent* (in connection with Hasse's appointment there). Dr. Ullrich left no favorable impression behind here, neither with his then-supervisor, Prof. Dr. Neumann, nor in the leading circles of our *Dozentenschaft*. At his departure Göttingen obtained no report on him, which at that time pleased us greatly, since we gladly wished "to be free" of him. Dr. Ullrich is an Austrian and a Catholic, which in and of itself is unimportant, were it not that his political position and his behavior have a Jesuitical-Center-Party-like character. Before the revolution he had sung hymns of praise to Brüning[33] and in Marburg had sought connections only with the Jewish-democratic circle of the then *Privatdozenten*. His mind and endeavors are solely and singly set on self-advertisement, which explains the fact that, when there was an election, he always named himself as a candidate. His ambition indeed

[30] BAK R21 (Rep. 76) 10417, Tübingen Rektor to the education minister in Berlin (and the local Württemberg official in Stuttgart), July 30, 1936.

[31] Ibid.

[32] BAK R21 (Rep. 76) 10417, Schwenk to Rektor, Sept. 4, 1936. Uwe Adam says (1977: 69 n. 131) that Schwenk took over from Schönherdt on September 2; however, in this letter he still signs himself as "acting leader." This is the same letter that complains about Knopp and Kamke.

[33] Heinrich Brüning was the chancellor of Germany from March 30, 1930, to May 30, 1932. Though toward the end of this period he found it necessary to rule by "emergency decree," he was, in some sense, the last "parliamentary" chancellor.

> went so far that in the voting in spring 1933,[34] he put his name on an election list and claimed that the list had been approved by the leading National Socialists of our university. An inquiry on my part indicated the opposite. His striving for a leadership role etc. [sic] is perhaps a reason that before January 1933 he sought proximity to the group named above, while, after the revolution, he looked to a connection with the colleagues with whom he had been up until then distant; he had then also found his National Socialist heart. I must leave it undecided how much the first or second phase of his time in Marburg corresponds to his inner being. At least he always exhibits a certain tacking behavior, first sounding things out, in order only then to follow suit. His thirst for power, and his pressing himself forward, have brought him no recognition; even if his disciplinary knowledge has received approval from his own opponents, he represents a disciplinary direction whose chief exponent is a Finn.[35]
>
> Dr. Ullrich has understood he should not shove his ultramontanist connections into the foreground in Marburg. Insofar as I however understand and can judge him, he will do that unhesitatingly, just so soon as it can be advantageous to him. His major goal remains the featuring of his own self, for which for him all means are holy. So he would, as my informant tells me (he will attest to it with his name), not be frightened of denouncing other people etc.

The anti-Catholic bias is not the least striking thing about this report and others—this atmosphere may make it somewhat easier to understand the extreme transformation of people like Doetsch.

The other side of character assassination is panegyric, and this selection closes with two examples among mathematicians of how these reports might be used by Nazi activists to promote their like-minded colleagues to the posts becoming vacant. In March 1935, Gerhard Kowalewski, a distinguished geometer, was under consideration as Rektor at the technical university in Dresden (and therefore, in the Nazi interpretation of the Rektor's position, effectively *Führer* of the university). An old acquaintance (and apparently old Nazi), the engineer Hermann Alt, wrote to the Nazi leadership in Saxony (after validating Kowalewski as scholar and colleague):

> Especial weight is to be given the question of K[owalewski's] relationship to the National Socialist movement. Indeed, only after the *Führer*'s seizure of power, did K[owalewski] become a member of the NSDAP. Nevertheless, in his personality structure (*menschlichen Struktur*) he is already a born National Socialist. In all questions of National Socialist thinking he is completely sure, unquestionably reliable, loyal, and not inclined to compromise. In his great human maturity he is free of a personal desire for self-promotion. His inner stance is characterized by the thought that he feels the ability to serve the *Führer* and National Socialism as providential (*als eine gütige Fügung*). He would accept appointment as Rektor and perform this office in this spirit.
>
> . . . Solely and alone through the appointment of Kowalewski as Rektor would a guarantee be offered that the National Socialist spirit will prevail at the Technical

[34] Hitler held an election on March 5, 1933.

[35] Rolf Nevanlinna.

University of Dresden. Against this winning out of the National Socialist spirit are the circles that inwardly are not able to be satisfied with National Socialism, and that, naturally, reject party-member[36] Prof. Dr. Kowalewski as Rektor on account of his clear and goal-centered National Socialist bearing.

While there were probably "circles" at the technical university that were not wholeheartedly Nazi, they were equally probably in 1935 not much worry to the powers regnant. They might have opposed Kowalewski's appointment, but the complaint that the National Socialist spirit yet needed to prevail at the university is likely also an example of the continual dynamism of the Nazi movement.[37] The enemies, old or newly discovered, were always there, always *had* to be there. In any case, Kowalewski did become Rektor, and was in fact the first one of "Nazi temperament."[38] As to that temperament, Kowalewski himself says in a report on a certain Dr. G. Wiarda:

I am in the position of being able to send along with this an autobiographical description by Wiarda. I wish to emphasize the indication therein of the pure racial genealogy of Wiarda and his wife. What a benefit, to find a pure German[39] in such a hopelessly judaized discipline. Such a man, the more so since he justly meets all the requirements scientifically, should not have to wait longer for a professorship.

Wiarda obtained his doctorate in theoretical physics at age twenty-four at Marburg in 1913. Two of the World War I years he spent studying with Landau in Göttingen, and in 1920, he "habilitated" in mathematics at Marburg. In 1923, he moved to Dresden as Kowalewski's *Assistent*, and worked in the physics institute there as well. In 1935, at age forty-six, with Kowalewski's support, he finally obtained his *Ordinariat* at the technical university at Stuttgart, after serving briefly as Felix Hausdorff's replacement in Bonn following Hausdorff's forcible retirement.[40] Wiarda's late professorship seems to have been because of his unextraordinary mathematical gifts, from which politics and Kowalewski rescued him. This can be contrasted with the late professorships of truly famous mathematicians mentioned in chapter 3.

The important point is not the effect that such reports may or may not have had. Rather, it is that mathematicians, like other academics, were operating in

[36] In German, *Pg.* for *Parteigenosse*. Address of a Nazi party member as such was both the "highest" and a leveling title. Thus Kowalewski was not so much Prof. or Dr. as Pg.

[37] Cf. Werner Weber, above in chapter 6, "Hasse's Appointment at Göttingen." For the role of "Will" in Hitler's (and so National Socialist) thought, see particularly chapters 7 and 8 of Stern 1975.

[38] Klemperer 1987: 276. Kowalewski wrote an autobiography, *Bestand und Wandel* (Munich: R. Oldenburg, 1950), in which he devotes one paragraph to the years 1933–45, claiming that he could not prove he was "fully Aryan" (p. 309). In any case, in 1939 he, at his own initiative, transferred from Dresden to the German university in Prague (presumably after Czechoslovakia ceased to exist on March 15) where he had also taught from 1912 to 1920 (see the brief obituary by J. Heinhold in *Naturwissenschaftliche Rundschau* 3 [1950]: 284).

[39] One might note ironically that while Kowalewski and Wiarda were not Jews, their names are hardly "Nordic."

[40] An obituary can be found in Hans Butler, "Georg Wiarda," *JDMV* 72 (1972): 105–106.

an atmosphere of subjective judgment of character and political stance that had nothing to do with ability and sometimes a great deal to do with personal animosities. The whole system of *Dozentenschaft* reports on faculty was purposely subversive of the university faculties as they existed prior to 1933. Leadership in the *Dozentenschaft* was primarily for the young and politically active, who were usually not yet disciplinarily established. Recall that established faculty were invited to join *if* they could cooperate in the National Socialist work. The importance of the *Dozentenschaft* reporting on their elder academic superiors was intentionally destructive of the whole hierarchical university system. As might be expected, state officials and party officials could well give different judgments on individuals who were candidates for positions. Consequently, in July 1935, a decree was issued authorizing only the highest-ranking local party officials to give such recommendations. This last came from the office of the *Führer*'s deputy (Rudolf Hess).

Foreign Contact and Travel

Already after January 1934, university faculty could not accept speaking invitations outside of Germany unless approved by the local Nazi political district leader—in order to ensure that people whose "personality and worldview development" did not permit them to be seen as "called representatives of the new Germany" would not represent it. On February 1, the university Kurators were asked to provide lists of suitable faculty, subject matters of talks, and possible countries of visitation.[41] One might think mathematicians would not be affected by such regulations, since obviously their talks were "unpolitical" in nature, and indeed, Erich Hecke, the distinguished Hamburg mathematician who not only was openly no ardent Nazi, but apparently was secretly connected with academic circles of passive resistance,[42] did make a U.S. trip in 1938. How that trip came about, however, is instructive in the regulations that could be faced even by mathematicians.[43]

In late 1936, Hecke received an invitation from Oswald Veblen in Princeton to visit. On November 25, 1936, the Hamburg Rektor wrote a letter recommending that Hecke be allowed to accept, since Veblen's invitation was "of an unpolitical nature" and Hecke "belongs to the most distinguished German mathematicians." However, by February 4, the ministry had not answered, so the Rektor wrote again, calling the matter urgent. On March 1, general permission was granted, but when Hecke knew the exact time of his absence from Germany and his exact itinerary for the trip, he needed to apply again for permission for these specifics. In May, Hecke inquired whether officials in

[41] Personalakten Bessel-Hagen, Bonn. Dr. Stuckart for ministry to Kurators, Jan. 12, 1934. At Bonn, the Kurator passed this on on January 25, 1934. Achelis for education ministry to Rektors, Feb. 1, 1934, passed on the next day to Dekans and by them to faculty.

[42] Wilhelm Flitner, "Das war unser Weg," *Deutsches Allgemeines Sonntagsblatt* 46 (Nov. 16, 1986).

[43] All material on Hecke's trip from Personalakten Hecke, Universität Hamburg.

Hamburg or those in Berlin were to approve his time and route and was told Berlin, and by late June his plans to travel to New York in January 1938 on a Hapag-Lloyd ship, returning in May, were approved. However, in August the Hamburg education officialdom raised the issue of whether, since Emil Artin was leaving Hamburg,[44] the university could afford Hecke's absence for five months. The Rektor raised the issue with Hecke, who, naturally enough, said that Artin's departure was no bar. Hecke also had to obtain individual permissions to lecture other than in Princeton, as well as make a separate application in September to receive his full pay. (Princeton was only covering his travel costs and stay there, and his household costs in Hamburg would, he said, be only slightly smaller.) In late September, however, it appeared that the Hamburg school authorities (*Kultur- und Schulbehörde*) also required Hecke to submit an official application for leave (despite Berlin's approval months earlier). So Hecke did, and permission was granted; however, Hecke was forbidden to speak at Harvard, and was required to contact U.S. branches of the Nazi party. He was awarded his full pay, and the university would have to cover both Artin's and Hecke's teaching duties. Hecke left for New York on January 18, 1938, and on June 4, after his return, filed an official report on his trip. Not only official bureaucracy caused Hecke difficulties with this trip. On May 28, 1937, shortly before the official permission from Berlin came through, Hecke complained to the Rektor that his (equally distinguished) mathematical colleague, Wilhelm Blaschke, was spreading rumors that Hecke was really planning to leave Hamburg and emigrate to the United States in connection with his lecture trip. Hecke emphatically denied this, which could cause him "official difficulties," and wished people not to consider anything Blaschke said as true—he didn't even speak to Blaschke.[45] Hecke declared that he had never had the plan nor made the attempt to get a position in the United States.

This story illustrates several aspects of life in Germany at this time: the overlapping and potentially contradictory bureaucracies; the potential role of rumor and denunciation in the Nazi atmosphere; and the tight tether placed on university faculty. Similarly, in March 1939, when Hecke wished to go to Copenhagen to give a talk, he was again required to contact official Nazi organizations there (who were no doubt apprised both there and earlier, in the United States of this requirement).

Such difficulties seem to have been typical, rather than a particular misfortune of Hecke. For example, Heinrich Behnke loved to travel, and felt obligated to do so, saying: "As an editor of *Mathematische Annalen*, doubtless the most respected German mathematical journal, I am anxious to undertake trips to

[44] On April 19, 1937, the education ministry issued a decree that deprived anyone married to a Jew of their position as a state employee (and so, of course, affected academics). Artin's wife was "half-Jewish." According to a law already promulgated on January 26, but which did not take effect until July 1, this would soon have been the case anyway. Just about a month earlier, Artin had been refused leave to visit Stanford because he was "indispensable." Personalakten Emil Artin, Hamburg.

[45] For Blaschke, see below, chapters 6 and 8, and above, chapter 4.

foreign countries."[46] Behnke traveled in March 1939 to England. The necessary correspondence and visits for this trip are detailed in his report on it. It reads:

A.	My application to the Minister	Jan. 17, 1938
	Permission	Feb. 17, 1938
B.	Inquiry to the German Academic Exchange Service in London	Feb. 25, 1938
	Reminder to the Congress Central Office concerning funds	July 1, 1938
	Interim decision of the Congress Central Office concerning funds	July 8, 1938
	Request (on my part) to the minister for postponement of the trip.	Oct. 2, 1938
	Cancellation with German Academic Exchange Service in London	Oct. 2, 1938
	Decision of the minister: New application needed	Oct. 26, 1938
	Approval of trip by the minister	Nov. 10, 1938
	Interim decision of the Congress Central Office	Nov. 14, 1938
	New notification to the Congress Central Office	Nov. 23, 1938
	Reminder to the Congress Central Office concerning funds	Jan. 5, 1939
	Further correspondence with the Congress Central Office	Jan. 25 and 29, 1939
	Award of funds of 250 RM by the official in charge in Berlin (*Oberfinanzpräsident*)	Feb. 1, 1939
C.	Day of initial travel	Mar. 1, 1939
	First visit to the German Embassy in London (Legation counsellor von Salzem)	Mar. 2, 1939
	Conversation with Dr. Krause of the German Academic Exchange Service in London	Mar. 1, 1939
	Appearance at the foreign office of the NSDAP (Lieutenant Commander Kalowa)	Mar. 5 and 7, 1939

[46] Hecke *Nachlass—Mathematische Annalen*, letter from Behnke to Hecke, Apr. 12, 1939, which contains a copy of the report cited below. When Erich Hecke died in 1947, some of his papers dealing mostly with the *Mathematische Annalen*, of which he, Behnke, and B. L. van der Waerden were editors, came into the possession of his student Hans Petersson, now deceased. Copies of these have kindly been provided to me by Hans Petersson's son Holger Petersson (also a mathematician) with the approval of his mother. Henceforth these are cited as HNMA.

Visit to the German Embassy (Embassy Counsellor Kunz)	Mar. 16 and 20, 1939
Day of return	Mar. 23, 1939

Division of Trip

London	1 day
Birmingham	3 days
Oxford	1 day
London	1 day
Cambridge	9 days
Oxford	1 day
London	5 days

It should be added that the 250 RM that Behnke received amounted to 21 English pounds, which would not have been enough to survive for twenty-three days had he not been frequently invited by his hosts. He comments somewhat acidly in his report that in 1930 the current edition of Baedeker indicated that a minimum amount for travel in England was 30 shillings per day (and so at least 34½ pounds would have been needed for Behnke's twenty-three-day trip).

Hitler's annexation of the remainder of Czechoslovakia took place while Behnke was in England. Consequently, although an additional lecture at King's College had been planned; Behnke noted, "Since in the meantime new political tensions had appeared and my very small allowance was exhausted, I did not accept the friendly invitation [since it would have meant staying for some additional days in London]."

Of course, with the war, difficulties increased. Thus Behnke, after returning from a trip to Switzerland, wrote on March 8, 1940, to Hecke:[47]

> The Paulis would be very happy if you paid them a visit again [in Zürich]. However, it is probably immensely difficult to obtain permission for this. I had an official invitation, and nevertheless had fabulously many difficulties. I assume that you have no inclination at all to subject yourself to these many negotiations, telephone conversations, telegrams and travel difficulties, negotiations over crossing the border, visits with police chiefs (here personally and in Zürich in his office). For me that is indeed a particular pleasure, and the more difficult it became, I applied all the more energy thereto. However, finally, on the day of departure, my nerves had had it.

Although by 1937 Hecke had troubles enough getting his leave and travel approved, it would appear that the ordinances of 1934 had some problems taking hold. In June 1935, Rust (acting through a subordinate) repeated that he had the "sole power of decision" as to who should be permitted foreign scientific trips, which could only be undertaken with permission. Further stressed was the necessity in many cases on foreign-policy grounds of consulting the

[47] HNMA.

Foreign Office or its officials in other countries prior to approval.[48] This was especially true for trips overseas or to the Soviet Union.[49] And in March 1936, Theodor Vahlen complained that applications for such leave were reaching him late, and late applications meant denial. "The start of a foreign trip before receipt of my approval is not permitted." Furthermore, a Rektor's approval of such trips, said Vahlen, was often too cursory, and he wanted a real recommendation. Also, signatures of faculty were too often illegible—typed names also, please.

With the war, regulations were stricter; the education ministry on February 24, 1940, wrote the Rektor in Münster an express letter confirming Behnke's trip mentioned above—he was to lecture in Zürich on February 27 and Basel on February 29. Fortunately, the Foreign Office, the NSDAP organization for foreign countries, and the congress central office had all been notified. Nevertheless, Behnke still had to obtain funds from the last, have his passport stamped at the Foreign Office to allow exit from Germany, and visit the NSDAP organization to verify permission to go. "Immediately" after arrival, he had to go to the appropriate German government representative, who would support and advise him, especially with respect to using the "Germanic greeting" (*Heil Hitler*) and the wearing of party insignia and uniforms in foreign countries. He also was "as possible" to contact the local division of the NSDAP organization for foreign countries, which would be notified in advance of his arrival. In addition to the regulations mentioned above about nonscientific political or religious speeches and general congresses (scientific or not), which were reiterated, any reception by a foreign official or "personality" was forbidden. If there were "pressing grounds" for such, approval had to be sought from the education ministry via the Foreign Office, or, alternatively, if already abroad, through the appropriate German legation and the Foreign Office. Thus Behnke had scarcely two days to complete all arrangements in Berlin—no wonder "his nerves had had it" at the end—and he was carefully kept track of in Switzerland.[50]

By April 1935, ministry approval had become necessary for individuals to belong to any foreign or international organization, for while on the one hand, such membership was certainly a good thing on "cultural-political" grounds, on the other hand, one had to take notice of how such organizations felt about National Socialism. The following year, in March 1936, faculty were warned that, given the German exit from the League of Nations, they must cease all activity with League organizations. Explicitly forbidden were not only organizations dealing with armed forces, or refugees, or tariffs, but also such organizations as those dealing with health, instructional film, the trade in opium and other forbidden drugs, the protection of children, and national statistics, to name a few. As a similar example, on January 23, 1936, Germans were forbid-

[48] Cf. Rudolf Weyrich's situation in 1938–39 as discussed in chapter 4, "The Winkelmann Succession." This represented an extreme case of an established policy.

[49] Ministerial decree of June 25, 1935, in Personalakten Bessel-Hagen, Bonn.

[50] Ministry to Rektor, Feb. 24, 1940, in Personalakten Behnke, Münster.

den to attend a February congress in Monaco dealing with the foundation of a new association dealing with reforms "necessary to ensure to each individual, liberty, dignity, and freedom from injury." The quoted matter was explicitly the reason for the ban on attendance. Also in April 1935, "as an express desire of the *Führer*," no discussion "written or oral" was to be permitted about the "reform of the Reich government" to prevent "unnecessary confusion and disturbance." Faculty were affected as civil servants.

Regulations such as those described served step by step to isolate the individual academic in his dealings with party and state. At the same time, these individuals were compelled to appear in artificial masses to support the state. May 1, 1934, was decreed the "Festival of German Work," and successive May Firsts were the "National Holiday of the German people." This and other compulsory celebrations followed. Failure to attend meant job loss. By the end of 1936, however, the voluntary number of celebratory events had risen to such a height that in November, a statement came directly "from the *Führer* and Reichschancellor," in the first person, complaining that the huge number of such gatherings placed unreasonable burdens on the leaders of party and state. For this and propagandaistic reasons, it was necessary to regulate such celebrations. Thus only those officially approved would henceforth take place.

Citing this, in January 1937, Werner Zschintzsch, writing for the education ministry, informed university Rektors of how academic professional meetings were to be regulated. Also in January 1937, the education ministry decreed that its permission was necessary for all foreign academic visitors to give talks in Germany, and furthermore, application had to be made sufficiently early so that, if necessary, the Foreign Office or other branches of German officialdom on foreign soil could be asked their opinion. A similar worry of corruption by inappropriate foreign contact caused the complaint around the same time that "German personalities, especially professors and students," undertaking foreign travel had not been diligent enough in contacting German officialdom on foreign soil. Thus, all academic travelers were henceforth required when on foreign soil to contact "immediately" a German consulate, a foreign branch of the NSDAP, and an appropriate branch of the German academic exchange service (if available). Such contact was "especially necessary" in countries "in which Jewry has a dominant place in culture and in which emigrants attempt to push questions of German cultural life into the foreground." This decree renews the complaint that academics were still not always following the regulations concerning foreign travel and the reiteration that this might lead to the trip's delay or its denial. Indeed, as late as February 1939, the Rektor at Bonn found it necessary to write his faculty, because for many there was "still lack of clarity about the procedure" for submitting applications. However, vacation trips were exempted from regulation.

These and many other, often very particular, regulations led to the habit of what became famously known as "inner emigration." Retreat into one's own work, holding all opinions to oneself, became a commonplace for some intellectuals and academics as a response to the ideological pressures compassing them about. For some mathematicians, whose work was naturally divorced from other external pressures in a way in which that of, say, historians, biologists, anthropologists, or architects was not, this became a way to live with themselves, even though they were not personally injured by the state—Bessel-Hagen and Konrad Knopp are examples. Nevertheless, the Nazi state was not satisfied with just having easily manipulable, atomized, and terrorized individuals; it needed to weld them into a unified whole whose only bonds to each other were those binding them to the state. This was true of academics as well, and such bonds were not limited to compulsory celebrations like the ones described above.

One compulsory activity already encountered was the "field sports camps."[51] These ten-week camps would serve for the "selection of university assistants and instructors according to their qualities of character and will. . . . Bodily deficiencies can be very well compensated for by joy in service and especially exemplary comradely behavior." Lest anyone misunderstand, the Bonn *Dozentenführer*, Dr. Schmidt, pointed out that only those people would be allowed *Habilitation* (i.e., entrance to the regular civil-service teaching faculty) who had completed a ten-week camp. In fact, "permission to begin the academic part of the procedure for *Habilitation* will be given by the minister on a case-by-case basis, if the applicant has proved prior participation in a camp."[52] If not enough faculty volunteered for a particular camp, they were drafted into it. The sports camp could be replaced by a work camp, and this had happened already in April 1934 in Prussia.

If initial enthusiasm was less than desired for the sports camps, the same was true for compulsory lectures on a political-ideological theme. In 1934 *Dozentenführer* Schmidt complained that at the first lecture in "National Socialist school teaching (*N.S. Schulung*)," the number of university teachers who appeared was "definitely miserable." Such poor attendance caused the impression that "a large part of the teaching faculty still have not comprehended the spirit of the times and also are fundamentally not ready for working with it." Furthermore,

> Precisely from those teaching faculty who have not yet done their *Habilitation* must I expect a substantially stronger participation, they represent the academic teachers of the future, wherefore it must be expected that they will perform their later university activity completely and fully in the National Socialist sense. Therefore, it is the duty of all these gentlemen, even more the duty of the older ones, to regularly participate in these lectures; for in these lectures actual directions and areas of activity of National Socialism will be handled in an absolutely competent way. I therefore expect that in the next lectures the teaching faculty will appear in a body, and in the case of continuing nonappearance I will make the corresponding report to Berlin.

No doubt, attendance improved.[53]

[51] Cf. "Hasse's Appointment at Göttingen" in chapter 4, and Franz Rellich's mistaken call to such a camp.

[52] Announcement dated May 14, 1934, in Personalakten Bessel-Hagen, Bonn. (See also the announcement of Oct. 30, 1933, ibid.)

[53] The documents cited above can be found in Personalakten Bessel-Hagen, Bonn.

Mathematical Camps

There were even mathematics camps. The longest-running such camps were those organized by E. A. Weiss of Bonn for students. These are worth closer examination because they are one example of how mathematics and Nazi ideology could interact.

In 1933, Weiss wrote a pamphlet called "Mathematics—For What?"[54] Beginning in 1934, he established Nazi student mathematics camps to realize the ideals of this pamphlet. Weiss's pamphlet is written in a hyperromantic tone, containing, for example, a girl with large black eyes, blackbirds flitting over the meadow and catching worms beneath tall elm trees, fishermen standing watch overnight when camping out, and similar images. Some of the problems raised by Weiss—that students need a purpose for studying mathematics; that insistence on mathematical originality for a new mathematical fact in doctoral studies demeans personal originality and discovery; that mathematics done for its own sake is difficult to justify intellectually to everyone—are perennial topics in mathematical practice and teaching in many times and places. Nevertheless, Weiss's discussion has a peculiar Nazi flavor to it. For example, when discussing the purpose of doing mathematics, he asks: "Is it really German, to do something—this something—for its own sake?"[55] While admitting that mathematics is in some measure an art form, he says, "I am afraid this art will never be properly a people's art." According to Weiss, an individual cannot answer to his people (*Volk*) that mathematics is done because of the yearning for knowledge.[56] Weiss's answer as to the purpose of pure mathematics was (all emphases in original):[57]

> *The cultivation and education to respectable behavior in public.* That is: to clarity[58] in speech and writing. Mathematics indeed teaches the communication of what one has only seen oneself in a short, brief, obvious, form so that the auditor or reader can have no doubts about it.
>
> *The cultivation and education to courage*, to the courage to say for oneself "No," if one doesn't know something or has not understood it, and truly and incorruptibly to say "Yes" if one has convinced oneself of something.
>
> *The cultivation and education to discipline.* That is, however: to give up something recognized as false immediately, and without contradiction, and not "to come back to the first statement when one has spoken reason for hours."
>
> That is, to have order in all thinking and as a good soldier to preserve beside all inspiration cool consideration and clear powers of judgment.

[54] "Wozu Mathematik," Bonn, 1933 (cited as WM).

[55] Ibid.: 8.

[56] Ibid.: 11.

[57] Ibid.: 12–13.

[58] *Sauberkeit*, i.e., "cleanliness."

That is: education to a worldview that is not like a many-sheeted Riemann Surface with many winding points that one can multiply cut without it falling apart.[59]

The cultivation and education to silence. That is: to learn to be silent and not to talk rot and twaddle about things one doesn't know, about matters that one doesn't investigate oneself, about books that one has not read, about people one has not looked in the eye.

Finally mathematics educates to *comradeship*, being the only discipline in which teacher and student from the very beginning stand at the same level. The teacher indeed is the leader (*Führer*). However, only if he goes hand in hand with the student, can both step forward together. For here nothing can be taught, and nothing learned. Here there can only be a joint working-out.

Weiss denies the ideas of mathematics for its own intellectual sake, and other common academic notions. He even has his sympathetic female student with the large black eyes say, "Or are our teachers men who, themselves busy out of fear of life, determine limit upon limit in order not to become conscious of their own emptiness?"[60] He rejects as inadequate all the usual ideas about the nature of mathematics and its practice, such as for applications, or its artistic quality. For Weiss, mathematics educates the personality. "Mathematical education is presupposition, is a forming mold."[61] Resisting the temptation to cite Weiss's overheated rhetoric further, suffice it to say that he sees the mathematically trained as those who can think reliably and clearly, who can reason from definitions to consequences and further consequences,[62] who can work diligently, cleanly, and honorably. Weiss sees his own mathematical area of geometry or algebra as providing more opportunity for this sort of mathematical education than analysis. For mathematical education should involve emphasis on all that brings order and system, and for Weiss this is not analysis, which he characterizes as full of devices. This mathematical education, of course, is in the name of the Nazi state—the closing words of the pamphlet are "Heil Hitler," which has a double meaning, both as a greeting in Weiss's framing story of discussants keeping watch over a campfire, and as an indication Hitler as the primary definition for Weiss's mathematical educational project, all being in his name.

Shortly after the appearance of this pamphlet, Weiss established "mathematics work camps" to realize its aims for students at Bonn. These will be discussed shortly, but first let us take a brief glance at the zealous pedagogue and convinced Nazi who created them. Indeed, Weiss's life seems in some respects almost stereotypical. Born May 5, 1900, in Strassbourg,[63] his father a retired

[59] As a simple example of a two-sheeted Riemann Surface, the reader may want to try to imagine two different planes, each cut entirely along the negative x-axis, with one lying above the other, and with the bottom cut edge of the upper plane glued to the top cut edge of the lower plane *and* the bottom cut edge of the lower plane glued to the top cut edge of the upper plane (this cannot be realized by a model in three dimensions without interpenetration, which is not allowed).

[60] WM, p. 5.

[61] Ibid.: 13.

[62] Cf. Arendt 1980: 469–479.

[63] "Unterlagen zu dem Nachruf für Prof. Dr. Weiss" (Mar. 26, 1942), in Personalakten Ernst August Weiss, Universität Bonn.

Prussian officer, and his mother a German Alsatian from Weissenburg,[64] Weiss came from a well-to-do family on his mother's side, as his maternal grandparents were the German owners of a hotel in Cannes. Weiss developed a gift for languages, and as an adult not only spoke French and German fluently but could carry on a conversation in English or Italian, and was trying to learn Russian while in army service.[65] In school one of Weiss's teachers was Hans Beck. Beck thought that the mother was "indescribably vivacious," that the father (whom he never met, and who apparently died in early 1914) was something of a pedant, and that E. A. Weiss himself showed both these characteristics in various succession. In 1917, Weiss (like Hasse) took the emergency school-leaving examination, went to war, won the Iron Cross (Second Class),[66] was captured by the Americans, and spent a year as a prisoner of war. In the same year, his mother returned to Weissenburg, from which she was then expelled at the end of the war by the French. She found a temporary home in Prerow (a North German resort hamlet), and by war's end, Weiss's former *Gymnasium* teacher, Beck, had become a professor at Bonn. Released from captivity, Weiss visited his mentor, and the latter recommended him to the famous (and famously cantankerous) geometer Eduard Study as a student.[67] Beck, twenty-four years Weiss' senior, himself had been Study's student, receiving his doctorate in 1905 at Bonn, but had been a *Gymnasium* teacher from 1901, except for two brief spates of military service in 1916 and 1917.[68] Beck had come to Bonn in early 1918 as Study's assistant with the effective rank of associate professor, presumably in part because of wartime exigencies. In the general promotion of 1920, he became an Ordinarius. According to the orientalist Paul Kahle, his mathematical work did not merit the promotion;[69] in any case, by 1926, Study and his former student and present colleague, Beck, had a falling out, and their differences affected Weiss. Weiss obtained his doctorate in 1924 under Study and, two years later, wished to "habilitate" at Bonn, where he had remained. Beck and Study agreed that the work Weiss presented to allow him to enter the regular faculty was of good quality. However, while Study wished to proceed with admitting Weiss to the regular faculty, Beck advocated delay, because he said Weiss was very narrow; his sort of geometry was unmodish; and he needed broadening at some institution other than Bonn. Letters went back and forth; Felix Hausdorff sided with Study, and this point of view won out.[70] Weiss's

[64] According to the extensive obituary of Weiss by Karl Strubecker (*Deutsche Mathematik* 7 [1943] 254–298, p. 255), she was born in Cannes. Pp. 264–298 of this are devoted to Weiss's mathematical work.

[65] Ibid.

[66] Ibid.

[67] All above material, not otherwise annotated, is cited from a letter of Hans Beck to Prof. Leick (Greifswald) on Nov. 4, 1936, in Personalakten Weiss, Universität Bonn.

[68] See Beck's curriculum vitae in the Nazi civil-service formulary in Personalakten Hans Beck, Universität Bonn. See also the obituary by Erich Salkowski in *JDMV* 53 (1943): 91–103. Beck was over forty at the time of this service.

[69] Kahle 1945: 16.

[70] Letters of Beck, Study, and Hausdorff in Feb. 1926, in Personalakten Weiss. Weiss was "habili-

work *was* narrow; his sort of geometry (Study's sort) was not pursued much at the time. To what extent Beck was truly motivated by his asserted concern for Weiss's future as a mathematician is unclear. What is clear is that also in 1926, Study used the journal of the German Mathematical Society to attack some of Beck's work "since 1917." He made Beck appear a clumsy, self-glorifying mathematical incompetent in conformity with the nonmathematician Kahle's view.[71] This was not the first time Study had taken Beck to task—an appendix to an earlier paper had made effectively the same charges. Nor was Study alone; in 1927, Ludwig Bieberbach's review of a semi-popular mathematics book by Beck also exhibits Beck as an incompetent mathematician.[72] Whatever the justice of this quarrel, Weiss was clearly caught between his erstwhile mentor and his present one. It is not impossible that Beck was jealous that the one person who was truly his protegé had become attached to his ex-teacher and present enemy Study.

Following his admission to the faculty, Weiss did study for three semesters in France, but then returned to Bonn, where, in May 1932, he became associate professor. With the help of Wilhelm Süss, himself a geometer, and the famous algebraist Wolfgang Krull, then director of the Bonn mathematics department, his financial situation improved somewhat in early 1939. However, Beck's earlier pessimistic prognosis for Weiss's career was not unjustified, even though Weiss's breadth of interest expanded considerably and he had publications and students of his own. It seemed impossible for him to advance anywhere to Ordinarius. His closeness to Study's own narrow, geometrical way of thinking and the booklet "Mathematics—For What?" were sometimes blamed for this.[73] For example, as late as July 1939, he was suggested in vain for a free chair at the University of Prague (at the same time that Rudolf Weyrich was angling for such a chair).[74] Weiss's difficulties were at least in part because "geometry [of the Weiss-Study sort] in institutions of higher learning [was] dying out" and because of his own lack of interest in applications.[75]

In December 1933, Weiss was called to one of the ten-week "sports camps" already discussed. He had shortly before joined the SA,[76] and the idea of these "sports camps" was apparently an inspiration to him. Weiss seems always to have been successful in a military or quasi-military context. Though a teenager

---

tated" on May 19, 1926, but only after, earlier in May, Beck continued to try to prevent it. See "Unterlagen," as in note 63, and Beck letter of May 12, 1926, in Personalakten Weiss.

[71] *JDMV* (1926): 1.

[72] For Study, see *Mathematische Annalen* 91 (1924): 249–251; for Bieberbach, see *Deutsche Literaturzeitung* (1927) (Heft 4): 176.

[73] See Dekan Pfeiffer to Rektor at Bonn, Aug. 1, 1936, in Personalakten Weiss.

[74] See chapter 4, "The Winkelmann Succession," and a letter of July 12, 1939, to Herr von Burgsdorff, in BDC file for Ernst August Weiss.

[75] Erich Salkowski to Hans Beck, Feb. 11, 1939, in Personalakten Weiss, turning down Weiss for a job at the technical university in Berlin. Various other failed attempts at finding him a job are alluded to in Personalakten Weiss.

[76] July 8, 1933, according to NSLB card in BDC file on Ernst August Weiss. According to Strubecker (1943), Weiss had had no political activity prior to January 30, 1933.

when he joined up in 1917, he was a lieutenant by war's end. In 1936, he was "brigade-adjutant" in his SA troop, less than three years after joining. According to a 1935 report on Weiss by a certain Herr Peters (presumably a Nazi student leader at Bonn), Weiss's attitude toward the SA was that because he had not joined prior to January 30, 1933, he had to make up for this by particular diligence and zeal.[77] The early death of Weiss's father, his own military imprisonment, and the expulsion of his mother provided cause enough for resentment and devotion to Hitler's "new Germany," to which were added the frustrations of his career as a mathematician. Nazis, of course, blamed these difficulties on Jews. Thus Peters:[78]

> W[eiss] . . . avoided every connection with his unfortunately in large part Jewish colleagues and thereby made his career much more difficult, since these gentlemen, as substantiated, still possess an always noteworthy influence on the distribution of academic teaching jobs, full professorships, etc. W[eiss] went so far that, for example, he avoided seminars organized by Jews in which the collective teachers of mathematics occasionally assembled. . . . Precisely in Bonn this had, in my opinion, already worked out very unfavorably for him and will in the future be even more unfavorable. . . . As I learned through indiscretions, this has already worked out unfavorably in Giessen.

There follows a description of how Ludwig Schlesinger (a Jew) had, in 1930, pushed forward his candidate for a position, while Weiss (in second place) was not pushed in the same way by his Giessen patron. Now in 1935, he was not even considered as successor to the forcibly retired Hans Mohrmann.[79]

In the "mathematical camps," Weiss's belief in National Socialism, the promotion of his sort of mathematics, and a quasi-military life, not to mention romanticism, were blended. The first such camp was held in spring 1934[80] and was explicitly a carrying out of Weiss's ideas set forth in his pamphlet.[81] Eight students (six male, two female) together with Weiss met for two weeks in a ruined castle in Kronenburg in the Eifel, low mountains in the vicinity of Bonn. They were going to "work in a National Socialist spirit," but their first act, after hoisting the swastika flag over the castle ruins, was to announce themselves to the local SA troop, in whose activities they participated. The day's regimen was (in the twenty-four-hour clock): 7:00, early sports; 8:00, breakfast; 8:30 to 12:30, mathematical work; 13:00, dinner, followed by hiking; 16:00, mathe-

[77] Gutachten über Prof. Dr. E. A. Weiss, Feb. 11, 1935, in BAK R21 838.

[78] Ibid.

[79] The great irony, of course, is that at Bonn, Weiss had been supported by a Jew, Felix Hausdorff, while opposed by his former mentor and future Nazi, Hans Beck.

[80] March 1 to March 15, 1934 (*Das Mathematische Arbeitslager Kronenburg* [Bonn, 1934], 5). Weiss had been ordered to the ten-week sport camp beginning January 7, 1934. March 1 thus fell in the middle of his eighth week of mandated camp activity. Whether Weiss was excused, his service postponed, or his mandated term shortened is unknown to me, but I would guess postponement (see letter to the Kurator at Bonn, Dec. 17, 1933, arranging for Weiss's absence). It is also possible that a "work camp" of shorter duration was substituted for the ten-week camp.

[81] *Das Mathematische Arbeitslager Kronenburg*, p. 7. Information below comes from this source.

matical review session; 17:00, lecture or singing; 19:00, supper and free time. The main morning mathematical session was devoted to geometry, the afternoon "review" (*Repetitorium*) to analysis. However, toward camp's end, the afternoon was geometrical as well. Some of the lectures dealt with Italo Balbo's book *March on Rome*, a diary of Mussolini's triumphal progression by one of its leaders. Balbo was Mussolini's air minister, and Hermann Goering wrote the introduction to the German edition. (The sort of geometry favored by Weiss had been promoted by a substantial school of Italian mathematicians, some of whom were fascist sympathizers.)

From the point of view of mathematical pedagogy, the camp is quite interesting, despite its clear political content and goal of "working in a National Socialist spirit." It built on a lecture series given the preceding semester, and its main idea (as already adumbrated in "Mathematics—For What?") was to give students a time to concentrate on an independent investigation, no matter whether "science was served" or not. Some new results might emerge, but the main purpose was to educate students to the meaning of mathematical activity and to do so in a context that would use mathematics to build "character," as Weiss had already polemicized in his pamphlet. Explicitly:[82]

> Pure mathematics (in the narrowest sense) has a meaning only as a means of education to a formal character-building that is consciously employed for service to the entire people (*Volksganzen*).
>
> Mathematical character-building: that is, cultivation of the masculine principal in spiritual life. We call this formal because the direction of its employment, seen from its own point of view, remains undetermined. Therefore, it is especially necessary to further this direction, and thus we have added the word "consciously" because we wish to distinguish ourselves from those mathematicians who are indebted to a different ideal for the spiritual impulse to their work. This is a setting of aims incompatible with the concept of science, perhaps even perceived as unworthy by science, and only after the fact can say of itself: "What splendid people we are though. People such as us can be used by the state!"
>
> More than ever we turn our backs on those who do not wish this avowal to emerge from their lips, on those who can find an answer to our question [of what is mathematics for] only in the sense: mathematics is for those who are so captivated by its beauty that the question therefore no longer exists for them.

Indeed, a purpose of the mathematics work camp was to "eradicate" the type of mathematician whose "instruction that is inexperienced in the ways of the world has brought and still brings mathematics ill fame and thereby endangers generally the possibilities of existence for pure mathematics."[83]

Paul Kahle says of Weiss: "Others [at Bonn] who became Nazis were not at all honest, either politically or socially, such as the mathematician Ernst Weiss, not

[82] Ibid.: 10.
[83] Ibid.: 8.

a cipher but most unpleasant in his whole character."[84] Certainly the photograph of Weiss in the Bonn archives seems to show a stern and humorless young man, and it would be understandable had he been full of resentment. However that may be, he seems to have been a gifted teacher. Although his was a narrow and unmodish mathematical specialty, he seems to have been successful with students, even though (because?) his pedagogy was emblazoned with sincerely felt Nazi rhetoric and political indoctrination. Thus, the aforementioned report on Weiss concludes:[85]

> I further remark that W[eiss] in my opinion has worked in all-too-close understanding with the student body, although it was a priori certain that he could not receive their undivided applause permanently, since he did not do one thing: he seldom *talks* about National Socialism, new science, companionship, rather he *behaves* as a National Socialist, does science, and while he achieves something in his specialty, furthers achievements among his students, and is a comrade even to the point of personal sacrifice that does not stop, as with most people, where the money pouch begins.

Weiss seems to have discovered for himself the efficacy of small-group learning in a concentrated atmosphere.

In any case, the mathematics work camp was so successful that a second one, of much the same format, was held in October 1934 in the same location, with two of the previous male students and six new ones (four male, two female). A third followed in February 1935 with again some old and some new students, some male and some female. The return of the Saar to Germany (March 1, 1935) occurred during the camp and was duly celebrated with the Kronenburg villagers. By this time, news of these camps had gotten around; this was, after all, the third within a year's time. A visit came from a representative of the *Studentenschaft*; on March 1, the leader of the university mathematics students nationally, Fritz Kubach,[86] visited as well, and discussed the possibility that mathematical work camps might be built in as an organic part of university studies. While this did not happen, Weiss's idea did spread to several other universities with seeming success.[87] Indeed, it was received so enthusiastically that in July 1938, by popular demand, so to speak, a mathematics camp for mathematicians from all over Germany was organized by Ludwig Bieberbach in Bernau for three days, and a second was planned for May 1939.[88] There was a

[84] Kahle 1945: 18. Kahle distinguished "honest" men who were Nazis from "dishonest" ones. Compare the "honest" mineralogist Karl Chudoba (ibid., and see note 105 below).

[85] Gutachten über Prof. Dr. E. A. Weiss, Feb. 11, 1935, in BAK R21 838.

[86] Fritz Kubach was a mathematics student who became national leader of the natural science (and mathematics) students in the Nazi national student directorate.

[87] *Deutsche Mathematik* 1 (1936) contains in addition to a report of the fourth Bonn mathematics work camp (the first three of which were already "widely known" [p. 12]), reports of such camps for students at Giessen (two in 1935), Hamburg (two, 1935, and over the New Year 1935–36), and Heidelberg (a general ecological emphasis to the scientific work, but involving also mathematics students), over two months.

[88] See *Deutsche Mathematik* 4 (1939): 109–139, for a report, and ibid.: 276, for an announcement.

fourth such camp for Bonn students from February 23 to March 7, 1936, and a fifth in October 1938. In all, twenty-five students (eighteen men and seven women) participated in the Bonn camps.[89]

Weiss's own thought underwent some development, as revealed by his commentary on these camps. Already in "Mathematics—For What?" Weiss had (anonymously) cited Évariste Galois and Jean Victor Poncelet as the sorts of mathematicians to be admired for bringing order and system to parts of mathematics. In this pamphlet, the romantic-heroic aspects of these mathematician's lives are used as tags to identify them—Galois was killed in an (illegal) duel; Poncelet was a soldier in Napoleon's army who, during the retreat from Moscow, was taken prisoner in 1812 and spent the next two years in a Russian prison camp, where, devoid of books, he first developed his views on projective geometry.[90] However, in the report on the fourth camp (1936), Poncelet is cited by name, as is Gaspard Monge, a distinguished French geometer who was active as a French revolutionary, had military, political, and educational posts under the Committee of Public Safety and Directoire, and became a friend and counselor of Napoleon, rising to much esteem under the Empire. Poncelet is cited for his development of evening courses for workers and is cited (in French quotation) for his desire "to make mathematics useful to the class of artists and engineers by writing for the multitude"; Monge, the "friend and counselor of Napoleon," for being instrumental in the creation of the École Polytechnique. "Men such as Poncelet can only arise from a school especially suited for the education of such types." Also, a volume of Monge's "shows what is possible through common education in the military and mathematical spirit." Thus Weiss is seriously concerned about school reform, and about making mathematics more understandable to nonspecialists. Though Weiss certainly had enough personal reasons to hate the French, it is striking that his one substantial sojourn away from Bonn[91] was at the École Normale Supérieure in Paris, that he makes French geometers into heroes, and that the report just cited has several quotations in French. The fact that he apparently grew up with both the French and German languages seems inadequate to explain this. Clearly, what counted for him was a system of mathematical education that he thought, in the spirit of his earlier pamphlet, created upright men, ready to serve the state, by dint of its insistence on rigor and logic. The French had produced such heroes. The German pedagogical system of 1936 was, in his eyes, wanting: teaching mathematics because of applications, or for the sake of knowledge, or to "advance the subject," but not for its character-building qualities.[92]

[89] See "Verzeichnis der Lagerkameraden," in "Das mathematische Arbeitslager Kronenburg 5. Lager" (1938).

[90] WM, pp. 15–16. According to Strubecker (1943: 255), Weiss spent part of his childhood in Metz, which had been Poncelet's birthplace and home, and after Weiss became a geometer (like Poncelet), he became fascinated with Poncelet's career and similarities between it and his own.

[91] Weiss also studied briefly at the technical university in Hannover and the University of Hamburg; see Strubecker 1943: 256.

[92] "Das mathematische Arbeitslager Kronenburg 4. Lager" (1936), 7–8.

Weiss's heroizing of some French mathematicians seems at first to contrast with the idea among some Nazi mathematicians that there was a peculiarly French, or peculiarly Jewish, or peculiarly German way of doing mathematics.[93] However, Weiss supported Ludwig Bieberbach's 1934 paper on styles of mathematical creativity, emphasizing exactly the issue of style as opposed to content. Thus,[94]

> That mathematics is indispensable "for the people"[95] is uncontested. That mathematics arises "from the people," more precisely, "that the doing of mathematics belongs to the development of the German nature," remains to be shown. We reject the criterion that mathematics arising from the German people must have characteristic features distinguishing it from the features of mathematics of other peoples. . . . A particular kind of mathematics cannot be characteristically German; on the contrary, only a particular spiritual attitude to the doing of mathematics, a valuation of the doing of mathematics emanating from the whole personality, can be particularly German.
>
> In the same way the soldiers of different armies are distinguishable less through their type of warring and only outwardly through their uniforms, but actually through the thoughts that lie at the base of their soldierhood.
>
> Therefore we do not ask: "What is characteristic for German mathematics?" but which spiritual attitude corresponding to the German nature lies at the base of our doing mathematics. It is the thought of breeding, of order, of cleanliness that in the realms of science reveals itself most purely in mathematics. . . . We have thoughts about the higher meaning of mathematics only on special holidays [and not while doing it]. The National Socialist breakthrough introduced such a holiday for us.

Indeed, Weiss also gave an example of mathematics he disliked, again linking it politically and stylistically:[96]

> The geometer Poncelet had an opponent in the analyst Cauchy,[97] who was active politically in the sense of political Catholicism. . . . In the scientific conflict between Poncelet and Cauchy we see not a conflict over the objective truth-content of certain mathematical theorems, but a struggle between two different styles conditioned by the opposite characters of their advocates.

For Weiss, the "spiritual attitude" of different mathematicians was not simply determined by nationality. Poncelet was a republican who, in addition to founding projective geometry as a new mathematical discipline, had done significant work in applied mathmatics. Cauchy was a royalist and devout Catholic (and mathematical opponent of Weiss's hero Poncelet). Poncelet insisted on the primacy of geometry, Cauchy on that of analysis.

[93] For "Deutsche Mathematik," see below, chapter 7. Weiss became an active collaborator on this journal.

[94] "Das mathematische Arbeitslager Kronenburg 3. Lager" (1935), 5–6.

[95] "People" in this citation especially inadequately translates the German *Volk*.

[96] As in Weiss 1936: 8.

[97] For Ludwig Bieberbach also, Cauchy was an example of mathematical style to be avoided; see below in chapter 7. Cauchy attacked the "principe de continuité" of Weiss's hero Poncelet.

Weiss would probably have been an excellent pedagogue in any era, but in his own, the "times were out of joint." His mentor-nemesis Beck, on the one hand, kept striving to get him a job away from Bonn and, on the other, treated him anything but collegially. For example, in 1936, Beck was the mathematics department head; on March 27, he sent Weiss a sharply worded note complaining that as an *Assistent*, Weiss was obligated to have asked him for leave to hold the mathematics camp three weeks previously.[98] That same November, in a letter of recommendation for Weiss, he somewhat slighted the camps, at the same time reluctantly admitting their success.[99] Nevertheless, Weiss's idea was undoubtedly successful, and a sixth camp was planned for October 1939. This, of course, never took place. Weiss did finally receive his professorship on October 1, 1941. After nearly eighteen years as *Assistent* at Bonn, he was appointed Ordinarius at Posen. However, by this time Weiss was in the army, and so never went to Posen.[100] On February 22, 1942, his wife was sent the battlefield news:[101]

> Your husband, Captain Ernst August Weiss, came into our field hospital on February 4, 1942, with a bullet wound in the belly accompanied by the destruction of his spleen and injury to his small intestine. Despite every possible immediate medical aid, his general condition worsened. An inflammation of the lungs caused by his wound developed. With increasing circulatory weakness he died on February 9, 1942, at 6:30. The dead man did not suffer during his time in the field hospital.

Not only are Weiss's mathematical camps an example of mathematical pedagogy and National Socialist ideology accommodating one another, but also Weiss's "program" for mathematics in the Nazi state is one of several formulated or sketched by Nazi mathematicians.[102] The camps were probably a good pedagogical idea even without all the ideological overlay;[103] however, they were inspired by Nazi ideology and were immersed in it. It is not unlikely that the enthusiasm of working for the *Führer* by doing mathematics, and thereby building Nazi character, helped provide the enthusiasm necessary for successful mathematical work.

[98] Beck to Weiss, Mar. 27, 1936, in Personalakten Weiss.

[99] Beck to Prof. Leick, Nov. 14, 1936; Dekan Pfeiffer to Kurator, Aug. 1, 1936; Wolfgang Krull to Dekan, June 21, 1939; *Nachruf* for Ernst August Weiss, Mar. 31, 1942 by Prof. Carstens, Rektor at Posen, all in Personalakten Weiss.

[100] C. Kröger, Dekan at Posen, to Dekan at Bonn, Mar. 21, 1942, in Personalakten Weiss.

[101] Dr. Thunn to Eva Renate Weiss, Feb. 22, 1942, copy in Personalakten Weiss.

[102] For Ludwig Bieberbach and his followers, see below, chapters 6 and 7; for Udo Wegner, above, chapter 4, "Hasse's Appointment at Göttingen," and below, chapter 7; for Max Steck, below, chapters 6 and 7. Also note the uses of mathematics by those concerned with primary and secondary education at the end of this chapter.

[103] André Weil, in *The Apprenticeship of a Mathematician* (1992) (translation of *Souvenirs d'apprentissage* by Jennifer Gage), says, "Somewhat later [than the first Bourbaki congress in July 1935], Nazi mathematicians in Germany had the idea of organizing 'Work camps' on the model of the *Arbeitslager*. . . . Since then, the institution has spread throughout the world . . . and has become one of the most common ways of channeling government subsidies to scientific activities" (p. 105). Weil only has the date of precedence wrong, since Weiss' first camp for Bonn students preceded the first Bourbaki congress.

Students and Faculty before and during Wartime

Of course, the war brought other difficulties—for example, on July 11, 1942, Erich Bessel-Hagen asked the Rektor if there were some way he could obtain necessary writing paper.[104] Bessel-Hagen was asked to contribute, and did, to the sending of a little book about the University of Bonn and two lectures on the war to the men at the front. There were also, of course, the annual contributions to the *Winterhilfswerk* for the less well-off in wintertime; during the war, these became spending for the war as well, and the amounts to be given were predetermined. Faculty were asked to donate books to students—Bessel-Hagen demurred, saying that unfortunately, he had no books in which students would be interested. By mid-1940, the Rektor at Bonn had become Karl Chudoba, a mineralogist, "an honest man, a good assistant, very helpful to the students, but no scholar," who progressed to the *Rektorat* by way of being *Dozentenführer*, and an appointment leading the mineralogical institute, largely on his political credentials.[105] Contrary to pre-Nazi academic practice, he held the post continually thereafter. In September 1944, Chudoba said he had no information as to what people should do if the situation on the west bank of the Rhine should become more acute, but that all faculty (except those sick or ailing) who were not members of the legal, medical, or political-science faculties, or engaged in vital war work, were expected to help man the western defenses.[106] Bessel-Hagen's chronic illness excused him, but he had already been compulsorily involved as an air-raid warden (under threat of arrest should he either act contrarily or reveal secrets learned while on duty—and negligence was no excuse).[107] Indeed, as early as January 1934, every German was to belong to the national league for air defense (*Reichsluftschutzbund*), as a "duty to honor and fatherland," and was consequently required to make a contribution thereto. On October 18, 1944, Bonn University was destroyed by an Allied bombing raid—ironically, it was the anniversary of its foundation in 1818. The mathematics department was formally transferred to Göttingen. The faculty such as remained, however, stayed in Bonn.

I use the phrase "such as remained" because gradually all able-bodied men who were not over-age (and university faculties were almost totally male) were drawn into the armed forces. Similarly, the student body was depleted. However, even before the war, National Socialist policies had led to a decline of the student body. The total number of German university students in 1938–39 was half what it had been in 1932–33,[108] while the proportion of students in natural

[104] Personalakten Bessel-Hagen, Bonn.

[105] Kahle 1945: 18. It should be noted that Kahle left Germany in 1939.

[106] Bonn is on the west bank of the Rhine.

[107] The official form was sent to Bessel-Hagen on October 12, 1941.

[108] Cäcilie Quätsch, *The Numerical Record of University Attendance in Germany in the Last Fifty Years* (1961), 42.

science (including mathematics) remained constant at about 9 percent.[109] Thus the number of students in mathematics in 1938–39 was presumably half what it had been in 1932–33. In 1934, compulsory service in the national labor force (*Arbeitsdienstpflicht*) was introduced, and in 1935, compulsory military service. These, of course, resulted in postponement (and perhaps abandonment) of studies by young people.[110] Under National Socialist policies, the number of female students dropped even more, the proportion going from about 18 percent to about 14 percent of all students. About 12 percent of these women were enrolled in mathematics or natural science. While clearly the decline in the German birthrate after World War I had something to do with the decline in student population in the 1930s, National Socialist policies and attitudes such as those already mentioned were its primary cause.[111] In addition, of course, the number of students tends to decline in economic boom times, and increase in moderately hard economic times.[112]

Similarly (correspondingly?), the 1930s saw an equivalent decline in university faculty prior to the onset of World War II. In 1938, the number of mathematics faculty (at all ranks) in German universities was approximately 75 percent what it had been in 1931.[113] However, about 50 percent of the 1931 mathematics faculty was no longer teaching in 1938.[114] Thus the onset of the war only accelerated a trend already begun. Though people like Ernst Weinel and Werner Weber would not be called up until 1941 prior to "Barbarossa," and enthusiastic veterans like Gustav Doetsch and E. A. Weiss might always volunteer, conscription guaranteed that, as the war went on, there were ever fewer students, and ever fewer faculty to teach them. Thus, though Bonn's mathematics department was left largely in the hands of Erich Bessel-Hagen and Hans Beck (and Beck died in 1942), it hardly mattered.

These statistics and the documentary evidence of the preceding sections have been cited in an attempt to provide the context in which students and faculty did academic mathematical work. Weiss's popular Nazi-inspired pedagogical innovation (which might even be considered to have positive aspects) was discussed as an example of how university teaching of mathematics was conceived by some dedicated National Socialists. However, documentation cannot provide the "feeling" of a period by itself; while personal narrative is subject to the strictures of subjectivity and the plague of an idiosyncratic point of view, it also provides the nuance of emotive detail that can only be inferred from documents. Thus the two kinds of source, when available, need to be used as com-

[109] Ibid.: 43.

[110] Ibid.: 18.

[111] For the above statistics and a discussion, see ibid.: 17–21.

[112] For discussion, ibid.: 23–26.

[113] Christian von Ferber, *Die Entwicklung des Lehrkörpers der deutschen Universitäten und Hochschulen, 1864–1954* (1956), 216.

[114] Ibid.: 143–145. For a discussion of the difficulties of assessing the exact loss through emigration and persecution, see p. 143.

plementary to one another, providing symbiotically a clearer picture than either alone would.

Heinrich Behnke was a famous mathematician who has left his picture of what life under the Nazis was like. Because some of these impressions are unpublished, and none have been translated into English, a partial recapitulation seems apropos.[115] Peter Thullen was Behnke's most brilliant pupil of that time.[116] Thullen emigrated to Ecuador in 1934 (where he began a successful nonmathematical career) and was in Rome on fellowship from September 1933 to October 1934. However, he kept notes of his impressions of the early Hitler years that provide an additional view on Behnke's impressions. Thullen was a serious and convinced member of the Catholic youth movement, and his own emigration was because of his principled opposition to the Hitler regime, rather than for any "racial" reason or because of prior political activism.

Heinrich Behnke was born into a Lutheran family in 1898 in Horn, a suburb of Hamburg; in the autumn of 1918, he became a mathematics student in Göttingen. On May 10, 1919, a new German university was opened in Hamburg. The first mathematics professor appointed was Wilhelm Blaschke,[117] and the second, in whose choice Blaschke participated, was Erich Hecke, then at Göttingen. By this time, Behnke had become Hecke's student, and when Hecke decided to accept the offer, he made it clear that Behnke would go along with him to Hamburg. Although reluctant to leave his freedom as a student away from home and return to the neighborhood of his family, Behnke naturally did so, and in May 1922 received his doctorate. Thereupon, he traveled for the rest of the summer semester to Heidelberg, where he met Aenne Albersheim from Frankfurt-am-Main, and studied philosophy (with Karl Jaspers among others) and mathematics. In August, he returned to Hamburg and, the following April, received an *Assistent*'s position there. In summer 1924, Behnke completed his *Habilitation* at Hamburg, thus becoming a *Privatdozent*. As a consequence, in 1925, he married Aenne Albersheim, who came from a Jewish family.[118] In 1927, however, she died of childbirth complications,[119] and Behnke was left widower and father to a son. His in-laws assisted at the time with the child. Later that same year, Behnke, not yet twenty-nine years old, received a call to become Ordinarius at Münster without having passed through the intermediate degrees of professorship. Münster was a small provincial town, in a very Catholic region of Germany, and Behnke, given his own religious background, was

[115] Material not otherwise annotated in this section is from Behnke 1978; see note 11 above on Behnke's memoirs.

[116] Behnke 1978: 110–111, 115. The book by Behnke and Thullen (ibid.: 80) is still a standard work in the subject of several complex variables.

[117] For Blaschke, see particularly chapters 6 and 8.

[118] Behnke avoids mentioning this directly in *Semesterberichte*. It does appear in a letter from Behnke to Hecke in HNMA, and was mentioned to me by both Karl Stein, also Behnke's student (interview, Feb. 26, 1988), and Peter Thullen (interviews in Fribourg, Switzerland, Mar. 21–23, 1988).

[119] Interview, Peter Thullen, Fribourg, Switzerland, Mar. 21–23, 1988.

somethat anxious about accepting the position. In fact, Münster's greatest historical fame perhaps is as the site of the development and destruction of the radical Anabaptist movement under Johann Bockelson (1534–35). The university had been refounded in 1902 (the first founding of 1771 had lapsed in 1818, though a philosophical-theological academy had remained for the next eighty-four years) and was the smallest Prussian university. Prior to his meeting with a ministry official, Behnke consulted three colleagues (all coincidentally assimilated Jews), two of whom had taught there.[120] Remembering a knowledgeable teacher of his who had declared, "Münster is better than its reputation," Behnke accepted the position, though still with some apprehension. His mother-in-law and brother-in-law accompanied him as far as Kassel, but refused to go farther.[121] In fact, as Barbara Marshall has shown, in those years of late Weimar, Münster's Catholic conservatism prevented it from becoming the hotbed of anti-Semitism and Nazi-leaning nationalism that the town of Göttingen was.[122] In 1932, Behnke married again. His second wife, née Elisabeth Hartmann, had been a mathematics student at Münster and came from an intellectual, but consciously Catholic, family.[123] Presumably sometime thereafter, Behnke's son came to live with his father and stepmother. With the coming of the Nazi regime, according to Peter Thullen, Behnke lived in constant fear that the Jewish antecedents of his son would be discovered, and this fear transmitted itself into a constant anxiety affecting his everyday living. Indeed, Thullen has remarked that already in April 1933, Behnke feared any sort of political talk.[124]

Although Behnke claims to have perceived a "subterranean unrest" in German academic life in 1931 and 1932, and to have heard of the inevitability of an impending war as early as 1931,[125] not until after January 30, 1933, was the lecture of a Jewish professor (a guest) disrupted at Münster.[126] As discussed in chapter 3, this was not necessarily the case elsewhere and again gives credibility to the thesis that the conservative Catholic atmosphere in which the university at Münster dwelt kept it from excesses prior to Hitler's accession to power.

While at a general university meeting in Göttingen during the Lenten holiday, Behnke recalled reading in the newspaper that eleven university professors had been dismissed, and that the general feeling was, "So few, that's not so bad."[127] By the time he returned to Münster a few weeks later, one of his col-

[120] Behnke 1978: 84.

[121] Ibid.: 84–85.

[122] Barbara Marshall, "The Political Development of German University Towns in the Weimar Republic: Göttingen and Münster, 1918–1930," Ph.D. thesis, University of London, 1972. This is one of the persistent themes in this thesis. E.g., p. xv: "The process of 'Nazification' in Göttingen was well under way by 1930. . . ." In contrast, in Behnke's Münster, the town was under the control of the Catholic Center party up until the very end.

[123] Behnke 1978: 113.

[124] Interview, Peter Thullen, Fribourg, Switzerland, Mar. 21–23, 1988.

[125] Behnke n.d.: 4–5.

[126] Ibid.: 5.

[127] Behnke 1978: 117. The Göttingen newspapers for April 15, 1933, reported the suspension of

leagues had sent him a letter saying he might only visit after 7 P.M., lest he endanger his host, and another letter warned him against openly buying the *Frankfurter Zeitung* (a liberal newspaper).

At the faculty meeting to choose the next Rektor in Münster, a candidate was put forward who formerly had been thought right-radical, and who had been a *Freikorps* fighter in the Baltic—plainly a conciliatory gesture to the government. However, suddenly, a man of nearly sixty with a great beard,[128] appointed only shortly earlier, arose and declared that according to the *Gauleiter* of North Westphalia,[129] he should be chosen Rektor, and asked for the votes of those devoted to the new state. Voting by secret ballot, the faculty rejected him for the *Freikorps* fighter, also a National Socialist. Some thought it a mistake; but the result was another one of those internal struggles among National Socialists that was seen in the story of Hasse's appointment at Göttingen in chapter 4. The rejected man, together with the activist students, made life difficult for their National Socialist colleague who was Rektor. An atmosphere of denunciation and dismissal permeated the university. Overheard remarks of an idle moment were recalled to the disadvantage of colleagues; one often learned of one's own dismissal from the newspapers. Students officially informed on their professors, assistants on their bosses. Behnke himself says he was under suspicion "because of unclear international connections and friendships with Jews."[130] The resulting isolation of individuals was accompanied by innumerable compulsory gatherings and associations.[131] Often, the leaders at such meetings were students. This demonstrates again the same inversion of faculty prerogatives met earlier. A student gave orders to the faculty; the "new youth" would free the university from its "calcification."[132] The Nazi regime understood very well indeed that in the universities as elsewhere, the quickest way to penetrate traditional structures that might offer resistance was to overturn them with the aid of natural internal allies.

In the mephitic atmosphere described by Behnke, no one spoke meaningful opinions, and gossip was rife.[133] Each sought advantage over his neighbor, and to keep the party's "interest" far away. Common ground was lost except for the artificial connections described and compelled and the injunction "to think and

---

sixteen professors at other universities. See Becker et al. 1987: 26 (introduction by Hans-Joachim Dahms).

[128] Behnke avoids his name. According to Prof. Karl Stein, his name was Baumstark.

[129] Behnke (1978: 119) claims there was only one *Gau* (or provincial district; its Nazi leader was called a *Gauleiter*) for all of Westphalia. Münster is in the northern part of this traditional region.

[130] Behnke 1978: 120; cf. Behnke n.d.: 8, where Behnke says "pacifist leanings and friendship with Jews."

[131] Cf. the beginning of this chapter.

[132] Behnke 1978: 123.

[133] Throughout his book, Behnke is especially hard on "wives" as purveyors of gossip, pro-Nazi sentiments, and denunciation, including the wives of those denounced. A possible reason for this misogyny is unclear. A specific case is alluded to in HNMA, Behnke to Hecke, Dec. 30, 1941 (though the wife in question was divorced at the time she denounced her ex-husband).

feel National Socialistically."[134] Behnke knew students who had been religiously Roman Catholic, only to "convert" to being party members with the same inward spirituality.[135] He knew of a sensible introverted student who marched before his *Führer*, one among many, many tears of excitement flowing down his cheeks. However, Behnke's student, Peter Thullen, was not such a one. A very active member of Catholic youth organizations (which were dissolved in July), he possessed a study grant for Italy awarded in 1932,[136] and felt fortunate to be able to leave Germany for Rome in October 1933, but it was clear to him (and he made it clear to Behnke) that he could never return to live in Germany while Hitler was in power. When his research stipend was over, Thullen refused German offers of jobs. Protected by a "German-living-abroad" passport with an Ecuadorian visa, he met his wife-to-be in Germany, they were married in Maria Laach,[137] and shortly afterward they left for Ecuador, which was where he could find a suitable job, though he didn't even properly know the location of Quito at the time. A man like Thullen felt constrained to leave because he understood the Nazi regime meant a realignment of all values in the name of a strident and oppressive nationalism—that it meant the end of culture. His picture of the early Nazi days is much like Behnke's: sudden conversions among students and others; parades; students feeling compelled to join Nazi organizations to be assured of passing their examinations; the attacks on the Jewish spirit that must be combatted; in general, the deliverance of the universities into Nazi hands. This seems to have happened much more easily than even the politicians and certainly the bulk of the faculty had imagined beforehand. In Thullen's view, this same basically ill German nationalism that Hitler had easily perverted to evil was still present in the emergent Adenauer government in West Germany. Thullen remembered the infamous call of the German students' organization against the "un-German spirit": "A Jew cannot write German. Were he to write German, he is lying. We demand that Jews write only in Hebrew; German editions are to be considered as translations from the Hebrew, and to be denominated as such."[138] Thullen thought such sentiments laughably ridiculous in 1933, but clearly many of his fellow students thought differently. According to him, also, Eduard Spranger was the only German professor who seriously "dared" to oppose such attitudes.[139]

What did a man like Behnke think of Thullen's asserted inability to live under Hitler's government? He wrote,[140]

[134] Behnke 1978: 125.

[135] Cf. above, Gustav Doetsch, in chapter 4, "The Süss Book Project."

[136] Thullen had had a research stipend to study in Italy beginning October 1932; however, as the result of illness, he could not take it up, and so his assumption of the stipend was postponed until October 1933. Interview with Peter Thullen, Fribourg, Switzerland, Mar. 21–23, 1988 (also the source of all personal material about him below).

[137] Maria Laach is a famous Benedictine abbey founded in 1093 and beautifully situated on the shore of a volcanic lake.

[138] This is from a leaflet that appeared at universities throughout Germany.

[139] For Spranger's lack of success, see above, chapter 3.

[140] Behnke 1978: 125.

> Had I known in 1933 all that still awaited us, I would not have found the courage to avoid flight abroad. Also in the following years of the brown storm I still frequently played with the idea of leaving for my own sake, until Jewish friends made it clear to me that places [abroad] must be held open for them.

Yet four pages later we find:

> Thus was the dignity of office and human dignity taken from us. Weren't those who had to lose their positions luckier? That was frequently said, and yet most colleagues defended their position. Naturally it makes an impression if someone simply leaves with a head held proudly high before the brown bureaucracy regulates and then perhaps dismisses him.[141] In certain cases such vanishing of one's own free will doesn't cost much, for example, when one is a successful young mathematician.[142] However, for me such freedom from ties (*Ungebundenheit*) was weird. For me those people were more sympathetic who had difficulty leaving their German university. If they had taken their profession seriously, so necessarily they left a substantial piece of their own life behind in the old place.

Behnke was thus, on his own testimony many years later, like many, conflicted as to the countervailing claims of staying to protect as possible one's university, and leaving. As mentioned earlier, there were mathematicians like Carl Ludwig Siegel who at first thought to stay, but then managed to leave when they realized the true situation.[143] Behnke's statement about leaving posts open for Jews also rings true. For example, the well-known mathematician Wilhelm Magnus told me that as a young man, when he was on a research grant at New York University in 1934 and wished to stay in the United States, Richard Courant, by then already emigrated, told him that he *had* to go back to Germany, because he *could* go back to Germany, to make room for those who could not do so.[144]

As to Behnke's personal experiences, he (like others) found notices on his door asking why he had not yet disappeared,[145] yet his lectures were not boycotted; a mathematically incompetent student (though politically very competent) affected an appointment of a student of his;[146] he was twice denounced for "foreign contacts," and a list of these was consequently demanded of him;[147] a drunken student who was a member of the SA boasted in a pub that he exam-

[141] For example, the physicist James Franck, who resigned from Göttingen prior to any dismissals there, or the famous aerodynamicist Theodor von Kármán, who left Aachen voluntarily shortly after Hitler's accession to power.

[142] This is clearly a reference to Thullen, who felt justifiably injured and misunderstood by it. This slighting reference to Thullen is not present in the corresponding passage in Behnke n.d.

[143] See chapter 2.

[144] Wilhelm Magnus, interview, April 1982.

[145] Behnke 1978: 125.

[146] Ibid.: 129.

[147] Ibid.: 126.

ined all of Behnke's mail and then detailed to his hearers remarks about Behnke's personal family relationships.[148]

However, gradually the wave of denunciation and accusation in which each sought to protect himself at the expense of others abated. In a sort of local Ninth Thermidor, the "bearded man" of the voting for Rektor, who had been the bane of the faculty until then, gathering innuendo and fomenting suspicion, was forced from the university.[149] Denunciation became a "profession" whose activity was carried on by subordinates, occasionally wives interested in their husband's careers[150] and party functionaries. Cleansed of Jews, the remaining faculty need only worry about the students who, inspired by the "new order," might consider them "calcified."[151]

What of these students? The situation in Münster was like the general one described earlier. In pre-Nazi times, elementary mathematics lectures at Münster had averaged about 200 students; in 1933, there were fifty such students, and in 1934, only one.[152] Similar local information is available also for Hamburg and Göttingen. In winter 1940 in Hamburg, after the university reopened, Hecke had a constant number of about thirty students in his elementary calculus class, of whom about one-third were mathematicians. However, with few exceptions, "they had a frighteningly low level of thinking."[153] At Göttingen around the same time, just after the war began, there were about twenty beginning mathematics students, and no advanced lectures.[154] Yet at Münster, at least before the war, the older, more successful students stayed, and the mathematics faculty was not as decimated as at some other institutions, thus attracting other good students. Indeed, Behnke says: "The scientific work of the seminar ran from 1935 to 1939 better than at any other time. Here the trust between professors and students was not troubled in the slightest."[155] In other words, though living every day in constant fear, and openly harassed from time to time by Nazi functionaries or Nazi colleagues, but untroubled by beginning students, Behnke found more time for mathematical research in the company of his advanced students, and this went better than ever before. There has hardly been a better description of what "inner emigration" meant, and, as has already been noted, this "solution" was particularly available to mathematicians. Indeed, as Behnke himself says:[156]

> The possibilities of defending oneself remained limited. . . . Mathematics, as an apolitical area and at the same time a science for whose validity and utility there was no

[148] Ibid.: 129.
[149] Behnke n.d.: 18–19, and 1978: 132.
[150] Cf. note 133 above.
[151] Behnke 1978: 133.
[152] Ibid.
[153] HNMA, Hecke to Behnke, Jan. 23, 1940.
[154] HNMA, Hecke to Behnke, Oct. 14, 1939
[155] Behnke 1978: 133.
[156] Ibid.: 135.

> political limit anywhere in the world, was particularly suitable for providing an umbrella against the brown storm.

In reflection on "living in the shadow of the brown storm," he adds in the same vein:[157]

> Each one had to defend himself alone or in a small circle. Even today, looking back, one would still say that it was better for the well-being of the universities that each stood alone. There is hardly a professional group that necessarily consists so much of individualists as the professoriat. . . . Organized conspiratorial resistance was never a question given the omnipotence of the state, our inadequate insight into its organs of control, and professorial lack of know-how (*Ungeschicklichkeit*).

On November 8–9, 1938, in the aftermath of *Kristallnacht*, Behnke says a number of his colleagues "defended themselves atrociously. They knew nothing, they heard nothing, and they emphasized that as researchers, they were accustomed to mistrust all news and only to worry about what was proved objectively after the fact."[158]

This man, who believed that only the isolated individual could best defend himself and that organized resistance was impossible (as it certainly was in 1938), says he shouted after such a colleague in anger, "Saint Peter will record this sin in his great book."[159]

In the summer of 1939, the lecture halls in Münster were empty; where formerly there had been perhaps five hundred mathematics students in all years together (Münster was a small university), now there were only twenty-five.[160] On his usual Swiss holiday, Behnke was in Chexbres when the Molotov-Ribbentrop pact was announced, and he recalled for the rest of his life parting from hotel friends at the railway station, they on the train to Paris, he on the one to Frankfurt. Directly across the border, mobilization was in full swing, delaying the return to Münster. The university, like all in Germany, was closed, in Münster's case for a semester.[161] While the onset of the war does not seem to have been greeted enthusiastically, the early German victories quickly changed this view.[162]

The university itself became militarized, and faculty meetings were chiefly for the purpose of communicating supposedly confidential political information. That the professoriat still in office should be indifferent at this final stage of their impotence hardly seems surprising, especially in the face of the early Nazi military success. Apparently, shortly prior to the war, the local *Dozentenführer*

[157] Ibid.: 136.

[158] Ibid.: 139.

[159] Ibid.

[160] Ibid.

[161] Adam 1977: 188.

[162] For Germany as a whole, e.g., Frei 1987: 132–133, 137–138. This accords with Behnke's impressions at Münster in 1978: 140–142. Cf. Siegel's view of Hasse in chapter 4, "Hasse's Appointment at Göttingen."

asked Behnke whether maps of English harbors that he had been given in confidence for scientific purposes should be made available to the military because of the "coming hostilities." Behnke replied no, because they had been given to him as confidential. After the war began, Behnke was informed that he was considered unendurable because of "morbid objectivity," and with a peace treaty he would be dismissed.[163]

The terror at the university, which had declined after the initial wave of 1933–34, increased. Already by 1939, as remarked earlier, the number of students was a small proportion of those in 1932; mostly only women were left as potential students, but under industrial and military pressures for trained people, this number began to rise.[164] Thus the earlier National Socialist policy that had especially discouraged female university study was reversed—not the only alteration the exigencies of war made on National Socialist ideology.[165] Thanks to a German-Swiss cultural exchange agreement, Behnke still managed to travel to his beloved Switzerland during the war, as well as to Romania. Most German universities were eventually severely damaged or destroyed by the Allied bombing raids, though the pretense of "carrying on" was maintained. Behnke's Münster was in fact the target of the first daylight Allied raid on a sizable German city in October 1943. At this time, Behnke was the only mathematics professor left. Similar situations existed at the other German universities (as has already been noted in Bonn).

A comprehensive picture of one mathematics department during the war is provided by a report covering 1939–43 that Behnke was asked by the Münster Rektor on March 10, 1944, to provide. This comprises ten pages followed by another three-page report devoted solely to the academic year 1943–44.[166]

In 1939, wrote Behnke, an active mathematical life still prevailed at Münster, though the number of students was a tenth of what it had been prior to 1933, and the accomplishment of the students was generally less. His report contains the following table of the rough numbers of students in mathematics courses.

[163] Behnke n.d.: 32, and 1978: 141.

[164] Behnke 1978: 145. This is also borne out for the country as a whole by Hartmut Titze, *Das Hochschulstudium in Preussen und Deutschland, 1820–1944* (1987), 142–143, tables 52 and 53.

[165] For example, Herman Göring on June 20, 1941, decreed that all building with armament purposes should be devoid of any particular aesthetic consideration, and an engineer named Skerl in 1943 denied that there were any aesthetic considerations against flat roofs, citing their greater resistance to air pressure and fire. Similarly in 1941, the Romanian Red Cross, by this time under Nazi domination, announced it would accept donations of "Jewish blood" because "Aryan blood" was in short supply. For the first, see Lothar Suhling, "Deutsche Baukunst Technologie und Ideologie im Industriebau des Dritten Reiches," in Herbert Mehrtens and Steffen Richter, ed., *Naturwissenschaft, Technik, und NS-Ideologie* (1980), pp. 273–274. For the second, see Proctor 1988: 156, citing the *British Medical Journal* for November 15, 1941.

[166] This is *not* in the Personalakten Behnke in Münster, which are amazingly sparse on the Nazi period, but was preserved in Personalakten Herrmann in Münster. I am grateful to the Münster librarian for this information. I have been told that much of the Personalakten Behnke for this period became the possession of his widow, but I have not verified this, nor have I seen them. Citations below, unless otherwise indicated, come from Personalakten Herrmann.

Summer Semester	*Beginners*	*Total Number*
1928	200	300
1930	200	600
1932	100	500
1934	5	200
1936	2	80
1938	10	30
1940	60	80
1943	20	50
1944	5	30

Behnke estimated the average achievement of individual students "as a first approximation" roughly proportional to the number of students; that is, the former larger numbers produced higher average achievement than the present smaller ones.

Despite the small numbers, prior to the war, activity continued unabated. Behnke had a strong interest in mathematical pedagogy, and on May 13, 1939, the twelfth successive conference sponsored by him at Münster on the connection between the schools and the university took place, with about 120 participants and well-known speakers. In July, a conference was devoted to the latest results in the mathematical theory of several complex variables. (Though Behnke had written a dissertation in number theory under Hecke's supervision, he had later turned to the then very new subject matter of several complex variables, and founded a long-lasting and important school now in its fifth generation.)

In the autumn of 1939, while the university was shut, people "worked with particular success," and, among other things, two dissertations were completed. Because a group of mathematicians interested in several complex variables existed in Münster, a seminar in the subject was set up; it had to meet from 5 to 7 P.M. because of other obligations of two participants. The seminar continued until one evening when the head janitor expelled them (including Behnke, the institute head), threatening to turn out the lights, and remarking that they should be working at the war front and not in the lecture room. The Rektor at the time supported the janitor, at least in part because of the lighting costs; however, Behnke was not allowed to pay these out of his own pocket. The seminar continued through the help of a physicist, in a room of the physics department. Nevertheless, Behnke told the Rektor in charge in 1944:

> The open scorn, the deep and lasting injury to people and dignity of office, which were connected [with the Rektor's action], have caused the undersigned to lose every interest in the total situation (*Gesamtgeschehen*) of the University of Münster. Since then, the undersigned remains concerned to trouble the leadership of the University of Münster as little as possible with the concerns of Mathematics.

Indeed, said Behnke, since then he had leaned on Wilhelm Süss, "president of the German Mathematical Society, leader of the mathematical division of the

National Research Advisory Board, and Rektor at the University of Freiburg," as well as the prominent industrialist Carl Still, for support. While Still and Süss[167] were certainly important figures, Behnke's insubordinate tone seems astonishing. This was, after all, a Rektor of whom he was speaking, and another Rektor to whom he was writing in March 1944, and the Rektor was officially the appointed *Führer* of the university. Either the present Rektor was sympathetic to Behnke's attitude, or he didn't care.

With the university open again in 1940, gradually the number of students increased until one could again lecture to, say, sixty students, although one could not place the same demands on these students as previously, because of their lack of prior preparation. Here again, Behnke quite openly blames the Nazi school reform for this situation. Again, either the Rektor was sympathetic or else mathematics did not matter much to him.

As to the faculty at Münster, one lowly helper began teaching school with the 1939 closure. In 1940, three faculty were called to wartime duties, including Karl Stein, who was drafted as a soldier. The following year, Helmut Ulm was called to service in the Foreign Office (as Gottfried Köthe had been the year before). Ulm, however, managed to lecture in Münster two days a week until 1942, when he, too, became a soldier; but after a year, because of his special knowledge of the mathematics of high-frequency electricity, he became a division leader in an institute for high frequency research.[168] Thus, it was 1943 before his knowledge of this research specialty came to the attention of officialdom, although already sometime during the winter of 1940–41 Behnke had told the authorities that the Russians had been asking about Ulm's papers in this area. Of the other faculty, a certain Dr. Daniel was killed in an air raid in 1943, and Ludwig Neder was relieved of his duties in 1942.

This last event has some intrinsic interest.[169] Neder, who was a minor mathematician, was born in 1890 and had been a volunteer in World War I, acquiring several medals and finishing the war as a lieutenant. He became a member of the Nazi party (#3,565,969), though clearly not an early member. While on the battlefield, he apparently contracted epidemic encephalitis. One consequence of this was a gradually developing "post-encephalitic Parkinsonism," which by 1941 had reached the state that a doctor's examination found him unfit for almost any occupation—except, said that doctor, as a mathematics professor. Indeed, Neder could and did continue some research activity; however, "he is so hampered by the suffering in his speech and movements, that for students, it is precisely torture to follow his demonstrations." Neder had, when he came to Münster in 1922, explicitly insisted on the right to eventual emeritization in his contract. The point of this is that emeritization meant that he

[167] For Süss, see, in particular, above, chapter 4, and below, chapters 6 and 8.

[168] While Ulm apparently considered himself an applied mathematician with also a strong interest in questions where algebra and topology intersect (see his *Lebenslauf* of November 24, 1947, in his Personalakten at Münster), it is as an algebraist that he is most remembered today, "Ulm's Theorem" being an important result in the theory of infinite abelian groups.

[169] Material on Neder is all cited from his Personalakten at Münster.

could still continue to receive income, as opposed to simply being retired.[170] Had Neder sought to be released from his duties before the civil-service law promulgated January 26, 1937[171] (which took effect on July 1 of that year), there is no question he would have been emerited. Nevertheless, despite the contract and a four-page letter from the Rektor, he was simply retired. After the war, the West German government refused to change his status in any way, since in his case, procedures were judged to have been regular. Neder lived to be seventy years old.

In any case, by 1943 Behnke was left as the sole competent mathematics lecturer. The mathematician Guido Hoheisel[172] at Köln was seconded to give guest lectures at Münster to help Behnke out. Hoheisel was also made to do the same thing at Bonn. Thus he actually gave over twenty hours of lectures a week. Having lectured in Bonn in the morning and Köln in the afternoon, Hoheisel would often arrive in Münster around midnight, then stay two days there before returning. Nor was he paid well for this effort. The situation didn't last long, for in February 1944, Hoheisel was ordered to Berlin (though not drafted). Behnke was thus alone again; in early March he fell ill, and so there were no lectures for about a week. While the number of students was small, there was still enough difference among them to require several different sets of lectures. Such a situation, said Behnke, could only be endured "from the viewpoint of the demands of total war."[173] Then he continued in his outspoken manner:

> If cultural life, and therefore also scientific development, is not to be lastingly damaged, then the new endeavors of the minister for the continuing education of teachers in the secondary schools must be one day expanded by an unusual amount. Badly educated secondary-school teachers cause in the long run just as much damage as badly educated doctors.

Hamburg had been leveled by the air raids of July 24–August 3, 1943, and a year later, in July 1944, Behnke requested the appointment of Dr. Ernst Magin of Hamburg to Münster to help him out.[174] Interestingly enough, Magin had been the gifted secondary-school teacher with an American wife who had decisively influenced Behnke toward mathematics in the years 1915–17 (when he was drafted) as well as stimulating his philosophical and belletristic interests.[175] I do not know whether the appointment was made.

Throughout the war, other mathematicians gave occasional colloquia at

[170] As an emeritus, he would also be allowed to give classes occasionally, to have students, and to be listed among the institute's faculty.

[171] See Rektor Münster to Kurator Münster, June 15, 1942 (p. 2), in Personalakten Neder.

[172] Hoheisel became well known mathematically for the first "breakthrough" in a very old number-theoretic problem in a paper published in 1930.

[173] Joseph Goebbels's famous "total war" speech was given in the Berlin Sportpalast on February 18, 1943. However, the idea of "total war" was not new. In 1935, Erich Ludendorff published a book, *Total War*, that sold 100,000 copies in two years. See the article by Günter Multmann in Holborn 1973: 298–342.

[174] Behnke to Dekan, July 9, 1944, in Personalakten Behnke, Münster.

[175] Behnke 1978: 17–22.

Münster (though in July 1941, those of Hubert Cremer were disturbed by the first air attack on Münster), until January 1943, when the conditions for holding such guest lectures had become too difficult. By spring 1943, over 30 percent of all lecture time was lost as a result of air raids.

As to deaths due to the war, in addition to Dr. Daniel, Behnke's first Ph.D., a certain Helmut Welke, died as a captain on the Russian front.[176] A member of the youth movement, Welke "greeted with joy the daybreak of the contemporary time [i.e., the coming to power of the Nazis]." Another (the most recent) doctorate also died as a result of the war, and there was more or less certain news about a number of former students dead on the field of battle.

The bombing raid of October 10, 1943, was the fifth on Münster, and it destroyed portions of the university; the mathematics department became completely unusable for scientific work. In March, the library was sent for safekeeping to Nordkirchen.[177] There were practically no mathematical textbooks at the beginning level to be had (though doctoral students were allowed extended loans of the books shipped to Nordkirchen), so that six or eight books had to suffice for twenty students. Similarly, it was no longer possible to make hectographed copies of lectures.

Behnke, true to his pedagogical interests, was unremitting in his complaints about student preparation:

> The preparation with which students begin the study of mathematics has sunk uncommonly in the past ten years. The number of instructional hours at the boys' and above all the girls' schools has been substantially constricted. In addition there are the deficits in instruction created by the many extra-scientific demands on the students. I could get a good picture of the helplessness of the secondary-school teachers with respect to these restrictions. . . .[178]
>
> There remains on all sides nothing else but the way out of giving up logical deduction in school instruction and instead, to bring forth the results as laws of nature and allow them to be so learned. That, however, has very bad consequences. Thus, young students lack logical schooling and thereby lack completely any security in their thought processes.

Behnke suggested that age twenty may be too late to remedy this deficit; however, one must try. He believed the intrinsic quality of the students was no different, it was just that their situation was so rotten.

Even Göttingen was not spared wartime incursion. Though the town and university were spared from bombs, within the first few months of the war its mathematics faculty was reduced to Gustav Herglotz and Theodor Kaluza (who

[176] See also Behnke 1978: 110, where, in what often seems Behnke's evasive style, Welke's name is not given.

[177] Similarly, the mathematical library at Göttingen was placed for safekeeping in a mine in Volpriehausen. However, this contained a much smaller percentage of applied mathematical books (important for the war effort) than others. After the war was over, an accidental explosion in the mine destroyed most of the books there. See Schappacher 1987: 360.

[178] Cf. Quätsch 1961: 30–31.

in the end was Courant's replacement) and the former student of Carl Ludwig Siegel, Helene Braun, as *Assistent*. Hasse, in the first part of 1940, returned to a naval position in Berlin but came back to Göttingen from time to time to carry out what institute business there might be; the others at Göttingen were simply drafted into the armed forces. Hasse's naval position was as the leader of a research group dealing with curves of pursuit, and he even assigned tasks in this line to people in Göttingen.[179] Kaluza became the operational director in Hasse's absence.[180]

The war also saw in Göttingen the attempt to use university facilities for other than university purposes, and this no doubt happened elsewhere as well. For example, in early 1943, it was proposed that the mathematics facilities be used for an installation from a dye factory. This did not happen, primarily due to the efforts of the Kurator, Justus Valentiner, who felt it would be "vandalism."[181]

Later in 1943, rooms at Göttingen were taken over by the navy; on October 31 of that year, entrance to these rooms, except by Kaluza as acting director and the janitor, was forbidden.[182] Although apparently work and experimentation were carried out in these rooms, nevertheless, their use by the navy seems to have been arranged by Hasse in order to prevent more deleterious use, such as for a hospital, and was apparently at first something of a pseudo-requisitioning.[183]

Thus, the war brought an unsurprising decline in faculty and students at universities; this, however, in fact was an acceleration of declines that had already begun under Nazi regulations. University work suffered considerably, and this was only exacerbated by the introduction of compulsory service in the national labor force and in the military. Adding to faculty troubles was local political infighting and an initial atmosphere of "sauve qui peut," though this gradually ameliorated. Yet, despite these troubles (and in part, because of them), mathematical research with the few well-trained students was able to go on, even better than before. Still, it would seem that even party membership

[179] UAG 3206b, Hasse to Dames, Mar. 16, 1940; MI, letter to Hasse, Nov. 4, 1944; Schappacher n.d.; and idem 1987: 360.

[180] As borne out, for example, by UAG 3206b, Kaluza to Kurator, Mar. 19, 1941.

[181] MI, Kaluza to Kurator, Jan. 14, 1943, Kurator to Gnade (Oberbürgermeister Göttingen), Jan. 21, 1943. This is mentioned in Schappacher 1987: 360, with details in Schappacher n.d. Valentiner seems to have been a Nazi who was, nevertheless, academically oriented; he also supported the attempt to prevent Courant's dismissal (Schappacher 1987: 351). A similar orientation toward some traditional academic values, despite one's political persuasion, may partly explain the racial scientist Karl Astel's behavior toward Ernst Weinel discussed in chapter 4, "The Winkelmann Succession." A stunning example of the occasional preservation of such values is the case of Dr. Siegmund Rascher. Rascher was the notorious doctor who did the infamous Dachau medical experiments. He dearly wanted to "habilitate" so as to become a professor of medicine at a German university. However, the universities of Munich, Frankfurt, and Marburg apparently all refused to accept the records of the Dachau experiments as a *Habilitationschrift* on the grounds that they could not be published openly. See Michael Kater, *Doctors under Hitler* (Chapel Hill and London: University of North Carolina Press, 1989), 125–126.

[182] MI, OKM (Naval Office) to Kaluza, Oct. 31, 1943.

[183] Schappacher 1987: 360, with more details in Schappacher n.d.

was not enough in 1942 to save salary, which tended to be returned in various forms to the state coffers. In the same way, university buildings that were endangered could be requisitioned for the war effort. Yet the universities carried on (even when "bombed out") with smaller numbers of poorly trained students and a sharply reduced faculty, both of which were whittled down even further by wartime deaths.

The Value of Mathematics in the Nazi State

Erich Jaensch, the Marburg psychologist and Nazi supporter, was cited in chapter 1 for his insistence that mathematics was a rational subject. However, Jaensch (whose general work was inspirational to Ludwig Bieberbach and the "Deutsche Mathematik" movement discussed in chapter 7) also was worried about the wave of irrationalism sweeping over Germany. Jaensch wished, through the especially illuminating case of mathematical thinking, to show that a "present-day, out of place, and racially contrary (*artwidrig*) type of rationalism that rules far and wide must be replaced by another racially suitable (*artgemäss*) [type of rationalism]." Indeed, mathematics will show the way to other disciplines in this matter to avoid the "abysses of irrationalism."[184] Jaensch also worried about the fact that many educated Germans considered their mathematical education (and so future mathematical education) in large measure superfluous.[185] This estimate leaned on a paper of Heinz George[186] that described the results of a questionnaire sent in 1933 to 1,154 people, with 360 usable responses and 13 late ones. The statistical validity of the procedures used does not concern us. Rather, what are important are the conclusions drawn and their interpretation by such as Jaensch. With the exception of mathematical professionals,[187] most of the respondents found only elementary-school mathematics useful within or without their profession. As to whether mathematical education itself was valuable, 35.8 percent gave either no answer or no usable answer, and only 37.2 percent found it in itself useful. Many respondents, though, commented strongly on the value of mathematics for teaching how to think clearly and correctly, and stated that mathematics instruction taught values like decisiveness, cold-bloodedness, self-discipline, self-criticism, consequential thinking, a sense for order, reliability, and the like. However, these comprised only 63 percent of the usable responses, and most of these could not give experiences

[184] Jaensch and Althoff 1939: ix.

[185] Ibid.: 9.

[186] Heinz George, "Laienurteile über den Lebenswert der Mathematik" (Lay judgments concerning the value of mathematics for life), *Zeitschrift für angewandte Psychologie* 53 (1937): 80–112. (Shortened form of a doctoral dissertation at the University of Königsberg.)

[187] This item is used for mathematics, physics, engineering, and architecture. See ibid.: table 12, p. 93.

justifying their belief. Dr. George concluded that, among other things (emphasized in original),[188]

> the unconditional acceptance of a formal influence of mathematics [in other areas of human activity] is a dogma or a matter of blind faith. The judgment that mathematics has a high educational value is indeed widespread, nevertheless, it is a purely traditional and uncritically accepted prejudice, that one may not simply trust—rather, in this form of generality, it must be unconditionally rejected.

George was writing in 1937 about work (in fact statistically naive) done in 1933; Jaensch wrote in 1938 about accumulated impressions over many years. Neither was a mathematician; however, as has been already seen, E. A. Weiss argued that, at the university level, these same "layman-perceived" virtues of mathematics were character-building for a good Nazi. Weiss seems to have been arguing out of his own solid political and intellectual convictions; however, Jaensch seems truly distressed that mathematics is threatened by popular irrationalism and studies like George's.

Further evidence of this distress and attempt to justify mathematics in the Nazi context—to provide it, in Jaensch's terms, with a "racially suitable rationalism"—is a speech given by the Jena mathematician Robert König in 1940 and later published in a mathematical educational journal. König had been Behnke's predecessor at Münster and, as has already been seen, would later have some sort of run-in with Ernst Weinel. The abstract prefaced to the published version[189] begins with the remarkable sentence: "Mathematics is not foreign to life; rather, it is one of the strongest biological functions of the human organism." The paper begins (emphasis in original):

> The intent of my lecture is to develop a new biological conception of the nature of mathematical science: *Mathematics is a, and perhaps indeed the, strongest-orienting function of our consciousness and, as such, is necessary for the effective success of our existence* (Durchsetzung des Lebenbestandes) *as a people.*

What are König's examples?

> 1. The necessity of introducing complex numbers into number theory to effectively treat cubic and biquadratic residues. This is not necessary for quadratic residues, and leads to a "biological drive to a new adaptation, a *new* means of orientation must be found, the *complex number*."[190]
> 2. The idea of a mathematical group.

[188] Ibid.: 112.

[189] Robert König, "Mathematik als biologische Orientierungsfunktion unseres Bewusstseins" (Mathematics as the biological function for orientation of our consciousness), *Zeitschrift für mathematischen und naturwissenschaftlichen Unterricht* 2 (1941): 33–47, original talk held September 19, 1940. All citations of this speech below are from this published source.

[190] A positive integer x is a kth-power residue of another positive integer q, if there is a perfect kth power such that it minus x is divisible by q.

3. The creation of analytic geometry, particularly using vectors. "We remain, however, always conscious that in Euclidean space and the Cartesian coordinate system we have only a biological means of orientation before us."

4. Spherical trigonometry, enabling navigation and the determination of time and location (at least if the earth is regarded as a sphere): "Who thinks, however, when he makes a statement about time or the distance between places, of this achievement of mathematics in orientation. . . . This is only a small example; however, one sees how difficult the sacrifice in quality of life (*Lebensgut*) is that, on the whole, the suppression of mathematical instruction and its general shortening costs."

5. Classical differential geometry. "The biological drives, i.e., in our case the question of the true form of the earth, do not rest."

6. *n*-dimensional manifolds, which make Euclidean space an "orientation form" instead of something with "absolute, hypostatic existence." "That is biological thinking in an extremely modern manner [by Riemann]."

This is perhaps enough to give the flavor of König's talk, which goes on in the same vein—a solid exposition of mathematics for nonmathematicians coupled with "biological words," and pleas for recognition of the importance of mathematical achievements and instruction. Among König's other examples are calculus and the idea of limiting processes, functions of one complex variable, probability theory, statistics, and topology. In general, the theme is "*science—especially mathematics, is building for life.*"[191] The example given for topology is arresting both in the Nazi context and in the light of biological developments fifty years later:

> *The substrate of life is bound up with structure, that is geometric form*—according to Frey-Wissling, "The cell nucleus represents a wonderful morphological building with a hugely complicated sub-microscopic structure"—so the mathematician in particular is inspired by *this* knowledge; he hopes with the most recent branch of geometric research, *topology*, also here this once to be able to work along substantially on the problem of particularly the greatest secret and wonder of life, the sub-microscopic morphology of inheritance.

With the war beginning its second year, and being in Jena, the center of the German optical industry, König briefly, but pointedly, discussed the importance of mathematics for the military effort: "In any case, mathematics preserves its

[191] König 1941: 43; "Science is building for life" is also the epigraph to the paper, paraphrased from E. G. Kolbenheyer. Erwin Guido Kolbenheyer was a writer and philosopher who believed (as early as 1925) that philosophical questions should be handled from an exclusively biologistic standpoint. His *völkisch*, anti-individualistic, and anticlerical positions made him someone whose ideas were much appreciated by the Nazis. After 1945, there apparently was much argument in Germany over his Nazi affiliation. He died in 1962 at the age of eighty-four. König was apparently a lifelong friend of Kolbenheyer and his philosophy. When König was eighty-six in 1971, he published a 728-page book on Kolbenheyer's philosophy. Nevertheless, König was no friend of Nazi ideas. See in the Archiv der Universität München a notarized statement by Hermann Weyl (May 10, 1946) and an obituary of König (Sudetendeutsches Gedenkblatt—König was born in Zattig [Zatec]). Both items are in the file for Robert König.

proud power, which nothing can replace, not only in peace as a cultural act, but also in war as a weapon."

König's conclusion is also remarkable. After summarizing the need for more attention to a solid and systematic school instruction in mathematics as well as at the university, and the need for adequate educational time, he returned to a startling biological analogy:

> Two biological laws cast in brass further that [there is a need for more educational time]!
>
> One is the fundamental biogenetic law discovered by our great Haeckel, according to which each individual must repeat the development of his kind, i.e., in our case the young mathematician must run through all the substantial steps of differentiation on the long and hard path of accommodation to his science *himself*—a *biological* process that cannot be arbitrarily compromised or interrupted.[192]
>
> And the second biological fact emphasized by Kolbenheyer: "An organ develops itself only with the possibilities of having a function," or expressed somewhat differently, the presence of a mathematical inheritance by itself helps nothing if the possibility of activity is not provided.

In the Third Reich, everything was biological, and so even a plea for improved mathematical instruction, respect for mathematics, and adequate pedagogical time (so sharply disturbed by Nazi innovations) was cast in biological terms to give it a chance of success. König's last paragraph contains more quotes from Kolbenheyer and the almost obligatory appeal for the confluence of mathematical work and National Socialist *Weltanschauung* and a "new historical consideration of mathematical creativity in which the different racial shares are comprehensible."[193]

The denigration of mathematics and science was perhaps predictable. Hitler had been at best an indifferent secondary-school student,[194] and as well-known passages in *Mein Kampf* have it (emphasis in the original):[195]

> *The folkish state must not adjust its entire educational work primarily to the inoculation of mere knowledge, but to the breeding of absolutely healthy bodies. The training of mental abilities is only secondary. And here again, first place must be taken by the development of character, especially the promotion of will-power and determination, combined with the training of joy in responsibility, and only in last place comes scientific schooling.*
>
> Here the folkish state must proceed from the assumption *that a man of little scientific*

[192] "Ontogeny recapitulates phylogeny" is Haeckel's famous "fundamental biogenetic law." However, the idea thereby expressed goes back at least a hundred years before him to Buffon. Haeckel was very important on the continent, and politically became a conservative nationalist. To what extent his ideas as expressed by his followers fertilized the German intellectual soil for the growth of Naziism is discussed in Gasman 1971. Haeckel was certainly Jena's internationally most famous scientist. He died in 1919.

[193] Comparison should be made to the efforts of Ludwig Bieberbach and his collaborators on "Deutsche Mathematik" to distinguish a truly Aryan mathematics, as discussed in chapter 7.

[194] Deuerlein 1974: 68–69.

[195] Hitler, *Mein Kampf* (1943 translation): 408 and 423.

> *education but physically healthy, with a good, firm character, imbued with the joy of determination and will-power, is more valuable for the national community than a clever weakling.*

In addition,

> It is the characteristic of our present materialized epoch that our scientific education is turning more and more toward practical subjects—in other words, mathematics, physics, chemistry, etc. Necessary as this is for a period in which technology and chemistry rule—embodying at least those of its characteristics which are most visible in daily life—it is equally dangerous when the general education of a nation is more and more exclusively directed toward them. This education on the contrary must always be ideal. It must be more in keeping with the humanistic subjects and offer only the foundations for a subsequent additional education in a special field. Otherwise we renounce the forces which are still more important for the preservation of the nation than all technical or other ability.[196]

As Behnke and the student numbers make clear, the degeneration of scientific education began very early in the Third Reich, long before the anxieties of Jaensch or König, however different their politics may have been. On November 22, 1933, the mathematician Gerhard Thomsen, a gifted student of Wilhelm Blaschke at Hamburg, gave a talk at the University of Rostock, where he was an associate professor, entitled "Concerning the Danger in the Repression of the Exact Sciences at Schools and Universities."[197] This talk was a plea for the importance of mathematics and natural science in the "new Germany." Using the ideological language of the Nazis as much as possible, it rejected the consequent reduction of the importance of learning, and especially the detrimental effects for mathematics. Curiously enough, with the ideological dressing stripped away, many of its points seem relevant to other places and other times. Because of this and its early date, as well as the lecture's outspokenness and its necessary attempt at an ideological guise, some of it is here excerpted as a flavor of the times. As he summarized at the end, Thomsen's lecture really addressed three things: the actual tasks natural science needed to fulfill for the state; the tasks of intellectual-spiritual education and schooling of future mathematicians; and how the spirit of mathematics could contribute to the inner psychic transformation of Germans. It is also clear that Thomsen's intention was to further the traditional and threatened aims of mathematics education within the ideological context of the Nazi state. Appeals to Nazi themes were used to support traditional educational values. Thomsen seemingly did not understand that the Nazi seizure of power truly meant a reordering of all values, especially educational

[196] Of course, men like E. A. Weiss thought that such forces could be fostered by mathematical training.

[197] Gerhard Thomsen, "Über die Gefahr der Zurückdrängung der Exakten Naturwissenschaften an der Schulen und Hochschulen, *Neue Jahrbücher für Wissenschaft und Jugendbildung* (1934): 164–175. The original talk was given November 22, 1933. It was reprinted in 1943 in *Physikalische Blätter*.

ones, and thought he could combat the antiscientific tendencies he saw *within* the Nazi context. Thus:[198]

> Impressive in their multiplicity are the tasks that students have today. In the foreground stand the tasks of political schooling and education, as they are foreseen in a widely ramified program within the framework of the new political university. Who can remove himself from the urgency and importance of these tasks! The facts of our history teach us that political incapacity has all too often destroyed the fruits of surpassing achievements in peace and war. . . .
>
> Who can further deny the urgency of other tasks as they should be solved above all in work- and defense-camps, and how they represent the hardening and defensive preparation of body and spirit or the education to a social conception of life tied to the people!
>
> Now, we are not done with the intensive affirmation of all these tasks; on the contrary, the *actual organizational problem is the quantitative division of the available time to these individual tasks*. For the students' time is bounded and finally also their will-power and working power are limited, despite the extraordinary readiness of the contemporary generation of students for sacrifice. Especially difficult here is the situation of students in the scientific and mathematical disciplines, for these indeed it is just within their disciplines that the most important tasks of the fatherland are to be solved.
>
> The real revolutionary epoch of the National Socialist movement is past. One of the most important tasks of the second period, now settling in, is the education of that part of the German people which is usable for disciplinary and special work for the goals of the fatherland, to intellectual concentration on this work. . . . We must with all means available recruit and work for natural scientific and technical education. . . . Political schooling is indeed a thing more of intuitive experience than of patient work. *In contrast, much time, work, and patience is necessary for the study of natural scientific disciplines*; here everything depends on the amount of time. I might formulate the situation this way: beside the necessary political schooling, disciplinary studies should remain the nucleus of university education.

If this were not sufficiently outspoken, consider that earlier, Thomsen had deplored the reduction of attention given to mathematical and scientific education at secondary schools and the resulting lower educational level of those continuing their education. He continued:[199]

> On the whole, the contemporary atmosphere seems to me to be more inimical to natural science than friendly to it. Certainly natural science was brought into disrepute by the fact that it was all too much monopolized by Marxism. In Soviet Russia indeed the humanities (*Geisteswissenschaften*) are as good as eliminated and only the natural sciences play a role. In contrast thereto in National Socialism one speaks much about a primacy of the humanities. The political university is chiefly oriented in this humanities direction (*geisteswissenschaftlich*) . . . political naturally in the general sense of a

[198] Thomsen 1934: 167–168.
[199] Ibid.: 166.

> patriotic political setting of goals and nothing lies farther from this idea [of a political university] than the erection of a politicizing university.

Like E. A. Weiss, Thomsen saw pedagogical values, but somewhat different ones, in mathematical education: "educating to level-headed clarity"; "compulsory development of inner energies"; the experience of insightful solution to a problem. He combatted the idea of mathematics as "intellectual acrobatics," or as dry and spiritually cold. He cited David Hilbert as an example of the "intellectual power of the best East Prussian race." He coined the term "disciplinarily trained soldiers [*Fachsoldaten*] of the Third Reich." He defends the fact that it takes perhaps decades for theoretical mathematics and science to find practical utility, by citing the length of time it took racial theory to develop, and "What is right for racial theory, for the benefit of our state, is also right for theoretical natural science and mathematics."[200]

Thomsen's main message always remained the need for more attention to mathematics and natural science in school and university, though he clothed this need in ideological rhetoric.[201]

> I consider that one of the most important tasks of National Socialist education of the people is training intellectual (*geistiger*) concentration and one should begin with this, naturally beside physical education, when children are still small, when the intellect (*Geist*) is still completely fresh. . . . If today students complain that because of the multiplicity of tasks set them they have no possibility for adequate intellectual concentration in their disciplinary work, then, on the one hand, the leaders must give instruction that this burdening with other things will be lessened, and on the other hand, one must say in the contemporary spirit, he who does not have intellectual concentration, must be educated thereto with the same reckless and almost brutal energy with which today's German also will be educated to other things which lie in the interest of the fatherland.

As his penultimate paragraph put it:[202]

> That which in the years after the war despite all agitation against us had still always preserved respect for the German spirit, is in the very first place, indeed almost exclusively, the reputation of German natural science.[203] Let us therefore not saw off the branch on which we sit, in that we substantially suppress natural science.

Thomsen's talk shows how, early on, mathematicians saw mathematics as threatened, and represents an appeal for rectification largely on the basis of traditional values dressed up in Nazi rhetoric. Much later would come Jaensch's new "healthy" kind of rationality or König's "biologizing." As we have seen, Thomsen's plea was not heard—indeed, it is hard to believe that the Nazis

[200] Ibid.: 165.

[201] Ibid.: 169.

[202] Ibid.: 175.

[203] Although, immediately after World War I, the French and British vengefully excluded German participation in international scientific affairs, this did not prevent a tremendous flowering of German science. This was particularly true in mathematics and physics.

would give credence to a man who said, "More important than education to energy-forms completely restricted to the physical is education to the higher energy-forms of intellectual concentration,"[204] or "We need today a student youth that has ever so little fear of a murderous weapon in wartime as of the monster of a five-fold integral with singularities in the integrand."[205] As to Thomsen himself, though in his lecture, which apparently attracted considerable attention, he said he was concerned about the next ten or twenty years, six weeks later, on January 4, 1934, he committed suicide.[206] One conjecture is that the Gestapo had taken an interest in him; it has also been said, not inconsistently, that he did this in public as an attempt to evoke protest against the regime.[207]

The difference between Weiss and Thomsen, who in some respects used similar rhetoric, is that Weiss sincerely claimed mathematics could contribute to the creation of the Nazi ideal type and created a pedagogical program on this basis, whereas Thomsen wished to derive benefits for mathematical education from Nazi ideology and rhetoric. Either Thomsen was a naïf or he was attempting a clever rhetorical manipulation. In either case, the early Nazi state had little use for him.

Secondary and Elementary Mathematics

Thomsen, König, and Jaensch, although primarily talking about the university, all indicate, as does Behnke, the declining state of secondary-school instruction. Already in 1933, mathematicians and mathematics instructors attempted to establish a place for mathematics in the National Socialist educational program using similar rhetoric. At the same time, the education ministry defined a "Nordic spirit" in mathematics: "Racial conditioning is also recognizable in their [mathematical] work. Nordic spirit corresponds to conquering [a problem] in a wealth of forms born in inner intuition, as with the creative hand, so also with pondering reason."[208]

Three days after Thomsen's talk in Rostock, at the other end of Germany, in Stuttgart, the "Mathematical Scientific Union [for the furtherance of instruction]

[204] Thomsen 1934: 168. Cf. *Mein Kampf*, Hitler 1943: 408.

[205] Thomsen 1934: 169.

[206] Pinl 1972: 206. A brief obituary appeared immediately in *Abhandlungen des Mathematischen Seminars, Universität Hamburg* 10 (1934).

[207] The Gestapo conjecture is from Pinl 1972: 206; the public protest conjecture is a personal communication from Christoph Maass. Reinhard Siegmund-Schultze has suggested that Thomsen's personal file at Rostock indicates a connection between the talk and the suicide. See his "The Problem of Anti-Fascist Resistance of 'Apolitical' German scholars," in Monika Renneberg and Mark Walker, eds., *Science, Technology and National Socialism* (Cambridge: Cambridge University Press, 1994).

[208] Jürgen Genuneit, "Mein Rechen-Kampf," in K. Fuchs, ed., *Stuttgart im Dritten Reich, Die Jahre von 1933 bis 1939* (1984), 211. The title of this article is a pun on *Mein Rechenbuch* (or *My Counting Book*), a popular elementary-school arithmetic book of the time, as well as Hitler's *Mein Kampf*.

in Württemberg"[209] met to "joyfully greet" "the national rebirth of Germany," yet worried about the danger of insufficient mathematical and scientific instruction. The "right to existence" of mathematical instruction in National Socialist schools was defended. One speaker saw the "Faustian spirit"[210] at work in German mathematics. Using rhetoric that was a cross between Thomsen's and Weiss's but somewhat more elevated intellectually, he commented on how in the National Socialist age, men were needed "who in their youth had learned to traverse steep, stony, and thorny paths and were always capable and willing to do so." This was true not only physically but also intellectually, and who better a guide for the latter than mathematics instructors? For in mathematics, one learned to distinguish knowledge from conjecture, and not to state what cannot be supported: "[In mathematics] one learns that only inexorable, serious, self-forgetting, and self-denying work, endurance of a strong will, leads to the goal. These, however, are exactly the qualities that we need in the Third Reich." In conclusion, the "Faustian mathematics of the northern race of men" must be taught in all schools.[211]

Examples of such rhetoric could be multiplied many times. Even Thomsen's speech has striking examples not cited. While each such individual image is fascinating, their cumulation ends up being boring—one quickly gets to know the terms of discourse and the expected flourishes.

What is more interesting is how this rhetoric was translated into texts. In the name of an "instruction tied to reality," that is, mathematics problems that related to real-life situations and, therefore, were presumably more interesting to students and showed them more clearly the value of learning the material, these books from elementary arithmetic on were militarized, racialized, and politicized. It should be noted, moreover, that the militarization and politicization of such texts was not some new creation of the Nazis. Already in the Weimar Republic, popular arithmetic texts were militarized (for example, an illustrated comparison of the airplanes and tanks of several countries, and the per capita outlay of funds for the army and navy in 1914, 1925, 1927, of Germany, Great Britain, France, Italy, the United States, and Japan, appear in a 1929 text).[212] Furthermore, these Weimar texts did not contain problems encouraging thoughts about parliament and representative democracy, as they well might have. The Nazis insisted on new arithmetic books, not out of opposition to such tendencies, but because they thought they were not sufficiently thoroughgoing, and certainly because they were not explicitly National Socialistic in tenor. The Nazis continued emphasizing this military resentment, to which were added "biological," racial, National Socialist, and anti-Semitic problems; and all was brought to such a pitch that a mathematics instructor, even if it were so wished, could not escape inserting such ideas in the mathematics curriculum. As one

[209] This was a local division of a national organization.

[210] "Faustian spirit" was a well-known Spenglerian term.

[211] Genuneit 1984: 209–210. The similarity to Weiss's argumentation is striking.

[212] Ibid.: 207.

slogan had it in 1934: "Let us take off our seven-league boots with which we sought to measure every discipline (*Wissenschaft*) in the school. Rather, let us put on SA-boots; then young men can follow better and with love and inspiration as well."[213] The task of a National Socialist arithmetic book, another author declared in 1935, was to be "a book of living for the German child, given to him so that it helps him to announce and transfigure his fate and his task," and "a book whose power outside the school, in particular in the family, must be a demonstrative advertisement (*werbend zeigen muss*) whose content finds its deepest meaning primarily beyond the school."[214] Two such rewritten texts, one primary and one secondary, were studied in detail by Jürgen Genuneit.[215]

In 1936 there appeared a booklet, "Mathematics in the Service of National-Political Education,"[216] directed toward mathematics instructors, but also "to the general public. It should help to dispatch the old misunderstanding that mathematics consisted in consulting tables of logarithms and the rote learning of ununderstandable formulas with which nothing sensible could be begun."[217] The content of the illustrative problems suggested for teachers' use ranged from the innocent to the tendentious. For example, in the section on probability theory and mathematical statistics, one question (p. 30) investigated (in the manner of Francis Galton) the heights of grown children versus those of their parents, but another (p. 25) began, "If a race (white) has on the average three children and another (black) four children and both races at the start have the same numbers . . ." and ended (after much detail) with "the white race is completely exterminated in 264 years. Represent this graphically." The problems could even be incipiently deadly (p. 42): "An insane person costs daily about 4 RM, a cripple 5.50 RM, a criminal 3.50 RM. In many cases a civil servant earns daily only about 4 RM, an employee scarcely 3.50 RM, an uneducated person not even 2 RM. . . ." It does not seem necessary to finish the details of this problem.

The exhortations to teachers of mathematics to teach National Socialistically began almost immediately. For example, the 1933 volume of the *Journal for Mathematical and Scientific Instruction at All School Levels*,[218] one of whose editors was the mathematical educator Walter Lietzmann, contained such articles mixed in with the usual fare. One Bruno Kerst began a two-page article on the importance of mathematics and physics for the German school with:[219]

[213] Ibid.: 212, citing G. Woeste in *Nationalsozialistische Erziehung* (1934): 332.

[214] Genuneit 1984: 212, citing Wilhelm Ebel, "Über Sinn und Gestaltung des Rechenbuchs," in *Deutsches Bildungswesen* 3 (1935): 761–768, p. 768.

[215] Genuneit 1984: esp. 220–230.

[216] Alfred Dorner et al., "Mathematik im Dienste der nationalpolitischen Erziehung," 3d ed. (1936).

[217] Ibid.: 1 (from foreword by Prof. Georg Hamel).

[218] *Zeitschrift für mathematischen und naturwissenschaftlichen Unterricht aller Schulgattungen*. It was founded in 1869 and had been continuously published since then. All citations below are from the publication year 1933.

[219] Ibid.: 149–150.

> If the pedagogy of the age of liberalism concerned itself with the goal of "general education," so the new era (*Zeit*) sets every German educational organ (*Erziehungswesen*) the task of educating the *German citizen* [emphasis in original].

He asserted *in medias res* that "an education to forceful working in our community of the [German] people (*Volksgemeinschaft*) is only possible in connection with a sufficiently extended mathematical-physical education," and ended,

> And in the lasting values that German culture constitutes in the present and future, mathematics and physics enter in particularly large measure. Therefore already in 1924 it was said with justice: "The largest part of all mathematical instruction is instruction in German."

Lietzmann himself had written in 1924 a paper entitled "Mathematical Instruction and the Homeland" (cited with several others by Kerst) and contributed to this volume an article, "Mathematical Instruction and Military Science," that ended with the hope that the close connection between mathematics and military science would be convincing, so that "also in the new German Reich, mathematics deserves an important place in secondary-school instruction."[220]

There was good reason for such statements, as, for example, Lietzmann himself as editor reports on the order of September 13, 1933, that racial science was to be introduced in secondary schools, the extra time, if necessary, to be taken from mathematics and foreign-language instruction. German, history, and geography explicitly were to be in the service of racial science. Lietzmann added: "One could easily have expressly also mentioned mathematics since the whole theory of inheritance depends on the application of statistical methods."[221]

The same volume, true to Lietzmann's assertion, has a ten-page article by the statistician Hans Münzner on the "Mathematical Foundations of the Theory of Inheritance," which remarks: "The task of research on inheritance is to give precise foundations to racial science." Münzner had been a pupil of the extremely broad mathematician and statistician Felix Bernstein, who was a Jew, had been active in liberal politics, and had made early investigations of the distribution of the Landsteiner bloodgroups. It was still possible in 1933, even in the context of racial science, for Münzner to mention his teacher's name positively.[222]

Most typical, however, was the article by H. Wolff in Lübeck on "The Renewal of the New German School from the Spirit of Biology."[223] This is an article of exhortation, welcomed by prefatory remarks of the editors. Indeed, it does not mention mathematics except in the line, "What does a bad achievement by an upper-level student in Latin or mathematics trouble him, if he shows himself in the service of the idea of freedom as 'brave and loyal' to his [SA] division

[220] Ibid.: 249–253.

[221] Ibid.: 394–395.

[222] Münzner eventually succeeded to the chair of statistics at Göttingen.

[223] *Zeitschrift für mathematischen und naturwissenschaftlichen Unterricht aller Schulgattungen* (1933): 305–311.

leader and his comrades."[224] In general, the article is full of the usual Nazi slogans: "The great German geniuses Wagner, Lagarde, Möller van den Bruck, and Rosenberg" as precursors of National Socialism; "the lordly will of Adolf Hitler leads us again up out of the abyss of degradation and destruction"; "today the humanistic educational ideal is *dead* [emphasis in original], empty of blood"; "the weakness and misery of the past age, which was called liberalism but which in its effects was only Bolshevization"; "the [Kantian] categorical imperative could only have been born from the Nordic feeling for life"; "the Jewish question is the question of a people's parasitism within other nations"; "pacifism is to be exterminated as untrue to life and cowardly"; and so forth and so on, dealing with art, religion, and political history among other disciplines to be reborn from the spirit of biology.[225]

Although Wolff's article is pure "consciousness-raising," it is the rhetoric that König later tried to use to the benefit of university and secondary-school mathematics. Immediately following Wolff's article, though, was an article directly relevant to mathematics by E. Tiedge.[226] Originally delivered on September 20, 1933, to acclamation at the annual meeting of the *Mathematische Reichsverband* (MR), an organization of German mathematical groupings concerned with pedagogy, this talk linked mathematics and the "mathematical natural sciences" as important to the tasks set by the new German school—albeit prefaced by the customary genuflection: "Biology has won a completely unchallenged importance for national education."[227] It is through technical achievements that this importance is most directly felt, and Tiedge took the "reactionary modernist" position: "Technical creativity has nothing to do with mechanization and despiritualization, with shallowness and acquisitive greed. Technology is in itself neither bad nor good, it is an accomplishment previously unheard of, which can be applied equally as well to the well-being of humanity as to its calamity."[228] He also explicitly rejected the idea that mathematics and mathematical science is in the service of "a growing internationalism" and has "little to do with the German nature and essence."[229] Tiedge's lecture-article was printed in the NSDAP organ for Lower Franconia on the next day and distributed by the MR to all interested mathematical circles, as well as sent to the education ministry. In addition to the printing from which I cite, it was also distributed as a pam-

[224] Ibid.: 305–306.

[225] The citations above are, in order, from ibid.: 307, 305, 306, 307, 309, 310, 310. The whole article has this tone, is dense with names of German historical heroes, and seems to cover every slogan of the Nazi revolution. For example, one not mentioned above is the "turning away from the emancipation of women" and the creation of a "manly state."

[226] Ibid.: 311–320.

[227] Ibid.: 313.

[228] Ibid. Jeffrey Herf, in his book *Reactionary Modernism* (1984), discusses how some reactionary literati in Weimar Germany as well as a number of like-minded engineers moved away from the traditional conservative abhorrence of technology to the argument that it could and should be adapted to their purposes.

[229] *Zeitschrift für mathematischen und naturwissenschaftlichen Unterricht aller Schulgattungen* (1933): 319.

phlet.[230] It is worth noting that Tiedge could cite the famous chemist Fritz Haber (an assimilated Jew who would shortly lose his position) and the physicist Heinrich Hertz (a "half-Jew") positively in this lecture. Tiedge's theme is carried out concretely in Lietzmann's article already mentioned.[231]

As to the MR itself, its leader was the already-mentioned, well-known Berlin (technical university) mathematician Georg Hamel, who was its president in 1933, when it met in Würzburg and heard Tiedge, among others. Here is a part of Hamel's report on that meeting:[232]

> The mighty revolution in Germany has not only presented each individual, but also organizations, with important decisions. The MR has two major tasks: simultaneously university-pedagogical questions of mathematical instruction, and the connections between university, school, and public life, insofar as these questions concern mathematics. Therewith, the MR had never considered itself as a one-sided disciplinary representative, it was always anxious to formulate its claims in the framework of the whole, as a claim for the education and rearing of young German people, *i.e.*, with regard to a general national educational goal. In this respect, it has nothing to change. However, while it earlier, often in a defensive posture, fought for mathematics, it can today no longer demand, but only play an advisory role. We must subordinate ourselves. In this sense the MR has directed an application to the ministers. We wish to work along upright and loyal in the sense of the total state. For this, however, it is unconditionally necessary that the MR, even if it only wishes to be heard as an advisory organ, must adapt itself to the demands of the time. Therefore, three axioms are presented for acceptance: 1. The leadership principle (*Führerprinzip*). The leader to be chosen anew is the sole bearer of responsibility. 2. The leader's advisory council (*Führerrat*). The leader determines his coworkers, in particular his advisory council. He is in this bound by the Aryan principle, and indeed in its strict form as it has been declared for civil servants in positions of leadership. 3. The system of advisory committees will be dissolved. It is left to the leader to organize this arrangement anew.
>
> These axioms were accepted and the previous president, Prof. Dr. Hamel, was chosen as *Führer*.

The mathematics teachers clearly not only had no problems accommodating themselves to the new state, but were eager to do so.

The National Socialist concept of intellectual education in general and at the university in particular was summarized in 1939 by Rudolf Benze. Benze divided education into (a) physical, (b) spiritual-character-building, and (c) intellectual. To the last, he said that National Socialism wished to set aside[233]

> that portion of intellectual schooling that has made so many good and clever Germans into polymaths and pallid theoreticians, who chase after goals estranged from life and

[230] Ibid.: 396.

[231] Ibid.: 249–253. Lietzmann (who had an adjunct position at Gottingen) had written articles as early as 1914 advocating the military-scientific justification of mathematics (ibid.: 249 n. 2).

[232] Ibid.: 395. Cf. *JDMV* 43 (1934): Abteilung 2, pp. 81–82.

[233] Rudolf Benze, *Erziehung im Grossdeutschen Reich* (1939), 6–7.

> finally themselves become estranged from life and useless for meaningful and daring action . . . it is senseless and unendurable for a people struggling hard for its life if its members squander the search for truth on things that perhaps attract the individual, but are worthless for his own people, and so also do not further human culture. . . . The selection of material therefore follows according to fundamental laws of the biology of the people, as then generally biology has become the nucleus of all schooling and all of education is built up on the theories of inheritance and race.

This is what Thomsen did not understand, and König came to.

THE WARTIME DRAFTING OF SCIENTISTS

One cannot leave the general picture of the academic situation in mathematics without a description of the so-called *Osenberg-Aktion*. In addition to the "Association for German Research" (*Deutsche Forschungsgemeinschaft*, DFG), which under the Nazis had first Johannes Stark and then Rudolf Mentzel at its head, in 1937 the education minister, Bernhard Rust, established a *Reichsforschungrat* (RFR), or "National Council on Research" (also with Mentzel as its head).[234] At its celebratory establishment, Hitler set foot in the science ministry for the only time in his life—another mark of how little he cared for scientific research is that he refused to shake hands with the divisional leaders when they were presented to him.[235] As noted previously, much later, during the latter part of the war, Wilhelm Süss became an RFR leader for mathematics. An indication of the low esteem in which mathematics was held is that there had previously been no mathematics division, and indeed, Zierold's definitive book on the DFG never mentions either mathematics as a research discipline or Süss.

As the war dragged on, more and more scientists were drafted, and there was growing complaint about the lack of research workers. Mentzel could not manage to regulate who among them should be granted deferment as "indispensable." One of his sharpest critics was Werner Osenberg, an Ordinarius and leader of the machine-tool group at the technical university in Hannover. Even at the beginning of the war, Osenberg had managed to preserve some individuals from the draft in the interests of naval research, and he indefatigably wrote numerous memoranda about the importance of such deferment. These finally found resonance with Albert Speer. Speer decided to establish a planning office for scientific matters, which would also solve the problem of deferments for scientists. On June 29, 1943, Hermann Göring established such a planning office *within* the RFR, and Osenberg became its chief. Osenberg was directly responsible to Göring and not to Mentzel, some of whose functions were thus taken away.[236] Mentzel apparently was sufficiently distressed that he started to

[234] Zierold 1968: 215–224.
[235] Ibid.: 223.
[236] Ibid.: 248–249.

assemble material against Osenberg with a view to denouncing him.[237] Osenberg went immediately to work determining the research necessities of the armed forces and obtained an order within six months (December 18, 1943) recalling 5,000 scientists from service. One estimate is that Osenberg either recalled or deferred more than 15,000 scientists.[238] The war went on; the disorganization in scientific research funding grew. Not only did Mentzel and Osenberg act independently, but also each of the three divisions of the armed forces, and a "National Office for Economic Growth" (*Reichsamt für Wirtschafsausbau*), independently gave money to scientists. Some of this overlap in action appeared in the Süss-Doetsch controversy of chapter 4. Following the June 6, 1944, Allied landing on the Normandy beaches, the Allied armies finally broke through on July 31; on the same day, Osenberg finally called for a unitary organization overseeing all these various competing bureaucracies. On August 24, Göring complied, creating a "Defense Research Community" organization *within* the RFR, with Osenberg at its head. Also in the summer of 1944, Göring established yet another organization: "The Office for the Coming Generation [of scientists]," *within* the RFR. These, of course, became just two more bureaucracies. As Zierold remarked: "The organization of the RFR became completely impenetrable."[239] Meanwhile, as the war went worse for Germany, some of those deferred or recalled through Osenberg's actions were redrafted.[240] Osenberg complained, with mixed success; however, on July 23, 1944 (three days after the attempted assassination of Hitler), Heinrich Himmler issued an order stopping these recalls, "since I consider the demolition of our research madness." A very minor mathematician, H. J. Fischer, then working for the "National Security Office," claimed to have been responsible for this by making Himmler aware of Osenberg's difficulties.[241] Fischer estimated that Osenberg saved nearly 20,000 (rather than 15,000) scientists, but this number is obtained by adding the original 5,000 to the number 14,600 in Himmler's order. How effective that number was or how much overlap there was is unclear.[242] In any case, a number of mathematicians were among those recalled, including Helmut Ulm and Martin Barner,[243] a recent director of the Oberwolfach Institute. One can agree with Fischer that many of those scientists recalled would certainly have fallen otherwise.

This brief summary of Osenberg's attempt at wartime deferment of scientists and mathematicians demonstrates again the lack of political concern for most

[237] Fischer 1985, 2:28–29.

[238] Zierold 1968: 250.

[239] Ibid.: 253–255.

[240] Fischer 1985, 2:29.

[241] Ibid.: 86–87.

[242] Beyerchen 1977: 190, citing British intelligence, claims only 4,000 were actually recalled and 2,000 had already died in the war. On the other hand, Hans Ebert (n.d.: 2) claims 25,000 were recalled. Unfortunately, Ebert's source is missing from my copy of his manuscript (obtained from Herbert Mehrtens).

[243] Personal communication, March 24, 1988. For some other mathematicians among those recalled, see below, chapter 6, The Creation of the Oberwolfach Institute.

science and mathematics until the war situation began to become desperate. But it also shows how the polycratic nature of the Nazi state interfered with meaningful action, even after the concern was taken.

Having dealt with some of the vicissitudes of the Third Reich for mathematicians as teachers of mathematics, as scientific researchers, and simply as members of a university community, it is time to turn to some of the corporate bodies mathematicians established, and how these were affected.

Heinrich Behnke. Courtesy of the Mathematisches Institut Oberwolfach

Wilhelm Blaschke. Courtesy of the Mathematisches Institut Oberwolfach

Harald Bohr. Courtesy of the Mathematisches Institut Oberwolfach

Luitzen Egbertus Jan Brouwer. Courtesy of the Mathematisches Institut Oberwolfach

Richard Courant. Courtesy of the Mathematisches Institut Oberwolfach

Felix Hausdorff. Courtesy of the Mathematisches Institut Oberwolfach

Erich Hecke. Courtesy of the Mathematisches Institut Oberwolfach

Erich Kamke. Courtesy of the Mathematisches Institut Oberwolfach

Hellmuth Kneser (and Mrs. Cramer), Oberwolfach. Courtesy of the Mathematisches Institut Oberwolfach

Konrad Knopp. Courtesy of the Mathematisches Institut Oberwolfach

Edmund Landau. Courtesy of the Mathematisches Institut Oberwolfach

Emmy Noether. Courtesy of the Mathematisches Institut Oberwolfach

Heinrich Scholz. Courtesy of the Mathematisches Institut Oberwolfach

Wilhelm Suess, ca. 1954. Courtesy of the Mathematisches Institut Oberwolfach

Theodor Vahlen. Courtesy of the Mathematisches Institut Oberwolfach

Bartel van der Waerden. Courtesy of the Mathematisches Institut Oberwolfach

Ernst Zermelo. Courtesy of the Mathematisches Institut Oberwolfach

Otto Blumenthal. Courtesy of Gerald L Alexanderson

Mathematisches Insititut Oberwolfach. Courtesy of the Mathematisches Institut Oberwolfach

Ludwig Bieberbach. Reprinted, by permission, from Norbert Schappacher, "The Nazi Era: The Berlin Way of Politicizing Mathematics," in *Mathematics in Berlin*, ed. H. Begehr, H. Koch, and J. Kramer, p. 130. © 1998 by Springer-Verlag GmbH & Co.KG

Kurt Reidemeister, with his wife, and Gustav Doetsch. Courtesy of Gerald L. Alexanderson

CHAPTER SIX

Mathematical Institutions

THE lifeblood of any mathematical or scientific enterprise is communication, which is why some historians of science have devoted so much attention to citation analyses and "invisible colleges." To sift out chaff, scientific communication has come to mean by publication, and in this way the science is verified by colleagues. This is still true today, and was even more true sixty or seventy years ago, before the advent of xerography, "preprints," electronic mail, and the like. It is natural to ask how mathematical journals were affected by the political pressures of Nazi Germany and how they responded. The three leading German mathematical journals of the 1930s were the *Journal für die Reine und Angewandte Mathematik*, founded in 1826 by August Crelle and known fondly as "Crelle," since no other convenient short name existed; the *Mathematische Annalen*, founded in 1868 by Alfred Clebsch and Carl Neumann; and, perhaps slightly behind, the *Mathematische Zeitschrift*, founded in 1918. During the Nazi period, the effective chief editors of these publications were Helmut Hasse ("Crelle"), Erich Hecke (*Annalen*), and Konrad Knopp (*Zeitschrift*). Whatever the differences in their attitudes toward the Nazi regime may have been, they all were intent on upholding mathematics and protecting it from chicanery and political interference. Furthermore, even mathematicians, like Ludwig Bieberbach or Ernst August Weiss, who were ardent Nazi supporters or party members distinguished between a mathematical fact and the "style" that led to it—the fact participating in mathematical truth like any other, however one might consider the formulation thereof flawed or deplore its style. Thus the way in which mathematical journals might be affected seems problematic at first. To thoroughly survey the interactions of the regime with mathematics publications is obviously impossible in a book of this sort, even if all the relevant documents were available. However, as mentioned in chapter 5, copies of some of Hecke's correspondence concerning the *Mathematische Annalen* during those years have come into my possession. The incidents described in these letters not only are interesting but also might well be taken as indicative of the sorts of pressure standard mathematical journals faced. The journal *Deutsche Mathematik* ostensibly devoted to an explicitly Aryan mathematics as distinct from other ethnic sorts, will be studied in detail in chapter 7.

In addition to issues directly facing journals, like the publication of ideological articles, the publication of Jewish authors (or even the dedication of articles to Jews), the treatment of non-German authors, and the financial support of Jewish former coworkers, this Hecke correspondence touches on two figures who were principals in other mathematical projects. One was the logician (and theologian) Heinrich Scholz, who created a prominent school of mathematical logic—strikingly, at the same time as some devotees of an Aryan mathematics were decrying "logic-chopping" and axiomatics. The other was Max Steck, a minor geometer turned historian and philosopher of mathematics, who was central to a project (toward the end of the war) to enhance German cultural prestige through publication of the works of the famous Huguenot mathematician Johann Heinrich Lambert. Steck is an example of a figure attempting to climb politically in his profession through adoption of Nazi ideology and philosophy. Scholz is an example of the conservative nationalist who bacame disillusioned by Nazi behavior. Indeed, Steck wrote a book on the philosophy of mathematics that called Hilbert's ideas "Jewish," and condemned Scholz as well. Scholz reviewed Steck's book savagely and negatively. The tone of Scholz's review is not surprising; what is perhaps more surprising is that it was commissioned by Bieberbach and appeared in his *Deutsche Mathematik*.

A primary mathematical institution was naturally the German Mathematical Society and its journal. In 1934 these were involved in a contretemps that is one of the central events in a discussion of mathematicians in the Nazi period. The protagonist in this event was Ludwig Bieberbach, and its dénouement saw him resigning all society offices and founding *Deutsche Mathematik*.

Another mathematical society of import was the *Mathematische Reichsverband*, an organization of mathematical societies founded in 1920 and dedicated to promoting the value of mathematics to the general public as well as improving mathematics instruction at both the secondary and university levels. Throughout its existence, its head was the Berlin mathematician Georg Hamel. The MR adapted itself with alacrity to the Nazi situation (after all, one of its functions was to increase the popularity of mathematics).

A mathematical figure whose rise to prominence was occasioned by the 1934 crisis in the German Mathematical Society was Wilhelm Süss. As a leader of that society from 1937 onward, he seems to have been someone of fundamentally conservative principles with whom the Nazi authorities felt they could do business, but whose primary interests were always mathematics and mathematicians. In promoting those interests, he sometimes managed successfully to speak openly in ways that were generally politically forbidden. Nevertheless, he certainly enjoyed prestige and perquisites under the Nazis as a consequence of his position. Süss was also the creator, late in the war, of the (soon) international mathematical center at Oberwolfach.

Applied mathematics might be thought to have had some sort of privileged position in a country first rearming, and then at war. However, it wasn't until late in the war that any general effort to mobilize German academic expertise militarily was made. Partly this was because of the general Nazi suspicion of academics. Thus a brief survey of the position of applied mathematics in Nazi Germany seems to ask for consideration.

Finally, an "institution" peculiar to Nazi Germany was the concentration camp. Somewhat surprisingly, there were attempts, again only late in the war, to make rational use of some inmates for mathematical purposes.

Thus this chapter covers a great variety of institutional establishments, from the usually ordinary faced suddenly with extraordinary pressures, to the attempt to use slave labor rationally mathematically, to a present-day gem of international mathematical collaboration.

The Case of Otto Blumenthal

In addition to being among the older mathematical journals in any language, the *Mathematische Annalen* had a long association with Göttingen. Rudolf Friedrich Alfred Clebsch in 1868 had just succeeded to the Göttingen professorship associated with the names of Gauss, Dirichlet, and Riemann. Clebsch died young (at thirty-nine in 1872); the man who would make the journal a premier mathematical publication was his former assistant, Felix Klein, who became one of two chief editors in 1876, while still at Erlangen. Klein was succeeded by David Hilbert as chief editor. Thus Göttingen's golden age as a mathematics center and the *Annalen*'s as a mathematical publication were closely linked.

In 1928, the *Annalen* was involved in the mathematical-political struggle between L.E.J. Brouwer and Hilbert, a struggle that was not only about intuitionism versus formalism[1] as philosophies of mathematics but also about nationalism in science, personal animosities,[2] and academic politics, including the rivalry between Berlin and Göttingen as mathematical centers. The *Annalen* became involved when in 1928, Hilbert attempted to remove (eventually successfully) Brouwer as a member of its editorial board. Because of Brouwer's own somewhat naive politics in the Nazi period,[3] the struggle became one more in a sequence of events that led to Brouwer's characterization as a dangerous reactionary. The tone of the struggle is exhibited by the fact that Brouwer suggested Hilbert was "of unsound mind,"[4] and Albert Einstein, who attempted to stay neutral, apparently said Brouwer was "an involuntary proponent of Lombroso's theory of the close relation between genius and insanity." Einstein also, with reference to a somewhat obscure Greek parody, gave this struggle the name with which it has been characterized in the literature—the "war of the frogs and mice."[5]

[1] Cf. above, chapter 2.

[2] See, e.g., Fraenkel 1967: 160–162; and Alexandroff 1969: 122–123.

[3] Above, chapter 2; see also below, chapter 7.

[4] In 1925, at age sixty-three, Hilbert developed pernicious anemia. Fortuitously for Hilbert, the same year saw the experiments of George Hoyt Whipple, and in the following year, the therapy developed by George R. Minot and William Parry Murphy, for which the three would share the Nobel Prize. In 1948, it was found that pernicious anemia was caused by an inability to absorb vitamin B_{12}. Hilbert's illness had made him worry about the future of the *Annalen* should he die. See Reid 1970: chap. 21. In 1928, he was still suffering some of the effects of the disease, according to his own testimony. See van Dalen (next note).

[5] The whole story is told in Dirk van Dalen, "The War of the Frogs and the Mice, on the Crisis of the *Mathematische Annalen*," *Mathematical Intelligencer* 12, no. 4 (1990): 17–41. See also Norbert Schappacher and Martin Kneser, "Fachverband—Institut-Staat," in *Ein Jahrhundert Mathematik*,

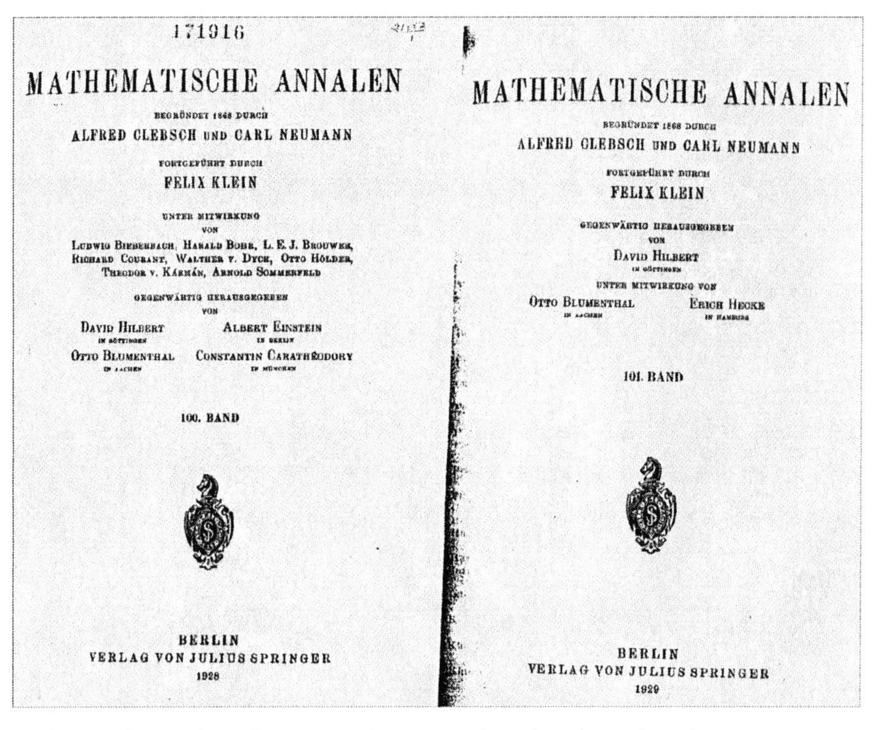

171916

MATHEMATISCHE ANNALEN

BEGRÜNDET 1868 DURCH
ALFRED CLEBSCH UND CARL NEUMANN

FORTGEFÜHRT DURCH
FELIX KLEIN

UNTER MITWIRKUNG
VON
LUDWIG BIEBERBACH, HARALD BOHR, L. E. J. BROUWER,
RICHARD COURANT, WALTHER V. DYCK, OTTO HÖLDER,
THEODOR V. KÁRMÁN, ARNOLD SOMMERFELD

GEGENWÄRTIG HERAUSGEGEBEN
VON
DAVID HILBERT IN GÖTTINGEN
ALBERT EINSTEIN IN BERLIN
OTTO BLUMENTHAL IN AACHEN
CONSTANTIN CARATHÉODORY IN MÜNCHEN

100. BAND

BERLIN
VERLAG VON JULIUS SPRINGER
1928

MATHEMATISCHE ANNALEN

BEGRÜNDET 1868 DURCH
ALFRED CLEBSCH UND CARL NEUMANN

FORTGEFÜHRT DURCH
FELIX KLEIN

GEGENWÄRTIG HERAUSGEGEBEN
VON
DAVID HILBERT
IN GÖTTINGEN

UNTER MITWIRKUNG VON
OTTO BLUMENTHAL IN AACHEN
ERICH HECKE IN HAMBURG

101. BAND

BERLIN
VERLAG VON JULIUS SPRINGER
1929

Mathematische Annalen title pages, volumes 100 (1928) and 101 (1929)

A result of this contretemps was that the title page of volume 101 (1929) of the *Annalen* looked very different from that of volume 100 (see figure). The only editors appearing in volume 101, besides Hilbert, were Erich Hecke (for the first time) and Otto Blumenthal. Blumenthal had been Hilbert's first doctoral student and effectively handled the nontechnical managing of the journal. In 1934 (vol. 109), the Dutch mathematician B. L. van der Waerden, who was a professor in Leipzig (and married to Franz Rellich's sister), was added to the main editorial board, and with volume 115 (1938), Behnke (who had been Hecke's student) also joined this group. Blumenthal was Jewish, but (probably through Hilbert's insistence) his name remained on the editorial page until volume 116 (1939), when it perforce vanished. With volume 117 (1940–41), Hecke officially replaced Hilbert as chief editor, though Hilbert had ceased to have much actual input in the journal some time before.[6] Next to Hermann

---

1890–1990 (1990), 54–57. "The War of the Frogs and the Mice" (*Batrachomyomachia*) is a brief parody of the *Iliad*, now attributed to the Carian Pigres, but formerly ascribed to Homer. A scholarly edition was made in 1896 by the German scholar Arthur Ludwich. It is often classed with the so-called Homeric Hymns. The *Batrachomyomachia* was translated into English by both George Chapman and William Cowper, among others, along with the *Iliad* and *Odyssey*. A charming contemporary blank verse English translation was made by Daryl Hine (Atheneum, 1972). The poem is the height of ridiculousness.

[6] Hilbert's seventy-seventh birthday was January 23, 1939.

Weyl, Hecke was perhaps the most important mathematician to have been a student of Hilbert.

These last events were not without some attempt, even in 1939, to retain Blumenthal's name and also Hilbert's official editorship. While men like Doetsch might be very careful about who his fellow editors might be, others were not so concerned. Ferdinand Springer was the publisher of the *Mathematische Annalen*, as well as many other scientific journals (including the *Mathematische Zeitschrift*). Apparently in 1938, he came under some pressure because of non-Aryan members of editorial boards of his journals, and he believed that if Hilbert, though seventy-six, were to cease being officially chief editor, Blumenthal could not remain on the editorial board, even though Springer recognized Blumenthal's managerial importance. Blumenthal, of course, had to leave (after thirty-two years of work on the journal, many as the functioning managing editor). It would appear that in 1938, the Nazi government attempted to expunge any last traces of Jews in German cultural life and any last traces of resistance to that expulsion. Thus Springer's complaint about pressure. Among Nazi government officials, Springer's publishing house had in fact long been characterized as "Jewish."[7] A similar story took place with the mathematical-pedagogical journal founded jointly by Behnke and Otto Toeplitz. Toeplitz, as a Jew, was forcibly retired at Bonn in 1933. In 1938, the Kurator at Münster asked the Dekan (who asked Behnke) why Toeplitz's name still appeared on the journal's title page.[8] As to Blumenthal, already in September 1933 (after an initial forcible furlough on May 10), he had been dismissed from his position in Aachen on political grounds as the ultimate result of a student initiative that began before the law of April 7.[9] Around this time, van der Waerden attempted to see whether a position might be found for Blumenthal in the United States, as appears from a letter of Oswald Veblen:[10]

> Dear van der Waerden:
>
> A quite considerable number of displaced German mathematicians have found temporary or permanent places in the United States. . . . We have a large number of native unemployed mathematicians, and there are also some signs of a growing anti-Semitism.
>
> On this account I have hesitated to suggest that Blumenthal should be invited to this country, but I have written to the "Emergency Committee," which has been helping in this matter. . . .
>
> The question also arises whether Blumenthal does not receive a large enough pension so that it might be better to use such funds for others who are in greater need.

In 1939, Blumenthal emigrated to Holland. With the German occupation of the Netherlands, he and his wife were arrested for deportation—apparently Dutch

[7] See HK, Theodor Vahlen to Hellmuth Kneser, Sept. 5, 1933.

[8] HNMA, Behnke to Hecke, May 9, 1938; see also Blumenthal to Springer, Jan. 9, 1938.

[9] Blumenthal had been an Ordinarius at Aachen since 1905, and thus fell under the "exceptions" clauses of the April 7 law. See Schappacher and Kneser 1990: 36–37 for more details.

[10] VP, Veblen to van der Waerden, Dec. 18, 1933. For the anti-Semitism referred to by Veblen, see Reingold 1981, and above, chapter 3.

friends wished to hide him, but Blumenthal refused because "he didn't want to endanger friends." His wife died in the collection center prior to deportation, and Blumenthal himself died, at age sixty-eight, in Theresienstadt on November 13, 1944.[11]

Editors of professional journals felt political pressure, even after Jews were eliminated from their midst. Thus Behnke wrote Hecke on January 27, 1939, concerning his becoming the official chief editor of the *Annalen*:[12]

> With your taking over . . . there must also be a new division of burdens, and indeed not in the practical work but in the political responsibility. I must be freed from my position as "sitting editor." The statement . . . at the end of every volume that I am responsible for the text, today means something quite different from earlier years.[13] Earlier it meant a potential liability in civil law, which, with the usual management practice, practically speaking, could scarcely occur. Today it is a political obligation, which at the moment is not at all urgent. In my opinion, however, it was especially strong in the second half of the year just past. I did not write you about that, however already at the time of the Baden congress [September 11–16, 1938] I had had many worries. Neither the publisher, nor the other editors can protect me from the fact that I am the first who could be made liable for Blu[menthal] working with us. At the moment, I see no danger, even with the sharpest supervision. However, what can happen once again is naturally not to be overlooked.
>
> I thoroughly wish to accept my portion of risk . . . however, please free me from this one-sided burden. . . .
>
> It would naturally be very good, if with the change in chief editor, an Englishman or some other foreigner would become a member of the editorial board.[14] . . . [If that should be successful] there also arises still again the question whether we are not then compelled also to add a German editor, so that we are not subject to the reproach that we made the old German scientific organ more and more into an international journal.

To the end of this typewritten letter is appended a handwritten note: "If it is in any way possible, I would like to visit you before my England trip.—We must attempt to help Ann[alen]—Otto."

The Lachmann Paper Incident

Kurt Lachmann was a mathematician, apparently something of an independent scholar,[15] in Berlin. In 1928, he published in the *Mathematische Annalen* a paper

[11] Pinl 1969: 168–170 (Aachen); Heinrich Behnke, "Otto Blumenthal zum Gedächtnis," *Mathematische Annalen* 136 (1958): 387–392; Arnold Sommerfeld and Franz Krauss, "Otto Blumenthal zum Gedächtnis," *Jahrbuch der Rheinisch-Westphälischen Technische Hochschule Aachen* (1951): 21–25.

[12] HNMA, Behnke to Hecke, Jan. 27, 1939.

[13] Cf., in HNMA, Behnke to other editors, Apr. 2, 1938, where he accepts such responsibility.

[14] On his trip to England discussed above in chapter 5, "Foreign Contact and Travel," Behnke attempted to discover such a suitable person.

[15] His name does not appear in membership lists of the German Mathematical Society.

concerning approximate solutions of a certain differential equation connected with pendula.[16] He was also a Jew. In November 1939, he sent another paper, also in ordinary differential equations, to the *Annalen*. In fact, Lachmann had apparently previously written two letters to Behnke, the first prior to July 14, 1935, asking whether he could submit a paper to the *Annalen*. Behnke had discussed the matter with Hecke, and Hecke wrote Lachmann on November 9, 1939, inviting the submission.[17] Van der Waerden recommended as referee his brother-in-law, Franz Rellich, then at Dresden. Earlier, at Göttingen, Nazis like Oswald Teichmüller and Erhard Tornier had considered Rellich part of the "Courant-clique" and had driven him out.[18] Rellich willingly undertook the job of reading the handwritten manuscript,[19] and as soon as he did so, Hecke wrote Lachmann, asking for his patience, indicating his paper was being read, and stating circumspectly:[20]

> I am sorry that I still cannot give you any final decision. The decision does not depend only on the editorial board, since, because of the particular nature of the question, the publisher also desires to become involved, and he must treat with higher officials (*höheren Instanzen*).

Within a month, Rellich had prepared a report that found the paper interesting, containing a nonobvious new result, but the arguments as presented insufficient, and so the paper needed considerable revision; in addition, he was unsure whether the paper was important enough to be published in as prestigious a journal as the *Annalen*. As to its authorship, he remarked sarcastically, "I would handle the case of L[achmann] confidentially. According to Blaschke, it ought to be a question of an old American."[21] On February 28, 1940, Hecke wrote Lachmann, conveying Rellich's criticisms and suggesting that even when reworked, the paper was perhaps more suitable in a journal for applied mathematics. He ended,[22]

> As to the personal difficulties for a printing in a German journal, so according to our latest information, if it is at all possible, the addition to your name is not necessary; this would also correspond to the wishes of the editorial board, which does not intend to introduce any defamatory additions to the recognition of authors.

What Hecke undoubtedly refers to here is the law of August 17, 1938, according to which Jews were only allowed to have distinctively Jewish first names, and those who did not had to append the name Israel if a male, or Sara if a female, to their names.

[16] "Beitrag Zum Schwingungsproblem von Duffing," *Mathematische Annalen* 99 (1928): 479–492.

[17] HNMA, Lachmann to Hecke, Nov. 14, 1939.

[18] MI, Tornier to Rellich, June 18, 1934, and Weber, Anlage 64.

[19] HNMA, Hecke to Lachmann, Jan. 27, 1940.

[20] Ibid.

[21] HNMA, Rellich to Hecke, Feb. 22, 1940. For the meaning of the sarcastic reference to Blaschke, who in an anti-Semitic article had excused the appearance of Old Testament names among Americans, see chapter 8.

[22] HNMA, Hecke to Lachmann, Feb. 28, 1940.

While this all seems straightforward, given the circumstances, much else had happened and would happen in connection with Lachmann's paper. At the time that Lachmann's paper was received, van der Waerden, one of Hecke's "coeditors," had suggested that, if the paper were of sufficient quality, then it should be sent to press without informing the publisher, Ferdinand Springer.[23] Behnke, the other "coeditor," had opposed this action as unfair to Springer, since he would have had ultimate responsibility for what was published in the journal. Behnke also warned of the serious consequences that might befall them, and the journal as well, should a paper by a Jewish author be published.[24] Hecke decided to talk to Springer, and did so sometime around December 20, 1939.[25] Of course, Hecke had not yet received Rellich's referee's report, and so had no idea whether he would actually face the eventuality of publishing a paper by a Jew. On January 3, 1940, he wrote his coeditors: "Fundamentally in such a case Springer insists on participation as publisher in deciding on the procedure. As he communicated to me, nowadays in all situations the publisher is made primarily (*in erster Linie*) responsible."[26]

No later than January 15, Springer's apparent refusal to risk such a publication had caused Hecke to want to resign his chief editorship of the journal, even though Rellich (who had been ill) would not send his referee's report until February 22.[27] Clearly, it was the principle that mattered to him. Hecke naturally let van der Waerden and Behnke make their own decisions about what to do. Behnke was dubious about Hecke's proposed resignation:[28]

> Naturally, I have grave doubts whether your position is not unnecessarily rigid. However, I also naturally have moods in which I completely agree with you. Then, however, I always say to myself that we, as long as we live, still must also hope. And for this there are now several grounds.

It is hard to see what grounds for real hope Behnke had on January 15, 1940.[29] In any case, his letter is filled with this tone of "maybe you're right, but on the

[23] HNMA, Behnke to van der Waerden, Oct. 8, 1940.

[24] Ibid. It might be noted that in 1938 Lachmann had managed to publish a brief note in the *Journal of the German Mathematical Society* (*JDMV* 48 [1938]: part 2, pp. 28–29). The editor at the time, E. Sperner, either did not know or did not care that Lachmann was a Jew—the latter seems unlikely. Rather strangely, up until April 15, 1940, there was no juridical ban on publications by Jewish authors. See Michael Knoche, "Wissenschaftliche Zeitschriften im nationalsozialistischen Deutschland," in Monika Estermann and Michael Knoche, eds., *Von Göschen bis Rowohlt* (Wiesbaden: Otto Harrassowitz, 1990), 260–281, p. 271. Thus Lachmann's paper was not forbidden *de jure*. In any case, presumably, after the start of the war, publishing a paper by a Jew was more dangerous. Strangely, neither Hecke, Behnke, van der Waerden, nor Rellich seem to have noticed this 1938 publication.

[25] HNMA, Hecke to Springer, Dec. 14, 1939.

[26] Cited in HNMA, Behnke to van der Waerden, Oct. 8, 1940.

[27] Hecke's intention to resign may be in his January 3 letter, which is not in HNMA. See HNMA, Behnke to Hecke, Jan. 15, 1940.

[28] Ibid.

[29] January 15, 1940, was in the middle of the *Sitzkrieg*, or "phony war." The *Graf Spee* had been scuttled in Argentina a month earlier (December 13, 1939), and Operation Weser, the German invasion of Norway, would begin on April 19. Presumably, Behnke hoped for a settlement of the

other hand." Among other problems he raised was the future of the *Annalen*, which also served as a connection to the non-German world. In discussing the possible constitution of an editorial board without Hecke, Behnke rejected Hellmuth Kneser as someone subservient to Nazi officialdom. Behnke also raised the specter that if he and van der Waerden were also to resign, then Springer would continue the *Annalen* "in a sovereign fashion" and was likely to engage Bieberbach and Blaschke as (safe) new editors, and they would be likely to accept such a position. Ludwig Bieberbach, as already remarked, as well as being a prominent mathematician, was a vocal supporter of the Nazis, while Wilhelm Blaschke, equally prominent, seems to have been an opportunist during the Nazi period, flattering the state and its regulations at every turn.[30] In addition, Hecke and Blaschke detested each other.[31]

There appears to have been another exchange of letters prior to a letter from Hecke to Behnke on January 23, 1940. Behnke, concerned with maintaining the *Annalen* even if Hecke should withdraw, had suggested the possibility of adding Herbert Seifert and William R.R.H. Threlfall as editors. These were two topologists and friends who had collaborated on an outstanding textbook in topology. Threlfall was nineteen years Seifert's senior; nevertheless, they were known in the mathematics community as "inseparable twins." They had known each other in Dresden, where they had "habilitated" seven years apart. At the time of this letter, Seifert was in Heidelberg, having been chosen to succeed Heinrich Liebmann in 1935 when the latter was forcibly emerited;[32] Threlfall was in Frankfurt, having succeeded to Siegel's chair when he left for Göttingen in 1938. Hecke applauded Behnke's desire to keep the *Annalen* going, and to involve the two topologists—though there was some question in Behnke's mind whether this was possible, since Threlfall was being very retiring because of his English ancestry.[33] In any case, Hecke seemed willing to put off his resignation for several weeks and apparently to handle the issue of the Lachmann paper by making a ministerial inquiry.[34] Concerning this, Behnke opined in March that

---

war favorable to Germany. Perhaps he thought (no doubt wrongly) that if that were to happen in early 1940, then ideological pressures would be relaxed.

[30] For more about Bieberbach and Blaschke, see below, "The Bieberbach-Bohr Exchange and the 1934 Meeting of the DMV," as well as, respectively, chapters 7 and 8.

[31] Personal communication from the late Hans Zassenhaus. Nevertheless, they jointly ran the Hamburg mathematics department for twenty-six years. It would seem they never spoke except about mathematics. Zassenhaus told me that Hecke was named correspondent in Blaschke's divorce from his first wife. Dr. Rotraut Stanik, who currently cares for Hecke's *Nachlass* in Hamburg, told me that Blaschke's second wife confirmed this to her. German law has prevented me from confirming this from the legal documentation of the divorce (at least in Hamburg, such documents are sealed to everyone except those, or the attorneys of those, who have a legal interest in them, a law that long antedates 1933). On the other hand, this secondhand information seems widespread. In any case, Hecke preserved his own marriage. Clearly also, Behnke knew of Hecke's detestation of Blaschke.

[32] See chapter 5.

[33] HNMA, Behnke to Hecke, Jan. 15, 1940, and Hecke to Behnke, Jan. 23, 1940. Threlfall's father was English. His mother was a niece of the microbiologist Robert Koch.

[34] Compare HNMA, Hecke to Behnke, Jan. 23, 1940, and Behnke to Hecke, Mar. 8, 1940.

they never would hear from the ministry, and that if they spoke of it again to them, it would redound on their heads.[35] A remark should be made about the opening line of Hecke's January letter. It reads, "Concerning your letter of January 20: 'Ha Ha!'" The German letter "H" is pronounced "Ha." "Ha" as an exclamation is used in German as in English. H.H. are, of course, the initial letters of the "Germanic greeting": "Heil Hitler," whose compulsory use was sought by the Nazis. One of the small ways in which Hecke safely expressed his opinion of the regime was by the punning exclamation "Ha Ha!"[36]

On the twenty-fourth of June, Hecke sent his resignation letter to Springer, after asking Behnke and van der Waerden to comment on its final form.[37] A revised manuscript from Lachmann following Hecke's February letter was received in June, and Hecke planned to tell him that acceptance of his paper would depend upon a decision of a new editorial committee.[38] Behnke, at least, did comment before the letter was sent. He still tried to dissuade Hecke from resigning:

> I must tell you again, that I find your withdrawal dangerous. The ministry can become very angry and cause injuries that go far beyond what one could imagine earlier. So in my opinion at this time, van der Waerden also generally no longer has the possibility of deciding according to his own judgment.

He asked Hecke also to at least modify a sentence in his letter, and worried about the future of the *Annalen*.[39]

This last theme appeared again in another letter a month later.[40] Behnke, as the editor responsible for the printing, suggested further that Hecke's and van der Waerden's leaving the editorial board should not happen until after the conclusion of volume 117.[41] He claimed the journal had a responsibility to publish the papers coming to it during the six months to a year it would take to put together a new editorial board.[42] Shortly thereafter, Hecke wrote Lachmann telling him that his paper could not appear before volume 118 and that there would be a substantial change in the composition of the editorial board for that volume:[43] "In a principally so important question as the acceptance of your paper we cannot anticipate now the decision of [this] future editorial board. So, to my regret, the decision as to the fate of your paper must be shoved off till

[35] Ibid.

[36] Both Horst Tietz (interview, Apr. 5, 1988) and Hans Zassenhaus (interview, Nov. 17, 1987) told me of this Hecke usage.

[37] HNMA, Hecke to Behnke and van der Waerden, June 16, 1940. Also HNMA, Tönjes Lange to Hecke, June 28, 1940, acknowledging receipt of Hecke's letter [of June 24] and putting it aside until Springer returned from vacation at the end of July.

[38] HNMA, Hecke to Behnke and van der Waerden, June 16, 1940.

[39] HNMA, Behnke to Hecke, June 19, 1940.

[40] HNMA, Behnke to Hecke, July 11, 1940.

[41] Ibid.

[42] Though there were many fewer papers than had been usual in pre-Nazi times. Cf. Behnke 1978: 134.

[43] HNMA, Hecke to Lachmann, July 14, 1940.

then." Springer wished, if at all possible, to prevent Hecke's resignation, and tried to get him to come to Berlin. Hecke refused, a position supported by Behnke:[44]

> You could come into an awkward (*peinliche*) situation there, for Springer can bring forth all possibilities in order to keep you on the editorial board. It seems to me not excluded that Springer might simply bring a ministry official to a discussion, who will represent himself there as your superior. Those fellows now do that frequently.

Springer, however, had a different card to play. Two days later, he suggested to Behnke that if Hecke and van der Waerden withdrew from the editorial board, he would stop paying Blumenthal's pension. Springer had agreed to pay this to Blumenthal when he was forced from the *Annalen* board because Hecke, Behnke, and van der Waerden had insisted on it as a condition of continuing to put out the journal. Behnke immediately wrote van der Waerden, but not Hecke, who was "terribly tired of life and I am afraid to torment him further."[45] He asked van der Waerden to "consider please whether it is right to create this new difficulty for Blumenthal."[46] The whole tone of Behnke's letter is to keep Hecke, and secondarily van der Waerden, on the editorial board, as he had been insisting for some time.

Behnke's letter made van der Waerden angry. He reminded Behnke that first, "Blumenthal's pension" was no longer paid to Blumenthal, but to his sister, who could not transfer it to him.[47] Second, in any case, after Hecke's withdrawal, Springer had every right to cancel the pension, whatever he, van der Waerden, did. Third, van der Waerden thought that his original idea of a surreptitious publication of Lachmann's paper (should it prove of suitable quality) would have avoided the present crisis, and only Behnke's opposition prevented this solution. Finally, he says, "In no case do I wish to be the hindrance if Springer wants to create a National Socialist editorial board for the *Annalen*. Therefore the matter must rest in that I allow him a free hand in forming a new editorial board."[48] Indeed, van der Waerden had earlier thought Behnke's willingness to continue the *Annalen* a sign of a "lack of character," an opinion that Hecke did not share, though he did say he would find it "refreshing" if van der Waerden were also to resign.[49]

Dismayed himself by van der Waerden's reply, Behnke sent him a long letter that was partly defensive—"Your shoving the guilt for this development on to me . . . has greatly wounded me"—and partly conciliatory—"However I think we both already have it hard enough and that in these times of need, under all circumstances, we must avoid further personal quarrels in our ranks." As

[44] HNMA, Behnke to Hecke, Sept. 18, 1940.

[45] HNMA, Behnke to van der Waerden, Sept. 30, 1940.

[46] Ibid.

[47] Blumenthal had emigrated to Holland in 1939. Needless to say, transfer of funds out of Germany in 1940 was impossible.

[48] HNMA, van der Waerden to Behnke, Oct. 5, 1940.

[49] HNMA, Hecke to van der Waerden, Jan. 24, 1940.

Behnke recalled, Hecke and van der Waerden had both been told in November 1939 that there might be problems with Blumenthal's pension, though it was true Behnke had then shoved these issues into the background. According to Behnke, van der Waerden's original plan for the eventuality of a Lachmann publication was useless, and he had opposed it in part because he thought it "unfair" to subject Springer unwittingly to such a risk.[50] The whole correspondence was copied by Behnke to Hecke, with the hope of smoothing things over.[51] However, all went well without Hecke's intervention, and van der Waerden replied in a way that reestablished his cordial relations with Behnke.[52] In the meantime, Springer wrote Hecke directly threatening Blumenthal's pension. As Hecke then wrote the others:[53]

> The threat regarding Bl[umenthal's] pension at the end [of Springer's letter] is indeed very deplorable. One should be clear, however, that if he means it seriously, then it is pure extortion and one should not let our procedure be influenced by such an action. Otherwise Sp[ringer] could, for example, arbitrarily repeat this action!

The story of this pension for a Jewish mathematician, which had become involved in the issue of whether a mathematical paper by a Jew was publishable in 1940, is itself somewhat complicated. Blumenthal had a sister who had been married to the son of the famous German Romantic writer, Theodor Storm. When the Nazi racial laws came into effect, her husband divorced her, and Blumenthal had since been her almost sole source of support. When Blumenthal decided in 1939 to emigrate to Holland, he, Behnke, and Springer agreed that the pension should be paid to his sister, and he had impressed this point on his friend Behnke. As things turned out, five years later she as well as her brother would die in the concentration camp at Theresienstadt.[54]

Hecke, anticipating the end of his activity on the journal, stopped accepting papers for its next volume.[55] Because of the *Annalen*'s long association with Göttingen and Hilbert, Springer (according to Behnke) wished to maintain a leading role in its editorial board for people who had had close associations with Hilbert, thereby maintaining its prestige. Hecke, as one of Hilbert's two most distinguished students, fulfilled this desideratum admirably. In a sense, Behnke did also, since he was Hecke's student, and had taken over the operational management of the journal from Blumenthal, Hilbert's first student. Who could replace Hecke in this sense? In Behnke's opinion, only Carathéodory, already sixty-seven, but he would only need to lend the *pro forma* use of his name. Behnke made this suggestion on October 17, 1940; however, Carathéodory was a Greek national (though a German citizen), and on October 28,

[50] HNMA, Behnke to van der Waerden, Oct. 8, 1940.

[51] HNMA, Behnke to Hecke, Oct. 10, 1940.

[52] HNMA, Behnke to Hecke, Oct. 12, 1940.

[53] HNMA, Hecke to Behnke and van der Waerden, Oct. 15, 1940.

[54] HNMA, Behnke to Hecke, Oct. 17, 1940; and Pinl 1969: 168–170, under Aachen.

[55] HNMA, Hecke to Behnke and van der Waerden, Oct. 15, 1940.

Greece entered the war against the Axis, thus eliminating the possibility.[56] In this eventuality, and with Hecke and van der Waerden resigning, Behnke could only suggest the possibility that Hecke "suffer" his name remaining on the title page, and Behnke alone, or possibly together with Kurt Reidemeister,[57] carry on the editorial work: "With the war's end we will be able to see further, and be better able then to make final decisions."[58] Behnke clearly was more optimistic about these possibilities than Hecke, who four months earlier had written: "Besides, I believe that after a victorious [for Germany] end to the war, for many years, mathematical journals will be of no great consequence. It will be rather irrelevant what occurs in them."[59]

For the next several months, there was "neither an editorial board nor a temporary substitute editorial board," in van der Waerden's words.[60] On February 5, Hecke followed up his letter eight months previously to Lachmann with:[61]

> Substantial changes have been undertaken in the editorial board of the *Annalen*, in particular, from the next issue on, I will no longer belong to the editorial board. Besides, I now think on the known grounds that it is out of the question that your manuscript will be accepted. And I leave it to you, not at all to send in your manuscript in the first place, in order to avoid unpleasantness that could perhaps result.

Lachmann survived the war, and presumably the monograph appearing under his name in 1946 with a much more minor mathematical publisher is the article that he originally submitted to the *Annalen*.[62]

As to the editorship of the *Annalen* during this hiatus, conversations were carried on between the three editors and Ferdinand Springer, and the resolution was reached that the title page that had read:

with presently as chief editor (*Herausgeber*)
Erich Hecke
in Hamburg
with the collaboration of

would now read:

with presently as acting chief editor
Erich Hecke
in Hamburg
with the collaboration of

[56] This is the date of the Italian invasion intent on conquest. Greece had attempted neutrality in vain.

[57] For Reidemeister, see above, chapter 3.

[58] HNMA, Behnke to Hecke, Oct. 17, 1940 and Oct. 31, 1940.

[59] HNMA, Hecke to Behnke, June 27, 1940.

[60] HNMA, van der Waerden to Springer, Mar. 6, 1941.

[61] HNMA, Hecke to Lachmann, Feb. 5, 1941.

[62] In *Veröffentlichungen Math. Inst. Tech. Hochschule Braunschweig* (1946), ii + 38 pages. A brief review by W. E. Milne is in *Mathematical Reviews* 11 (1950): 138.

and the back page, on which instructions to authors had been signed

> The editorial board of the *Mathematische Annalen*

would now read:

> The editorial board at this time of the *Mathematische Annalen*.

This arrangement would last for the next six months.[63] Hecke apparently agreed to this primarily to preserve the Blumenthal pension.[64] However, he insisted that after October 1, he should not be counted on even as a "substitute editor."[65] Be that as it were, on July 14, Behnke sent Hecke a handwritten letter in which he said that Harald Geppert had told him Ludwig Bieberbach had approached Springer with the plan of retaining Hecke on the editorial board and expelling both Behnke and van der Waerden from it on grounds of political unreliability. This would be the only way the public would learn of the changes in the *Annalen*. Apart from the *Annalen*, he and van der Waerden could not be indifferent to such a possibility. Thus Hecke, in any future negotiations with Springer, which were sure to occur, should bear this in mind.[66] Clearly, Behnke was trying to suggest that his (and van der Waerden's) personal safety would be involved in Hecke's not continuing as chief editor, even if only provisionally.

Hecke, however, was adamant. He was willing to let the present situation stand as agreed until the first of October; in April, both he and Springer had hoped that by July "the situation would be clearer," and that obviously was not the case.[67] Nevertheless, October 1 was the end of the present arrangement,[68]

> since I believe that no change of the situation will make the future clearer.
>
> The action of Bieberbach's described by you runs completely independently of what I do or undertake. His intentions and the pretended basis for them are naturally grotesque. If he should succeed, then that is the best sign that there is no place in the editorial board for people of my sort, and my intention to withdraw is fully grounded. If he doesn't reach his goal, then you have no basis for fears. Besides I can't take these attempts of his very seriously. Also Spr[inger] has up till now told me nothing of them.

On September 10, Springer, having returned "refreshed" from his vacation, did write Hecke, asking him to continue the provisional chief editorship beyond October. If Hecke rejected this absolutely, then he wouldn't press him further. In that case, however, the only person who seemed suitable as Hecke's replacement was Blaschke.[69] As Hecke surely recognized, and van der Waerden (who had received a copy of Springer's letter) pointed out, the mentioning of

[63] HNMA, Springer to Hecke, Apr. 1, 1941. See also HNMA, van der Waerden to Springer, Mar. 6, 1941.

[64] HNMA, Behnke to Hecke, Oct. 31, 1940.

[65] HNMA, Hecke to Springer, July 7, 1941, in response to Springer to Hecke, July 4, 1941.

[66] HNMA, Behnke to Hecke, July 14, 1941.

[67] "Operation Barbarossa," the German invasion of the Soviet Union, began on June 22, 1941.

[68] HNMA, Hecke to Behnke, Aug. 5, 1941.

[69] HNMA, Behnke to Hecke, July 14, 1941.

Blaschke was "certainly intentional, as he knows quite well that you cannot stand Bl[aschke]." Van der Waerden urged the continuation of the present arrangement also because: "What will our friends in foreign countries say if, after the war, scientific connections to Germany should be taken up again, but suitable structures and intermediaries thereto are lacking?"[70]

On September 15, Hecke agreed to continue provisionally as editor, presumably because of his desire to maintain the possibility of future international contact, as well as his detestation of Blaschke. In fact, only two volumes were published under the provisional arrangement: volume 118 for 1941–43, and a much shrunken volume 119 for 1943–44. By the time the *Annalen* reappeared after the war, Hecke had died, and he and Blumenthal both appear with Klein and Hilbert as distinguished previous editors in volume 120 for 1947–49. Behnke and van der Waerden remained on the editorial board, which was enlarged to six, with no "chief editor."

This "Lachmann paper incident" shows how difficult it was in the Nazi atmosphere for decent people to uphold various professional attitudes they thought appropriate, not just because of the pressure of the state and its ideology, but also because the mephitic social atmosphere brought ordinary decent actions into conflict with one another in impossible ways. Earlier Behnke and Hecke had thought of involving a foreign national in the editorial work of the *Annalen* as potentially a partial protection for their journal. This was also supported by Ferdinand Springer, the publisher, and "as soon as possible."[71] Behnke pursued the idea actively on his 1939 trip to England. One possibility for Behnke was the well-known mathematician G. N. Watson. However, G. H. Hardy, perhaps the leading figure in contemporary British mathematics, suggested privately that "from a purely scientific point of view," E. C. Titchmarsh might be preferable. Titchmarsh, whom Hecke also preferred,[72] was Hardy's student, and would indeed become one of the leading mathematicians of his generation. However, Behnke was at least a partial realist:[73]

> I am now not certain whether, given the stormy development we are living through at present, all our considerations don't come much too late. If the opposition between political forces remains as sharp as in the last 10 days, we can naturally not count on an English cooptation at all. I emphasize, however, that on my [English] trip things looked quite otherwise.

Behnke seemed to acquire an *idée fixe* about Watson, though Hecke insisted that if a foreigner were to be added to the editorial board, it had to be the very best available, hence Titchmarsh, "even if he understands no German" (which

[70] HNMA, van der Waerden to Hecke, Sept. 12, 1941. Van der Waerden was anticipating that the situation would be different than it had been after World War I.

[71] HNMA, Springer to Hecke, Apr. 26, 1939.

[72] HNMA, Hecke to Behnke, Apr. 5, 1939 and Apr. 25, 1939.

[73] HNMA, Behnke to Hecke, Mar. 27, 1939. On March 15, Hitler had marched into the remainder of Czechoslovakia.

seems to have been the case).[74] Of course, all such speculation (which involved Dutchmen as well), while unlikely in May 1939, was totally foreclosed in September.

Max Steck and the "Lambert Project"

The Lachmann incident was not the only instance of the *Annalen*'s editors having to worry about seeming "pro-Jewish." Heinrich Liebmann lost his position at Heidelberg in 1935 as the result of a student boycott and a putative Jewish grandparent.[75] A geometer, he had published a book called *Synthetic Geometry* in 1934, and a paper in the *Mathematische Annalen* in 1935. Of course, at the time of these publications, Liebmann was still in office, having fallen under the exceptions clauses of the April 7 law (he had habilitated in 1899, though he did not come to Heidelberg until 1920).[76] He even published his last paper in 1937 in the *Proceedings of the Bavarian Academy*.[77] But 1937 was obviously not the same political situation as in November 1939, when Lachmann's paper was submitted to the *Annalen*. Shortly after Liebmann's death on June 12, 1939, the geometer Max Steck sent a paper to the *Annalen* dedicated to Liebmann's memory.[78] This brief paper was in effect an improvement on Liebmann's 1935 *Annalen* paper. Max Steck was not trying to make any anti-anti-Semitic point. In fact, as will be seen, he wrote an anti-Jewish article for Ludwig Bieberbach's *Deutsche Mathematik*, and later wrote a book laced with Nazi rhetoric that attacked Hilbert and his logical school of formalism in a scurrilous fashion. Steck seems to have been simply a geometer wishing to honor posthumously another geometer, his teacher, who had had over a hundred publications. Steck was also not very prominent as a mathematician. Hecke, who had no personal objections to the dedication, though he recognized that complaints might come from some active Nazi sympathizers, asked Behnke:[79]

> Do you know what sort of a man the author Steck is, whether a civil servant or something similar? In this case, his professional prospects would certainly not be improved by the dedication. Is he perhaps so naive that he does not notice at all such connections?

At this time, Steck had already published several geometric articles in *Deutsche Mathematik*. He had been a doctoral student of Liebmann's at Heidelberg, receiving his degree in 1932. Apparently in 1935 the *Annalen* rejected a paper of his (because of similarity to work he had already published), and he asked his fellow geometer Wilhelm Süss for help in getting it published.[80]

[74] HNMA, Hecke to Behnke, Apr. 28, 1939, and May 25, 1939.
[75] See above, chapter 5.
[76] Pinl 1972: 162–163, under Heidelberg.
[77] Ibid.: 167.
[78] HNMA, Behnke to Hecke, July 23, 1939.
[79] HNMA, Behnke to Hecke, Oct. 24, 1939.
[80] HK, Süss to Kneser, Jan. 11, 1935.

He "habilitated" at the technical university in Munich in 1938, and his *Habilitationsschrift* was published the next year.[81] This brought him to Hecke's attention, and he wrote Behnke that he should be made aware in any case of the risk he ran through the dedication.[82] Steck's little paper appeared in volume 117 (1939–41) of the *Annalen*, without the dedication to Liebmann, but with a first paragraph devoted to Liebmann's work, which was slightly refined. It seems honestly to have been a homage from a newly fledged geometer to his recently deceased teacher, despite Steck's affinity for the ideas of *Deutsche Mathematik*.

Max Steck was also the central figure in a curious incident revealing another aspect of mathematical policy and activity in the Third Reich. Johann Heinrich Lambert (1728–77) was one of the most important of eighteenth-century German-speaking mathematicians. Almost completely self-educated, he was the first person to prove that π was irrational, and his posthumously published book *Theory of Parallels* was an important precursor of the non-Euclidean geometry of Bolyai, Lobachevsky, and Gauss; he also was a notable proto-Kantian philosopher. Lambert was born to a family of Lorraine Huguenots who had fled to Mülhausen (Mulhouse), which is presently in the southeastern corner of France but in 1728 was Swiss and, as an Alsatian city, at the end of June 1942 was, of course, German.

At the end of 1943, a learned institution in Zürich called the Schnyder von Wartensee Foundation had apparently announced a prize for work concerning Lambert's mathematics. Steck, who in 1943 had published an edition of Lambert's collected *Papers on Perspective*, as well as the first *Bibliographia Lambertiana*, submitted these for the prize and, as winner, was awarded on March 11, 1944, an amount of 700 Swiss francs. His work had contained a plan for publication of a collected works of Lambert, and he was invited by the foundation in May to visit for the purpose of planning such an edition.[83] A week after receiving the prize, Steck wrote Rudolf Mentzel, then president of the *Deutsche Forschungsgemeinschaft* (DFG), pointing out that in his work, he had explicitly claimed Lambert as German, "which he without doubt is," and especially advised the DFG and others to put forth an edition of Lambert's papers from a "*German authoritative point of view*" (emphasis in original), especially since Switzerland had for some time claimed Lambert for its own.[84] Thus it was a matter of German prestige that there should be German and not just Swiss participation in such a venture, and Steck was asking for DFG support of it. There was also the matter that, with German participation, Swiss money would come to German institutions and research. Most of Steck's letter is devoted to building himself up as the right person to be the German leader in such a

[81] Poggendorf (1960), vol. 7A: 4, p. 504, *Mathematische Zeitschrift* 45 (1939): 609–634 (submitted June 17, 1939).

[82] HNMA, Hecke to Behnke (handwritten note), Nov. 5, 1939.

[83] BAK R73 14903, Steck to Mentzel, Mar. 18, 1944. All material on this "Lambert project," unless otherwise annotated, is from this BAK file.

[84] In order to escape the Burgundians, in 1515, Mulhausen allied itself with the Swiss, until 1798, when it became French. From 1871 through 1918, it was German, then French again.

venture, mentioning not only his mathematical and historical qualifications, particularly with respect to Lambert, but also, "I myself work together with the *Führer*'s chancellery, with the Foreign Office, and with the National Student Leadership, and through them indirectly with the OKW."[85]

It may be worth noting that on March 18, when this was written, Mentzel apparently asked the physicist Walter Gerlach for an opinion,[86] and a month later, Gerlach replied positively to Dr. Karl Griewank, who was the DFG referee for humanities (*Geisteswissenschaften*). On April 18, Gerlach wrote that "the suitability of such a work as a cultural-political measure cannot be at all contested." He mentioned how more scientists should know something of the history of their subject, and the value for contemporary research of the working through of previous ideas.

It is perhaps not astonishing that Gerlach promoted the project and Steck, nor that he supported its sponsorship by the RFR (*Reichsforschungsrat*, or National Council on Research), so that Steck could tell the Swiss that the Germans already had such an effort in train (whereupon Gerlach was sure the Swiss would want to cooperate). But it does seem rather astonishing that in the middle of a war going increasingly badly for the Germans, though still far from decided, Gerlach talks about making Steck comfortable for this research as though it were in idyllic peacetime:[87]

> To begin with one must in some way clearly make Herr Steck, who is still a *Privatdozent*, secure so that he is absolutely sure of his income for many years, and indeed in such a way as corresponds to the importance of his task. Thus Herr Steck must receive the necessary aid. As I imagine, a small office and above all certainly arrangements for the making of photocopies.[88] However, he must then also be given the possibility of still drawing in one or another coworker for the working over of special questions. In essence, however, the whole project must be done by Herr Steck himself. One man must solely plan and carry out such a task, for if it is not carried out in a unitary fashion, it has only limited value.

Gerlach was convincing. Griewank declared that a "fundamental decision has become pressing" in order to prevent Switzerland from taking over the plan and the associated "culture-political prestige." Mentzel approved, so informing Steck on May 9, though mentioning for the first time the war, and promising oral communication about "especially the question of how far [the project] can be attacked during the war."

It was inevitable that Wilhelm Süss should hear of the project. Süss was a

[85] *Oberkommando der Wehrmacht*, or Armed Forces High Command; in 1938 it replaced the old German war ministry.

[86] Walter Gerlach succeeded Abraham Esau as head of both the physics section of the RFR and the nuclear physics project in late 1943. See Walker 1989: 129, which contains more information about both men. On Gerlach, see also Beyerchen 1977: passim.

[87] Gerlach to Griewank, Apr. 18, 1944.

[88] One should bear in mind that in 1944 this literally meant "photographic copies" with appropriate cameras, tables, and lighting arrangements.

geometer; he was president of the German Mathematical Society, and Rektor at Freiburg. This university was the closest to the Alsatian city of Mülhausen and, jointly with it, had established (in April 1943) its own Lambert Prize, to be first awarded in 1944. Süss was explicitly mentioned in the announcement of this award. On June 1 (when Field Marshall Kesselring was retreating in Italy), Süss wrote a Dr. Hagert in the *Führer*'s chancellery, pulling out all the stops already heard played about Lambert as a *German* cultural figure, and the importance of German involvement in the Lambert project, including that the Swiss could not do it alone, as well as the value of the situation of the University of Freiburg for such tasks. "As to your confidential question about the suitability of the instructor Dr. Steck of the technical university in Munich, I must indicate to you that Dr. Steck will be almost unanimously rejected by his disciplinary colleagues for such a task." Süss's reasons for rejecting the only obvious Lambert scholar in Germany are interesting. Although, says Süss, Steck had been concerned with the history of mathematics for many years, he had made many blunders therein.

> In particular, in a book on the main problem of mathematics, he had so pilloried the recognized leading German mathematician Hilbert,[89] not only in an unbelievably injudicious, but also in an especially clumsy and tactless manner, that one can only hope that this book, in view of the scientifically small importance of its author, will become as little known in foreign countries as possible.

On the other hand, Steck was an expert on Lambert, and had connections to Swiss mathematicians, and so it would be useful to involve him in the effort of arranging Swiss participation in a German edition of Lambert's works. "I might only ask that, for this involvement of Dr. Steck, we do not pay the price of his participation as a coeditor of the publication."

Actually, Süss had had earlier knowledge of Steck as an incompetent. In 1938, Constantin Carathéodory retired from his mathematics chair at the University of Munich. The position was not immediately filled, and in 1941 various supporters of *Deutsche Mathematik* pushed Max Steck for the position. When he had an opportunity, Süss objected to Wilhelm Führer of the education ministry (and temporarily the mathematical referee there) and to Führer's close friend, one of Steck's promoters, the astronomer Bruno Thüring, even though Süss had no official say in the matter.[90] He also refused to let the matter rest and privately asked Hellmuth Kneser to give him an opinion of Steck's papers. Kneser's report was devastating. Among the phrases are "a serious lack of mastery of even those mathematical aids which must be very obvious" and "stupidly dull" (*stumpfsinnig*)." Süss sent a copy (without, at Kneser's request, mentioning him) to the mathematical Ordinarius Oskar Perron at Munich with the hope that it could be used by the faculty to prevent Steck's succession to the chair in

[89] Hilbert died on February 14, 1943.

[90] The physicist (and SS man) Führer and the astronomer Thüring were both strong supporters of "Deutsche Physik" and, in Thüring's case, of Bieberbach's *Deutsche Mathematik* as well. See Beyerchen 1977, under their respective names, and *Deutsche Mathematik* 1 (1936): 10–11 and 705–711.

Munich. If necessary, he planned to go to the Dekan, Rektor, and even the national leaders of the *Dozentenschaft*.[91] In the event, the chair was not then filled, but Steck was given an adjunct position teaching geometry at the university.[92] Eventually, in 1944, a mathematician of distinction, Eberhard Hopf, succeeded to Carathéodory's chair. It may have been that the Nazi education ministry was particularly desirous of making Munich an example of "Aryan culture," and this may account for both the succession of Wilhelm Müller to Sommerfeld's chair in physics and the attempt to promote Steck to a distinguished chair in mathematics.

Steck's book mentioned by Süss, entitled *The Main Problem of Mathematics*, was published just a few weeks after David Hilbert's eightieth birthday (January 23, 1942).[93] For Steck, the "main problem of mathematics" is "*the nature of the mathematical in itself*,"[94] and "in epistemological systematics—the main problem of mathematics reveals itself in two completely different aspects."[95] These turn out to be: (1) the problems of axiomatics,[96] and (2) connected with the idea of isomorphism, the problem of "*giving meaning to the formal*," "so that every empty form of that axiom system fills itself with material content, with sense and meaning, with being and existential nature."[97] These citations are perhaps sufficient to indicate the bloated and pedantic philosophical tone of Steck's book, a tone already noted in Steck's work some years earlier by Kneser. As to the content complained about by Süss, Steck launches a bitter attack on Hilbert's formalist philosophy of mathematics, complete with attacks on Georg Cantor and his famous credo, "The essence of mathematics lies in its freedom,"[98] and on Hilbert's philosophic ideas as "Jewish."[99] Steck seeks a true German sense of the nature of mathematics in intuition and concrete concepts. Hilbert and his program, attacked throughout the book, are called solipsistic, intellectually dishonest, decadent, sophistic, believing in the relativism of truth, and dictatorial. It should be emphasized that neither Russell's logicism nor Brouwer's intuitionism suited Steck's philosophy either—it seems to be a philosophy of mathematics *sui generis*. Further, some of the common academic infighting can be seen in Steck's 1942 rejection of the "logicist" Heinrich Scholz as the philosopher on

[91] HK, Süss to Kneser Apr. 26, 1941; Kneser to Süss, May 2, 1941, May 9, 1941, May 29, 1941. The Dekan in Munich at the time, though, was Wilhelm Müller, the contemporary physics incompetent who had succeeded Arnold Sommerfeld.

[92] As in note 81; also Scharlau et al. 1990: 231.

[93] The copy I have seen once belonged to a training school for teachers in the "Adolf-Hitler Schools."

[94] Max Steck, *Das Hauptproblem der Mathematik* (1942), 11, "das Wesen des Mathematischen an sich."

[95] Ibid.: 17.

[96] Ibid.: 17–24.

[97] Ibid.: 24–30. Citation from p. 24.

[98] Among many other places, Ibid.: 106 and 178. Cantor's phrase was "Das Wesen der Mathematik liegt in ihrer Freiheit." Ludwig Bieberbach (see below, chapter 7) would also find the phrase offensive.

[99] Ibid.: 109. Hilbert's association with Jews is mentioned on pp. 184–185.

the Lambert commission and praise for the philosophical viewpoint of one Theodor Haering, and Süss's 1945 rejection of Haering and suggestion of Scholz for the position.[100]

Although Süss rejected Steck, on August 17, the distinguished Swiss mathematician Andreas Speiser, together with the Swiss director of an institute for foreign research, wrote Mentzel under the letterhead of a Swiss commission for the publication of Lambert's works, suggesting Steck as the guiding genius of the project. They also had plenty of money for it.

Two letters went to Süss six weeks later, on September 27 (by which time the first incursions of enemy forces over the German border in the west had occurred). One was official, from Mentzel, appointing Süss head of a German commission that would work together with the Swiss on the publication of Lambert's collected works. Other members of this commission were the historian of mathematics Joseph Hofmann, Ludwig Bieberbach, two philosophers, the physicist Max Caspar (who had been editor of Kepler's collected works), and Griewank. "Doz. Dr. Steck, who will come into consideration in the first place as a German worker [on the project], ought, on a case by case basis, to be brought into the discussions of the commission."

The second letter, from Karl Griewank, remarked on Steck's connections with the Swiss, his preparatory steps toward the project at the DFG's behest, and the fact that his outline suggestions would by and large be followed. Although Griewank recognized that details of Steck's work had been sharply criticized, he was the obvious German worker, and a task of the commission would be to give him direction and control his work so as to forestall objection. Griewank's letter contained the briefest allusion to the war: "Work on the German side, under present-day circumstances, can only be set in motion slowly."

Süss replied, accepting, on October 16 (by which time the Germans had evacuated Athens; the Hungarian leader Admiral Horthy had signed an armistice with the Russians, leading to his kidnapping by the Germans; and the battle of the Hürtgen forest had reached its bloody conclusion, to German detriment). Süss's letter contained a warning about the necessity of undertaking only obligations that could be fulfilled vis-à-vis the Swiss, but in essence its page plus might have been written in 1934, and, absent the "Heil Hitler!" in the close, almost any time. The one hint that all is not as usual is the remark that the letters of September 27 "unfortunately because of present traffic situations came very late into my hands." The next day, Griewank (not yet having received Süss's acceptance) wrote Steck telling him of the commission, lamenting his inability to be definitive because he had not heard from Süss, and suggesting an oral discussion prior to the start of the winter semester.

The tone of academic business as usual continued in the remaining correspondence. Steck thanked Griewank on October 30 and said:

[100] Süss to Griewank, Jan. 24, 1945. This letter also repeated the rejection of Steck for a leading role in the project as well as the rejection of the philosopher Weinhandl mentioned below. For Scholz, see below, "Heinrich Scholz, Logician," where his relations with Steck are also discussed.

> Daily I have waited for the fixing of oral discussion in the Lambert matter before the winter semester begins, so that I can also arrange myself somewhat in the work. . . . I can however, now after the research community [DFG] has made its choice for a president of the German Lambert commission, think that the delays are based therein. As that person who finally must do the work, I would not wish that the delays to it and thereby also the execution of the whole matter in the same tempo as up till now persist, since Switzerland. . . . I would therefore also have been more for at once working, first, before we set up commissions and the like, whereby indeed only advertising notices (*Aushängeschilder*) are created, but no research work is carried out.

Griewank also informed Süss that he had a free hand in picking commission members, and that the first communication was only a suggested list, who had not even been officially informed of their participation. He also urged him to take the matter in hand soon so that the joint work could be directed toward a realizable first goal.

On January 24, 1945 (after the Ardennes offensive ["Battle of the Bulge"] had failed), Süss was still writing bureaucratic letters, though from the "Lorenzenhof" in Oberwolfach,[101] to which he had removed for safety's sake. In a letter over three pages long, he objected to the two suggested philosophers, Weinhandl and Häring (who had been Weinhandl's suggestion). He suggested instead (but not definitively) the mathematician and theologian Heinrich Scholz[102] (whom, unbeknownst to Süss, Weinhandl had explicitly rejected the preceding June). As to Steck, while there was no question he would concern himself in extraordinary measure about the project, nevertheless, he would need careful watching if it were not to "swarm with remarkable novelties." Steck could begin work (following a consultation with Süss). However, "the naming of Dr. Steck as 'editor' (*Herausgeber*) in my opinion is at the moment out of the question" (though that seemed to be the Swiss desire). The only touch of reality in Süss's letter was, "I had hoped to be able to be in Berlin on February 7 . . . in the present situation, I admit it is not forseeable whether my trip will then take place."

Griewank's letter had taken about a month to reach Süss (Berlin to Oberwolfach in the Black Forest), and Süss's reply, apparently six weeks. On March 12, 1945, Griewank complained to Steck about such delay. He then communicated almost verbatim from Süss's letter those parts of it relevant to Steck's beginning work, and asked him what financial support he would need so an appropriate sum could be relayed to him.

By this time, Steck had also removed himself from the large city of Munich to the town of Prien on the Chiemsee, a lake southeast of the Inn. However, Griewank's letter reached him fairly promptly—postal service in eastern Germany was apparently less disturbed than in the west. On March 19, 1945, the very day of Hitler's infamous "scorched earth" decree, Steck replied in a manner completely oblivious to the German circumstances: "The assurance of the work

[101] See below for the Oberwolfach Institute.

[102] And also possibly the Hamburg philosopher, Heinrich Sauer.

for the long term should be guaranteed"; "the contractual connections to the publisher of the volumes appearing in Germany . . . should be clarified"; "for the first period of my work in Basel I would request a monthly allowance of at least 400–500 Swiss francs"; "the DFG must arrange with the Rektorat at the technical university in Munich for my leave as Dozent during the time of the work"; "the interruption by this work of my activity as Dozent should in no way negatively influence my further professional advancement to an *Ordinariat* or the like, nor my pay and its incremental increase with years of service"; and "I assume the DFG is ready to compensate me after the fact for the expenses and outlay of cash already incurred for the publication of Lambert's works"; also his family (which could not go to Basel) needed to continue to receive money from the technical university in Munich.

The story might end here, but attached to the letter is a "*Secret Matter.*" If the rest of Steck's letter is the usual academic insistence on proper credit, proper respect, and proper payment, the "*Secret Matter*" is truly fantastic. Steck wrote:

> On the occasion of the meeting of scientists and mathematicians who belong to the *Dozentenbund* and its national leadership in Salzburg at the beginning of December 1944, the question was raised with the local *Gauleiter* by the scientists present, chiefly physicists and chemists, of the procurement of foreign scientific literature important for the war. Since Herr Dr. Kubach,[103] who led the meeting, knew about my commission for Lambert's collected works, he suggested that I take over this task on my trip to Switzerland, which works especially well since I am well initiated into the Swiss libraries, and besides everything can be camouflaged by [the rubric] "Lambert's Collected Works project." Photocopies of the desired foreign literature must then be sent by courier, by way of the German consulates in the diplomatic pouch, into the hands of the leadership of the national university teachers' organization, and from there come to the physicists or chemists involved. Already for this reason, beginning work in Switzerland should be arranged as early and as quickly as possible.

Steck also suggested that personal messages to his family as well as the genuine "Lambert-mail" go via diplomatic pouch.

Needless to say, nothing proceeded prior to Hitler's suicide on April 30 and the final German collapse on May 8. After the end of the war, Steck apparently lost his position at the technical university in Munich, presumably through the denazification procedures of the Allies. In 1951, he finally did travel to Basel and worked on the Lambert manuscripts, among other things, updating his 1943 work, and this new edition was published in 1970. Steck died in 1971. The new catalogue he prepared was itself published in Switzerland on the 200th anniversary of Lambert's death in 1977. In 1952, Steck managed to obtain regular employment again, as a professor at an academy for applied technology in Nuremberg, moving to a similar institution in Munich for architectural technology in 1957.[104] An unremarked irony of the whole story is that the

[103] Kubach was the national leader of the mathematics students under the Nazis.

[104] Poggendorf (1960), vol. 7A: 4, p. 504.

unpublished Lambert manuscripts, now in Basel, had been sold in 1799 by Johann Bernoulli III, together with those of his famous mathematician-grandfather, to the Duke of Sachsen-Gotha for his library, and they had remained undisturbed in these German hands until 1936, when the then-director of the ducal collections sold them back to Basel for much-needed cash. Furthermore, on their way to Basel, they had been in the library at Freiburg (where Süss had been on the faculty since 1934) for Swiss examination prior to their taking possession.[105]

In telling the story of the Lambert commission project, I have made a counterpoint with the progress of World War II in order to emphasize the seemingly incredible business-as-usual academic atmosphere that surrounded it. This can be interpreted in several ways. It might be argued that many German mathematicians truly lived in an "ivory tower," remote from reality, and this was just a fantastic extreme case. While this suits popular prejudices about mathematicians, it hardly seems valid. Süss and the physicist Walter Gerlach, for example, not to mention education ministry officialdom, were far from inexperienced in the ways of the world. It also might be argued that a fanatic belief in an ultimate German victory allowed ignorance of the true military-political situation. This may have played something of a role, especially for Steck, but hardly seems adequately explanatory. It is probably true that the educational bureaucracy, faced with an issue couched in terms of German prestige, simply found it easiest and safest to proceed in a bureaucratic way. However, for Steck, a minor mathematician with an inflated style, whose philosophical-cultural fulminations about the nature of mathematics won the respect of Nazi true believers but the disapproval of his colleagues, the issue was probably personal advancement along a quasi-political ladder. One should note again his offer on March 19, 1945(!) to use the diplomatic pouch to smuggle photocopies of technical books out of Switzerland under the cover of the Lambert project as a reason for proceeding with it. The same letter, after all, contains detailed demands about salary and rank. It was not patriotism, but manipulation of a given situation for personal advantage that motivated Steck. This is not to deny his genuine interest in Lambert, nor the harmlessness to others of these activities. Indeed, the innocence of his devotion to his teacher, Liebmann, whom he did not blame for his putatively Jewish grandfather, is perhaps even somewhat touching, given the times and behavior of others.

As for Süss, as seen in the book-manuscript project, and as will be seen again in the creation of the Oberwolfach Institute, he continually used his position and prestige to do business with the Nazis to the benefit of both mathematics and mathematicians whom he considered to have their discipline as their central academic concern. Steck, with his philosophical denunciation of Hilbert in the name of a "German realistic" philosophy of mathematics, did not fit in this category for him, despite Steck's acknowledged expertise in the Lambert matter.

[105] See preface by M. Steinmann, in *Der handschriftliche Nachlass von Johann Heinrich Lambert* (Basel: Universitätsbibliothek Basel, 1977).

Heinrich Scholz may well have. One of Süss's basic concerns seems to have been the acquisition of much-needed funds for mathematics and mathematicians he deemed worthy, whether the excuse for that were the military effort in the war (as in the controversy with Doetsch) or some putative necessary (and, given the times, somewhat fantastic) enhancement of German prestige (as in the Lambert project).

Resistance to Ideological Articles

Articles by and dedications to Jews were not the only problems to bother the *Annalen* editors. Hecke, Behnke, and van der Waerden also wished to prevent any articles containing tendentious Nazi ideological argumentation from appearing in the *Mathematische Annalen*. Thus, when van der Waerden, in March 1940, received a paper entitled "Problems of Strict Scientific Thinking in the Light of the Study of Inherited Characteristics" for consideration for publication in the *Annalen*, he initially had some doubts. Later, when he had looked at the paper more closely and consulted a psychologist friend about the paper's presuppositions, he seemed much more positive:[106]

> With respect to scientificality and exactness of expression, it is far superior to the average of the literature appearing today on psychological typology and researching inherited characteristics. Also, concerning the foundational questions of mathematics that the author discusses, he is quite suitably in the know. He understands nothing about physics, however the paragraph about physics can be omitted without injury to the paper.

Nevertheless, the question remained whether the *Annalen* wished to indulge in the "contemporary fashion of handling questions of inherited characteristics in mathematical journals" provided the level was high enough. Van der Waerden sent Hecke the paper (by Friedrich Requard at Köln, who was apparently not a mathematician[107]) so he could make his own judgment, and four days later, Hecke replied. In general, he was against allowing the current "bad fashion" in the *Annalen*. However, the essay by Requard seemed to him "completely in order and even interesting; above all, however, a connection with research on inheritance is indeed brought in only artificially . . . if one therefore deletes every reference to inheritance, there remains a clean paper interesting to psychologists as well as mathematicians, which perhaps in a reasonable way should be published in a mathematical and not a psychological journal." Therefore, he was in favor of returning the paper to Requard to rework in this sense, acknowledging its value.[108] Van der Waerden was generally in agreement—his psychologist friend, who was "very critical" of this sort of inheritance research,

[106] HNMA, van der Waerden to Hecke, Apr. 11, 1940.
[107] At least, he was not on the mathematics faculty.
[108] HNMA, Hecke to van der Waerden, Apr. 15, 1940.

thought some of Requard's statements established and so should not be deleted—though presumably the emphasis and title could be changed.[109]

From Hecke, the paper went to Behnke, who was not at all so accommodating. He was completely disparaging, saying that its concepts "were completely unclear" and that it belonged in an "entertainment journal."[110] Van der Waerden was offended, for he had considered himself and Hecke "sufficiently schooled in philosophy" to distinguish nonsense from proper psychology; however, as a convenience, he called in a philosopher at Leipzig whom he trusted for consultation.[111] This man, Professor Gadamer, opined:[112]

> [Behnke's] judgment concerning the inadequate precision of concepts was unwarranted and the paper, namely in the first four sections, was undoubtedly good and scientifically justifiable; however he could understand [Behnke's] fears concerning the undesirable consequences of accepting such papers in disciplinary journals . . . and that Requard's paper was again not so overwhelmingly superior that one had unconditionally to take the risk [of receiving masses of inferior similar papers] into the bargain. Also he thought that it would be very difficult for the author [Requard] to achieve a reworking in the sense required by Hecke.

Thus van der Waerden returned the paper to Requard suggesting he approach a psychological or philosophical journal with it. Requard's paper did, in fact, so appear.[113] The paper takes as its starting point a paper by the mathematical logician Gerhard Gentzen.[114] It attempts to associate with Brouwer's constructive and Hilbert's *an-sich*[115] standpoints[116] two different kinds of thinking, without judging the value of either. Hecke seems correct that the notion of inheritance is somewhat arbitrarily dragged in through the work of the Tübingen psychologist G. Pfahler. Pfahler, in a 1932 book, *Inheritance as Fate*,[117] apparently distinguished antipodal poles of inheritable "basic functions," which Requard associated with the two kinds of mathematical thinking.

In his annoyance at Behnke's judgment, van der Waerden remarked, "The real reason for his [Behnke's] rejection seems to be that he is afraid of Heinrich Scholz!"[118] Two years later, Scholz gave Behnke a manuscript, "What Is a Mathematical Theory," by his student Karl Schröter for consideration of publication in the *Annalen*. Behnke recused himself from judgment.[119]

[109] HNMA, van der Waerden to Hecke, Apr. 19, 1940.
[110] HNMA, van der Waerden to Hecke, May 3, 1940.
[111] Ibid.
[112] HNMA, van der Waerden to Behnke, May 3, 1940.
[113] *Zeitschrift für Angewandte Psychologie und Charakterkunde* (1941): 351–370.
[114] See below, chapter 8.
[115] "In itself": an echo of the famous Kantian concept of a "thing in itself" (*Ding an sich*).
[116] Compare discussion of these in chapter 2.
[117] Gerhard Pfahler, *Vererbung als Schicksal* (Leipzig: J. A. Barth, 1932).
[118] HNMA, van der Waerden to Hecke, May 5, 1940.
[119] HNMA, Behnke to Hecke, Mar. 4, 1942.

> I ask you and perhaps Herr van der Waerden to decide the question of acceptance of this manuscript without my collaboration. Given the personal and local connections you will understand that I must declare myself biased, if not intellectually, nevertheless in my function [as an editor]. The manuscript indeed deviates in style from the usual mathematical paper. In my opinion, by its tone, the essay would be better suited to the *Jahresbericht* [of the German Mathematical Society]. On the other hand, already in the discussion of Requard's manuscript, van der Waerden had seen such a distinction as irrelevant. On the negative side, besides, one could perhaps throw in that presumably no new results appear.
>
> What carries very positive weight is that this essay will be read on all sides, because it is accessible to every mathematician and touches on things that concern us all. For a usual journal some such would be decisive, but how would Hilbert have judged the matter?

Within a month, Hecke had decided:[120]

> I have long thought about Schröter's manuscript, however, in the end I find it unacceptable for the *Annalen*. It is a report about a new and difficult theory, in which references for the most important assertions are to the future or other publications. . . . To say this naturally expresses not the slightest demeaning of its content.

Schröter's paper did appear later in the *Journal of the German Mathematical Society*,[121] a possible place of publication mentioned also by Hecke.

Heinrich Scholz, Logician

Hecke's letter went not to Behnke, but to his second wife, Elisabeth (Lisa), who, prior to marriage, had been a mathematics student at Münster, and who helped her husband out in some of the *Annalen* work—Behnke was on a few weeks' vacation in Switzerland.[122] She asked to be allowed to communicate directly with Scholz, as contact with her husband would take too long, and "I know however how impatient Scholz is for news, presumably is already somewhat irritated that he! [*sic*] must wait for something."[123]

Who was the impatient Heinrich Scholz, promoted by Süss, attacked by Steck,[124] colleague and friend of Behnke (but of whom Behnke was possibly afraid)? Heinrich Scholz was a remarkable figure in the history of mathematical logic. Trained as a religious philosopher by the famous Adolf von Harnack in Berlin, he was a professor of philosophy at Kiel at the same time Otto Toeplitz was a mathematics Ordinarius there, and was influenced by Toeplitz, as well as by Russell and Whitehead's *Principia Mathematica*, to take up the philosophy of mathematics. Fourteen years Behnke's senior, he arrived at Münster from Kiel as

[120] HNMA, Hecke to Lisa Behnke, Apr. 9, 1942.

[121] *JDMV* 53 (1943): 69–82.

[122] HNMA, Lisa Behnke to Hecke, Mar. 27, 1942, and Behnke 1978: 147.

[123] HNMA, Lisa Behnke to Hecke, Apr. 14, 1942.

[124] E.g., Steck 1942: 126–127, among many other places.

a professor of philosophy the year after Behnke's arrival.[125] The fact that Scholz was the son of a Lutheran minister and that Behnke still perhaps felt himself somewhat at sea in Catholic Münster may have drawn the two men together. Also, they were both widowers. Scholz's first wife died in 1924, while he was still in Kiel. At his wife's grave, he told the pedagogical philosopher, Eduard Spranger, a life-long friend from school days, "You (*Du*) understand how I am *not able* to work on things with content."[126] Scholz's feeling for structure was no small thing. He apparently felt that when having guests for dinner: (1) no more than six people should be invited; (2) there must be an excellent menu; (3) a discussion theme must be planned; and (4) the guests should have prepared themselves as much as possible beforehand on this theme.[127] His wife's death, Toeplitz's influence, Russell and Whitehead, and Scholz's own love of truth and feeling for structure all, no doubt, influenced him toward mathematical logic.[128] Scholz's humanism as well as love is illustrated by the inscription he placed on his wife's grave:

> Alles Vergängliche ist nur ein Gleichnis,
> die Liebe aber ist das grösseste unter ihnen,

which literally reads:

> Everything transitory is only a similitude,
> Love, however, is the greatest of these.

The first line is a famous quotation from the closing lines of Goethe's *Faust*, Part 2, which ends five lines later with the even more famous, "The eternal feminine draws us upward."

When Scholz arrived at Münster in 1928, he regularly gave two courses: one on mathematical logic, and one on the great philosophers.[129] Though an autodidact, his work in mathematical logic became so respected that, for example, on September 26, 1939 (three weeks into World War II), Haskell Curry, then president of the (American) Association for Symbolic Logic, wrote him asking that he allow his name to be nominated as a foreign member of the association's Council. On December 5 (three months into World War II), the education ministry gave its permission, despite all the caveats seen earlier about joining foreign bodies.[130] Indeed, the other foreign member of the association was to be

[125] Behnke 1978: 107. Article on Scholz's seventieth birthday in *Münstersche Zeitung*, Dec. 20, 1954, in Personalakten Heinrich Scholz, Universität Münster. Hans-Günther Bigalke, *Heinrich Heesch* (1988), 20–23. Toeplitz, it may be recalled, also stimulated the young Helmut Hasse to mathematical activity.

[126] In a letter from Spranger read December 20, 1957, at Scholz's obsequies. In Personalakten Heinrich Scholz, Universität Münster.

[127] Bigalke 1988: 22–23.

[128] See summary of remarks by the mathematician Herbert Seifert and mathematical logician Hans Hermes (Scholz's most distinguished student) at his obsequies. In Personalakten Heinrich Scholz, Universität Münster.

[129] Behnke 1978: 107.

[130] Curry to Scholz, Sept. 26, 1939, and education ministry to Scholz, Dec. 5, 1939, in Personalakten Heinrich Scholz, Universität Münster.

a Briton.[131] Earlier, in 1937, Scholz had received permission to join the association in the first place.[132]

Prior to the Nazi assumption of power, politically, Scholz, like his close friend Eduard Spranger, was a small-minded Prussian nationalist.[133] Behnke learned to avoid discussions with him that might touch on difficult topics. However, under the Nazis, Scholz developed into someone who was adept at obtaining resources for his own academic interests based on his reputation. On the other hand, he also was someone who attempted to help as possible, without endangering his own life, people in trouble with the government, particularly Polish mathematical logicians. Scholz was also a man who seems never to have publicly uttered a word in the tones of Nazi propaganda—a man who "in critical times acted according to his conscience."[134] For example, Scholz might mention gratuitously in an official report on a 1943 lecture trip to Zürich that it was the first time since the collapse of the University of Warsaw that he had spoken of his work on foreign soil, that the Warsaw logicians were connected with the "great German master" Gottlob Frege, and that their leader was Jan Łukasiewicz,[135] who had received an honorary doctorate from Münster in 1938.[136] In November 1939, after the collapse of Poland in the war, Scholz wrote Karl Griewank at the DFG that he was going to Berlin "in order to worry personally about the help that must be directed toward our unhappy Warsaw friends before it is too late."[137] Such pro-Polish sentiments, or at least pro-Polish academician sentiments, occasionally caused him trouble with the authorities. For example, on October 2, 1940, the education minister wrote Scholz (with a copy to the Münster Kurator) as follows. The letter reveals very clearly the attitudes expected of academics, even those whose stock in trade, like theology or symbolic logic, was far removed from Nazi concerns.[138]

> As I have first learned now, on March 14, 1940, you sent a petition to the education division of the *Generalgouvernement* [i.e., the occupied region of Poland] in which you advocate the liberation of the former Cracow professor of theology Dr. Jan Salamucha, now in protective custody. I must object as to both form and content of this petition in the sharpest way. In form it is no longer a question of a request, as possibly would

[131] Curry to Scholz, Sept. 26, 1939, in Personalakten Heinrich Scholz, Universität Münster. No doubt in part this was an assertion of the association's neutrality.

[132] Scholz to education ministry, Mar. 24, 1937, in Personalakten Heinrich Scholz, Universität Münster.

[133] "Small-minded" (*engstirnig*) is Behnke's adjective (1978: 108). For Spranger, see above, chapter 3.

[134] Article on Scholz's seventieth birthday in *Münstersche Zeitung*, Dec. 20, 1954, in Personalakten Heinrich Scholz, Universität Münster.

[135] Jan Łukasiewicz is perhaps best known for the invention and study of "non-Aristotelian" many-valued logics. He was also a groundbreaker in many areas of formal logic. See the article on him in *Dictionary of Scientific Biography* 1970.

[136] See Scholz's official report on January 3, 1944, of a lecture trip to Switzerland, Personalakten Heinrich Scholz, Universität Münster.

[137] BAK R73 15934, Scholz to Griewank, Nov. 26, 1939.

[138] In Personalakten Heinrich Scholz, Universität Münster. Salamucha died in 1944 during the Warsaw uprising.

> have been compatible with the duties of a German vis-à-vis the responsible officials for the citizens of an enemy state, but of a demand raised with particular emphasis. In content, in scarcely veiled form, there is raised in it the reproach of life-threatening treatment applied by official German agencies, and the feared consequence of the death of Prof. Salamucha as a "stain," if not for Germans generally, then indeed set down for German research. You have thereby heavily injured the national honor in a matter that is exclusively capable of being treated from the standpoint of the collective interests of the German people now at war, and not at all from the viewpoint of the individual person. I pronounce to you my sharpest disapproval on account of this behavior. In order to exclude once and for all similar occurrences, which can serve propaganda inimical to Germany as a dangerous weapon against the political leadership of the German people, I hereby forbid to you every further petition in matters concerning foreign scientists, unless by the official channels through me. If you should go against this order, then I will apply against you the measures of the civil-service criminal law.

Nevertheless, three years later, Scholz did try to help Łukasiewicz, albeit in a tangential manner, in a report that, of course, was officially addressed to Education Minister Rust. This report also contained mention of meeting with, among others, Paul Bernays, who was among those expelled from Göttingen in 1933 as a Jew, and who had emigrated to Switzerland. Bernays's name appears once in the context of being a prominent mathematical logician without further mention—but it is not omitted.

Even Ludwig Bieberbach, perhaps the primary propagandist for Nazi ideology among mathematicians, seems to have been offended by Max Steck's attack on Hilbert mentioned earlier, and he commissioned Heinrich Scholz to write a lengthy article for his journal *Deutsche Mathematik* entitled "What Does Formalized Study of the Foundations of Mathematics Aim At?"[139] This article in part is a counterattack on Steck that not only is substantive but also is filled with sarcasm. Thus, after a citation from Steck, Scholz remarked: "For the grammatical or syntactical quality of this sentence I am as little responsible as I am for its content."[140] Similarly, he supposed that Steck had only ever read some reports about formalized foundational research, and never the *Principia Mathematica*, since he consistently misspelled Russell's name with one "l."[141] Scholz managed in his report on his trip to Zürich also to mention this article, Bieberbach's name, and Steck's "distorted picture of our Göttingen senior master Hilbert" "in an embarassing book."[142]

Scholz used his connection to Bieberbach to his (and mathematical logic's) good advantage as well. Within the funds provided by the DFG to support Bieberbach's journal, *Deutsche Mathematik*, discussed in detail in the next chap-

[139] *Deutsche Mathematik* 7 (1943): 206–248. For Bieberbach's "commission," see p. 3 of Scholz's 1943 report.

[140] *DM* 7 (1943): 208–209 n. 3.

[141] Ibid.: e.g., end of footnote 4, pp. 211–212.

[142] See note 136 for the report, and *DM* 7 (1943): 206–248.

ter, were funds for producing a set of monographs on logic.[143] These very respectable publications began in 1937. Scholz obtained funds in 1935 and 1937 to travel with two of his students, as well as his (second) wife, to international congresses on the unity of science.[144] It is perhaps worth remarking that one of the organizers, as explicitly mentioned by Scholz, was Rudolf Carnap, who had emigrated to the United States in 1935.[145] Scholz founded the first German institute for mathematical logic with a considerable library. Already at the first international congress in 1935, the phrase "the Münster school of mathematical logic" was used.[146] In 1938, Scholz's professorship of philosophy was converted into one for the "Philosophy of Mathematics and Science."[147] In 1943, it was converted again into one for "Mathematical Logic and Research on Foundational Questions [in Mathematics]."[148]

The rage of Steck at Scholz, who obtained money during mid-war for monographs in mathematical logic, which Steck thought un-German, was expressly stated:[149]

> What he [Scholz] has understood is doubtless this, to obtain from the German state and make fluid huge amounts of publication money for this logistic production. We fundamentally reject this logistic which prizes the English empiricists and sensory philosophers as the *great Englishmen* Locke, Berkeley, Hume *and by now find it really time to speak for once of the "Great Germans"* [emphasis in original].

He also characterized Scholz as a follower of the Polish logistic school of Tarski, Łukasiewicz, et al. and the "Wiener Kreis": Carnap, Reichenbach, Schlick, Hahn, Menger, et al.[150] This was doubtless true, but Steck's implicit point was the far more pointed one in 1943 Germany that, with the exception of the Pole, Łukasiewicz, and Carnap, all the others were Jews.

There was therefore perhaps more than a little malice against Steck in Süss's promotion of Scholz as the philosopher for the Lambert project. Scholz seems to have been the sort of person who, though a conservative nationalist, soon realized that this was not what the Hitler government was about, and, while inwardly resistant to Nazi beliefs, attempted to use his contacts with those ideologically dedicated thereto for whatever benefit he could. He also was ideologically dedicated; but his ideology was one of religious truth. Like Ludwig Prandtl

[143] BAK R73 15934, Karl Griewank to the publisher S. Hirzel Verlag, June 28, 1938; marginal note apparently written by Griewank on Nov. 6, 1939; Scholz to Griewank, Mar. 14, 1942.

[144] See his application in 1938 to attend the fourth such international congress in Personalakten Heinrich Scholz, Universität Münster.

[145] See his application to attend the third congress and a philosophical congress as well in Personalakten Heinrich Scholz, Universität Münster (Scholz to ministry, Apr. 2, 1937).

[146] Ibid.: 2.

[147] Ministry to Scholz, May 21, 1938, in Personalakten Heinrich Scholz, Universität Münster.

[148] Ministry to Scholz, October 25, 1943, in Personalakten Heinrich Scholz, Universität Münster. The letter says the change takes place "immediately"; thus there is a small error in Scharlau et al. 1990, which dates this professorship from 1944 (p. 236).

[149] Steck 1942: 148–149 n. 127.

[150] Ibid.: 176.

and Wilhelm Süss, he seems to have been the sort of man who could do business with the Nazis without necessarily being personally tainted thereby, his business being for the objective good by any standards of his discipline and its practitioners. Like Prandtl, his original conservative nationalism and his political silence in most matters seems to have protected him—as did, in Scholz's case, the extremely unpolitical nature of his academic interests.

Miscellaneous Non-German Authors

Some things with which the editors of the *Mathematische Annalen* concerned themselves, and which in ordinary circumstances would be trivia, took on exaggerated importance under the Nazis. For example, in early 1942, a Romanian mathematician named Michel Ghermanescu wished to publish a paper in the *Annalen*. Ghermanescu knew very little German, with the result that the German text of his paper was "unreadable." Lacking other possibilities, he had rewritten the paper in French, which was an international language he did know. If we look at the situation, on the one hand, there was a war on, French was an enemy language, and the author was a non-German; therefore his writing in French might be considered political. On the other hand, France had fallen, the Vichy government was well installed in the part remaining under French control, and Romania was an ally. Could the *Annalen* publish such a paper in French?[151] Hecke's decision: nothing against a French text, especially from an allied national, besides "one with such a beautiful name"—provided, of course, that the mathematical content was adequate.[152] Ghermanescu's paper did appear in French as the very last paper published in the *Annalen* prior to the end of the war. Indeed, even its author's name was in French, as "M. Michel Ghermanescu."[153]

Rather more serious was a letter van der Waerden received from the famous Czech topologist Eduard Čech sometime around the turn of the year 1942–43. Čech's student, Bedřich Pospíšil, had written him that he had certain results in quantum theory that he would like to publish in the *Annalen*. There was, however, a difficulty. Pospíšil, who in ordinary times was in Brünn (Brno), at the time of the letter was in Wohlau prison in Silesia. He had asked Čech to ask the *Annalen* editors to write the prison authorities for permission to send a long letter to the *Annalen* containing these results. Van der Waerden was unwilling to write to a prison warden because he himself was a foreigner.[154] Van der Waerden had good reason to be concerned. In May 1940 he was temporarily forbidden to hold lectures (the German conquest of Holland began on May 10 and took only five days), and he was worried that he might be required to become a German citizen: "In itself I have nothing against German citizenship, however, at this moment, since Germany has occupied my homeland, I would not gladly

[151] HNMA, Lisa Behnke to Hecke, Mar. 27, 1942.

[152] HNMA, Hecke to Lisa Behnke, Apr. 9, 1942. *Ghermanescu* means "German" in Romanian.

[153] *Mathematische Annalen* 119 (1943–44): 288–320.

[154] HNMA, van der Waerden to Hecke, no date, but sometime in late 1942.

give up my previous neutrality and throw myself in a certain measure publicly on the German side."[155] In a second letter to Hecke on the same day in 1940, van der Waerden had had the idea that even if he did not have police permission to travel from Leipzig to Hamburg, nevertheless, "It occurred to me that it perhaps might be very favorable for me to be invited to Hamburg"; he also indicated that he had telephoned Behnke, "who was constantly concerned about me," and asked him to use what influence he might have to effect his reinstatement as a lecturer. Behnke had apparently boasted of his "connections" to van der Waerden.[156]

Whether Behnke attempted to help van der Waerden is unclear; certainly he was unwilling to intervene himself on behalf of Pospíšil, and neither Pospíšil nor van der Waerden are mentioned in this context in his autobiographical memoirs.[157] Behnke, as discussed previously, had reason not to want to call attention to himself (whatever boasts he may have made to van der Waerden). His letters to Hecke reveal his justified concern about his "mixed-race" son. By 1942, he was writing: "I now worry a great deal about my son. The first cases are occurring in which miscegenants (*Mischlinge*) will be treated as full Jews. That is no accident. As I also learn from the ministry, that will be developed further according to plan."[158]

Hecke replied that the treatment of half-Aryans seemed to be regionally variable: in Hamburg a year before, they were forbidden to study at a university, but in Leipzig, van der Waerden had managed to obtain a doctorate for such a person, and there was no question of limiting his study.[159] However, three months later, Behnke wrote:[160]

> You have certainly already heard of the new measures against miscegenants. According to them my son must now leave school. Süss wishes to inquire at the ministry whether I will be allowed to send my son to Switzerland. That is thoroughly uncertain. And experienced colleagues have warned me. An accusation could still be brought against me on account of paragraph 71 of the civil-service law, if I tried to send him abroad. That could also be seen as an action inimical to the state. Sometimes one must simply doubt the sense (*Sinne*) of this world.
>
> Are we all only there in order to reciprocally torment one another?

Hecke, however, had none of the inhibitions their situations imposed upon van der Waerden, Behnke, and others, and so on January 11, 1943, he wrote the prison management in Wohlau:[161]

[155] HNMA, van der Waerden to Hecke, May 16, 1940 (first letter of two written on this date to Hecke).

[156] HNMA, van der Waerden to Hecke, May 16, 1940, letter 2.

[157] In Behnke 1978, van der Waerden is mentioned briefly three times; on two of these occasions, Behnke mentions that he had received the highest Prussian decoration, the "Pour le Mérite." See pp. 31, 240.

[158] HNMA, Behnke to Hecke, May 22, 1942.

[159] HNMA, Hecke to Behnke, May 27, 1942.

[160] HNMA, Behnke to Hecke, Aug. 19, 1942.

[161] HNMA, Hecke to Wohlau prison management, Jan. 11, 1943.

A certain Bedřich Pospíšil is at present an inmate of the prison there in Division B IV, Cell 268. He is not known to me personally, but as the author of mathematical papers in German scientific journals, and communicates via one of his earlier university professors that he has developed a mathematical-physical theory that might well be suitable for publication in the mathematical journal edited by me.

I make the prison management aware of this and leave to it granting the prisoner the permission for him to send the manuscript of this paper to me for checking.

Presumably the request fell on deaf ears, and Pospíšil in fact died in prison. No mathematical paper of his appeared after 1939.[162] As to Hecke's willingness to publish a paper on "relative quantum mechanics," by 1943 unpoliticized physicists had managed to reassert themselves disciplinarily, and Johannes Stark, the Nobel laureate and Nazi physicist who had earlier attempted to dominate physics politically, was in eclipse.[163]

If Hecke could not help Pospíšil, as a prisoner, to publish, sometimes he was asked to publish Czech authors. In July 1939, Ludwig Bieberbach sent Behnke a paper by Frantisek Wolf, an instructor at the technical university in Prague. Only three manuscript pages long, Bieberbach thought it unsuitable for the *Berlin Academy Proceedings*, and "I cannot accept it for *Deutsche Mathematik*[164] since fundamentally we print only contributions of German authors."[165] Behnke, not knowing the subject matter, thought the paper was only a sketch, but sent it on to Hecke with Bieberbach's letter, on which he commented sarcastically. Hecke sent it to Carathéodory, who by return mail announced it surprising and worth publishing.[166] The paper, only one printed page long, appeared in volume 117 of the *Annalen*, by which time Wolf was in neutral Stockholm.[167]

One aspect of the Nazi government was its officially anti-elitist stance; after all, it was a "workers' party." For the *Mathematische Annalen*, this egalitarian emphasis, coupled with the denigration of learning in favor of feeling, resulted in the submission for publication of scores of inadequate amateur papers. These were mostly sent to the publisher, Ferdinand Springer, who then forwarded them to Behnke, the journal's "copy editor," the person who composed each volume and received corrected galley proofs. Thus he wrote: "The many lay contributions of papers are extremely unpleasant. It is always a torture to study these manuscripts. And, however one makes it out, afterwards, there is trouble."[168]

And a month later:[169]

Indeed, [Ferdinand] Springer always has the tendency to protect you from irritation. Therefore, I also indeed receive the whole dungpile of German and foreign manu-

[162] The latest appearing paper by Pospíšil is in *Fundamenta Mathematica* 33 (1945): 66–74. However, this was in press and already printed in 1939 (Ibid.: ix).

[163] For the history of this struggle, see Beyerchen 1977.

[164] For *Deutsche Mathematik*, see chapter 7.

[165] HNMA, Bieberbach to Behnke, July 1, 1939.

[166] HNMA, Behnke to Hecke, July 4, 1939, and Carathéodory to Hecke, July 20, 1939.

[167] *Mathematische Annalen* 117 (1939–41): 383.

[168] HNMA, Behnke to Hecke, Nov. 10, 1941.

[169] HNMA, Behnke to Hecke, Dec. 11, 1941.

> scripts of such lay persons who send their mostly lousy[170] manuscripts to the publishing house and not to us. Correspondence with such people sometimes ends with fat letters and naturally also with political threats. Unfortunately, contributions of this sort, even in wartime, have not yet decreased very much.

Professional contributions from mathematicians were, however, much reduced. Thus: "I surmise that, if the war should not now suddenly end, we will enter a quieter time at the *Annalen* editorial offices. Till now most [professional mathematicians] could still work, now that holds only for a few."[171]

June 1940 saw the German military at its wartime zenith, and so perhaps the war might have soon ended. In fact, volume 117 of the *Annalen* covered 1939–41; 118, 1941–43; and 119, 1943–44, but in a reduced size and number of pages. Volume 120 appeared in 1949, covering 1947–49. Of course, the lack of suitable manuscripts meant the lack of paper was not felt as an undue hardship.[172]

The Bieberbach-Bohr Exchange and the 1934 Meeting of the DMV

Though the *Mathematische Annalen* was one of the major mathematical organs, responsibility for it devolved not upon the mathematical community as a whole, but only on the editors and publisher. This was obviously not the case of the *Journal of the German Mathematical Society* (*Jahresbericht der Deutschen Mathematiker Vereinigung*).[173] In 1933, the editors of this journal were not all of the ilk (however different among themselves) of Hecke, Behnke, and van der Waerden. These journal editors were Helmut Hasse (since 1932), Konrad Knopp (since 1934, replacing Otto Blumenthal), and Ludwig Bieberbach (since 1921). Hasse has been discussed in chapter 4; Knopp seems to have been an upright man who had a conservative distaste for the Nazi hegemony, and who attempted to do his best to maintain his discipline both nationally and as an Ordinarius at Tübingen in difficult times. Bieberbach, however, was not only a conservative German nationalist, but perhaps the most vocal propagandizer of Nazi ideology within the mathematical community.[174] Certainly there were others, but Bieberbach, unlike Tornier, for example, or even Hamel, was one of the most distinguished German mathematicians of his generation. For mathematics under the Nazis, as a very highly respected professional, he was somewhat analogous to the Nobel Prize-winning Nazi physicists Phillip Lenard and Johannes Stark. In June 1934, Bieberbach, against the wishes of his fellow editors, published an open letter attacking the famous Danish mathematician Harald Bohr and up-

[170] *Schmierig*; literally "greasy."

[171] HNMA, Behnke to Hecke, June 8, 1940.

[172] On lack of suitable manuscripts, cf. Behnke 1978: 134. In late 1942, Süss had the DFG send required formularies to prospective authors of his book project so they could obtain paper (BAK R73 12976, Süss to Griewank, Nov. 5, 1942), as discussed in chapter 4.

[173] For the rest of this chapter, referred to as "the journal" or *JDMV*.

[174] For Ludwig Bieberbach's activities in discerning a "Nordic style" of mathematics, see below, chapter 7.

holding the twin notions that only "Aryan" mathematicians were suitable to teach German youth and that such national awareness was essential to international scientific cooperation. This letter was a primary focus of attention at the annual business meeting of the German Mathematical Society, held in 1934 at the spa of Bad Pyrmont.[175] The other focus of attention was the attempt to introduce the *Führerprinzip*, or "leadership principle," into the German Mathematical Society. Before turning to these matters and how the society conducted itself, it is necessary to understand the origins of this letter and why it was as late as autumn 1934 that the society was considering adoption of Nazi principles and *Gleichschaltung*, or assimilation (literally "coordination"), into the reigning system.[176]

Although the law of April 7, 1933, caused immediate turmoil in the universities as a consequence of the numerous dismissals, the German Mathematical Society seemed surprisingly unaffected as an institution. In 1931, the German university mathematics faculties contained 200 teaching faculty, of whom ninety-eight were *Ordinarien*.[177] As a consequence of the April 7 law, in 1933–34, these faculties lost thirty-five people (of whom thirty were Jewish and fifteen *Ordinarien*). That is, 17.5 percent of the mathematics faculties, including 15.5 percent of the full professors in mathematics, were dismissed in the first year of Nazi hegemony. Of these dismissals, 85.7 percent were of people who

[175] This is perhaps the most frequently discussed occurrence among mathematicians during the Nazi hegemony.

[176] I am unable to address such matters from the original documents in the archive of the German Mathematical Society, as Prof. Martin Barner, who controlled access to them, sealed them from view and refused me permission in 1988 to see them. Nor is this some idiosyncracy of Barner; in 1969, a preface to Pinl's publication series *Kollegen in einer dunklen Zeit* said that the archival bases for Pinl's report existed in the German Mathematical Society Archive and were available for historical investigations. However, in 1983, when Ralf Schaper asked for such viewing permission, he was told by the then-editor of the *JDMV*, Walter Benz, "what was meant however are historical investigations for some later time period, since [the archive contains] private letters concerning persons who are still living and could damage them." (See Ralf Schaper, "Mathematiker im Exil," in preprint 7/83 of Gesamt-hochschule Kassel, 26 pages, p. 23. This has apparently been published in *Die Künste und die Wissenschaften im Exil, 1933–1945* [1983]). However, material on this crucial incident appears in the personal papers of Hellmuth Kneser (cited, as before, as HK) to which his son, Prof. Martin Kneser, has kindly allowed me access. These and other sources were also used by Herbert Mehrtens to discuss this incident in an article appearing in *Jahrbuch Überblicke Mathematik* (1985): 83–103. Apparently Prof. Barner allowed Norbert Schappacher and Martin Kneser access to some of the material in the archive for use in their 1990 article, already cited several times, which appeared in the centenary celebration volume of the German Mathematical Society. Thus the main sources for what follows are *nolens-volens* Schappacher and Kneser 1990, Mehrtens 1985, and HK. More recently the archive has been opened to historians, but not in time to be used in the composition of this book. Citations that follow are amply documented, often from more than one source, though unfortunately not the primary one.

[177] Ferber 1956: 216 (or p. 198). The *Ordinarien* included two honorary professors. In addition, there were nineteen emeriti (who, after all, had the right to teach), five guest lecturers, and fifteen "teaching fellows" (*Lehrbeauftragter*), for a total teaching strength of 239. These figures include teachers of the history of technology. Cf. Schappacher and Kneser 1990: 21.

were Jews or "Jewish-related."[178] Yet the German Mathematical Society seems to have accepted placidly this initial upheaval. It was as though mathematicians as a professional group were maintaining their traditional apolitical stance—mathematics, after all, had nothing to do with politics—much in line with the traditional German academic attitude discussed earlier, and despite the political upheaval affecting their membership.[179] The official report in the *JDMV* of the annual meeting in September 1933 held in Würzburg contains as its only nonroutine item a discussion over the difficulties caused by the existence of two refereeing journals in mathematics.[180] Indeed, in 1933, the members of the German Mathematical Society's executive board who were responsible for publishing its journal were Ludwig Bieberbach, Helmut Hasse, and Otto Blumenthal. Bieberbach had been the secretary of the society since 1921. Already forcibly furloughed at Aachen, Blumenthal resigned this office in early June. While people like Hecke, van der Waerden, and Behnke might have worried about Blumenthal,[181] the society as a whole seemed not at all disturbed by maltreatment of one of their most universally known colleagues, and David Hilbert's first pupil. On June 25, Bieberbach wrote Hasse that he had explicitly discussed with Blumenthal the necessity of his resignation and "could not dissuade him from this purpose (*konnte ihm von seinem Vorhaben nicht abraten*)."[182] Given Bieberbach's own attitudes, this would seem at best disingenuous. It is clear that Bieberbach would have liked his non-Aryan colleagues simply to accept the new order of things and vanish from the scene without much fuss, but to attribute such feelings to Blumenthal (who appeared on the *Annalen*'s title page until 1938) seems rather much.

In July 1933, Bieberbach gave a lecture on mathematical intuition at the Prussian Academy of Sciences in which, to judge by an abstract, he connected such intuition with race, leaning on well-known remarks of Felix Klein.[183] Ear-

[178] Schappacher and Kneser 1990: 21. "Jews," of course, means "racially," that is, descended from people who had been Jews. Most of these followed either Christian religious practices or none at all. For example, Matthias Landau, Edmund Landau's last surviving child, told me (interview, Düsseldorf, Jan. 26, 1988) that his father was the only one of the many "Jewish" mathematicians at Göttingen with any religious practices at all, and these were rather weak. "Jewish-related" in the context of this sentence means at least one Jewish ("racially") grandparent—a few years later, it would include being married to a Jew.

[179] The notion that professional mathematicians should have nothing to do with politics as a society is also common today. Such attitudes were passionately expressed, for example, in business meetings of the American Mathematical Society over issues like the Vietnam War, "Star Wars," and the Equal Rights Amendment, with varying resolutions.

[180] *Zentralblatt der Mathematik* and *Fortschritte der Mathematik*. The former was founded because the latter was years behind in its reviews. For a fascinating view of the politics involved (which also involved Ludwig Bieberbach), see Reinhardt Siegmund-Schultze, "Das Ende des Jahrbuchs über die Fortschritte der Mathematik und die Brechung des deutschen Referatenmonopols," *Mitteilungen der Mathematischen Gesellschaft der DDR* (1984): 91–102.

[181] See above, chapter 4, "The Case of Otto Blumenthal."

[182] Schappacher and Kneser 1990: 52.

[183] *Sitzungsberichte der Preussischen Akademie der Wissenschaften* (Berlin, 1933): 643. For Klein's remarks, see below.

lier, on April 1, the day of the proposed boycott of Jewish business and professionals,[184] and six days before the law under which academics would be dismissed, Bieberbach apparently opened his lecture by publicly saying that he deplored the fact that his "dear friend and colleague Schur today may no longer be among us."[185]

Rather than protest the dismissal of Jewish colleagues, the German Mathematical Society meeting in Würzburg in September 1933 adopted the principle that someone selected by vote for a society office needed to be of "purely Aryan descent and wishing to work in the sense of the national state."[186] Things took their inevitable course so that on November 14, Bieberbach could tell the other council members of the society that Schur "as a voluntary decision relinquishes his office in our council. In agreement with our president [Oskar Perron] and with Herr [Richard] Baldus [the preceding president], I recommend Prof. Conrad Müller of Hannover . . . Herr Müller is (as is his wife) of purely Aryan descent."[187]

The society thus seems to have made easily the "apolitical" adjustment to the Nazi regime. There was worry on the part of some that the departures and dismissals would be deleterious to German mathematics,[188] and vain individual efforts were certainly made to keep individuals like Blumenthal, Courant, and Noether, but, as an organization, the German Mathematical Society was content to be left alone to organize mathematics, accommodating political-ideological demands as necessary.

In 1934, under pressure to be more actively engaged for the National Socialist state, this passive accommodation would change. The agent for this change was Ludwig Bieberbach. Although Bieberbach's July 1933 lecture was not immediately published, on Easter Tuesday, 1934, he gave an address whose published version was entitled "Personality Structure and Mathematical Creativity,"[189] which attracted international attention, even before its complete publication. Bieberbach published a summary of his ideas in *Forschungen und Fortschritte* (Research and progress)[190] that prompted a letter to *Nature*[191] by G. H. Hardy. Hardy was not only one of the leading British mathematicians of the day, but also a man with strong personal contacts to Edmund Landau, whose forced resignation had taken place the preceding autumn.[192] Hardy's letter, "The J-type and S-type among Mathematicians," contained mostly citations from Bieber-

[184] The official boycott lasted only two days. Somewhat ironically, April 1, 1933, was a Saturday; Monday's boycott was half-hearted, and by Tuesday it was called off, officially because it had become "superfluous." Hermann Graml, *Reichskristallnacht* (1988), 118.

[185] Biermann 1988: 198, quoting Kurt Hirsch, who was a student in the class. Bieberbach and Schur had coauthored a paper.

[186] Schappacher and Kneser 1990: 52.

[187] Ibid.

[188] An extract from a letter by Erich Kamke to this effect is in Schappacher and Kneser 1990: 53.

[189] "Persönlichkeitsstruktur und mathematisches Schaffen."

[190] *Forschungen und Fortschritte* (June 20, 1934): 235–237.

[191] Vol. 134 (1934): 250.

[192] Above, chapter 4, "Hasse's Appointment at Göttingen."

bach's summary demonstrating its virulently anti-Semitic and anti-French bias, and its quasi-mystical attempt to distinguish "racial" (i.e., ethnic) differences in mathematical style.[193] It concluded:

> But perhaps I have quoted enough; and I feel disposed to add one comment only. It is not reasonable to criticise too closely the utterances, even of men of science, in times of intense political or national excitement. There are many of us, many Englishmen and many Germans, who said things during the War which we scarcely meant and are sorry to remember now. Anxiety for one's own position, dread of falling behind the rising torrent of folly, determination at all costs not to be outdone, may be natural if not particularly heroic excuses. Prof. Bieberbach's reputation excludes such explanations of his utterances; and I find myself driven to the more uncharitable conclusion that he really believes them true.[194]

A report of the oral lecture appeared in *Deutsche Zukunft* (The German future),[195] a journal with apparently a chiefly academic audience.[196] This report, entitled "New Mathematics," is by a certain "P.S.," who seems to have had some knowledge of mathematics and mathematicians. His article took issue with Bieberbach on racial influences affecting C.G.J. Jacobi as contrasted with Gauss, as well as the greater intelligibility of James Clerk Maxwell to Aryans than to non-Aryans. However, basically "P.S." accurately represented favorably the tone and substance of the published article:

> However, here Bieberbach's conclusions are more important. . . . There is no independent realm of mathematics, independent of intuition and life; the struggle over the foundations [of mathematics] that now rages is in reality a racial conflict: "Political rootedness gives thinking its style!"
>
> . . . Since German mathematics is rooted in blood and soil (*Blut und Boden*), the state ought to and must support and cultivate it, the great accomplishments of German mathematics in the past and present reveal the science as a powerful manifestation of the people (*Volkstum*); therewith it needs no further justification.
>
> Perhaps Bieberbach's exposition, through this sort of foundation, does a service for the maintenance of mathematical activity in Germany and its salvation through the [present] unquiet time.

This review began by calling Bieberbach's speech of historic importance for the development of mathematics in the Third Reich because it "places the most abstract of sciences under the total state," and ended by deploring that he did

[193] For further discussion of Bieberbach's article, see below, chapter 7.

[194] Contrary to widespread belief among mathematicians, Hardy was no pacifist. While he thought that the desire in Britain to "crush" and "humiliate" Germany in 1914 was "monstrous," and he wished for peace on reasonable terms, he nevertheless applied for army service and was rejected, three times (in 1914, he was thirty-seven). Though himself not a conscientious objector, Hardy fought against Bertrand Russell's expulsion from Trinity College, Cambridge, as a pacifist. See Russell correspondence, Russell Archive, McMaster University, Hamilton, Ontario, Hardy to Bertrand Russell, Sept. 16, 1914 and July 1916.

[195] Vol. 14 (Apr. 8, 1934): 15.

[196] According to Schappacher and Kneser 1990: 58.

not drive his conclusions to a decision about tasks and problems that were racially suitable. It is possible to read parts of this article as lightly ironic—for example, this last sentiment, or the article's title. However, despite minor differences with Bieberbach, they seem in context to be sincerely meant and wishing well to a German mathematics stripped of French and Jewish influences. "P.S." himself is surprised that Bieberbach did not mention Hermann Minkowski and Georg Cantor as Jewish "S-types." One should not misinterpret the genuine ideological passion of a bygone time through the ironic spectacles of contemporary rationality.

The famous Danish mathematician Harald Bohr[197] did read the article in *Deutsche Zukunft* and published a rejoinder called "'New Mathematics' in Germany" on May 1 in the *Berlingske Aften* (written in Swedish). Bohr talked about the stimulating mutual influence of different mathematical schools, and the destruction of international traditions through a "little fanatical circle with Berlin professor Ludwig Bieberbach at their head." In addition to forcefully rejecting Bieberbach's reported arguments, Bohr also commented that Bieberbach's editorial work on the (international) journal *Compositio Mathematica*[198] contradicted his nationalist mathematics. Further, thinking (like "P.S.") of Georg Cantor as Jewish, Bohr seemed to believe that Cantor was omitted by Bieberbach because of his absolutely fundamental insights.[199]

Bieberbach was incensed. At the same time, perhaps, he saw an opportunity to bring the German Mathematical Society more actively into harmony with the National Socialist state. Thus on May 21, 1934, without telling the other editors, he inserted an "open letter to Harald Bohr" in the society's journal. His coeditors, Helmut Hasse and Konrad Knopp, saw this letter for the first time in galley proofs a little later and protested to Bieberbach, who was, after all, using the official society journal to carry on what was essentially a personal quarrel.[200] Nevertheless, Bieberbach's open letter did appear when the journal issue was sent to subscribers in early July. It was entitled, "On the Art of Citation."[201]

Bieberbach's open letter is not without interest. On the one hand, he condemned Bohr for responding to a journal's report of his lecture, a report that he called a "malicious caricature" in a journal of "the eternally bygone" (*ewig gestrigen*). But why such a claim? Bieberbach remarked that "P.S." in *Deutsche Zukunft* quoted him as saying, "A people that has come into self-consciousness (*zu sich selbst gekommen ist*) cannot abide such teachers [as Edmund Landau], and must reject foreign thinking," but what he really said was, "A people that

[197] The brother of the physicist Niels Bohr, perhaps the best known of all Danish mathematicians, and also in his youth an Olympic-class soccer player.

[198] Founded by L.E.J. Brouwer after his expulsion from the editorial board of *Mathematische Annalen*.

[199] I am indebted to Ruth Ohman for translating Bohr's article. A German translation appears in the Stanford University Archives, Papers of Gabor Szegö (GS), box 9, folder 15. This translation has some handwritten corrections (by Bohr?).

[200] Schappacher and Kneser 1990: 59.

[201] "Über die Kunst des Zitierens."

has realized how foreign appetite for dominance gnaws at its marrow, how those foreign to the people (*Volksfremde*) work to force their foreign style on it, must reject teachers of a foreign type." This, claimed Bieberbach, is nearly the opposite of what "P.S." and Bohr accused him of saying. His rather intricate and somewhat disingenuous point is that he does not reject the foreign style of doing mathematics, and certainly not foreign mathematics, but he does reject foreign style for Germans. Thus the cited sentence contextually indicated the "inner basis" for the "manly appearance of the Göttingen students [that] put an end to the further activity of Herr Landau as a teacher of German youth." Indeed, Bieberbach maintained that only through the recognition of the differences of different peoples could proper international cooperation take place. "A people that wishes to develop its own talents needs teachers of its own sort." He also condemned Bohr's apparent ignorance of racial science and psychological typology.

Deutsche Zukunft was not the only German journal to take note of Bieberbach's lecture with the same interpretation. On April 4, after Bieberbach's lecture, the *Deutsche Allgemeine Zeitung* reported it in a straightforward fashion:[202]

> Bieberbach underlines and approves of this standpoint [of the students who boycotted Landau's class]; if mathematics may be in its results international and abstract, so it may not be in its sort of thinking and exposition. Mathematical style is racially determined [*bluterfüllt*; literally, "filled with blood"] and an appropriate subject of contemporary racial and typological science. . . . The result is that political rootedness gives thinking its style, practical research the rejection of non-Aryan mathematics, which ought not to be looked upon as German. . . . With this lecture by Bieberbach, mathematical science has integrated (*eingeordnet*) itself into the Third Reich. The reproach by the state of an intellectualism that is foreign to reality will indeed no longer strike it.

Among mathematicians protesting privately to Bieberbach was Gabor Szegö, in a lengthy letter (basing himself on the *Deutsche Zukunft* report). Bieberbach had distinguished the way Landau introduced π as Jewish; among other things, Szegö pointed out that the possibility of this analytic nongeometric introduction of π appeared also in G. H. Hardy's *Pure Mathematics*, where it was termed "satisfactory," though Hardy did not prefer it for his book. Bieberbach answered Szegö in the same disingenuous fashion in which he answered Bohr. He even used the same phrase, *ewig gestrigen*, to characterize *Deutsche Zukunft*, and again called the review by "P.S." a "malicious caricature." He also pointed out that he had never used the word "Aryan" in his lecture.[203]

Why should Bieberbach attempt such a convoluted defense of ideas that fitted so naturally and simply—for example, in the interpretation of "P.S."—

[202] VP, Bohr folder, contains a typed copy of this article, as does the Personalakten Bessel-Hagen, Universität Bonn, though this last does not identify the journal.

[203] For these letters, see GS, box 9, folder 15. Szegö was not "fully Aryan"; Hardy, of course, was. This is implicitly part of Szegö's point, although such considerations are not mentioned by either correspondent. For Szegö's reference to Hardy, see G. Hardy, *A Course of Pure Mathematics* (Cambridge: Cambridge University Press, 1908), section 224.

into the ambient political atmosphere? This question is even more salient since there was considerable fear that in the Nazi atmosphere, mathematics would be under attack as a subject matter. If Bieberbach's lecture, when interpreted in an obvious way (as by "P.S," Bohr, Hardy, and others), rescued mathematics from such ideological pressures, why should he seemingly refuse to acknowledge this, and indulge in such a complex and "intellectual" defense?

Parts of the answer may be the following. There is a long tradition of German scholarship, going back at least to Johann Gottfried von Herder, that individual peoples can best thrive by separate development of their individual creative gifts, which then may complement one another. Herder, as is well known, believed in the tolerance of each nation-state for others, but he was not the only famous German thinker whose ideas would be perverted by Nazi ideologues to their own purposes. No doubt in the Herder tradition is the well-known statement about ethnic differences in mathematical abilities made by Felix Klein (in English) at Northwestern University in 1893:[204]

> Finally, it must be said that the degree of exactness of the intuition of space may be different in different individuals, perhaps even in different races. It would seem as if a strong racial space intuition were an attribute presumably of the Teutonic race, while the critical, purely logical sense is more fully developed in the Latin and Hebrew races. A full investigation of this subject somewhat on the lines suggested by Francis Galton in his research in heredity might be interesting.

For Klein, though certainly a conservative nationalist, was also certainly no anti-Semite—he had helped bring first Hermann Minkowski and then Edmund Landau as well as Karl Schwarzschild to Göttingen, and spoke favorably of the emancipation of the Jews in Prussia in 1812 (by Napoleon Bonaparte): "With this action a large new reservoir of mathematical talent was opened up for our country, the powers of which, coupled with the increase therein achieved by the French emigrants, very soon proved itself fruitful in our science."[205] He also had good relations with a number of Jewish mathematicians and had an extensive correspondence with his good friend Max Noether (Emmy's father): eighty-nine letters from Noether and 129 from Klein, often addressed "Lieber Noether."[206]

Bieberbach was strongly influenced as a young man by Klein,[207] and used his Northwestern remarks as one of the anchors for distinguishing racial types in

[204] Felix Klein, Evanston Colloquium Lectures, lecture 6, delivered at Northwestern University, September 2, 1893, published in Klein's *Gesammelte Mathematische Abhandlungen*, (Berlin: Springer, 1921–23), Vol. 2, p. 228, p. 4 of the original. The original is in English.

[205] F. Klein, *Vorlesungen über die Entwicklung der Mathematik im 19. Jahrhundert*, vol. 1 (Berlin: Springer, 1926), 114.

[206] In *Nachlass* Felix Klein, Handschriftenabteilung der Niedersächsischen Staats- und Universitätsbibliothek, Göttingen. Cf. also Rowe 1986, where the fact that Klein attempted to encourage Jewish mathematicians and resisted anti-Semitism in the faculty is thoroughly documented.

[207] Herbert Mehrtens, "Ludwig Bieberbach and 'Deutsche Mathematik,'" in Esther Phillips, ed., *Studies in the History of Mathematics*, Mathematical Association of America (1987), 197, 207–209. See also below, chapter 7.

mathematics. Bieberbach may even have himself believed that his lecture was still in this tradition of racially distinguishing different but equally valid mathematical abilities, when it clearly had a cast that was pervertedly different: the purging of "German mathematics" or German teaching of mathematics of all Jewish or French influence. This was roughly the opposite of Klein's intention, as his later historical remarks demonstrate. Bieberbach's attack on Edmund Landau may even have been motivated by personal animosities.[208]

In any case, Bieberbach, in the same pseudo-Herder style toward the end of his "open letter," and after calling Bohr a "well-poisoner," condemned him as "a harmful pest (*Schädling*) for all international cooperation. This can only thrive on the basis of the inner strength and self-consciousness of peoples, borne by the reciprocal respect of all." How did the usually acquiescent German mathematicians react to Bieberbach's lecture and to his consequent attack on Bohr? As to the former, Oskar Perron, then president of the German Mathematical Society, wrote Bohr:[209]

> I completely understand your excitement over Bieberbach's talk. I am myself horrified, must however, for today, confine myself to telling you that these absurd views are certainly only shared by very few mathematicians, and that in any case, the German Mathematical Society, which has at all times reckoned the large number of foreign members to its honor, stands completely apart from these views.

Perron went on to discuss the article by "P.S.," which he said took a position rather against Bieberbach. He also remarked that in the May 13 issue of *Deutsche Zukunlt*, one of Bieberbach's students had attempted to mollify the matter; and so "he [Bieberbach] seems therefore also to see his mistake himself, even if open acknowledgment of this naturally is difficult for him."

As events would prove, on this second point Perron was wrong. On the first, the occasional lightly ironic tone of "P.S." seems directed rather toward the idea that mathematics should fit so smoothly into the ideology of the Third Reich, seemingly incredible to him, rather than that Bieberbach's ideas were necessarily spurious, or that it should not be so fitted in. As noted, another newspaper also saw the same straightforward intent in Bieberbach's ideas. Perron may also initially have found it hard to credit that Bieberbach believed the things he said he did.

The reaction of Helmut Hasse is interesting.[210] Not only was he officially one of Bieberbach's coeditors of the *JDMV*, but this was exactly at the height of his trouble in taking over in Göttingen.[211] Bohr had written him on May 13 about Bieberbach's lecture. Hasse did not reply until June 3,[212] excusing the delay with

[208] Personal communication from the late Alexander Ostrowski. Cf. below, chapter 7.

[209] VP, Bohr folder, Perron to Bohr, May 22, 1934.

[210] Schappacher and Kneser 1990: 59–62; Schappacher 1987: 355–356; Segal 1980.

[211] Above, chapter 4, "Hasse's Appointment at Göttingen." Werner Weber refused Hasse the keys to the Göttingen Institut on May 29, 1934.

[212] VP, Bohr folder.

an allusion to "difficulties" in Göttingen that had "exhausted (*mitgenommen*) [him] outwardly and inwardly." He went on:

> I have discussed his lecture with Herr Bieberbach both orally and in writing. I myself have had very many thoughts about the attractive theme of different kinds of modes of thinking and sorts of style in mathematics. . . . However, I cannot follow Herr Bieberbach so far that I see the main source for these differences in blood and race. For my part, environment and education are far more the main factors. As to the valuation of different sorts of styles, so I take the standpoint: *De gustibus non est disputandum* [Latin in original].
>
> . . . The lecture itself in fact has another aspect [from that given it in *Deutsche Zukunft*]. I have had Bieberbach's manuscript sent to me and inspected it. *I would certainly assume that the whole matter will be discussed at the autumn meeting of the DMV in some form or other* [emphasis added]. I do not wish to take a personal position.

At this time Hasse knew of Bieberbach's intention to publish "The Art of Citation" in the *JDMV*, and he and Knopp had already attempted to prevent it. Knopp wrote on May 28 to both Bieberbach and Hasse about the matter, on June 1 again to Hasse, and June 13 again to Bieberbach. Hasse also apparently complained in some measure to Bieberbach, as would seem to appear from a letter of Bieberbach to Hasse on June 7.[213] On June 6, Bohr replied to Hasse:[214]

> Naturally, what you write about Bieberbach's arguments—whose impression on the listener indeed comes out unambiguously in *different* [emphasis in original] newspaper reports—interests me. I would however not consider it right, not to say to you plainly that your communication that you "do not wish to take a personal position" astonishes me and injures me.

He went on to say that it is one thing to compare different mathematical styles and another to justify thereby the approval of Landau's removal. Hasse's reply on June 11[215] asked Bohr not to think ill of his request to take no personal stance on the Bieberbach lecture, his reasons are difficult to explain in a letter and are connected with the Göttingen difficulties. He repeated his unwillingness to draw Bieberbach's conclusions from objective stylistic differences. He also, however, said,

> Because of the grounds indicated above, I must ask you in a most heartfelt fashion in no way to make use of this letter of mine publicly, and generally to leave my name out of the whole matter. What I could personally do, I have done, namely repeatedly indicated to Herr Bieberbach, both orally and in writing, the impression that his lecture has evoked among many foreign and even German mathematicians. Bieberbach perceived this, however blamed it completely on the distorted resumé of his lecture in the review in *Zukunft*, which alone has come in further measure to the public. Besides

[213] Cited in Schappacher and Kneser 1990: 59.
[214] VP, Bohr folder.
[215] Ibid.

I must myself assume the standpoint that it is Herr Bieberbach's own business what he does. I can make no prescriptions about that for him.

Hasse also said that an oral conversation with Bohr would clear the air. Perron's letter and Hasse's thus contrast in the possible attitudes that might then be taken among German mathematicians toward Bieberbach's lecture, even after allowance is made for Hasse's difficulties in Göttingen at the time. And even Perron (who learned of Bieberbach's lecture from a personal letter from Bohr, and Bohr's newspaper article) emphasized the somewhat ironic attitude of "P.S.," and the belief that Bieberbach had seen his error.

Oswald Veblen, one of the leading American mathematicians of the day and a good friend of Bohr, also wrote a letter to Bieberbach objecting to the content of his lecture, with respect both to his defense of the boycott of Landau's classes (Veblen called Landau "one of the great mathematicians of our epoch") and to his distinction between a German and a Jewish mathematics. Veblen admitted, however, that his information was from the *Deutsche Zukunft* article, and invited Bieberbach to repudiate such statements. Veblen sent copies to, among others, Harald Bohr, G. H. Hardy, and L.E.J. Brouwer.[216] However, at least in early June, Veblen tended to think that the brouhaha over Bieberbach's lecture was something of a "tempest in a teapot."[217]

In July, Bieberbach's open letter to Bohr appeared, against the wishes of Hasse and Knopp. Bohr apparently received a number of private letters of support from, among others, Konrad Knopp, Erich Kamke, Hans Hahn, Karl Menger, Georg Pólya, Arnold Sommerfeld, Erich Hecke, and Oskar Perron.[218] Knopp was seemingly so incensed by Bieberbach's action that he insisted that the next number of the journal should make clear that his "open letter" was published against the wishes of his coeditors, and threatened to resign his position.[219] Hasse, Bieberbach's other coeditor, took quite another tack. Apparently he wished to put everything into Perron's hands and have a general decision of the governing council banning ideological articles in the journal; but he wished to avoid any further publication concerning the conflict with Bieberbach, including Knopp's proposed disavowal of it.[220] Bohr and others had written him, and he suggested on July 4 that he and Knopp jointly sign a letter of reply. Hasse, as we know, had just come to Göttingen after considerable difficulty, and with the "strong Nazi" Erhard Tornier as an accompanying appointment. Hasse also remarked: "After consultation with Herr Tornier, for myself I must stress that this [letter] of Herr Bohr will be considered as a purely personal communication, which should never be used publicly, therefore I have especially emphasized this in the letter [to Bohr]."[221] Among the people writing Hasse about the open

[216] A copy is in Personalakten Bessel-Hagen, Universität Bonn.
[217] VP, Bohr to Veblen, June 6, 1934.
[218] VP, Bohr to Veblen, Aug. 11, 1934.
[219] Schappacher and Kneser 1990: 60.
[220] Ibid.
[221] Schappacher and Kneser 1990: n. 112.

letter was Carl Ludwig Siegel, who encouraged him: "Give a signal to everyone with your resignation from the DMV."[222] Hasse, on the contrary, thought his duty was "now to come to an agreement and to prevent the DMV from being further injured by a continuation of the quarrel."[223]

On the same July 9 that Hasse was writing this to Siegel, Bieberbach told him that his proposal to ban ideological articles was against the by-laws of the society, which had "the position of deservedly raising the status of the mathematical sciences in the spiritual life of the nation," and he considered it "in the long run intolerable that about 40 percent of our membership does not belong to the German people and yet should contribute to the fulfilling of this task. I therefore consider necessary a change in the by-laws as soon as possible. This must also clearly express the position of a scientific society in the National Socialist state."[224]

Hasse did decide to join Perron and Knopp in a public declaration that Bieberbach's letter had been published against his fellow editors' wishes. It would have been natural to attempt publication in the society's journal. However, they worried that Bieberbach, as editor, might simply delete it from the galley proofs, and so they proposed to publish an extra single page at their own cost and send it to the membership. On July 23, Hasse wrote the society's publisher, Teubner, with this intention, mentioning that Bieberbach apparently had further intentions stemming from some Nazi party office that stood behind him. Unsurprisingly, without a formal order from the president [Perron] or perhaps a decision of the executive board (which included Bieberbach), the publishing house declined to print and send the membership this explanation of the Bieberbach letter.[225] Consequently, this plan was abandoned, and so the ground shifted to the annual meeting of the society to be held that September in the spa of Bad Pyrmont, as Hasse had originally predicted prior to Bieberbach's "open letter."

Hasse's personal version of the events in Bad Pyrmont was apparently made available to a number of mathematicians.[226] The official report appearing in the *JDMV* was apparently Knopp's edition of Bieberbach's version.[227]

Bieberbach had invited a number of students to come to the meeting,[228] and

[222] Mehrtens 1985: 87.

[223] Schappacher and Kneser 1990: n. 113.

[224] Schappacher and Kneser 1990: 60–61.

[225] Ibid.: 60.

[226] According to Schappacher and Kneser (1990: 63). Among the people they mention as receiving this report is Harold Davenport, but no copy of the report appears in the Davenport-Hasse correspondence as given by Davenport's widow to Cambridge University. Direct quotation from this report appears in Schappacher and Kneser 1990, and the following remarks follow their reportage. Presumably the copy Hasse made available to Blaschke is in the DMV archives in Oberwolfach and this was made available to Schappacher and Kneser (it is not in the Personalakten Blaschke in Hamburg, though relevant letters are). See above, note 176.

[227] Schappacher and Kneser 1990: 63. *JDMV* 44 (1934): Abteilung 2, 86–88.

[228] Among them, according to B. L. van der Waerden, was Fritz Kubach, the mathematics student who had become a Nazi national student leader. Mehrtens 1985: 87–88.

Erhard Tornier had invited the retired Captain Weniger, presumably on the excuse that he was a friend of Theodor Vahlen, as well as presently associated with the DFG. Vahlen himself, mathematician and minister, was not present. Weniger's impeccable Nazi credentials and his former view of himself as an official spy on Göttingen mathematics no doubt also recommended him to Tornier.[229] The presence of students' and teachers' organizations was purportedly justified by the political questions to be discussed. However, Oskar Perron, in the chair, ruled that this was not a teachers' meeting, but a meeting of the membership of a scientific society. He also rejected any particular reason for Weniger's presence, despite his close association with Vahlen, in particular refusing to allow Weniger to exercise a proxy for Vahlen. Perron was here acting completely consistently with an earlier refusal to allow reporters into the meeting since they were not members. When the majority of members present also expressed the wish that nonmembers not be present, they left the room.

The first order of business was the report of the secretary (Bieberbach), in which context the issue of the Bieberbach-Bohr exchange arose.[230] One of the speakers against Bieberbach was Egon Ullrich,[231] who had come from Marburg at Hasse's suggestion to rejoin him in Göttingen as Otto Neugebauer's replacement. Somewhat naively, Ullrich said that in Scandinavia he had personally received the impression that a number of respected Scandinavian mathematicians saw the Bohr article as unjust.[232] Perron and others had been attempting to keep the discussion at the level of Bieberbach's inappropriate behavior with respect to the *JDMV* and his refusal to countenance the opinion of his fellow editors. Ullrich's remarks unfortunately allowed the grounds of discussion to shift to general political ones. Hecke apparently blamed Perron's insufficiency as chair for this situation.[233] In any case, after lengthy discussion, Conrad Müller, the "full Aryan" mathematician at Hannover who had been Schur's replacement in the governing council, submitted the following motion:[234]

> The membership condemns most sharply the attack of Herr Bohr on Herr Bieberbach, insofar as therein an attack on the new German state and National Socialism is to be seen. It deplores the form of Herr Bieberbach's open letter and his procedures for its publication, which took place against the will of both other coeditors and without the knowledge of the president.

[229] For Weniger and Tornier, see above, chapter 4, "Hasse's Appointment at Göttingen."

[230] *JDMV* 44 (1934): Abteilung 2, 86.

[231] A *Dozentenschaft* report on Ullrich appears above in chapter 5.

[232] Ullrich worked in the field founded by the path-breaking work of the Finnish mathematician Rolf Nevanlinna. Nevanlinna had the reputation of being sympathetic to Germany (e.g., MI, Hasse to secretary of the mathematics and physics division of the Göttingen Academy of Sciences, Nov. 5, 1936, recommending Nevanlinna for a corresponding membership, and UAG, memorandum on Nevanlinna, n.d.). Nevanlinna's mother was German, and his (first) wife a Nazi sympathizer (Weil 1992: 130). In 1936–37, he spent a year as visiting professor in Göttingen (Schappacher 1987: 358, 369, and n. 141). See also UAG, Neumann to Nevanlinna, Sept. 3, 1935, Nevanlinna to Neumann, Sept. 6, 1935, and Neumann to Bacher, Sept. 10, 1935.

[233] Mehrtens 1985: 82.

[234] *JDMV* 44 (1934): Abteilung 2, p. 87.

Van der Waerden attempted to change the word "deplores" (*bedauert*) to "disapproves of" (*missbilligt*), but this was rejected.[235] Müller's motion was accepted on a secret written ballot by a vote of 50 yes, 10 no, 2 abstentions.[236]

An additional motion was presented by the mathematician Erich Schönhardt:

> The membership recognizes that Herr Bieberbach in the Bohr matter was concerned to protect the interests of the Third Reich.

This motion just obtained a bare majority of those voting (again by secret written ballot): 31 yes, 20 no, and 10 abstentions.[237]

Bieberbach stated: "I have always declared, and also declare today, that I have answered Herr Bohr's newspaper article in a sharp form in the *Jahresbericht* because I saw and see in it an indeed partly hidden but infamous attack on the Third Reich." Egon Ullrich proposed that by unanimous consent, this declaration would appear in the official report in the *JDMV*, which was granted.[238]

It seems clear that the attendant membership of the DMV had no desire to allow political intrusion into professional matters; at the same time, they were anxious to ward off pretexts for such intrusion by accommodating themselves to the regime. Thus, by reaffirming their loyalty, they would maintain their independence. This was not necessarily a conscious calculation for some—nationalist feeling coupled with professional responsibility provided grounds enough. Nor were Bieberbach's supporters necessarily insignificant mathematicians. For example, the distinguished mathematician Hellmuth Kneser wrote Bieberbach on July 14, 1934, at the close of a mathematical letter:[239]

> In your feud in matters of personality-structure, composition, etc., I take an active (*lebhaften*) part. May God grant German science a unitary, powerful, and continued political position.
>
> Heil Hitler!
> Yours

The attempt to walk a political straight-edge would appear again in the business meeting, when the third item of the agenda, the introduction of the *Führerprinzip*, had a hearing. However, first came elections.

On August 3, 1934, Perron had met with Hasse and Knopp, and they united on Wilhelm Blaschke as candidate to be Perron's successor:[240] whether as *Führer*, or just as president, remained of course to be seen. In the elections, Blaschke *was* chosen to enter the governing council and was then chosen by them, as provided in the by-laws, as president.

There followed the crucial political issue: should the current organization of

[235] Schappacher and Kneser 1990: 63.

[236] According to *JDMV* 44 (1934): Abteilung 2, p. 86, sixty-three members were present at the meeting. The vote is given on p. 87.

[237] Ibid.

[238] Ibid.

[239] HK, Kneser to Bieberbach, July 14, 1934.

[240] Schappacher and Kneser 1990: 62. For more on Blaschke as a personality, see below, chapter 8.

the society, with its officers chosen in a representative-democratic fashion, be replaced by the *Führerprinzip*? This had a prehistory. Seven weeks earlier, on July 17, Bieberbach had already suggested to the governing council that the "new times" demanded the introduction of the *Führerprinzip*, as well as the selection of Theodor Vahlen, by then already director of the scientific office in the education ministry, and Erhard Tornier, the "strong Nazi" who had just accompanied Hasse to Göttingen, for the two seats to be vacated on it in the usual rotation. Five days later, Perron rejected both the preliminarily determined selection of council members and the *Führerprinzip*, stressing the scientific goals of the society and its success, which had led to the large number of foreign members. This epistolary war continued, with Bieberbach suggesting Tornier as new *Führer* of the society and telling at least Perron and Hasse that Vahlen desired Tornier as German Mathematical Society *Führer*. Perron despaired and thought of dissolving the society; however, Hasse thought they might be able to obtain a viable counter-candidate to Tornier.[241]

With this background, Bieberbach proposed at the business meeting that Tornier become *Führer*, with the power to change accordingly the by-laws of the society in consultation with the education ministry. Tornier then proposed the same, only with Bieberbach as *Führer*. Erich Hecke then proposed that the by-laws be altered so that the president be chosen "by the assembled members either through acclamation, or through a written secret ballot with majority rule for two years" (instead of one year, which had been the rule until then). The president would appoint and dismiss the members of the governing council and executive board,[242] and the president would name the executive board in accordance with the April 7 laws for "reform of the civil service."[243]

For Tornier as a new *Führer* of the society, there were 11 yes, 40 no, and 3 abstentions. For Bieberbach as a new *Führer*, there were 10 yes, 38 no, and 5 abstentions. For Hecke's "moderate *Führerprinzip*" proposal, there were only fifty members of the original sixty-three present and voting, and the vote was 38 yes, 8 no, and 4 abstentions, thus barely attaining the three-fourths majority of those present and voting required by the by-laws for a change in them.[244]

Blaschke was then chosen by acclamation "as president in the sense of this change in the by-laws." At this point in the officially published report appears a footnote: "For formal-juridical reasons, the entry of this change in the by-laws in the official index of associations could not yet take place."[245]

[241] For these preliminary skirmishes, see Schappacher and Kneser 1990: 61–62.

[242] I have translated throughout *Ausschuss* as "governing council" and *Vorstand* as "executive board."

[243] *JDMV* 44 (1934): Abteilung 2, p. 87.

[244] Ibid.: 87–88. Since there were sixty-three members originally present at the meeting (above, note 236), one wonders whether the others had absented themselves, or simply refused to vote. It might be noted that there was always a minimum of thirty-eight voting members against any kind of *Führerprinzip*, but that Hecke's motion only carried the day because the number of voters had shrunk to fifty.

[245] Ibid.: 88.

Indeed, Bieberbach, as secretary of the association, was charged with seeing that the change in the by-laws was appropriately entered by the appropriate juridical body in Leipzig that oversaw the by-laws of registered associations. He hesitated to do so. Apparently as early as September 23, he had told Hellmuth Kneser that he had deep reservations about the change in the by-laws.[246] However, on September 27, he told Knopp he would proceed forthwith with the change.[247] He did not. Letters went back and forth throughout October.[248] Finally, on November 7, Bieberbach took the proposed changes to a lawyer, then, on November 13, asked the juridical official in Leipzig for an opinion, which turned out to be that the changes in the by-laws as approved in Bad Pyrmont could not be accepted. There really was a problem, which has no doubt escaped the reader, as it did the mathematicians at Bad Pyrmont. The difficulty was that in the original by-laws, the governing council consisted of six mathematicians elected for three-year terms with two elected each year; this council then elected three others (a secretary, a treasurer, and a managing editor of their journal) to form the executive board and also to join them to make up a council of nine. The latter three also apparently acted as coeditors of the journal. The executive board had terms of indeterminate duration, so long as they were members and wished to serve. The governing council elected from among its other members the president, who served a one-year term in this capacity.[249] With the change as proposed by Hecke, on the one hand, the president would consitute the sole executive, and on the other, he was charged with naming the members of the executive board. These "formal inconsistencies" caused a formal difficulty that the notarizing official in Leipzig pointed out[250] (and that Bieberbach may have earlier realized, though on reflection, he thought not).[251] In between his consultation of a lawyer and his consultation of the Leipzig official, Bieberbach wrote Blaschke on November 10 that there were points in the changes in the by-laws "which, in my opinion, give occasion for juridical second thoughts." He therefore wished to wait until this matter was cleared up by a legal opinion, since "I wish, in any case so long as I am responsible, to spare the DMV the disgrace of a juridically denied application for inscription [of the changes in the by-laws]."[252] Blaschke and the other members of the executive board realized that these formal problems were easily handled by striking the word "executive board" in one place, by striking a sentence relating to contingent election of council members, and by replacing the word "elections" by "election." The man in Leipzig decided (prodded by Bieberbach?) that for these changes to be effective, they would have to be approved by a new business meeting of the society (by, of course, a

[246] HK, Bieberbach to Kneser, Jan. 19, 1935, alludes to such a communication.

[247] HK, cited in *Rundschreiben* I (the first of two circular letters), to be discussed below.

[248] HK. Excerpts from these letters appear in the same *Rundschreiben* I as in the previous note.

[249] Schappacher and Kneser 1990: 51 n. 95.

[250] HK, *Rundschreiben* II.

[251] HK, Bieberbach to Kneser, Jan. 19, 1935.

[252] HK, *Rundschreiben* I, p. 2.

three-fourths majority).[253] There is no little irony in this quarrel in the Nazi state of 1934 over quasi-parliamentary legal technicalities.

At Bad Pyrmont, many of the thirty-eight mathematicians voting for Hecke's proposal no doubt thought that they had again managed to walk the straight-edge, rejecting Bieberbach and Tornier's attempt at overt politicization while accommodating themselves to the regime. They could go on being apolitical mathematicians. Furthermore, there was something of a snare for Bieberbach in the Hecke proposal—for if the new "president" could appoint and dismiss other officers at will, then, in fact, he could dismiss Bieberbach from his office in the society. That this may in part have been Hecke's intention, and was certainly assumed to be by some who opposed Bieberbach, appears from a letter of Emmy Noether, already emigrated to the United States and installed at Bryn Mawr College, to Oswald Veblen. Her opinion regarding the Bohr letter was that all that could be achieved under the circumstances had been, which also speaks to this feeling of relative success by those mathematicians opposed to Bieberbach-style politicization.[254] This feeling of success was now entrammeled in legalistic technicalities.

Nor was this all. In November 1934, Blaschke, Hasse, and Knopp sent a letter to Bieberbach, with copies to 131 other members of the society, including many foreigners.[255] Virtually all of this circular letter consisted of quotations from letters between Bieberbach and Blaschke, Hasse, and Knopp, concerning Bieberbach's delay in seeing that the change in the by-laws be appropriately registered. The last letter cited, from Hasse and Knopp to Bieberbach on November 18, 1934, contained the following forceful statement:[256]

> With respect to us we must see in this [Bieberbach's delay] a renewed insincere and uncollegial behavior and procedure on your part. Therefore, it is no longer possible for us to work together with you in the same way as up till now. We both, since we constitute a two-thirds majority on the executive board of the DMV, have therefore decided to reestablish the conditions as they appear in the by-laws, which have been violated for a long time by the union of the offices of secretary and publisher in your person,[257] and thereby take into our control all essential business of the executive

[253] HK, *Rundschreiben* II.

[254] VP, Noether file, Noether to Veblen, Oct. 4, 1934. She cited a letter from Emil Artin on the matter, attributed the plan to censure Bieberbach to Hasse, and mentioned that Hecke was on his way to Copenhagen to give a personal report (to Bohr).

[255] HK contains two lists. To judge superficially by the typewriter, one list appears to be a copy made by Bieberbach and sent to Kneser—Bieberbach seems frequently to have used an inferior typewriter with a decidedly used ribbon. The other list is presumably a copy of this (clearer and more neatly arranged) in which five names (including four consecutive G's) have been accidentally omitted.

[256] HK, *Rundschreiben* I, p. 2.

[257] Bieberbach was officially secretary of the society and had been since 1921. However, practically, he had also assumed primary responsibility for publication of the *JDMV* for some years, though technically this was Otto Blumenthal's position; see Schappacher and Kneser 1990: 51. Bieberbach in 1934 wrote on stationery denominating him managing editor (*Schriftführer*). See Bieberbach to Szegö, Apr. 18, 1934, in GS, box 9, folder 15.

> board. Blaschke has approved our decision. We have sent a corresponding message to Teubner [the publisher of the *JDMV*].

The circular letter itself concluded with the reiteration that a further collegial working together with Bieberbach in the leadership of the DMV was impossible, and asked him to draw the consequences. It was signed by Blaschke, Hasse, and Knopp, and dated "in November 1934."

Clearly, the three signatories thought they had found a way to dispose of Bieberbach, and for the society to continue in a relatively unpolitical manner. Although the letter was apparently not sent to the membership until November 26,[258] Bieberbach had received his copy on November 23, and replied six days later. His somewhat disingenuous complaint was that he had indeed acted in the interests of the society since the change in the by-laws was improperly formulated. Furthermore, the collection of citations failed to contain citations from his letter of November 22 explaining the juridical situation. He requested a retraction. He sent copies of this letter to fourteen other people (who presumably he thought of as potential supporters).[259] More correspondence followed. A technical point that should be noted is that Bieberbach's letter of November 22 arrived in the hands of at least some of his three opponents on the twenty-third, the same day on which he received the original of the circular letter. At least Konrad Knopp believed that the circular letter had been sent to the other society members prior to receipt of Bieberbach's letter explaining the position of the registry official. Knopp also said he did not agree with the sending of the letter to foreigners.[260]

Receiving insufficient satisfaction in the exchange of paper, Bieberbach repeated his demand for exculpation, mentioning not only the omission of his letter of November 22, but also one of November 14 to Blaschke where he had indicated there might be legal trouble. He called his letter of November 22 "surpressed." While Bieberbach's tone is one of an angry man whose honor has been falsely impeached, he nowhere discusses in either this or the previous letter his procrastination since the end of September. At least one of Bieberbach's supporters, Hellmuth Kneser, saw the matter similarly:[261]

> I come to your fight with the Triple Alliance[262] Bl[aschke]-Ha[sse]-Kn[opp]. I find your expression "defamation" for the procedure of the three right on the mark. . . . I wish,

[258] The date appears from a letter from Bieberbach to Blaschke, Hasse, and Knopp, on Dec. 11, 1934, a copy of which is in HK.

[259] These were thirteen mathematicians and the representative of the publisher Teubner. Bieberbach to Blaschke, Hasse, and Knopp, Nov. 29, 1934. A copy is in HK. Names of those receiving copies appear in Bieberbach to Blaschke, Dec. 11, 1934.

[260] Bieberbach *Rundschreiben*, Jan. 4, 1935; copy in HK. Blaschke and Knopp were especially well-traveled and somewhat cosmopolitan; Hasse had spent some time in England—the issue of foreign members may not initially have occurred to any of them, though Knopp's objection may be taken as genuine.

[261] HK, Kneser to Bieberbach, Dec. 15, 1934. This is Kneser's response to a letter Bieberbach sent him on Nov. 25, 1934, and to Bieberbach's sending him copies of the correspondence.

[262] *Dreibund*. Whether there is intentional allusion to the Triple Entente of France, Britain, and Russia after 1907 is unclear.

however, to tell you my impression of what is in your disfavor. . . . You have unnecessarily and unhappily done yourself an injustice as you veiled your intention from Blaschke despite his questions. He certainly had a right to ask, and I do not see how your intention of allowing the rejection of the vulnerable alteration in the by-laws at the earliest possible stage could have been endangered through openness.

The letter closed with the assertion that, despite differences, Kneser believed he agreed with Bieberbach in the important things.

Feeling a need to rectify themselves with the membership, given Bieberbach's accusations, the "Triple Alliance" sent "at Christmastime" a supplementary circular letter that explained the simple alterations necessary in the change in the by-laws, but asserted that the necessity of these changes in no way affected their reproaches against Bieberbach:[263]

For these relate to the behavior of Herr Bieberbach against Herr Hasse and Herr Knopp in May 1934 at the printing of his open letter to Herr Harald Bohr, as well as against the three undersigned in the time after the Pyrmont meeting until our open break with Herr Bieberbach, which was produced by his missive of November 10.[264]

It also emphasized that, contrary to Bieberbach's implications, their reproaches against him related simply to his behavior prior to the unfavorable decision of the Leipzig registry official, but had nothing to do with this decision.

Blaschke, Hasse, and Knopp may have thought that Bieberbach's "uncollegiality" gave them an advantage in the dispute, and a means to force Bieberbach to resign his office, whether or not the change in the by-laws was duly registered immediately. In fact, their action was to prove their undoing.

Bieberbach's letter of November 29 demanding a retraction from the other three society officers had not mentioned the breadth of distribution of their first letter. His December 11 letter repeating his demand for a retraction did complain about being slandered "to native and foreign members." It also complained that Blaschke's response at being asked to whom the first circular letter had been sent was, "To all members they knew of." On December 28, Bieberbach finally received a list of the 131 other recipients of the circular letter. It should be emphasized that these 131 were far from "all" members of the society, whose number at the time of the Bad Pyrmont meeting was 1,100.[265] They included many important mathematicians, both German and foreign, the personal acquaintances of the officers other than Bieberbach, or presumably "active members" of the Society. Among the 131, the foreigners included, for example, an American (Birkhoff), a Dutchman (Brouwer), Englishmen (Hardy and Littlewood), a Japanese (Takagi), a Finn (Rolf Nevanlinna)—in all, thirty-six Foreigners by Bieberbach's count (or about 27 percent of the recipients) as well as a number of Jews, such as Landau, Tullio Levi-Civita (also Italian), Hausdorff, and Emmy Noether. Although Bieberbach clearly already knew that foreigners

[263] HK, *Rundschreiben* II.

[264] This is the letter in which Bieberbach first told Blaschke of the technical difficulties.

[265] *JDMV* 44 (1934): Abteilung 2, p. 86. About half of these received the journal.

had received the circular letter, the numbers on the list and their prominence apparently inspired an effective counterattack.[266]

> Unfortunately I find myself forced to declare that the distribution in foreign countries of such a circular letter directed against the integrity of a German colleague is a method of procedure that speaks of a deplorable lack of national feeling and national pride—all the more so in the present time of organized agitation against everything German.

This political point would give Bieberbach the advantage. This is not to say that there were not well-known mathematicians already supporting Bieberbach at this time. As previously noted, Hellmuth Kneser was one. Another was Kneser's very close friend (and Bieberbach's student) Wilhelm Süss. Both had received Bieberbach's November 29 letter, and on December 7, with reference to this letter, apparently not knowing who the other recipients were, Süss wrote:[267]

> In the interests of German mathematicians, as well as of the Third Reich, I can only greatly deplore the whole development. At the first instance I thought that one must write to Vahlen and ask him in the interests of all to intervene authoritatively (*mit einen Machtwort*). Bieberbach writes to me that for his part, he wishes to fight his battles through without government help. However, he indeed must hold out rather alone, since most colleagues see caution, always and again as the beginning and end of all wisdom. . . . Should I feel hurt? . . . Besides, among the students, Bl[aschke] is recognized as a good man of the Third Reich. I consider him far too clever for Bieberbach alone to be able to be a match for him.

Whatever Bieberbach may have told Süss, he did meet with Vahlen, and on January 19, acting as a sort of plenipotentiary for the government, issued demands to the governing council of the society:[268]

> Ministerial director Vahlen has commissioned me to notify (*eröffnen*) the governing council of the following: It is to be expected that the Ministry of Education will forbid those civil servants beneath it membership in the German Mathematical Society if 1) Herr Blaschke does not resign his office in the governing council, since he is responsible for the defamation of a fellow German (*Volksgenossen*) to foreigners, and if 2) an alteration in the by-laws is not brought about in which it is fixed that the responsible leadership of the German Mathematical Society needs the confidence of the state authority, i.e., its confirmation through the education ministry.
>
> I thus ask Herr Blaschke to make known his resignation by January 24 and to take care that this declaration comes to my knowledge.
>
> In the certain expectation of this resignation, I recommend to the council the cooptation of Herr Hamel, and simultaneously suggest that he be chosen as president for

[266] HK, Bieberbach *Rundschreiben*, Jan. 4, 1935.

[267] HK, Süss to Kneser, Dec. 7, 1934.

[268] HK, Bieberbach to the members of the governing council of the DMV, Jan. 19, 1935.

the current period of office, or respectively, till the time of the introduction of new by-laws. In case I do not hear to the contrary by January 24, I assume agreement.

For the purpose of carrying out the necessary change in the by-laws, I commission the notary Hartmann to work out juridically incontestable by-laws meeting the Pyrmont decisions as well as the requirement of the ministry. In case I do not hear to the contrary by January 24, I assume agreement. I will first of all then lay these new by-laws before the ministry and thereafter pass them on to the members of the governing council. Then a new meeting of the membership will be called for the purpose of passing a resolution about the change in the by-laws.

With the introduction of the new by-laws, my office in the German Mathematical Society naturally becomes extinct.

This letter has been cited in full for both its content and its tone. Except for the last sentence, Bieberbach is speaking as though he were de facto *Führer* of the society, at least *pro tempore*. The last sentence was in fact added at the request of Hamel, whom Bieberbach had called the day before with the news of the ultimatum he was about to deliver, and presumably to obtain Hamel's agreement to be the new president.[269] Vahlen's threat meant the dissolution of the society, since the large majority of active members were professors with civil-service status. As for Hamel, it has already been seen in the previous chapter how he adapted with alacrity the *Führerprinzip* for the *Mathematische Reichsverband* (MR), of which he had been president and was now *Führer*, declaring that subordination to the total state was the order of the day. Indeed, at Bad Pyrmont, Hamel had indicated that the society's political duties to raise the position of the mathematical sciences in the spiritual life of the nation could find fulfillment within the MR, of which it was a consituent. Commenting on the organization of the MR on the *Führerprinzip* which had taken place the previous year at Würzburg, he further remarked that since he was not a Nazi party member, he was only *Leiter*, not "*Führer*"[270] at the MR; however, if the DMV rejected the *Führerprinzip*, he would give up his office to the benefit of a National Socialist *Führer*.[271]

If Bieberbach's ultimatum of January 19 came as a shock to the members of the governing council, nevertheless in the two preceding weeks there were a number of mathematicians who wished to bring the government into the dispute to help him or to compel Blaschke's resignation. For example, Süss proposed to Kneser the organization of a group of mathematicians to demand Blaschke's resignation, even speculating whether Knopp might perhaps be asked to join. Further, says Süss, it would be good if Vahlen could be brought in, as he could declare that Blaschke could not be tolerated as president and could designate someone as acceptable:

[269] HK, Hamel to Kneser, Jan. 19, 1935.

[270] While both words may be translated "leader," Führer, under the Nazi regime, had, of course, the connotation of "maximum leader."

[271] Schappacher and Kneser 1990: 63–64, citing Hasse's report.

> That was the partial notion (*Idee*) of Doetsch for Bie[berbach]'s procedure in Py[rmont], which I found not bad. I emphasize that for me neither X nor Y [putative choices by Vahlen] need be identical with Bie[berbach]. At any rate, aside from yourself,[272] I know no one whom I could unreservedly recommend. For the cohesion of the DMV, a new name would be best, therefore also neither Knopp nor Hasse. Van der Waerden is a foreigner and was already in the fight. Tornier: ?? Back to Perron, who was already unsuitable, he unfortunately comes also not into consideration. E. Schmidt? . . . At the earliest by your side!

Also, according to Süss, Doetsch proposed writing Tornier about the matter, but Süss believed he would not.[273]

Three days later, Süss announced to Kneser that he and Doetsch jointly proposed to discover the names of those who found Blaschke unbearable and then, on all their behalfs, ask for Blaschke's resignation. He was also seriously considering a letter to Vahlen "with the request to make an end by decree first of all of the internal state of the DMV until the next meeting."[274] Kneser, who was an old acquaintance of Vahlen's, replied with his wish to speak to Vahlen personally about the matter. Süss also announced that he and Doetsch had composed a joint letter to be sent simultaneously to the education ministry and to Blaschke, Hasse, and Knopp, which he thought to send in first draft to Kneser, Vahlen, Tornier, and Erhard Schmidt, for their approval and possible signature, and then potentially to others who might be expected to sign.[275] Among other things, they demanded the resignations of Blaschke, Hasse, and Knopp "in a time when the German people stand in heaviest battle for honor and dignity with respect to foreign countries (*das Ausland*)."[276] This feeling is also reflected in the opening line of Süss's January 15 letter: "Before everything we congratulate ourselves on account of the Saar." Similarly, Bieberbach had added a postscript to his January 4 letter: "Only eight days to the Saar plebiscite."[277]

In two letters, Kneser demurred. He thought that the text brought in the arguable matter of whether Bieberbach had been dismissed from the executive board by Hasse and Knopp. In his view, there was no relevance in any of the issues involving whether the reproaches against Bieberbach were justified. The sole issue was the distribution to foreigners of attacks on a German colleague. He would be in accord with a declaration about the impeachment of national dignity, and he would send it to all "German (that is, neither born in a foreign country nor Jewish)" recipients of the first circular letter that had requested

[272] *Du*. Kneser and Süss were very close friends.

[273] HK, Süss to Kneser, Jan. 8, 1935.

[274] HK, Süss to Kneser, Jan. 11, 1935.

[275] HK, Süss to Kneser, Jan. 15, 1935.

[276] A copy is in HK.

[277] After World War I, a League of Nations commission administered the Saar until a foreordained 1935 plebiscite determining whether it should be part of France or Germany. In the meantime, the French had control of the well-known coal fields there. This plebiscite was held on January 12, 1935, when over 90% of the Saarlanders voting opted for union with Germany. Cf. also E. A. Weiss's study camp, above, chapter 5, "Mathematical Camps."

Bieberbach's resignation. Kneser wished also, if possible, to avoid bringing in Vahlen and the ministry.[278]

> I am not for forcing coordination (*Gleichschaltung*) on the DMV from without. In Pyrmont I voted for Bieberbach's proposals and will act correspondingly in the future; however, it must come from within. I mention this because V[ahlen] according to my impression is little interested in a change of personnel without full *Gleichschaltung*.

Kneser was also in correspondence with Hamel, and wrote him in the same vein.[279] Vahlen had told him in effect that if the ministry intervened, then it would do so in a radical fashion, and Kneser felt that "I myself also would belong just as gladly to a DMV transformed in such a radical way as to the earlier one, and more gladly than to the present one." Still, he did not wish to encourage Vahlen to such action. Foremost, this was because the DMV should bring its own house in order so long as it was able, something for which Kneser had not given up all hope. And "second, the transformation would, if it came from outside the DMV, perhaps endanger the role of the DMV as an active outpost of Germany in foreign countries."

While in Berlin to see Vahlen, Kneser also spoke to Erhard Schmidt and Bieberbach. All agreed nothing could be done while Blaschke was still president. Kneser announced his plans to bring together demands for Blaschke's resignation, written as energetically as possible, and to keep Schmidt informed of progress. He asked Hamel to join this effort.

The next day, Hamel replied with the news of Bieberbach's ultimatum, but saying that this did not make actions like those contemplated by Kneser superfluous:[280] "It now seems to me to be essential that the meeting of the DMV in March should be prepared, especially morally. There ought not to be lengthy controversies there again."[281] He asked Kneser to communicate with and elucidate other mathematicians.

> For it will be good that besides the ultimative form of Bieberbach, they hear from another side, what opinion is, and what caused me to agree to B[ieberbach's] action. Namely, it is not merely love of B[ieberbach], but rather it is worry about the DMV.

In fact, though, all these planned actions against Blaschke were made unnecessary by Bieberbach's stroke. However, while unimportant from the point of view of the succession of significant events, they reveal the attitudes of an important group of active mathematicians at a crucial moment in the society's history. They may have been a minority among the mathematicians active in the society, but they were not necessarily more nationalist in their feelings than

[278] HK, Kneser to Süss, Jan. 19, 1935. The notion of "German" here is not one of residence (*ius solis*), but the traditional German one of genetic descent (*ius sanguinis*). This is still true today, as witness, for example, the official reception in Germany of Silesian Germans as sort of displaced natives.

[279] HK, Kneser to Hamel, Jan. 18, 1935.

[280] HK, Hamel to Kneser, Jan. 19, 1935.

[281] An extraordinary March meeting, which in the eventuality was never held, was contemplated to ratify a legally correct version of changes in the by-laws.

mathematicians on the other side—Hasse's strong nationalist feelings have already been discussed. They truly saw an affront to German dignity in the sending of the first circular letter to foreigners. They also exhibit how many people at first genuinely saw Hitler's accession to power as a means of reestablishing German honor and respect among nations at a time when they felt Germany was still unfairly disdained and still unfairly placed under a burden of guilt, both financial and moral, for the World War. Chapter 3 discussed how these attitudes sometimes expressed themselves in extreme forms among academics during the Weimar period. The events of the Bohr letter, Bieberbach's response, and the circular letter *contra* Bieberbach provided concrete local instances around which such feelings could coalesce. Hellmuth Kneser's desire for an "inner-directed" coordination of the German Mathematical Society into the German state, indeed as an "active outpost" spreading German culture abroad, was the genuine feeling at the time of a sincere, honest, and upright conservative. Moreover, the "sides" exhibited in the 1934 controversies had no durability, nor did they even reflect the continuing attitudes of the participants. As was seen in chapter 4, by 1943, Süss and Doetsch were enemies both in fact and in attitude toward politics. Bieberbach had already become an ardent propagandist among mathematicians for Nazi ideas, and his efforts are discussed in chapter 7. By 1936, Süss (and Kneser) had fallen out of sympathy with Bieberbach. For example, Bieberbach planned a visit to Freiburg (where Süss was Rektor), but then had to call it off. Süss felt on the one hand a good deal easier "that we have remained spared [from Bieberbach's visit]. I would have in any case also very gladly for once determined what in the meantime has become of the man, from whom I myself indeed always received only help and friendliness. It is completely ununderstandable to me that he now behaves [the way he does]."[282]

In fact, Süss, Kneser, and Erhard Schmidt, while certainly initially inspired by the National Socialist revolution, seemed to become three mathematicians who lived through the Third Reich as decent and upstanding individuals, whatever their initial pro-Hitler disposition. It is tempting to ascribe this, at least in part, to an inner-directed conservatism that at bottom had almost nothing to do with right-wing radicalism. While Kneser and Erhard Schmidt essentially restricted themselves to their local scene, Süss became, from 1937 to 1945, the representative of the mathematical community to the government. Kneser, although he felt that Vahlen "had exercised through his personality real influence on me," also seems not to have been able to follow Vahlen deep into the Nazi thicket.[283] Indeed, on the contrary, Kneser and Erich Kamke became good friends.[284] Erhard Schmidt was socially prominent and was listed (as, for example, was

[282] HK, Süss to Kneser, Oct. 29, 1936.

[283] HK, Kneser to Süss, Apr. 17, 1939, p. 2, top.

[284] HK, Kamke to Kneser, Aug. 1, 1937, announcing Kamke's pending dismissal, and Apr. 15, 1958, in celebration of Kneser's sixtieth birthday. Also see Kneser's memorial article for Kamke in *JDMV* 69 (1968): 206–208.

Edmund Landau) in the *Reichshandbuch der Deutschen Gesellschaft* for 1931, the German social register.[285] Not unlike Hellmuth Kneser, he was[286]

> of nationalist persuasion before the [political] breakthrough [of the Nazis] and after the takeover of power let it be known that he approved the goals of National Socialism. He has participated in the social measures of party and state in ample measure up until now. Sch[midt] is a subscriber to National Socialist periodicals (*Schriften*).

Yet gradually that nationalist conservatism, unbeknownst to the powers that were, took him (like Kneser) into a silent rejection of the government, about which he could do nothing, and an attempt to be as personally upright as possible.[287] Süss was a somewhat similar figure, though his career was complexly interwoven with mathematical life in the Third Reich, as will be discussed later. On the other hand, Wilhelm Blaschke, who will also be discussed in more detail later, seems to have become an opportunistic manipulator, who, in Horst Tietz's words, "worked on both sides [of political issues]."[288] These developments are mentioned here as a caution that the behavior of individual mathematicians in September 1934 only reflects their feelings at that time and should not be taken as a necessary predictor of later attitudes. Some, in their vastly different ways, like Hecke, van der Waerden, or Doetsch, whether anti- or pro-Nazi, seem not to have changed their minds much; others, like Kneser and Süss, seem to have been driven by firmly held inner-directed principles in a direction they had not thought at first theirs.

Another caution is also in order. The way I have presented events, the date of January 19, 1935, seems like some sort of apogee. It was not. On January 19 alone, there is in the Kneser correspondence a letter from Kneser to Süss, from Hamel to Kneser, from Bieberbach to Kneser, and from Bieberbach to the governing council of the Society. Presumably others involved were also engaged in similar correspondence at the time. Agitation was promoting action like the Süss-Doetsch initiative. Bieberbach's involvement of Vahlen and the ministry sharply cut all this off.

Blaschke had been told to resign by Thursday, January 24. On Friday, January 25, a special meeting of the governing council of the society, arranged telegraphically, took place in Berlin. Eight of the nine members were present. (Leopold Vietoris, who was Austrian and employed at Innsbruck, did not appear.) Hamel was also present. Blaschke took full blame for sending the November circular letter even to foreigners and resigned as president. Bieberbach resigned as secretary (as he had promised Hamel). Both left the council as well. Hamel was coopted into the council and chosen by them (this in the old-fashioned way) as president. Emanuel Sperner was chosen as secretary to re-

[285] Page 1656.

[286] BDC file on Erhard Schmidt, Hauptstellenleiter Kuhn of NSDAP to Gauleitung München-Bayern and repeated by them to Bavarian ministry of education, July 10, 1941, Aug. 14, 1941.

[287] For more on Schmidt's attitudes, see below in chapter 7.

[288] Horst Tietz, interview, Apr. 5, 1988.

place Bieberbach. Hasse and Knopp agreed that "they would assure themselves of Hamel's agreement for all important decisions."[289] Furthermore, no special meeting of the society was to be called, as no change in the by-laws would be demanded, and Vahlen himself did not wish this; it sufficed that the person selected as president had the approval of the ministry.[290] Hasse and Knopp remained members of the executive board, as neither Blaschke nor Bieberbach nor Vahlen requested their resignations. Also, as already noted, Hamel was suitably National Socialistically inclined, and, as he wrote Kneser, "the role as *Führer* was given me only as a matter of protocol."[291]

Thus, by the end of January 1935, the society was operating under its old laws of an annual change in the president who could only be reelected after a three-year hiatus, except that the president was subject to the approval of the ministry—a very weak *Führerprinzip*. By choosing sufficiently "national" people, the society could be ensured of ministry approval. It seems clear that Theodor Vahlen and the ministry wanted a thriving and active professional organization providing only that it was suitably "coordinated" into the state. As Oskar Perron had apparently remarked at the start of the whole trouble, the interior ministry wished to maintain the unpolitical character of purely scientific organizations. Indeed, he apparently read a statement to this effect, which he had personally obtained from the cultural affairs division of the Foreign Office.[292] Thus the ideological fanatics like Bieberbach were at the time somewhat sidelined in the name of order. One way in which Bieberbach reacted will be discussed in the next chapter.

The MR and the Content of University Mathematics Teaching

From its origin in 1920 until 1945, the president of the *Mathematische Reichsverband* (MR) was the mathematician Georg Hamel. Already noted is the speed with which the MR adapted itself to the *Führerprinzip* under his leadership. Indeed, as early as 1928, Hamel had prefaced a sketch of a neo-Kantian approach to problems in the foundations of mathematics with nationalist *völkisch* sentiments. While moderate in tone for a conservative academic of the time, they also reveal that his conversion to the *Führerprinzip* was not of the Damascene variety.[293] The MR began life as an organization dedicated to the interests of mathematical education at the secondary and university levels. Its potential

[289] *JDMV* 45 (1935): Abteilung 2, p. 1. No details appear in this report.

[290] HK, Hamel to Kneser, Jan. 27, 1935, and Jan. 31, 1935.

[291] HK, Hamel to Kneser, Jan. 31, 1935; literally "mir nur protokollarisch die Führereigenschaft zugelegt."

[292] Schappacher and Kneser 1990: 63–64.

[293] Georg Hamel, "Ueber die philosophische Stellung der Mathematik," *Akademische Schriftenreihe der Technischen Hochschule Charlottenburg* (1928). This was Hamel's inaugural address as Rektor of the technical university in Berlin. For much less moderate academic sentiments, see chapter 3, above. For the MR's adoption of the *Führerprinzip*, see above, chapter 5, "Secondary and Elementary Mathematics."

members were all the members of the mathematical organizations that officially belonged to the *Verband*. "Interests" meant both improving mathematics instruction and making political propaganda for its value.[294] Perhaps because Hamel was at the technical university in Berlin, perhaps because it was good publicity for the value of mathematics, the MR seems to have been especially interested in mathematical applications. It was certainly not the only organized attempt at improving mathematical pedagogy—for example, there were the long-running pedagogical seminars organized jointly by Heinrich Behnke and Otto Toeplitz mentioned earlier. The DMV always appointed two of its members as official delegates to the MR, though there was naturally a considerable overlap in membership, and both organizations usually met at the same time and place.

While the members of the DMV were still fighting over the *Führerprinzip* in Bad Pyrmont in 1934, the MR, having enthusiastically accepted the dictatorial principle a year earlier, had a series of lectures on the role of applied mathematics, extracts from and summaries of which were published in the *JDMV*.[295] As Hamel said in his introductory remarks, for the MR, "Future work shall be above all practical," and "a completely new positive relationship of the universities to applied mathematics" was required.[296] At this meeting, copies of speeches by Ernst Tiedge and (the late) Gerhard Thomsen, which have already been examined in chapter 5, were distributed.

Despite the early *Gleichschaltung* of the MR, ideological rhetoric in the manner of Thomsen or Tiedge was certainly not necessary in addressing it, as a talk the following year in Stuttgart by Erich Kamke on the education of mathematics students demonstrates. Kamke's address, reprinted in full in the *JDMV*, is somewhat pedagogical-technical and (as might be expected) devoid of any sociopolitical rhetoric. Its tone is no different than it might have been had it been given ten years earlier or twenty years later.[297] It even mentions favorably Otto Toeplitz, who had already, as a Jew, been forced from his professorship at Bonn.[298]

These 1934 talks reflect the attitudes and thinking of people seriously concerned with mathematical education at the time. Their rhetoric is not mere lip service to the reigning ideology. Many people believed the things that were said. The early Nazi government had reawakened in many a German national pride that was felt to have been demeaned since the end of the war. These longings have been discussed in chapter 3, and their genuineness needs to be respected. As one speaker put it with conviction, the new contemporary educational task was the "training of the German man."[299] All things considered, mathematics

[294] *JDMV* 29 (1920): Abteilung 2, p. 52. The MR was founded on January 6, 1920. In addition to Hamel and two secondary-school people, Issai Schur also signed the announcement of its creation.

[295] *JDMV* 45 (1935): Abteilung 2, pp. 2–21.

[296] Ibid.: 3 and 2.

[297] *JDMV* 46 (1936): Abteilung 2, pp. 14–19.

[298] Ibid.: 19. Toeplitz was forcibly furloughed at Easter 1933 and emerited in 1935. See Personalakten Toeplitz, Universität Bonn, and Schappacher and Kneser 1990: 40.

[299] *JDMV* 45 (1935): Abteilung 2, p. 20.

education somehow had better not be far behind. If one was devoted to teaching mathematics and devoted to the new state, it was apropos to see how these devotions might serve one another.

The rhetoric of these talks reflects some of the themes in mathematics already encountered. It could be, for example, a moderate allusion to mathematical rigor as serving the new state—an argument Ernst August Weiss was forcefully making at the same time. Such was the case with Georg Feigl, who, in discussing what the university desired in secondary-school mathematics instruction, could say:

> The responsibility with respect to the people as a whole (*Volksganzen*), for which consciousness has now again become awakened and active in our land, makes it a strict duty for the individual to cultivate his plans carefully and at the place where he is put to bring them to full development in the general interest.
>
> . . . No school subject holds one thus to concentrated intellectual work as mathematics. Mathematical instruction requires continual attention and concentration. In no other subject is there a similar closed development, progressing from fact to fact, at every step presuming the knowledge and understanding of the preceding one. . . . Mathematics also educates to honor in intellectual work: not only the result, but also every step leading thereto must be able to withstand testing, and the disjunction "right or wrong" leaves no place for connotations, interpretations, and evasions.

The rhetoric might be a defense of mathematics as a truly appropriate German subject, felt to be immediately necessary:[300]

> The lecturer opposes the view exposited by the ministerial advisor Benze in his book *Race and School* (1934) that "the mathematical-natural scientific disciplines are—except for biology—less called upon for the education of the people (*völkischen Erziehung*), but above all to serve intellectually formal schooling and preparatory training for certain professions." He explains how, in the course of the coming into consciousness of a people, the customs and usages originally present in legend and religion have developed manifest *connectedness to* nature, to *knowledge* of nature, and finally to the *science* of nature. Even today, this same ancient Germanic instinct that once lived in the Vikings, and strives, in opposition to dull devotion, for the mastery and domination of the powers of nature, shows itself in our youth when they, for example, despite all the restrictions of the Versailles Treaty and all the neediness of the time, press toward flying gliders and generally to flying . . . only through the total sum of mathematical-natural-scientific-technical knowledge is the human spirit placed in a position to dominate nature and to make it serviceable to the benefit of the whole people. It is therefore a fundamental error if one assumes the whole complex of mathematical-natural-scientific-technical work is inwardly foreign to our people and belongs to the outer precincts of a merely formal schooling. . . .
>
> Mathematics plays a fundamental role in this whole complex, since its connections to nature are deep-lying and penetrating. They go so far that every process in nature in

[300] Ibid.: 5, in a talk by W. Schmiedler of Breslau entitled "Mathematics Instruction in the Third Reich."

> principle has its image in the realm of mathematics—the possibility originates of replacing or predicting by mathematical computation what occurs or will occur. The research drive in mathematics, the true living element of this science, has in the deepest fundamental sense, the same sources as the research drive in the natural sciences. The cultivation of mathematical thinking is therefore necessary not only for reasons of expediency, but also it takes into account the same fundamental feeling (*Urgefühl*) of our people that from olden times has driven us to the researching and domination of nature.

This rhetoric might even be a call to what can only be described as illogical mathematics (for the use of jurists):[301]

> Earlier, mathematical thinking was of directly decisive value for jurists. Occupation with mathematics, as no other branch of education, sharpens rational thought (*Verstand*). It gave the ability to order occurrences in life according to strictly logical points of view in general lines of arrangement. . . .
>
> Today we can no longer work with this sort of jurisprudence and formal-logical mathematics. It is not only useless, but dangerous and injurious. *Jurisprudence and jurists are to be put much more into the life of the people.* Here the fundamentals and assumptions of thinking and operating are to be found. We must therefore declare battle with a mathematics that exhausts itself in logical conclusion-making and in mathematical computation. Instead, for young jurists, we need a thinking that first and always goes into the people-centered (*völkische*), not everywhere rationally apprehensible, conditioning of life. Teleology must step into the place of pure logic. Beginning and end of thinking and operating must be fulfilled by the people-centered aim.
>
> For this new juridical thinking, we need mathematics. It is not possible for legal life to be ruled by a dull feeling for legality. That were even as false as the other extreme of exclusively abstract thinking. . . .
>
> . . . The schooling of the intellect can be especially mediated by mathematics. This is much more suited thereto than pure logic because it has the advantage of being easier to intuit (*grösserer Anschaulichkeit*), and within mathematics again, applied mathematics deserves the advantage over pure mathematics. The mathematics that we wish for the beginning jurist is a mathematics rooted in the people, in the irrational, and not merely formally constructed. . . . Thus, from the standpoint of the jurist, the following requirements for mathematics instruction in the Third Reich result:
>
> I. In place of a formal-logical mathematics must step a mathematics that strongly emphasizes the limits of logic, the assumptions of all life, and is conscious at all times of the national (*völkischen*) conditioning of all thinking and living. . . .
>
> II. The connections between thinking and feeling, between science and national life are already to be brought near one another in school instruction. Mathematics is to be applied not only to the natural sciences but also to every sort of traditional knowledge of the people (*Volkskunde*).[302]

[301] Ibid.: 18, from an article by K. Siegert in Göttingen.

[302] *Volkskunde* can also be translated "folklore," or "the scientific study of folklore," but the notion of traditional knowledge seems more appropriate here.

As to mathematical as opposed to rhetorical content, there was much appeal to educating intuitive understanding of three-dimensional space and the motion of objects in it.[303] There was also an emphasis on teaching computational skills and on reducing the subjects covered in secondary school (one speaker suggested students go no further than the beginning elements of differential and integral calculus) while intensifying knowledge of them.[304] There was even a talk on "Soldiers and Mathematics," which, to judge from the extract, discussed ballistic questions, the pursuit of airplanes by anti-aircraft artillery, and estimation of the position of enemy batteries.[305]

In 1937, the meeting of the MR concerned itself with two issues: (1) the occupations other than teacher into which university mathematics students went and how university mathematics should serve them; and (2) the transition from secondary school to university.[306] The first was at the suggestion of Erich Kamke, who at the time of the meeting had already been dismissed as "Jewish-related."[307] His summarizing address was reprinted in full in the *JDMV*.[308] The second item consisted of talks by Feigl and Lietzmann.[309] One of the problems was that students did not go immediately from secondary school to university. Fully effective July 16, 1937 (though first ordered the preceding autumn), the education ministry had decreed that future schoolteachers had to spend a year at a "teachers' college" (*Hochschule für Lehrerbildung or HFL*) before going on to more traditional tertiary education. Given the compulsory work-service following secondary school, and perhaps two years of army service, the gap between secondary education and university might be as great as 3½ years, and was at least 1½ years. Lietzmann frankly says that the majority of students had a 2½-to-3½-year gap in their education. In addition, the decree of July 16 shortened the period that could be spent at university to no more than six semesters. Both speakers recognized the necessity of building in lectures in "elementary mathematics" or some equivalent thereto as a preliminary to real study.[310]

Feigl remarked diplomatically about the "teachers' colleges": "We must in any case attempt to configure the collaboration of the HFL in the disciplinary educa-

[303] E.g., *JDMV* 45 (1935): Abteilung 2, pp. 9–11, 13, 19–20.

[304] Ibid.: 14.

[305] Ibid.: 15–17.

[306] Here (and elsewhere) *Hochschule* has been translated as "university"; in German there was a distinction between the arts and sciences *Universität* and the *Technische Hochschule*, or technical university. Both are *Hochschulen*, i.e., they represent tertiary education.

[307] Above, chapter 4, "The Süss Book Project."

[308] *JDMV* 47 (1937): 250–256. There were three other talks on this theme reprinted in ibid.: 232–250.

[309] Ibid.: Abteilung 2, pp. 80–88 (Feigl) and pp. 88–95 (Lietzmann).

[310] That this (for different reasons) should have been necessary both in the Germany of the 1930s and the United States of the 1970s is an irony too sharp not to be worth remark. In Germany, the same thing had happened in the years following World War I, when no less distinguished mathematicians than Erhard Schmidt and Constantin Carathéodory had held such elementary mathematics lectures for soldiers returning from the front. See *JDMV* 47 (1937): Abteilung 2, pp. 86–87 (Feigl's address), and M. Zacharias, "Elementarmathematische Ergänzungsvorlesungen," *Zeitschrift für math.-naturwiss. Unterricht* 50 (1919): 277–280.

tion of our future teachers more effectively than it has been up till now." Furthermore, the hours for mathematics in the schools were restricted, a matter detailed and deplored by Lietzmann. Much that the universities could once have assumed as even heard by students (let alone known or remembered) could no longer be so supposed. Gone from these two talks is all the rhetoric of three years earlier (though Lietzmann, true to his colors, mentioned the military importance of mathematics). Instead we have a view of a mathematics education declining in quality, with no prospect that the future would be anything but more difficult at all levels. As the end of Feigl's talk somewhat desperately put it:

> Those areas [of mathematics], however, which are now too remote or too difficult for the future teacher and which are not of importance for the future practical user of mathematics . . . areas in whose development German research has always had a prominent part, ought not to be pressed into the background or disappear entirely from our plan of study. For the university indeed has to serve preparation for professions *and* pure research and the exclusive emphasis on the point of view of usefulness will quickly lead to an impoverishment and devastation of mathematical life at our universities.[311]

The Post-Crisis Mathematical Society and the Role of Wilhelm Süss

As to the German Mathematical Society, its crisis over, things proceeded normally. In September 1935, it met in Stuttgart, reaffirmed Hamel's membership in the governing council, and chose Erhard Schmidt as one of two members to replace those whose terms had expired; later, the council chose Schmidt as president.[312]

The following year, there was some concern again about selections being made by appropriate voting procedures, though all selections of new council members were in fact made by acclamation. One of these, the *Gymnasium* teacher Walther Lietzmann, who had an adjunct position at the University of Göttingen, was then selected by written vote to be the next annual president. Lietzmann, as was seen earlier, had long been a rattler of mathematical sabres.

At the end of 1935, Konrad Knopp resigned as editor of the society's journal and was replaced by Emanuel Sperner, Sperner's position as secretary being in turn taken by Conrad Müller, who had recently left the governing council.[313] These selections were made by the council. Sperner continued as editor until the last appearance of the journal during the Third Reich, at the end of 1943 (volume 53). Hasse remained as treasurer, though when, with the war, he re-

[311] The quotes from Feigl appear in *JDMV* 47 (1937): Abteilung 2, pp. 82–88. For the emphasis on mathematics (including statistics) as valuable (only) because of its applicability and the need for a general recognition of the utility of mathematics, see, for example, the talks cited in note 295.

[312] *JDMV* 45 (1935): Abteilung 2, pp. 720–721.

[313] *JDMV* 46 (1936): Abteilung 2, p. 86.

turned to military service, his functions were fulfilled first by Sperner, then by Ulrich Graf, and then by Georg Feigl as deputies.[314]

As to the presidency, the society had its usual September meeting in 1937 in Bad Kreuznach. A member of the council had died in January and, in accord with society rules, Wilhelm Süss was co-opted as a substitute replacement until approval by the next annual meeting. Süss's co-optation was approved (though this was *post factum*—the term had had seven more days to run). Süss was then chosen as one of the two people elected for a new three-year term (by acclamation). This did not contravene the statutes because he had not been previously elected. Then, by written ballot of the council as usual, Süss was chosen president for the forthcoming year.[315]

In 1938, at the annual meeting held in Baden-Baden, not only was the scientific program enriched by a number of talks from the "German Eastern region"[316] as well as by speakers from Romania, Italy, France, and Switzerland, but also changes in the statutory by-laws were undertaken in the business session. These were adopted "unanimously." One of them permitted the reappointment of the president without the heretofore three-year wait. Süss (in the second year of his council term) was then reelected president.[317] The planned meeting for September 1939 failed to take place, no doubt because of the war; however, the council provisionally elected (subject to the approval of the next business meeting, where approval was usually, before as well as after 1933, *pro forma*) two new members to replace those whose terms had ended, and reelected Süss as president.[318] Süss himself published a report talking about the increase in the work of the president and the tasks of the society in raising the position of mathematics in Germany, and its possible developing involvement in the manifold aspects of mathematics from secondary instruction to refereeing policies of journals to doctoral examinations. "In good knowledge of our duties within the politically rigidly aligned state, we stand at the disposal of the leading offices of national and political life as advisor and source of suggestions concerning our particular discipline." He concluded in words that clearly indicate the direction of the society's governance.[319]

> The position of the president within the framework of the society has naturally substantially changed with his activity. This change in the organization of the society, actually already operative and visible for some time, must someday also be given formal expression. Since the by-laws of our society stem from the '90s of the last century, it is not remarkable that they must be seen as antiquated in some points. In this way, it is understandable that on the agenda of the business meeting of the society

[314] *JDMV*: see "Mitteilungen an die Mitglieder," appended to Abteilung 2, for vols. 51 (1941) and 52 (1942).

[315] *JDMV* 47 (1937): Abteilung 2, pp. 76–79.

[316] "Deutsche Ostmark," i.e., Austria.

[317] *JDMV* 48 (1938): Abteilung 2, pp. 80–83. These "suggested changes" were announced in a *Mitteilung* sent to all members.

[318] *JDMV* 49 (1939): Abteilung 2, pp. 76–77.

[319] Ibid.: 77–78.

> during the planned gathering in Marienbad that has now failed to take place, one point was "the empowerment of the president and the executive board to make a new version of the by-laws." The sense of this empowerment and the intention that is connected with it is plain from the activity and attitude of the president and the governing council in recent years. The new version of our by-laws will have no other goals than to meet the determinations that have become necessary for the activity and position of the president and the executive board and according to which the society for ideal and practical reasons already has been and will be led.

While this sounds considerably closer to a *Führerprinzip*, it also corresponded to a political reality. On November 15, 1938, the ministry of education had issued the so-called academy decree (*Akademieerlass*), which, among other things, decreed the *Führerprinzip* for scientific academies, and which applied to organizations like the DMV.[320]

Ordinarily, on September 30, 1940, Süss's term as a member of the governing council of the society would have expired, and with it his statutory availability as president. Apparently, there was no annual meeting for 1940 as well, and Süss continued as president. However, the governing council did meet in March 1941. Conrad Müller wished to resign as secretary of the society effective April 1; the council appointed Süss, who had been continuing president, as secretary. This brought Süss into the executive board and thus still part of the official governance of the society, though his term as council member (and as president) had technically expired. This, as well as the appropriate changes in the council (Conrad Müller became one of the new members), was approved by the 1941 business meeting. Also approved was item (c) of the agenda, which was empowerment of the executive board to create new by-laws. Süss, now also secretary, was reelected president in the customary fashion.[321] Thus the society, while maintaining its parliamentary procedures, gradually and perforce accommodated itself to effective dictatorship of its affairs. The usual rota in the governing council took place, at least for 1942–43,[322] but there seem to have been no further standard annual meetings during the Nazi period.

There was, however, a meeting with the usual sorts of lectures on mathematical research held September 6–10, 1943, in Würzburg under the sponsorship of the RFR. There were about 140 participants.[323] The meeting contained an open public session and a closed session, where mathematics presumably directly related to the war effort was handled confidentially. Some people spoke at both sessions (e.g., Ernst Mohr and Martin Eichler).[324] This confidential session had earlier, on August 4, been cleared by Süss with Ludwig Prandtl, then president of the air force research organization, provided it was limited to people

[320] Cf. Schappacher and Kneser 1990: 70.

[321] *JDMV* 51 (1941): Abteilung 2, pp. 63–66.

[322] Ibid.: "Mitteilungen an die Mitglieder," following p. 76 of Abteilung 2.

[323] BAK R73 12976, Süss to Fischer, Sept. 17, 1943.

[324] Though Martin Eichler's name is well known for his "pure" work in algebraic number theory, at the time he was employed at the rocket research facility in Peenemünde.

who through their "other efforts for military research are trusted with the obligation to keep secrets."[325] Nevertheless, ninety minutes before the conference was to begin with a reception on Monday evening, September 6, and despite Prandtl's apparent explicit approval of the individual secret talks, Seewaldt at the air force research directorate communicated to Süss that ten proposed talks would be prohibited, under threat of criminal punishment should the prohibition not be respected. The ostensible reason was that the manuscripts had not been reviewed in advance and secrets of air force research might be revealed. Süss did not have friends for nothing—he immediately phoned Prandtl, who reaffirmed his approval and that the ban was "erroneous," communicating the same to his executive secretary, Seewaldt. In the end, eight of the ten forbidden actually held their confidential talks. Süss commented acidly that Gustav Doetsch was probably behind the "unfriendliness," which he bypassed through appealing to higher authority.[326]

Despite the pressures for industrially and militarily applied mathematics as a vindication of the importance of mathematics or even its right to existence, the man the DMV chose to guide it through most of the Hitler years was not an applied mathematician. Nor was he a seemingly ardent Nazi fellow-traveler, like the applied mathematician Hamel, nor a military-minded pedagogue, like Lietzmann. Wilhelm Süss was a geometer.[327] At the time he was an Ordinarius at Freiburg, having succeeded Alfred Loewy, who was Jewish. Loewy died as the consequence of an operation in 1935; his widow committed suicide in 1940 when about to be deported.

As seen above, shortly after Süss's arrival at Freiburg, he and Doetsch joined together in a conservative nationalist effort to "rectify" the attitudes of the DMV as personified, in their view, by Blaschke. Doetsch was three years older than Süss, but both had served in World War I and been decorated; both had "habilitated" at about the same time. They would seem to have naturally shared common conservative veterans' values, to have wanted to uphold a denigrated Germany to its rightful place in the community of nations, especially when seen from the perspective of their own profession. A significant number of prominent colleagues shared their views. Thus, for them, Blaschke's angry and thoughtless attempt to recriminate against Bieberbach's delay had to be rejected, and he must be made to suffer the consequences. Hitler's "new Germany" at its outset held forth to many the promise of a new and justifiable respect for Germany and Germans. Yet Süss and his friend Kneser were of a different temperament from Doetsch. Doetsch seems to have been quarrelsome and self-aggrandizing, for whatever reasons. Süss, though he guided the DMV from 1937 onward and was Rektor at Freiburg from 1940 to 1945, seems to have had "greatness thrust upon him." Unlike Malvolio, he seems to have been up to the

[325] BAK R73 12976, Prandtl to Forschungsführung, Sept. 7, 1943.

[326] For the above, see BAK R73 12976, Süss to Fischer, Sept. 17, 1943, Forschungsführung to Süss, Sept. 1943 (hand-delivered), Prandtl to Forschungsführung, Sept. 7, 1943. This BAK file contains a complete program of both the public and the confidential sessions.

[327] For more details about Süss and his career, see below, chapter 8.

task. It is unclear what caused the break between Süss and Doetsch, formerly politically close collaborators. It may have been nothing more than Doetsch's self-conversion into the "110% Nazi" described by Behnke, coupled with Süss's own growing disenchantment with ideological politics and his determination to obtain the most possible for his discipline in perilous times.[328] Doetsch followed the route of apparent adulatory devotion to the present state, whatever his earlier politics might have been. Süss, on the other hand, seems to have become *engagé* in the best sense, doing what he could to modify and mollify ideological dictates and attitudes in the interest of the personal lives of mathematicians and of mathematics, while never losing credibility with his political overseers.

An outstanding example of this outspokenness on the part of Süss is the speech he gave in the middle of the war to a conference of university Rektors as their president. This assembly met in Salzburg, and at the time Süss spoke, on August 26, 1943, the war was not going well for the Germans, though it was also not clearly lost.[329] Süss's speech is worth extensive discussion, both because of what it reveals about Süss and because of its seemingly extraordinary content. Though he is speaking as the Rektor at Freiburg to his fellow Rektors, and not as a mathematician, the themes he presents are those that he brought to his leadership of mathematics as well. It also shows his penchant for speaking plainly. As he said in its conclusion, his task was "to give a blunt picture of contemporary science and our universities."

As to research, Süss first made some remarks about how "since the first World War" there had been a relatively greater reduction in the importance of German science relative to that of foreign research. Also,

> since the first World War, sort of in the American fashion, a series of half-ripe sciences have acquired prominence among us, which, for the German university shaken out of its quiet development by the breakthrough [of the Nazi state], often represent rather more ballast and therefore a danger rather than a gain.

Such remarks blaming the "Weimar system" for a decline in German learning, and attacking disciplines like sociology, were certainly politically unexceptionable,[330] but when Süss turned to "the natural sciences, technical sciences, and medicine," which have a "great practical importance" in wartime, matters were somewhat different.

> One can here speak of a scientific war potential of both sides in the fighting, since there are certain standards of comparison that, as I will substantiate, unfortunately provide a not-pleasing picture for us. In passing, let it be mentioned that naturally, with the scientific emigrants into enemy foreign countries, we have delivered to the opposition a not-insignificant potential gain.

[328] But see below, chapter 8, "Ernst Zermelo."

[329] Mussolini was deposed and arrested on July 25, 1943. The famous tank battle at Kursk had ended to German detriment around July 13. The devastating saturation bombing of Hamburg took place July 24–August 3, 1943. A copy of Suss's address is in HK.

[330] Though some of his auditors may have silently considered "racial science" among the "half-ripe" as well, this, of course, could not be said.

This implicit criticism of political policy is somewhat startling in a "total state" carrying on a "total war." Süss went on to use the war as justification for improving natural science and mathematics research and teaching. He commented positively on the tremendous strides of American physics in the interwar years, and the success of the American National Research Council in coordinating science in wartime and in creating a militarily useful sort of "scientific general staff." As observed in chapters 4 and 5, the German situation, with its overlapping bureaucracies and invitations to personal politics, was not adept at this sort of coordination of mathematics. The same was true for natural science in general.[331] With allowances for Süss's overvaluation of the American effort in this regard, there is still some irony in the representative of a totalitarian system praising the organizational capacity of a republican one. Süss exhorted his university colleagues (emphases in original):

> *We must exploit the scientific potential of Germany in an optimal manner and engage it completely and fully for the direction of the war.* . . . In many respects we could *utilize the available forces more rationally* than has occurred up until now . . . the achievement of our German *university institutes* in the past is based on [the fame and preeminence of German natural science in the world] . . . [to maximize their potential] an *immediate palpable increase of* the staff *of coworkers* and the number *of assistants (Hilfskräfte), so that the equipment of the institutes will be used fully for research* . . . [necessary] will be beginning immediately with the education of the half-finished, and the training of a number of young coworkers. For we have to set our minds on a long-lasting war and to calculate therewith that *the longer the war lasts, all the greater can be the weight of scientific discoveries for its outcome.* . . . *The army must be able to survive the loss of (Ausfall . . . verschmerzen) several hundred researchers and two to three thousand assistants (Hilfskräfte) from its ranks. The scientific direction of the war, however, has a pressing need for these men. Possibly they will decide the war.* . . . The improvement of institutes, building them up in personnel and materially, ought not to be left to chance, to the favorable disposition or changing understanding of the most heterogeneous administrative offices of the state or the party. . . . [If the professional experts don't get everything they think is necessary] for the areas of military research,[332] then the German people will have to suffer the consequences (*büssen*)!

Having dealt with the fundamentals of the organization of German science and the political attitude toward academe in this sharply critical way, as well as demanding increased support for academic research in the name of the German people, Süss turned to some of the immediate needs of research. His criticism of state and party was no less sharp.

> Often the total work production in a discipline [to educate just medical students in science] depends more or less on one man or on two. . . . *One cannot repeat enough the warning to give immediately to these university teachers* [subjected to pressures to inten-

[331] Cf. Zierold 1968: passim.

[332] For Süss's attitude toward military research, see above, chapter 4, "The Süss Book Project."

sify instruction] *coworkers and assistants* (*Hilfskräfte*). Many stand in danger of ruining their health, and several have already broken down.

For another example of an antirational division of forces, I would like to refer to [the following]: In many new research institutes and research positions, for example in the army, young people are in leading positions. For their task they have their whole time and rich financing at their disposal. For this one could certainly rejoice, if one did not know that often their experience and their capability is not comparable with that of their academic teachers who remained at the universities and there will have their powers consumed by instruction and other tasks alien to research. . . . All in all, I speak plainly of a *winning back of the equilibrium between research and teaching lost in manifold ways*, and for some disciplines indeed a *movement of the center of gravity of the university institute in the direction of research.*

While deploring the fact that university research had been undermined, Süss saw it as necessary as well for the training of the next generation of scientists and the successful prosecution of the war: "The exhaustion of these numerically too small forces in only instruction is not responsible to the future of German science; in the militarily important disciplines, however, it is completely unbearable, and in the long run, simply suicide."

Süss pleaded for more positions for *Assistenten* "so that alternate occupation of the *Assistenten* in an institute in research and teaching will be possible." However, he also was not afraid to complain about political interference.

My observation is that also today, time and again, there are young students (*Akademiker*) at the universities, who are possessed by their work. They have, however, become less frequent, perhaps also they venture less gladly into the light of day than previously; some of them might also therefore find no avenue to the career of university teacher, because they deviate too much from the ideal (*Wunschbild*) that here and there political judges have set up from a standpoint often far distant from science. Gifted natures often are frightened away from the university exactly by this. Here a fundamental alleviation is first to be expected, if for once we can arrive at the standpoint that political issues go without saying (*sich das Politische von selbst versteht*). . . . Indeed, imagine for once a corresponding procedure for character judgment and selection, as here and there was used in the universities in the years of the breakthrough [of the Nazi state], in our army! How often in the discussion of political judgment is the disciplinary achievement of the affected person completely overlooked!

Süss admitted that things did seem to have gotten better as "youthful elements" incapable of making the sorts of balanced judgments necessary had lost influence.[333] However, much needed to be redressed:

One-sided or unjustified judgments are to be corrected in the political personal files. . . . The unlucky persons affected often have no knowledge at all what has been chalked up against them, and so are not able to defend themselves against it.

[333] This seems to have been basically true.

Individual rectifications were not enough; attitudes inimical toward academic learning needed to be fought:

> At the same time, we should more energetically (*nach Kräften*) again separate from our own ranks gravediggers of science, insofar as they are found among us or have snuck in among us or attempt to do so, and indeed without any tolerance or softness. Herewith belongs also proceeding against people who dare with venal formulations and suspicions to attack our truly great and unique scientific personalities and to befoul their life's work.[334] They act in their zeal nourished by ambition, hate, envy, or inferiority complexes, without noting what damage they thereby inflict on the German people both outwardly and inwardly.

Süss not only thought that the material and social status of university professors needed to be elevated so that theirs was a sought-after occupation, but even went so far as to say that "a scholar of decisive importance [like the mathematician David Hilbert] should be completely, publicly, put on a level with deserving leaders of the army or other leading men." He further deplored that supplemental food rations were in general not available for "intellectual workers," however deserving the case. Moreover, the elevation of the reputation of the academic was essential for the health of Germany,

> not only in decisive places in state and party, but the conception must be anchored widely among the people that intellectual work in all areas of our cultural life is first-rate work for the benefit of our people, and that in particular without German science, also Germany cannot live.

As for students, they should study and not be drawn off to irrelevancies:

> However, apart from service during catastrophes and the continual needs for air-raid protection, I consider that the best service of students lies in carrying on study that *in itself can be valued as a militarily important* employment. . . .
>
> In the interest of the reputation and the future of the university, we must promote *also for studies a more academic attitude in scientific work.*

Süss's speech would have been startling any time after the Nazi consolidation of power; that it happened at mid-war, when the struggle was beginning to turn decisively, as it proved, against the Germans, is even more shocking. Paradoxically, it would seem that the war situation (which made "defeatism" a capital crime) also enabled Süss to attack various policies and situations as deleterious to the war effort, whereas without the war, he would perhaps not have gotten a hearing.

Wilhelm Süss was certainly a German nationalist, and like most human beings, no doubt enjoyed the perquisites of power and influence when they came to him. But his interests always seem to have been primarily to benefit the universities, mathematics, and mathematicians. He tended to avoid the sort of opportunistic self-aggrandizement to which men like his colleague, Gustav Doetsch, fell prey; nor was he the reactionary ideologue his teacher, Bieberbach, became. Effectively the *Führer* of German mathematics from 1938 onward, he

[334] No doubt Max Steck and his attack on Hilbert is one incident Süss has in mind here.

used that position in the interests of his colleagues as best he could. While this inevitably involved compromises with the regime, these also allowed him a position of respect where, in his widow's words, "He would tell the Nazi big shots what they were doing wrong and they were so nonplussed (*verblüfft*), they would listen to him."[335] This in fact has the ring of truth, provided it is borne in mind that this was possible only if the ideological fundamentals of the regime were not transgressed. It would have been in vain and only have brought suspicion on one to defend Jews, but it was possible to save a distinguished "Aryan" mathematician who was "Jewish-related" (Erich Kamke) from being sent to a labor camp, because of the importance of his scientific activity. However, Süss was only enabled to behave in this way because he had established a reputation of being, in Nazi jargon, *politisch zuverlässig*, or politically reliable.

The Creation of the Oberwolfach Institute

One of Wilhelm Süss's greatest achievements on behalf of mathematics was the creation of the Oberwolfach Mathematical Institute. He is remembered primarily today not for his mathematics, or even for his guidance of the German Mathematical Society in perilous times, but for this accomplishment, perhaps even more remarkable for being achieved in the last years of the war.

According to his widow, Süss for some time had had the idea of creating a central institute for mathematics to avoid duplication of labor on repetitive discoveries. Naturally enough, he initially thought of Göttingen as the appropriate place for its establishment.[336]

On March 24, 1944, Helmut Hasse wrote Süss suggesting he leave Freiburg and take a position in Göttingen.[337] At the time, not only was Süss president of the German Mathematical Society and Rektor at Freiburg, and had recently addressed the annual Rektors' conference as its president, but he was also the leader of the mathematical working group (under the division of physics) in the RFR.[338] In short, he was as mathematically and politically important an academic as could be imagined.

Needless to say, both Freiburg and the local Baden government were reluctant to lose such a man, especially in the seemingly ever more perilous days of 1944. Nor was Süss reluctant to use a little false modesty as appropriate in promoting the idea of a mathematics institute. Thus he wrote Fischer at the RFR "confidentially" that in the moment when a general ministerial decree might free him from the burdens of being Rektor, Göttingen was talking about a pro-

[335] Interview, Irmgard Süss, Mar. 25, 1988.

[336] Irmgard Süss, *Beginnings of the Mathematical Research Institute Oberwolfach at the Country House "Lorenzenhof,"* trans. P. L. Butzer, Mathematisches Institut Oberwolfach (1967), 5–6. The original German edition is out of print. Much of this material also appears in *General Inequalities* 2 (Basel: Birkhäuser Verlag, 1980), 3–13 and *General Inequalities* 3 (Basel: Birkhäuser Verlag, 1983), 3–19.

[337] Ibid.: 47–50, reproduction of a handwritten letter from Hasse to Wilhelm Süss on March 24, 1944.

[338] Abraham Esau to Süss, June 4, 1943, reproduced in Süss 1967: 36.

fessorial appointment "which for me on the one hand would simplify the building of a national institute (if only I were the right man for it!); which, however, on the other, a positive decision [to accept] could rip apart my heart, which is strongly rooted here."[339]

According to citations by Irmgard Süss, Rudolf Mentzel (the head of the RFR) thought that war considerations were most prominent, and so did not want any university town as the site, as he thought such areas were liable to bombing.[340] "The Ministry of Education of Baden, wanting to keep me in Freiburg at least in the present difficult situation, has offered me a place of rare advantages in the Black Forest where I hope to start the most urgent work without delay and undisturbed." Therefore, Mentzel suggested the appointment to Göttingen be delayed and that "I ought to arrange for the foundation of the institute in the Black Forest place."

The Dekan at Göttingen responded that Freiburg and Baden would try to keep such an institute once the bombing had stopped; then both institute and Süss would be lost to Göttingen. Süss replied,[341]

> I unite the consciousness of the high honour and strong obligation of a call to Göttingen with the conviction that to conserve Göttingen as the internationally accepted stronghold has to be the aim of us mathematicians . . . the offer [of a "beautiful place" for the "temporary accommodation" of the planned institute] is explicitly made to me quite personally, not to the mathematical chair at Freiburg. There is not hidden behind this any ambition to conquer the Institute for Baden . . . it means the winning for science of a place favorable regarding air raids and quiet for work.

In the meantime, the *Oberbürgermeister* of Freiburg told the Baden education ministry that Freiburg was well aware of Süss's importance for the local university and that he had been granted the reservation for future purchase of a "splendidly situated" parcel of land. "Also, the necessary steps have already been initiated for an improvement of his present living conditions." Even in Nazi Germany, an appropriate academic offer could work the same magic it does in contemporary American universities.[342]

Süss had suggested the establishment of a central institute for mathematics, or one for applied mathematics, or a mathematical-technical institute (presumably this meant for work on calculational devices) many times to Mentzel.[343] However, the creation of such an institute was not authorized until August 3, 1944, when Mentzel told Süss that he could begin preparations for the erection of a "national institute for mathematics" and that Göring's official appointment

[339] BAK R73 12976, Süss to Fischer, June 14, 1944.

[340] The irony is that Göttingen was undamaged by the war.

[341] Süss 1967: 6–7, letters from Süss to Hans Kopferman, July 8, 1944 and Aug. 26, 1944, Kopferman to Süss, Aug. 1, 1944. Translation is as given in Süss. For more about the attitudes of Kopferman, who was a physicist, see Becker et al. 1987: 393, 395; Walker 1989: 53, 132; and Beyerchen 1977: 177, 220 n. 11, 228 n. 46.

[342] Letter of *Oberbürgermeister*, Aug. 12, 1944, in Personalakten Süss, Freiburg. The "splendidly situated" tract refers to the site of the Lorenzenhof (Oberwolfach). Earlier Süss had turned down an offer of a chair at Munich (Personalakten Süss, Freiburg, letter from Süss to Rektor, Feb. 22, 1943).

[343] BAK R73 12976, e.g., Süss to Mentzel, Sept. 19, 1942.

of him to lead such an institute would follow "as soon as . . . the necessary means and forces were available."[344] On August 29, Mentzel informed Süss that the means would be provided for the most necessary calculating machines and mathematical apparatus, and the mechanisms for transferring personnel to his institute. In particular, for civil-service faculty, it was only necessary to provide their names to the education ministry. Also, Süss's notion of making other mathematical libraries safe at the "Lorenzenhof" (which was to be the institute site) was affirmed, but he would have to carry out such negotiations personally.[345]

Throughout such negotiations, Süss had no independent standing as a division leader in the RFR. In fact, only on June 4, 1943, did Abraham Esau, then leader of the physics division, write Süss, "It seems to me necessary to found a mathematics working group [within the physics division] that comprehends the pure as well as the applied side," and ask Süss to be its leader, especially since "you have already successfully variously supported us with word and deed."[346] On November 3, 1944, Werner Osenberg[347] wrote Süss saying he had already a month previously suggested to Göring the elevation of Süss to a division leader for mathematics in the RFR, but this had not yet taken place for "administrative reasons"; nevertheless, Osenberg asked him to act as though so commissioned and sent him copies of relevant memoranda.[348] The letter contains the sentiment that "I hope that your rich disciplinary knowledge and your known activity will be successful in leading the research institutes under your supervision to as decisive as possible a contribution to victory in this war."[349] Göring's official appointment of Suss did come through on January 3, 1945 (at a time when the German offensive in the Ardennes had been halted and the Russian crossing of the Vistula was imminent). Süss was asked to take up his activity "as quickly as possible," a statement that in the eventuality is not without unintended irony.[350]

What was this "Black Forest location" or "Lorenzenhof" that so late in the war (and the history of the Third Reich) was to become an official national mathematics institute? Before World War I, a Baron Stoesser had built and landscaped a Black Forest hunting lodge with appurtenant buildings for gardening and pig-keeping, as well as a stable and a coach house. This was the "Lorenzenhof" at Oberwolfach. Actually, "Lorenzenhof" was originally the name of a farm in the valley below. In 1928, most of this property, including all the buildings, was bought by a Belgian banker, who also cleared some of the land. A Nazi law of 1936 forbade foreigners to have shooting licenses within fifty kilometers (about thirty miles) of the German border. Oberwolfach is actually

[344] Mentzel, Aug. 3, 1944, reproduced in Süss 1967: 50.

[345] Mentzel to Süss, Aug. 29, 1944, reproduced in ibid.: 52.

[346] Esau to Süss, June 4, 1943, reproduced in ibid.: 36.

[347] For the confused overlapping of Nazi bureaucracies, especially in the late war years, see above, chapter 5, "The Wartime Drafting of Scientists," and Zierold 1968.

[348] Reproduced in Süss 1967: 37–38.

[349] Ibid.: 38. While it may have been necessary for Osenberg to be officially hopeful (or he may actually have been so), by November 1944 many Germans realized that the war was lost. An open question, though, was how badly lost?

[350] Reproduced in ibid.: 39.

about forty-four kilometers from Strassbourg (then and now French). So the Belgian businessman sold the property to a Black Forest lumber merchant. In 1942, the Baden education ministry considered buying the place, possibly as an adjunct to the university in Freiburg. Süss, as Rektor at Freiburg, apparently supported such a purchase; however, rather than coming to the university, the building became adjunct to a teachers' training college at nearby Bad Rippoldsau (about fifteen kilometers, or nine miles, away). In fact, it became a training camp designed to imbue Alsatian teachers with National Socialist ideals—originally for men, it later became a women's training camp. With the Baden ministry anxious to keep Süss at Freiburg, they went looking at various places, including one on a nearby hillside, and a dilapidated convent near Birnau on Lake Konstanz. The idea of the Lorenzenhof then arose.

Süss was influential enough, and in the waning years of the war, science, even mathematics, had acquired enough stature as "militarily important" despite the earlier, somewhat negative party-political stance toward it, that the Alsatian women were dispersed to the benefit of the mathematics institute. However, this did not happen without a denunciation by their supervisor, which failed to find resonance.[351] Access to the former hunting lodge often involved a two-day hike or a sixty-five-kilometer (about thirty-nine-mile) bicycle ride. The initial mathematical library was that of the university at Strassbourg (which had been entered by the Allies on November 23, 1944).[352] Among the first mathematicians to find refuge at Oberwolfach, either from bombs and advancing enemy armies, or from political threats, or from both, were Emanuel Sperner, William Threlfall, Herbert Seifert, Gerrit Bol, Wilhelm Maak, Heinrich Behnke, Hermann Boerner, and Theodor Schneider. Threlfall (who had had an English father)[353] had already been rescued by Süss from the consequences of a denunciation caused by some dinner-table remarks in Frankfurt, and was at an aeronautical institute in Braunschweig. On the other hand, Sperner, Boerner, Maak, Bol, and Schneider were part of the "Osenberg action."[354]

With the war's end, the Allies attempted to collect and interrogate all German scientists having anything to do with rocketry or atomic research. Among the men sought was a rocketry expert named Hellmuth Walter. By mistake, the British in fact brought the Darmstadt mathematician Alwin Walther, an expert on the then-extant computing devices,[355] to London. They assigned, among others, the mathematician John Todd and his wife (the well-known mathematician Olga Taussky, who had left Germany in 1932)[356] to interrogate him. Todd

[351] Ibid.: 1–3 and 8–10.

[352] Ibid.: 13, 53.

[353] HNMA, Behnke to Hecke, Jan. 15, 1940, speaks of Threlfall's "English descent." From the British point of view, Threlfall had remained a British citizen; see John Todd, "Oberwolfach 1945," in *General Inequalities* 3 (1983), 21.

[354] In Süss 1967: 5, 13–14. See also chapter 5, "The Wartime Drafting of Scientists."

[355] For Walther (as opposed to Walter), see above, chapter 4, "The Süss Book Project," and below, "Applied Mathematics in Nazi Germany."

[356] An *Assistent* in Göttingen, following a holiday, she did not return there in the autumn of 1932 because of a warning from Courant about the political situation. See Pinl 1971: 184.

was in fact a computational mathematician like Walther, having set up an Admiralty Computing Service within the British Admiralty. From Walther, who was returned to Darmstadt, Todd learned of Oberwolfach. Together with the well-known mathematician G.E.H. Reuter[357] and others, Todd conceived the idea of an intelligence mission to investigate mathematics in Germany, particularly Oberwolfach. "We knew our way around Whitehall and before the week was out, we were officers in the Royal Navy Volunteer Reserve." Equipped with open orders and maps of Germany, after a number of adventures and a month on the road, Reuter and Todd arrived in early July 1945 at Oberwolfach. "For safety, we posted notices on the main entrance to the effect that the building was under the protection of the British Navy," and later Todd managed with the authority of a British naval officer to fend off a troop of French Moroccan soldiers, telling them "to leave the mathematicians and *même les poules* undisturbed." Oberwolfach ended in the French zone of occupation and, briefed by Todd, the well-known French mathematician Szolem Mandelbrojt paid a visit there and was able to legitimize the protection of the institute initially provided by Todd and Reuter. Since then, Oberwolfach has become one of the most important international mathematical centers, as originally envisioned by Süss.[358] Thus the French and British reaction to German mathematics (and science) following World War II was directly opposite to the reaction in the aftermath of World War I. Rather than boycotting German mathematics, they were interested in supporting German mathematicians. While gossip in the mathematics community makes Todd's appearance at Oberwolfach serendipitous, it was in fact intentional.

There is no room here for the many interesting stories of Oberwolfach's early days told by Irmgard Süss, but the institute's beginnings point out that bureaucratic business as usual went on even to the last days of the war and Reich with the authorization and expense of substantial sums of money for academic activities. Of course, these activities had to be "militarily important," but one has the impression that as the war went worse and worse for Germany, more and more activity became accepted by the educational bureaucracies as "militarily important," and someone as gifted as Süss at bureaucratic manipulation was continually able to obtain funds for his projects, whether the translation and composition of mathematical books, or the mathematics institute at Oberwolfach.

Though many Germans, including mathematicians, knew that Germany was irrevocably collapsing in late 1944 and 1945, and Süss was among them,[359] the bureaucracy behaved bureaucratically, simply carrying on in face of increased difficulties—the "your letter, alas, took five weeks to reach me" sort of thing—as though nothing were different. There are perhaps several reasons for this. One is simply that there may have been no choice: defeatism was a capital offense. Another is that no one knew what the termination of the war would

[357] His father was Ernst Reuter, who, after the war, became the legendary mayor of Berlin at the time of the "Berlin Blockade." Reuter had been a teenager when his family fled Germany because of his father's prominence as a Socialist party politician.

[358] Todd 1983. Cf. also I Süss 1967: 27–28.

[359] According to Süss 1967: 9.

bring—World War I, after all, had ended with German armies still on French soil. The *Wehrmacht* had been resisting strongly, and the *total* collapse and ensuing occupation were not part of the German vision. Then there were romantics, as was perhaps Max Steck, who believed in the Nazi message and that the war would turn around even at the last moment—that the *Führer* would produce *Wunderwaffen*. Indeed, the Ardennes offensive ("Battle of the Bulge") took the Allies by surprise in late 1944, V-2 rockets did rain on London, and the Germans might well have achieved an atomic weapon but for a number of miscalculations, including the belief that development could not come rapidly enough to affect the war.[360] With the bureaucracy and others carrying on as usual, figures like Walther and Süss, of course, did likewise, especially if they could obtain the funds necessary for projects beneficial to the mathematics community by making them *kriegswichtig*, or militarily important. Süss's book project, an improved differential analyzer (one of Walther's interests), the establishment of a mathematics institute, enhanced computational methods, are all examples of activities that, while declarable as "militarily important" in the context, were foremost benefits to the German and (certainly in Süss's vision of the institute) eventually the international mathematical community. Max von Laue's remark that "sometimes too the possibility arose of protecting political suspects from concentration camps or worse, by assigning them research work of more or less 'military importance',"[361] had application to other disciplines as well as physics and went far deeper than just saving lives.

Applied Mathematics in Nazi Germany

Süss's personal relationships with other mathematicians will be considered further in chapter 8, but mention again of Alwin Walther brings us to the applied mathematics community in Nazi Germany. One would have thought that in a militarizing and then warring Germany, appropriate applied mathematics, at least, might have been nurtured. But this did not happen until late in the war, when Süss took advantage of the situation.

An academic-military-industrial-political complex was forged during World War II in Great Britain and the United States. No such complex was built in Germany, however, despite some advantages: most German academics were state civil servants, and the state had an example of early academic-industrial cooperation involving mathematics at the Zeiss Werke in Jena.[362] This failure probably stemmed from the deep suspicion of academe held by the unify-

[360] See Walker 1989: passim.

[361] Max von Laue, "The Wartime Activities of German Scientists," *Bulletin of the Atomic Scientists* 4 (1948): 103.

[362] See above, chapter 4, "The Winkelmann Succession," for an example of how, despite this, the Zeiss Werke had relatively little influence in an issue of academic appointment at Jena during the war. As noted there, this had already proved true previously under Weimar in 1930. Jena was in Thuringia, and by 1930 Nazis had entered the Thuringian government.

ing political forces,[363] a suspicion that, as many studies have shown, was basically unnecessary. Not until Joseph Goebbels's speech of July 9, 1943—"The intellectual worker in the battle for the fate (*Schicksalskampf*) of the nation (*Reiches*)"—was a positive reevaluation of the role of academics publicly declared by the Nazi state.[364] Süss's address to the assembled Rektors cited above also had this theme of mobilizing academia in the war effort. Of course, aeronautics was an exception, both because of its obvious military importance and because of Hermann Göring's prominence in the Nazi hierarchy.

In 1904, the very first Ordinarius for applied mathematics in Germany had been created at Felix Klein's instance, and filled by Carle Runge. In 1904, also with Klein's influence, the twenty-nine-year-old Ludwig Prandtl became associate professor for technical physics at Göttingen. This was the same year in which he had made fundamental discoveries about fluid flow. By 1907, he was Ordinarius for applied mechanics at Göttingen; in 1908, he built the first Göttingen wind tunnel, and by 1909, he also was giving lectures on aerodynamics, the first such academic position in Germany. Prandtl and his most famous student, Theodor von Kármán, who in 1913 became (at age thirty-two) the Ordinarius at the newly founded Aerodynamical Institute at the technical university in Aachen, were two of the world's leading early aeronautical experts. Von Kármán, whose mother was a Hungarian Jew, was forced into emigration, which was in his case serendipitously easy.[365] During World War II, he was chief scientific advisor to the U.S. Army Air Force, later playing similar roles for the postwar U.S. Air Force and NATO. He is the only mathematician to be honored with a U.S. postage stamp (in 1992). Indeed, von Kármán was already so highly regarded by the Nazi hierarchy that on November 16, 1933, the air ministry made a request to keep the internationally recognized scientist in the Prussian civil service—the education ministry refused. However, Bernhard Rust's letter of April 21, 1934, expelling von Kármán from his position at Aachen effective

[363] See, for example, Gerd Rühle, *Das Dritte Reich, 1934* (1935), 221–222, for the initial Nazi attitude toward universities. I. Süss (1967: 3) tells an anecdote of a conversation between the education minister, Bernhard Rust, and Süss in 1942 in which Rust told Süss that in the early Nazi days, it was necessary for him to bring the professors into line, because otherwise "there was such a storm of hatred raging through Germany against the intellectuals, universities would have been simply swept away."

[364] Number 20 in Helmut Heiber's edition of Goebbels, *Reden*, vol. 2 (1972), 240–257. The speech was given at the University of Heidelberg.

[365] In October 1929, von Kármán accepted the offer of Robert Millikan to direct the newly funded Guggenheim Aeronautical Laboratory at the California Institute of Technology (for which he had been a consultant). He moved to the United States in December; however, he did not sever his ties with Aachen, but took a leave of absence and returned to Germany in both 1931 and 1932. After January 30, 1933, he merely remained in California. See Theodor von Kármán, *The Wind and Beyond* (1967), 124–125, 145–146; and Ulrich Kalkmann, "Die Vertreibung Aachener Hochschullehrer durch die Nationalsozialisten" (1991). See also folder 447 in the archive of the Technische Hochschule Aachen for details of von Kármán's attempt to extend his leave and his offer of resignation should that not be possible. The accounts of Kalkmann and von Kármán differ somewhat in nuance. As in all such cases, the documentary evidence of the historian must have precedence over the memory of the protagonist.

the preceding April 1, suggested he could find a position in the air ministry, and presumably this was with Göring's approval or even at his insistence.[366] Von Kármán refused the offer. If such were the case for the internationally famous von Kármán, one can imagine how hopeless the situation was for one of his best students, the important aerodynamicist Ludwig Hopf, who was Jewish.[367]

Prandtl, however, was *echt Deutsch*, having been born to a Lutheran family in Freising in 1875. He was also a dedicated scientist. The GAMM, or *Gesellschaft für Angewandte Mathematik und Mechanik* (Society for Applied Mathematics and Mechanics), was one of Prandtl's brainchildren; contrary to the usual order in such matters, it was preceded by its journal.[368] As a result of agitation within the "Union of German Engineers" (which had existed since 1856) to be better informed about applications of mathematics and theoretical mechanics, the *Journal of Applied Mathematics and Mechanics* (*ZAMM*) was founded in 1921 with Richard von Mises as editor; von Mises only the preceding year had become Ordinarius for applied mathematics at Berlin, a position for which Erhard Schmidt had been agitating for two years.[369] The object of the journal was perhaps put most succinctly in a letter of Felix Klein: "*The goal of theoretical natural science should be not only a passive understanding, but also an active domination of Nature.*" Though, as Klein remarked, this was not to be in contradiction to "ethical requirements."[370]

Prandtl, von Mises, and Hans Reissner (who was von Kármán's predecessor at Aachen and in 1913 had become Ordinarius for theoretical mechanics at the technical university in Berlin) had discussed among themselves the founding of a Society for Applied Mathematics. On September 21, 1922, effectively such a society was founded, with Prandtl as its president, von Mises as its secretary, and Reissner as one of the other members of the governing council. A year later, Reissner was vice president as well. Thus, while in mathematics there was and had been continuing competition between Göttingen and Berlin,[371] in "applied mathematics" (defined as theoretical mechanics and theoretical engineering science) cooperation reigned, with Prandtl at Göttingen, von Mises at the univer-

[366] Kalkmann 1991 and von Kármán 1967: 146. See previous note.

[367] According to von Kármán, Hopf was one of those academics, like Eduard Spranger (above, chapter 3), who had sympathy for the "national feeling" of the students in 1929 when they protested giving honorary degrees to members of what had been enemy states during World War I. See von Kármán 1967: 144.

[368] For the history of the GAMM, the following sources have been used for material not otherwise annotated. Helmut Gericke, "50 Jahre GAMM," *Ingenieur-Archiv* 41 (1972) (supplement); Mehrtens 1985 and 1986; Renate Tobies, "Die 'Gesellschaft für angewandte Mathematik und Mechanik' im Gefüge imperialistischer Wissenschaftsorganisation," *NTM Schriftenreihe* 19 (1982); 16–28; and Tollmien 1987. The *Akten* of the GAMM from its founding were destroyed in the Dresden air raid of February 13, 1945 (Gericke 1972: 7).

[369] Biermann 1988: 189–191.

[370] Gericke 1972: 7, emphasis in original.

[371] Partly this was a genuine rivalry between first-class mathematicians, partly it was the result of the enormous antipathy between Karl Weierstrass and later Georg Frobenius on one side and Felix Klein on the other. Klein was already at Göttingen in 1886. See Biermann 1988: 166–167 and passim.

sity in Berlin, and Reissner at the technical university in Berlin.[372] The new association intended to become recognized as a formal official scientific organization, but this continued to be put off, and in fact was never done.[373]

This proved fortunate in 1933. Up until then Prandtl, Reissner, and von Mises had been continually reelected to their respective positions. At this time, the GAMM had 444 members, of whom more than a third were foreigners; thus its membership was a little less than half the DMV's, indicating substantial growth, and substantial foreign interest, in less than eleven years.[374] In 1933, the "humanistically minded"[375] Prandtl was faced with serious difficulties—among other things, Reissner and von Mises were of Jewish descent.

On June 22, 1933, Prandtl wrote Erich Trefftz, a member of the GAMM's governing council, about the new situation. Trefftz also had an "Aachen connection," having "habilitated" there with von Kármán and remained, eventually as Ordinarius, until he moved to Dresden in 1922. He was what one might call a "concrete mathematician" who was interested in the development of numerical methods for problems coming from mechanics.[376] Prandtl wrote:[377]

> Our colleagues, von Mises and Reissner, have indicated to me that the Society for Applied Mathematics and Mechanics on the occasion of its next major meeting must become "coordinated" (*gleichgeschaltet*) and that they themselves on this occasion would wish to resign.

He suggested his own resignation as well and asked Trefftz to take over the presidency. Three days later, he wrote that von Mises remained as editor of the *ZAMM*, and that Reissner and von Mises no longer wished to work "publicly" for the GAMM so as not to bring it into difficulties. In Prandtl's view, there was no question that a scientific organization need not be "coordinated" with a National Socialist at its helm. Trefftz's reply was that the GAMM should not be endangered in this way, and Prandtl should remain president; however: "If we *must* exclude Jewish members, I would consider dissolution the worthiest (*würdigste*) action."

Against such ethical considerations, Prandtl raised the issue of the discipline, and its newly found discernment as distinct from both physics and mathematics, and that it ought not be politically involved. At the time Prandtl was fifty-eight, a man who had been raised in the strict Wilhelminian tradition of the loyal civil servant, faithful to the government that gave him his bread, even if he deplored some of its actions.[378] He was a man who believed science and politics

[372] See citations from GAMM, as in Gericke 1972: 7–10.

[373] Ibid.: 10.

[374] Mehrtens 1985: 96.

[375] The phrase is Renate Tobies's (1986: 22).

[376] Lothar Collatz, "Numerik," in *Ein Jahrhundert Mathematik, 1890–1990* (1990), 284.

[377] Mehrtens 1985: 96; all quotations below from the Prandtl-Trefftz correspondence are taken from citations in Mehrtens 1985.

[378] Cf. above, chapter 4. The varied and evolving attitudes of Süss, Hasse, Kneser, and Erhard Schmidt as described elsewhere are worth consideration in this evaluation.

did not mix, while ironically he was himself one of the most politically skillful of scientists in a politically difficult time. Thus, he stated,

> we ought allow no political moods to influence us, and therefore even if the exclusion of Jewish members should be requested, *must* further preserve the GAMM. In my opinion, this has nothing to do with considerations of what would be worthiest, since it is simply a question of a need of our discipline.

However, Prandtl not only had no desire to exclude Jewish members of the GAMM, but actively fought against such proscription. The earlier failure to register as an official formal society was used by Prandtl to circumvent official ideological pressures: the GAMM "had the sole task of holding scientific gatherings in our disciplinary area and therefore cannot count as a disciplinary society in the usual sense."[379] He also fought as possible for his Jewish or Jewish-related colleagues; for example, he was active in the attempt to prevent the forcible furloughing of Richard Courant.[380] He spoke up repeatedly for his colleague Reissner, and others.[381] As early as April 27, 1933, he wrote Wilhelm Frick, the Nazi interior minister, that "quarter-Jews" were after all "three-quarters German," attempting to prevent the dismissal of "quarter- and half-Jews."[382] In 1937, when those married to Jews were dismissed, he spoke up for a former *Assistent* who was so married.[383] All such efforts were, of course, in vain. Nevertheless, Prandtl continued the attempt to keep his disciplinary organization, the GAMM, clear of "political influence." For example, on June 15, 1938, he wrote the education ministry:[384]

> Mechanics, just as mathematics and the exact natural sciences, has not the slightest connection to politics according to its entire internal structure. Advances in these sciences rest on international cooperative work. Given the contemporary specialization, a single country no longer produces enough brains in order to do without this cooperative work. This viewpoint, which, since the breakthrough of 1933, has fallen somewhat in the background, must generally be helped to become again valid if Germany does not wish to suffer damage.

A month earlier, he had written Friedrich Willers, the editor of the *ZAMM*: "A book that represents an important advance in science must be reviewed, whether the author is Aryan or non-Aryan, and similar holds with respect to papers for publication."[385] This was written with respect to a demand from the Society of German Engineers (UDI), whose publishing house published the *ZAMM*, that no Jewish authors appear in it. Such representations had no effect;

[379] Gericke 1972: 13.

[380] Schappacher 1987: 351; Tollmien 1987: 472, 484 n. 42; Reid 1976: 145, 151–152. Courant's furloughing is discussed in chapters 15 and 16 of Reid 1976.

[381] Tobies 1986: 22, 26 n. 58.

[382] Tollmien 1987: 472–473.

[383] Ibid., where other activities of this sort by Prandtl are discussed.

[384] Gericke 1972: 414. Collatz, in *Ein Jahrhundert Mathematik* 1990: 276.

[385] Gericke 1972. Willers himself had trouble with Nazi students and officialdom; see below, chapter 8, "Wilhelm Süss."

but around the same time, Prandtl did refuse to have the GAMM join the UDI because the latter would have insisted on the exclusion of non-Aryan members.[386] Similarly, in December 1933, Trefftz made the decision that only his name would appear on the title page of the *ZAMM* as editor, "with the cooperation of well-known disciplinary colleagues," so as to avoid the demand that non-Aryan names be stricken.[387] Nevertheless, by 1940, no non-Aryan names appeared among the members of the GAMM (many of whom had not paid dues for five or more years); one of the last to be stricken was Reissner.[388] Thus, though it was true, as von Mises wrote Prandtl in 1933, that "when I draw a comparison with the procedures in other associations, I know very well to value in what worthy and high-minded fashion, here, in our society [the GAMM], the consequences of the contemporary circumstances were drawn," such decency was ultimately of little practical avail.[389]

Prandtl, though an applied mathematician interested in fluid mechanics and aerodynamics, took occasion also to help theoretical physics if it were a question of science being surpressed by politics. He took advantage of being Heinrich Himmler's table companion to speak of the unjustified attacks on Werner Heisenberg by *Deutsche Physiker*, including Johannes Stark, and wrote Himmler a detailed follow-up letter. Prandtl's letter seems to have been what moved Himmler to action, and he wrote Reinhard Heydrich, "We cannot afford to lose this man [Heisenberg], who is relatively young[390] and is able to produce students, nor to silence him (*tot zu machen*)."[391] Prandtl's letter apparently even contained the statement that Einstein was a first-class physicist and that most physicists thought relativity theory was physically correct.[392]

Even as late as 1942, Prandtl wrote Rust a strong protest about the appointment of Wilhelm Müller to succeed the famous mathematical physicist Arnold Sommerfeld at Munich. Müller was a Nazi fellow-traveler and, as a physicist, an expositor of the differences between German and Jewish physics. He had been Theodor von Kármán's successor in the chair of mechanics at Aachen. Scientifically, he was a competent mathematical expositor of classical mechanics with no interest in some of the theoretical developments that had intrigued von Kármán. Furthermore, he was no physicist, had never published in a physics journal, and was succeeding arguably the most famous theoretical physicist in Germany. Prandtl called it an act of sabotage.[393]

[386] Gericke 1972: 13–14.

[387] Ibid.: 13.

[388] Mehrtens 1985: 98; Gericke 1972: 13; cf. Schappacher and Kneser 1990: 69–71, where the similar situation in the DMV is discussed.

[389] Von Mises to Prandtl, Oct. 2, 1933, cited in Gericke 1972: 12; Collatz, in *Ein Jahrhundert Mathematik* 1990: 276.

[390] Heisenberg was thirty-seven.

[391] Tollmein 1987: 475–476; Beyerchen 1977: 162–163, for Prandtl. "The Aryan Physics Political Campaign" is chapter 8 of this last.

[392] Beyerchen 1977.

[393] Tollmien 1987: 476; Beyerchen 1977: 186. See also above, chapter 5, "The Winkelmann Succession," note 84.

Thus Prandtl was always unimpeachable when it came to issues of science versus politics; for example, Süss, in the controversy with Doetsch over mathematical publications, could have faith that Prandtl would see things aright.[394] A nationalist-conservative like many of his academic colleagues, he publicly supported the DNVP even in the elections of March 1933. Unlike Hasse, also a DNVP supporter prior to 1933, he never tried to become a member of the Nazi party, despite his tremendous importance as the German leader in aerodynamical research, which led to substantial expansion of his facilities at Göttingen. His prestige and importance seemed to obviate the need for political credentials, and he was clearly not anxious to provide them. Like Süss, he was "socially acceptable in National Socialist circles,"[395] and personally knew Himmler, Frick, Göring, and General Erhard Milch—the latter two naturally also in support of his aerodynamic research. He remained distant from the party as a member, while socially close to its topmost echelons. In another regard, however, he was like Hasse, in that he gradually became accommodated to the regime, though continuing to speak out on issues where, in his view, politics interfered with science. In correspondence with the British aerodynamicist G. I. Taylor, he even defended the German struggle against the Jews while deploring its effects upon Jewish scientists.[396] The correspondence of Hasse with Harold Davenport examined in chapter 4 shows a similar tone. After the war, the seventy-year-old Prandtl was unimpeached by the Allies, and could argue in 1947 for "all humanly valuable young researchers and teachers, who were not activists, but only wished to serve the state and their science," just as he had for the "half- and quarter-Jews" with Frick in 1933. In the postwar case, "they should now again be taken up in clemency (*Gnade*) and not come to injury because in recent years they had no other way than the one through the party, which became completely fused with the state."[397]

During the Weimar period, the state handed out grants to scientists in response to applications through the *Notgemeinschaft der Deutschen Wissenschaft* (literally, "Community of Need for German Science," founded October 30, 1920). One substantial recipient of funds was Prandtl's group at Göttingen, and one should remember this included not only the Institute for Fluid Dynamics (*Institut für Strömungforschung*) of the *Kaiser Wilhelm Gesellschaft*, but also the Aerodynamic Experimental Station (*Aerodynamische Versuchsanstalt*, or AVA). Prandtl taught at the university as well, being head of the Institute for Mechanics until 1934.[398] Two other applied mathematicians receiving substantial funds were Theodore von Kármán at Aachen and another student of Prandtl's, Wilhelm Spannhake, who led a fluid dynamics institute in Karlsruhe. In fact, this Weimar concentration on aerodynamics as applied mathematics in these three locations had the result that even highly distinguished mathematicians

[394] See above, chapter 6, "The Süss Book Project."
[395] "Sozusagen nationalsozialistisch gesellschaftsfähig," Tollmien 1987: 476.
[396] Ibid.: 476–477 and passim. See also von Kármán 1967: 38–40.
[397] As cited by Tollmien 1987: 482.
[398] Ibid.: 472.

had difficulty obtaining grants for other purposes. These included Richard Courant, Alwin Walther, and Richard von Mises—though the last, denied a grant for the expansion of applied mathematics at Berlin, did manage to receive one for the study of turbulence. This only reemphasizes the concentration on fluid-dynamic and aerodynamic problems.[399] Under the Nazis and Göring, the AVA expanded tremendously.[400] Yet, despite Prandtl's efforts, his subject was greatly injured by the proscription of many of the best scientists in the name of ideological rectitude. Though Prandtl benefited personally during the Nazi hegemony, he tried, in vain, to retard this proscription in the name of science.

If this was the situation in a subject as important to Nazi intentions as aerodynamics, one might imagine that the situation in applied mathematics in general was even worse. This was almost true. Alwin Walther, at the age of thirty, in 1928 assumed a chair at the technical university in Darmstadt. The position was nominally in either descriptive geometry or simply mathematics. Walther's *Habilitationsschrift* had dealt with theoretical aspects of the Riemann zeta-function. Nevertheless, from his coming to Darmstadt, he had built up an "institute for practical mathematics," which, in his own words, was

> a mathematical laboratory for research and instruction, with "hands-on" involvement (*Selbsttätigkeit*) of the students. Practical mathematics is a customary expression for numerical, diagrammatic (*zeichnerische*), and apparatus-related (*gerätemässige*) procedures of mathematics as a complement to the formulary methods, and besides emphasizes the close relationship between theory and practice, science and life.

Strongly influenced by Richard Courant, with whom he had spent several years as *Assistent*, and the well-known Danish mathematician Nils Nørlund, Walther's main interests involved the various branches of applied mathematics and their connection with "pure mathematics." Needless to say, the military had an interest in computationally intensive work. Under Weimar, Walther had been refused money for research on machine-generated solutions to mathematical problems, because "the disciplinary reviewing committee considers the provision (*Ausführung*) of the requested special equipment as not urgent and believes that its use is restricted."[401] Under the Nazi regime, he was able to obtain substantial funds for such machines, though not without help from Wilhelm Süss, and not until the war began going increasingly badly for Germany. Walther seems to have had an easy time getting military support for his efforts in "practical mathematics," but civil support was subject to lengthy bureaucratic delay, apparently due more to bureaucratic infighting than to disturbance by the Allied bombing raids.

On May 7, 1943, after a preliminary phone call with Rudolf Mentzel, Süss forwarded a request from Walther for funds to construct a differential analyzer. Such an analog computer had first been constructed by Vannevar Bush in 1927 at MIT, and similar machines existed in Manchester, England, and Oslo, Nor-

[399] Tobies 1986: 20, 25, nn. 42, 43, 44.
[400] Tollmien 1987: 468–470.
[401] Tobies 1986: 20 and 25 n. 3.

way. In Aachen, using funds provided by the army, Robert Sauer had already constructed such a machine devoted to military problems. Süss effectively requested funds for both Walther and Sauer to continue the development of such machines for questions of military research and technology, for "roughly our entire present-day technical research will have a tremendous advantage with the construction of such machines."[402] Walther already had an army contract for such work, but the products thereof would consequently belong to the army, whereas, as he explained in a further specification of his request, he wished such machines for scientific investigations, "—now in wartime, naturally, only for those of a militarily important nature—." He also wished to do further research on automated computation.[403]

Eight months after Süss's letter to Mentzel, Walther still had no RFR money; after another two months, Walther Gerlach, the RFR physics head, signed on to Walther's proposal, calling it "really urgent" and complaining about an apparent wish of the air force to monopolize all such efforts.[404] After Gerlach's letter, Walther finally received his money on April 1, 1944.[405] Things went easier. As late as February 1945, Alwin Walther was able to ask for and receive 44,000 RM for his "institute for practical mathematics" at Darmstadt, which at the time employed ninety-two people (including eighteen academics) in computational research.[406] Of course, Walther's "institute for practical mathematics" had other contracts,[407] including one from the RFR for the tabulation of Planck's radiation law (August 29, 1944). While the RFR in this case reserved the right to dissolve the contract, should it not have the funds therefor, Walther nevertheless was asked to report on progress by April 1, 1945, and semi-annually thereafter.[408] On March 13, 1945, Walther wrote the RFR saying he had used one-third of the money allocated to him for the differential analyzer, and wished the remaining two-thirds carried over to the next fiscal year. By March 13, 1945, Koenigsberg had been invested, and in the West, the Allies had reached the Rhine and seized the bridge at Remagen.

On June 18, this letter was stamped by someone as "not disposed of."[409] In addition to the glacial bureaucracy, what strikes attention here is once more the attitude of bureaucratic business as usual. It is again not that the bureaucrats did not know that the walls were falling in, but simply that they carried on their functions as routinely as possible. The money was there. Among both investigators and ministry bureaucrats were probably not many who believed in a sud-

[402] BAK R73 15487, Süss to Mentzel, May 7, 1943.

[403] BAK R73 15487, Walther to RFR (via Süss), Jan. 18, 1944.

[404] BAK R73 15487, Gerlach to Fischer, Mar. 20, 1944.

[405] BAK R73 15487, Walther to RFR, Feb. 15, 1945, and Mar. 13, 1945.

[406] BAK R73 15487, Walther to Mentzel, Feb. 15, 1945.

[407] Ibid.

[408] BAK R73 15487, Gerlach to Walther, Aug. 24, 1944 (this letter was "countersigned" with Mentzel's stamped name).

[409] BAK R73 15487, Walther to RFR, Mar. 13, 1945.

den reversal of fortune, but there probably were many, Süss and Walther among them, who attempted to prepare a suitable basis for their disciplines in the unknown postwar era.

Walther's institute at Darmstadt still was well staffed in early 1945. It was supported and successful. In 1939 it had taken over the calculations for the German rocket program at Peenemünde.[410] But except for it and Prandtl's AVA for aerodynamic experiments, applied mathematics, at least at universities, seems to have declined under the Third Reich. At Aachen, von Kármán had been succeeded by Wilhelm Müller. At Göttingen, in mathematics, Klein's dream of a theoretical mathematics integrated with a concern for applications died with the destruction of the faculty under the Nazis. Richard Courant had been Klein's successor in carrying out this endeavor—as it turned out, however, New York University was where he would ultimately create a version of it. Ironically, from this perspective, it was Carle Runge's successor, Gustav Herglotz, who was left in the mathematics department after the initial Nazi "cleansing." For while Herglotz was apparently a brilliant lecturer and an extremely broad mathematician, he was hardly the computationally concerned mathematician that Runge was, and that Prandtl, for example, might have wished as a colleague. No more so was Theodor Kaluza, who eventually succeeded Courant, and whose multidimensional ideas about space-time have achieved a renaissance in contemporary "string theory."[411]

An interesting case of the decline of applied mathematics is afforded by the University of Berlin. Although one of the chairs was held by Ludwig Bieberbach, one of the most pro-Nazi of mathematicians and a man who wielded considerable influence in Berlin mathematics, the applied mathematics institute effectively withered. This was not just because of the forced emigration of both Richard von Mises, its director, as a part-Jew, and some of his colleagues. Bieberbach (whose own very distinguished mathematical work prior to 1933 was quite "unapplied") simply had no interest in attempting to maintain a first-class applied institute. There *was* a problem in Nazi Germany of finding adequately trained, ideologically acceptable mathematical personnel, but it was not as severe as it has often been made out to be until the incursions of the war denuded the universities. When compared, for example, to Alwin Walther's institute for computational mathematics at Darmstadt (which was not even at a university, let alone a great and famous one), the Berlin effort seems paltry indeed.[412]

[410] Mehrtens 1986: 334.

[411] See Schappacher 1987: 357, for the fact that not only was Prandtl dissatisfied with Kaluza as an applied mathematician, but neither did Kaluza consider himself such. Cf. Mehrtens 1986: 324.

[412] Reinhard Siegmund-Schultze, "Berliner Mathematik zur Zeit des Faschismus," *Mitteilungen der Mathematischen Gesellschaft der DDR* (1988): 61–84; idem, "Theodor Vahlen—zum Schuldanteil eines deutschen mathematikers am faschisten Missbrauch der Wissenschaft," *NTM Schriftenreihe*. 21 (1984): 17–32; idem, "Einige Probleme der Geschichtsschreibung der Mathematik im faschistischen Deutschland—unter besonderer Berucksichtigung des Lebenslaufes des Greifswalder Mathematikers Theodor Vahlen," *Wissenschaftliche Zeitschrift der Ernst-Moritz-Arndt-Universität Greifswald* 32 (1984):

Although von Mises had served on the front in World War I and so technically fell under the exemptions of the law of April 7, 1933, it was also clear (if from nothing else, then from the situation of Richard Courant in Göttingen) that it was only a matter of time before he, too, was forcibly furloughed or dismissed. In addition, his student and future wife, Hilda Geiringer-Pollaczek, in September lost the right to teach by application of this law.[413] Intent on trying to preserve an applied mathematics institute despite his absence, when von Mises left for Turkey in November 1933,[414] he had actually recommended that his successor be the old Nazi (Karl) Theodor Vahlen, who was already sixty-four years old.[415] Both Erhard Schmidt and Max Planck voted for Vahlen; curiously enough, Ludwig Bieberbach, for all his Nazi sympathies and later cooperation with Vahlen, desirous of a "mathematical statistician," voted against him.[416] Vahlen, who already on March 15, 1933, had been called to a position in the higher education division of the Prussian ministry, took over von Mises's chair on January 22, 1934. In April, he became leader of his ministry division, and on June 1, "Head of the Scientific Office" in the newly established national education ministry. In October, Vahlen departed the university to devote all his energies to the ministry.[417] The faculty established in February 1935 the usual list of three for the new successor to von Mises's chair; these were, in order, Georg Hamel, Erich Trefftz, and Werner Schmeidler. Hamel's attitudes were clearly what the Nazis called "reliable." He was also an established applied mathematician of repute. However, if the university were to have a mathematician who would continue a strong applied mathematics tradition, instead of just a member of the political establishment,[418] Hamel, at fifty-seven, was perhaps too old, as well as too occupied with activities like the *Mathematische Reichsverband*. As to Trefftz, though mathematically exemplary, politically, he was hardly in sympathy with the Nazi state. Schmeidler, who was Hamel's professorial colleague at the technical university in Berlin, was similarly disqualified. The Berlin *Dozentenführer* said of him: "However, politically one does not have the

51–56; idem, "Zur Sozialgeschichte der Mathematik an der Berliner Universität im Faschismus," *NTM Schriftenreihe* 26 (1989): 49–68. This last and the first paper listed above have considerable overlap; the first being an invited address, the second a more detailed exposition of the same material.

[413] Biermann 1988: 237.

[414] Ibid.

[415] Siegmund-Schultze 1988: 67; 1984b: 55; 1984c: 26; 1989: 52.

[416] Siegmund-Schultze 1984b: 55.

[417] Ibid.

[418] While Vahlen was a competent and fairly broad mathematician who, after World War I, wrote on applied mathematics topics like ballistics and the magnetic compass, his original training was in number theory (he was a student of Georg Frobenius). Though already forty-five when World War I broke out, he served throughout the war on both fronts and was wounded. After 1923, he was a Nazi "*alter Kämpfer*." When Hitler came to power, he was nearly sixty-four. At age seventy-six, he starved to death in a Czech prison camp in Stechovice. For the above and further information about Vahlen, see *Deutsche Mathematik* 1 (1936): 389–420; *Sitzungsberichte Akademie Berlin* (1938): xcviii–xcix. (Antrittsrede des Hrn. Vahlen); and a handwritten *Lebenslauf* from 1937 in the BDC file on Vahlen. For Vahlen's death, see HK, copy of letter of Franz Krammer to H. Pinl[sic], Nov. 23, 1946.

trust that he has already really freed himself from his liberal and democratic ways of thought."[419] The much more prestigious position of Ordinarius at the University of Berlin would not be Schmeidler's.

Just as with Max Winkelmann and Ernst Weinel in Jena,[420] in the absence of an appointment, Vahlen's position was filled on a day-to-day basis by the assistant Alfred Klose. Klose was an astronomer who had started his career at Greifswald when Vahlen had been Rektor there. He went to Riga as a professor of astronomy in 1925, and four years later to Berlin in a similar position (at a lesser rank). Berlin clearly was a more desirable location than Riga.[421] In December 1933, Klose inherited in addition Hilda Geiringer-Pollaczek's position as *Assistent* in mathematics after she was dismissed and left for Istanbul with von Mises. On March 1, 1937, he became the new Ordinarius and institute director, even though in 1935 the faculty had considered him inappropriate for the position, despite his substituting in it, since he was a specialist in mathematical astronomy.[422] However, as early as May 1935, Bieberbach, acting as Dekan of the faculty, had written the ministry that, taking into account a telephone conversation with Vahlen at the ministry, although astronomy was not one of the disciplines to be fostered by the applied mathematics institute, nevertheless, "Herr Klose politically is unconditionally reliable and belongs to the number of really active National Socialists at the university."[423]

What was the applied mathematics institute of "active National Socialist" Klose like? In the academic years 1934–35 and 1936–37, after Vahlen's departure, when Klose was substitute director before being officially named, there were respectively seventy-two and twenty-three students actively in the institute for applied mathematics. On the other hand, between 1934 and 1936, twelve "unemployed academics" worked in the institute computing tables.[424] Indeed, after becoming director, Klose needed to defend the employment of additional teaching help in a way that has a familiar ring:[425]

> The number of students indeed has recently gone down considerably, so that the examination of the papers handed in [by students] no longer takes the time it did earlier; however, the preparation of the Praktikum is substantially independent of the number of participants. Therefore the expenditure of work in this respect has not become less. With this, however, comes the fact that the undersigned has undertaken the duty to convert the institute more and more to the requirements of the technical sciences and to work along in the framework of the Four-Year Plan [of 1936]. Thus under the rubric (*im Rahmen*) of help to academics, a set of mathematical tables was calculated. In the past year, among other things, a diagram for the location of radio stations was drafted and calculated. For such tasks, the present forces are insufficient;

[419] Siegmund-Schultze 1984b: 55; 1988: 67.
[420] Above, chapter 4, "The Winkelmann Succession."
[421] Scharlau et al. 1990: 41, 133.
[422] Siegmund-Schultze 1989: 52; 1984b: 55.
[423] Siegmund-Schultze 1984b: 55; 1988: 67–68.
[424] Siegmund-Schultze 1989: 53; 1988: 68.
[425] Siegmund-Schultze 1989: 53.

for the two *Assistenten* are completely busied with running the Praktikum and administrative work.

The elevation of triviality to importance has rarely been more pathetically expressed. It demonstrates how little could actually be expected of Klose's institute.

Von Mises's "last pupil in Germany" was Lothar Collatz, who would become an applied mathematician of international stature. Von Mises had provided the stimulus for Collatz's doctoral work, and the latter was scheduled to have an oral examination with his "Doktorvater" in the middle of November 1933. On November 2, he received a card from von Mises saying: "If you still want to do your examination with me, you must come to my house tomorrow at 10 A.M." Collatz did so, and (since von Mises was apparently a feared examiner) protested that he was not yet ready with his preparation for the examination. Von Mises answered: "Never mind. I must leave this country. I know you already sufficiently; exceptional times necessitate exceptional behavior. I would like in this hour to talk through with you how you might develop your dissertation; for you will now be placed alone with your paper."[426] Collatz completed his studies *pro forma* with Klose, though he received no help from him. In 1935, there was no place for this clearly coming mathematician in Klose's institute.[427] So Collatz went to the technical university in Karlsruhe to "habilitate"; as was seen in chapter 4, a few years later, Jena would be attempting in vain to give him a chair. However, Collatz would not leave Karlsruhe until 1943, when he went to the technical university in Hannover.

That Collatz should find a home in these places rather than in Berlin indicates how rapidly and how much the institute had declined. By late 1938, Klose had to fight to keep a second *Assistent*, and did so by invoking the importance of applied mathematics without any concrete evidence of the activities of his institute.[428] Klose publicized an unrealistic view of the capacities of his institute, perhaps unaware, perhaps simply as a means of self-aggrandizement. Thus on July 24, 1939, he wished to make a trip to examine Picone's applied mathematics institute in Rome. Though Prandtl did not think much of this institute in terms of applications, Doetsch had praised it.[429] Klose declared:[430]

Besides, I have been thinking for some time about the creation of an institute, which, like the Italian one, carries out mathematical commissions from research establishments, public institutions, and industrial concerns. I have frequently experienced that a need for such an institute exists. I have also, insofar as the small means of my

[426] Collatz, in *Ein Jahrhundert Mathematik* 1990: 282.

[427] Siegmund-Schultze 1989: 52–53; 1988: 68.

[428] Siegmund-Schulze 1989: 54; 1988: 69.

[429] Above, chapter 4, "The Süss Book Project."

[430] Siegmund-Schultze 1989: 54; 1988: 69; for another attempt at relations with Picone's institute, see Reinhard Siegmund-Schultze, "Faschistische Pläne zur Neuordnung der europäischen Wissenschaft, Das Beispiel Mathematik," *NTM Schriftenreihe* 23 (1986): 1–17.

institute both materially and with respect to personnel allow, already often in recent years carried out such commissions.

By 1940, Klose and his two assistants had both been called up, and the existence of the institute was largely a paper one. It clearly mattered little to anyone with any power; no one was even freed from service and seconded to the institute for "militarily important" work.[431]

Thus, an established applied mathematics institute might wither because ideology prevented appropriate appointments. Innovative individuals also had difficulty getting a hearing. In 1936 Konrad Zuse built a sort of computer, and in 1938 he completed his first binary calculator, but it was not until 1940 that he received a military contract. It is worth noting that Zuse used electromagnetic relays (instead of vacuum tubes) and by 1941 was using a punched tape for data entry. In 1945, in the United States, Prosper Eckert and John Mauchly's ENIAC used vacuum tubes, and had to be programmed by resetting the wiring.[432]

One modernizing idea introduced by the Nazi educational bureaucracy that proved of lasting positive value was the *Diplom*, a terminal degree prior to the doctorate. Mathematics and physics were among the last disciplines to have such a degree introduced, but the decree to that effect, including the establishment of new examinations, was issued August 7, 1942, effective November 1, after more than a year's discussions. Ludwig Bieberbach was involved in these last, but not Wilhelm Süss.[433] The new arrangement of examinations explicitly put applied mathematics on an equal footing with "pure" mathematics, thus giving the former a role in university study that it heretofore had never enjoyed. Nevertheless, given the "yawning, empty lecture halls" and wartime conditions, there was no change in positions available. Even Prandtl, explicitly referring to the new role of applied mathematics, could not obtain a new applied mathematics position in Göttingen.[434] Actually, as noted earlier, the role of applied mathematics had been discussed as early as 1935 among mathematicians; but this was at least partly a justification of the existence of their discipline and an attempt at more political influence.[435] Considerations of more status for applied

[431] Siegmund-Schultze 1989: 54; 1988: 69. One of Klose's assistants was Karl-Heinz Boseck, for whom see the next section. Boseck, who was physically unfit for soldiering, apparently returned to the university on a part-time basis.

[432] Mehrtens 1986: 334. The idea of using electronic instead of mechanical electromagnetic relays seems to have originated with Helmut Schreyer. For this and the history of Zuse's efforts, see Hartmut Petzold, *Moderne Rechenkünstler* (Munich: C. H. Beck, 1992), esp. chap. 5. Zuse, who was interested in automating a universal language, had contacts with the logician Heinrich Scholz (above, "Heinrich Scholz, Logician"), at first through his student, Hans Lohmeyer (Petzold, ibid.: 196–197), and Alwin Walther (ibid.: 197, 228–229). Walther was, however, more interested in speed than in an internal universal language, and also originally more inclined to analog technology along the lines of Vannevar Bush's differential analyzer than digital technology.

[433] Mehrtens 1986: 332.

[434] Ibid.: 332–333, 324; Siegmund-Schultze 1989: 54; 1988: 70. "Yawning, empty lecture halls" is Siegmund-Schultze's phrase.

[435] This is Herbert Mehrtens's judgment, which I share.

mathematics had apparently gone on for several years in the education ministry[436]—mathematics, even applied mathematics, apparently did not matter much to the ministry. Aerodynamics was different because of Göring's patronage. Similarly, at the University of Rostock, applied mathematics apparently had considerable influence, largely due to the status of the Heinkel aircraft factory.[437] However, the Zeiss factory did not enjoy similar influence in Jena. By 1942, the *Diplom* in mathematics or applied mathematics was an idea whose time had clearly come. It had first appeared in (the free city of) Danzig in the 1920s, and although the National Socialist ministry may have introduced it in the interest of cutting short the years to a terminal university degree, especially given the government's new requirements for ideological education as well as wartime conditions,[438] after the war, its worth was still recognized. Thus, at the meeting of the German Mathematical Society in Cologne in September 1949, it was decided virtually unanimously that the *Diplom* should be retained as a degree and that the examinations for it should be determined autonomously by the universities, rather than by the state.[439]

The experience of what had been von Mises's institute for applied mathematics only adds to the impression that applied mathematics mattered little more than "nonapplied" mathematics, even through most of the war. The quarrel between Doetsch and Süss concerning a book publication project was discussed in chapter 4. While both men probably wished for Germany to win the war, the role of applied mathematics therein seems not to have been the real aim of either one. Doetsch was interested in personal self-aggrandizement, and Süss in the protection and employment of mathematicians. Similarly, despite the interest of the Zeiss works, the Winkelmann succession in Jena, also discussed in chapter 4, dragged on and on. In neither case were there authorities who cared enough to put an end to the dispute rapidly in the interest of the war effort, let alone in the interest of science. The Nazi suspicion of academe, despite the "German national" position of most of the professoriat, and the Nazi emphasis on emotion as opposed to intellect, left traditional intellectuals of small actual import, even if their work might have been of value to the state. Hitler's emphasis on achievement through "will" in a mystic religious sense[440] also informally undermined any trust in rational processes. His attitude toward world events seems to have become rather like Humpty Dumpty's toward words: it was only a question of who was to be master. If "will" was the essential fact and only short-term research goals almost immediately convertible into products mattered, then long-term research, whether it was the development of atomic weaponry[441] or applied mathematical research, received slight, disparaging inattention.

[436] Mehrtens 1986: 332.

[437] Mehrtens 1986: 324; Scharlau et al. 1990: 239; *Geschichte der Universität Rostock, 1419–1969*, vol. 1 (1969), 287–288.

[438] Gerd Schubring, "Zur strukturellen Entwicklung der Mathematik an den deutschen Hochschulen, 1800–1945," in Scharlau et al. 1990: 264–278, p. 276.

[439] *JDMV* 54 (1950): Abteilung 2, p. 8.

[440] See, e.g., J. P. Stern 1975: passim.

[441] Walker 1989: passim.

Helmut Joachim Fischer, the Ph.D. mathematician who became a member of Heydrich's SD (*Sicherheitsdienst*, or Security Service) and whose degree had been forced on the faculty by Udo Wegner,[442] claimed he was among a group proposing a "scientific-technical leadership staff" to advise Hitler. Ernst Kaltenbrunner[443] presented the proposal to Hitler, who greeted it warmly. However, it was November 30, 1944. Fischer lamented the failure to create such a staff years earlier, but also quite accurately recorded the strong possibility that earlier Hitler would have rejected such an advisory staff: "First in 1944 as he [Hitler] had had sufficient bitter experience was [he] in favor of a better connection with science and technology."[444] Only late in the war, when such efforts were too late, were such actions possible: Süss's speech to the Rektors or the Osenberg action recalling scientists from the front are two other examples.

Mathematics in the Concentration Camps

In discussing mathematical institutions under the Third Reich, one that was peculiar to it should not be omitted: the concentration camp. Surprising as it may at first seem, there was organized mathematical activity in this context.[445]

At Plaszów, a suburb of Kraków, Poland, a concentration camp had been established. This camp is the one that figures in the story of Oskar Schindler's rescue of numerous Jews. Its commandant was one Amon Goeth, who was notorious for his brutality.[446] In connection with the ministry for armament, various contracts were assigned to Plaszów, including some in astronomy and mathematics under the general direction of a certain K. Walter. Professor Walter was director of the observatory in Kraków under the auspices of the occupying German government (*Generalgouvernement*). About fifteen Russian prisoners on

[442] See letter of K. Freudenburg et al., Nov. 8, 1947, in Personalakten Wegner, Heidelberg. This lengthy letter mentions *inter alia* other actions of Wegner at Heidelberg affecting mathematics, and attempts to evaluate his sometimes ambiguous behavior under the Nazis. It is partially excerpted in Vezina 1982: 155, in which the "keen National Socialist" Fischer is mentioned. Fischer is mentioned here with forename "Hans," however, this is an incorrect memory; see Fischer 1985, 1:37, for his version of the story.

[443] Ernst Kaltenbrunner was an Austrian lawyer and Reinhard Heydrich's successor as head of the SD and RSHA (*Reichssicherheitshauptamt*, or National Security Office) on the latter's assassination. He was the ultimate supervisor of the extermination camps. Tried for war crimes and crimes against humanity, he was executed by hanging at Nuremberg on October 16, 1946.

[444] Fischer 1985, 2:138–144.

[445] The following (sometimes overlapping) sources have been used in the ensuing discussion: Nuremberg Military Tribunal (NMT) documents labeled NO-1056 and NO-640 (copies obtained from the U.S. National Archives and, of the former document, also the Hoover Institution at Stanford, Calif.); the BDC file for Karl-Heinz Boseck (this file also contains NO-1056 in its entirety); a memoir by Hans Ebert entitled "Mathematiker im KZ 1944/1945," copied from a copy in the possession of Herbert Mehrtens; Fischer 1985, 2: esp. 97–104; U.S. National Archives, Captured German Documents, T-580/125/39. All material about concentration camps not otherwise footnoted can be found in one or more of these references.

[446] See Thomas Keneally, *Schindler's List* (New York: Simon and Schuster, 1982), a "novelized" version of Schindler's humanitarian efforts, which is essentially true. This has also been made into a renowned movie of the same title.

average worked on contracts for the air force, the navy, and the astronomical institute in Berlin (at Dahlem). By the end of July 1944, not only had there been military computations completed, but also a set of mathematical tables "needed by research and industry," and, at the request of Alwin Walther in Darmstadt, translations of some Russian mathematical papers.

However, by the end of July 1944, the Russians were less than eighty miles from Kraków, and the activities of the institute were moved to Germany. Some went to the concentration camp at Ravensbrück and the nearby town of Alt-Thymen.[447] In these places the work was continued in September and included, among other efforts, the numerical solution of partial differential equations connected with ballistic investigations under an industrial contract, as well as the computation of the ephemerides of Mars and Jupiter.[448]

The origins of the mathematical effort at Kraków-Plaszów are unclear but seem to have been at the initiative of the aforesaid Professor K. Walter[449] and were among several efforts of a military nature, including the successful extraction of sulphur from poor natural deposits and the combatting of agricultural pests. On the other hand, the origins of the other mathematical effort within the concentration camps are quite clear. On May 25, 1944, Heinrich Himmler wrote his SS subordinate, Oswald Pohl: "Among the Jews whom we have now received from Hungary as well as also among our concentration-camp prisoners, without doubt are a whole lot of physicists, chemists, and other scientists."[450] Pohl was delegated to establish in some concentration camp a "scientific research establishment" in which "the disciplinary knowledge of these people will be applied to the humanly stressful (*menschenbeanspruchende*) and time-consuming computation of formulas, the working out of individual constructions, and also, however, for fundamental research." The whole operation was placed under the aegis of Himmler's favorite *Ahnenerbe*,[451] with the scientific leadership assigned to its member Walter Wüst.[452] Himmler's order mentions the already extant operation with Russian prisoners.

[447] Many of the larger camps had numerous satellite camps associated with them. For example, the camp at Sachsenhausen discussed below grew out of a smaller camp at Oranienburg, the nearest city, and had sixty-one such satellite camps.

[448] NO-640, memorandum of Wolfram Sievers, Nov. 24, 1944. This is an extract from a report to the RFR that was apparently sent to Karl-Heinz Boseck around December 1.

[449] Not to be confused with Alwin Walther, the computational mathematician at Darmstadt. Such confusion may have existed in the SS hierarchy, as Walther's name is frequently misspelled "Walter." For more about Walter at Ravensbrück, see below.

[450] NO-640. Pohl was an *alter Kämpfer* who had joined the Nazi party as early as 1926, and the SS in 1929. In 1942, he was made head of the chief office for economic administration (WVHA = *Wirtschaftsverwaltungshauptamt*), and had since 1939 been in charge of all SS economic enterprises. He was executed for war crimes in 1951.

[451] *Ahnenerbe*, or "Ancestral Heritage," was an SS organization originally directed toward research on Germanic prehistory, and the collection of evidence of the value of pure Aryan bloodlines. Heinrich Himmler was one of its founders in 1935, and it came to be a place for investigation of anything of interest to Himmler, which included such extremes as the "World Ice Theory" and the skull and skeleton collection of Prof. August Hirt.

[452] According to Fischer (1985, 2:141), Wüst was a scientific incompetent.

Himmler's order envisioned the *Ahnenerbe* working together with the RSHA, where the quondam mathematician H.-J. Fischer was employed, and the matter landed on his desk. Fischer claims in his memoirs that he realized political opponents and prisoners could not be expected to have the requisite inner drive for scientific activity on behalf of the Third Reich, but, not wanting to disappoint Himmler, he hit on the idea of a computational institute connected with militarily important research.[453] Fischer seems to have given himself too much credit, given the existence of the effort in Poland as well as Himmler's explicit mention of a certain SS leader, Koppe, as providing the original suggestion for such a use of skilled concentration-camp labor. Nevertheless, it naturally fell to Fischer, as the possessor of a doctorate in mathematics, to make a selection of the prisoners to be used.[454]

Himmler's original order to Pohl had asked for monthly reports beginning August 1, 1944, and on June 15, Pohl held a meeting at which it was determined that the site for the concentration-camp computational institute would be the camp at Sachsenhausen (about fifty miles from Berlin).[455] By late July, Fischer had visited Sachsenhausen and promised to travel in early August to Dachau and Buchenwald to seek further suitable prisoners.[456] At Sachsenhausen, Fischer saw about seventy prisoners who[457]

> were foreigners and came from almost all European countries that at the time were occupied by the German army. . . . Throughout, they were political prisoners, recognizable through a red triangle . . . [other prisoners], say, asocial individuals, professional criminals, Jehovah's Witnesses, homosexuals, were not represented; also there were no Jews.

In about three hours, Fischer found about thirty of the seventy at first impression usable. However, only fourteen turned out to be so. In Dachau, Fischer found another handful.[458] Fischer also suggested the mathematician Karl-Heinz Boseck as direct supervisor of the operation.[459]

Karl-Heinz Boseck was an *Assistent* in Berlin at Klose's applied mathematics institute. He had earlier worked part-time for the RSHA (which seems to be how Fischer met him) and also had done work for the DFG. At nearly twenty-

[453] Ibid.: 99.

[454] Ibid.: 100–101; NO-640, Wolfram Sievers to R. Brandt, Aug. 14, 1944.

[455] NO-640, Sievers to Brandt, Aug. 14, 1944.

[456] NO-1056; Sachsenhausen and Oranienburg are used interchangeably in the correspondence for what was effectively the same concentration camp.

[457] Fischer 1985, 2:101–102.

[458] Ibid.: 104; Maurer to Sievers, in Aug. 16, 1944, in BDC file on Karl-Heinz Boseck. A memorandum, Fischer to RSHA, in NO-640 details these individually as three mathematicians and eight chemists. The latter were intended for a parallel chemical effort in the concentration camps, but presumably were capable by training for the computation of formulas and the use of logarithmic tables. Fischer, in his memoirs (1985), exaggerated when he put the number at another thirty. It should be added that Fischer was a Nazi apologist who maintained that prior to 1945, there were good conditions in the concentration camps (ibid.: 104–105).

[459] NO-1056.

nine years old, the apparent reason he was not in military service was that he suffered from severe varicose veins in both feet.[460] Boseck declared himself ready to supervise the prisoners beginning August 15, which would be after he had completed his examinations for the *Diplom* degree in mathematics. As of July 28, he had been successful in these. However, Boseck was only willing to undertake the supervision if, "for reasons of his authority," he were made a specialist officer in the SS. His varicose veins were here a difficulty. There were, however, no other suitable personnel, and so on August 30, Himmler agreed that "at least for the duration of the war," Boseck would be an SS officer, specialist class.[461] In fact, to the annoyance of the Sachsenhausen camp commandant, Boseck did not show up on August 15, and difficulty in obtaining suitable calculating machines also delayed the work.[462] Indeed, it was August 21 before Klose was asked to release Boseck from his duties as *Assistent*; Klose agreed "with heavy heart," and in the hope that Boseck's new organization would be helpful to him "in scientific military tasks."[463]

A vivid picture of Boseck at the institute in Berlin has been left by the mathematician Alexander Dinghas, then a student:[464]

> [There] came as [Nazi] disciplinary leader (*Fachschaftsleiter*) the man who for ten years would be unbounded master of the mathematics faculty: Karl-Heinz Boseck. Boseck was no ordinary man. At that time, he indeed was still half a child, displaying, however, already all the signs of the later Robespierre on a small scale. For I am certain; had Boseck been a member of the Convention in the French Revolution, so had he been *no* less than Robespierre or Saint-Just, for he possessed all the qualities through which a fanatic obtains external power: narrow vision, hunger for power, desire to rule over other people, and blind belief in ideas. Chiefly, [he believed] in the ideas of National Socialism, if here one can also speak with difficulty of ideas. . . . Later I came to know other members of his family and ascertained that they all possessed the same fanaticism and belief as Boseck, . . . my situation was weak. I could only then last if I had the support of Boseck, for a word from Boseck would have sufficed to remove me from my position. That this conjecture of mine was correct, I had later confirmed in 1944, . . . Boseck supported me until 1943 without reservation. In 1939, he rang up the ministry and requested that my receipt of a teaching post be expedited. I relate this here only to show what power Boseck had at his disposal in those years. Later, his power would be still greater and Bieberbach[465] sank ever deeper. It is sad that Bieber-

[460] *Ibid.* According to Fischer (1985, 2:104), Boseck had suffered a leg injury through an accident.

[461] NO-1056 and NO-640, Sievers to SS Personalhauptamt, Sept. 13, 1944. (This document is also in the BDC file for Boseck, which contains an amplification on September 19 as well.) See also in NO-640, Sievers to Brandt, Aug. 14, 1944.

[462] BDC, Boseck file, Maurer to Sievers, Aug. 16, 1944.

[463] BDC, Boseck file, Sievers to Klose, Aug. 21, 1944; Klose to Sievers, Aug. 25, 1944; OKH (Army High Command) to Sievers, Sept. 12, 1944.

[464] Dinghas 1998: 199–200. Dinghas wrote this in part during his internment in August and September 1945 in a camp at Luckenwald.

[465] Dinghas was a student of Erhard Schmidt; however, of the *Ordinarien*, Bieberbach was presumably the administratively dominant personality. None of the others (including Schmidt) had such interests, whereas Bieberbach did.

bach never made the attempt to free himself from Boseck. I believed for a long time that he wished to and just as us, grinned and bore it, until finally in 1944, I convinced myself of the opposite: Bieberbach had had respect for Boseck.

Boseck's appointment took place effective October 1 (a Sunday). By Tuesday, he had examined the physical situation in Sachsenhausen, and on Thursday (October 5) was clarifying his situation vis-à-vis the university, as he did not want to give up the opportunity of becoming a university teacher at Berlin, as well as questions of his pay and associated bureaucratic details. In particular, an agreement was reached with Sievers that only a very limited number of organizations could submit computational requests, including explicitly the SS weapons office, the army weapons office, the various branches of the air force research directory, Walther Gerlach, and Wilhelm Süss (by virtue of their positions in the RFR).[466]

Actually, Wolfram Sievers, the executive secretary of the *Ahnenerbe*, had already informed Gerlach on August 21 of the plans for the computational institute for the "humanly stressful and time-consuming computation of formulas, the working out of individual constructions, and also, however, for fundamental research," and Gerlach had replied enthusiastically on August 29, as well as informing Süss.[467] A week later, Süss himself wrote Sievers, mentioning the case of Ernst Mohr and suggesting him as a suitable participant.[468] Alwin Walther at Darmstadt was also enthusiastic. Fischer had suggested he might provide overall scientific supervision, and in fact, in late July, Walther was already sufficiently involved to declare himself satisfied with Boseck, but to suggest that it were necessary before the work began for Boseck or whomever would be in charge to have preliminarily two weeks' training at his institute in Darmstadt. He was also consulting Werner Osenberg[469] about a possible supervisor.[470]

Sievers, writing to Gerlach, repeated Himmler's original phrase requoted above as to the work to be done in Boseck's institute. Indeed, in almost a parody of the hierarchical transmission of commands, the phrase echoes in letter after letter in the files: it is also in Sievers to Klose on August 21 (where after "constructions" appears the weary "etc." (*usw*). It appears in Sievers to the commandant of the Dachau concentration camp (Weiter) on November 4, 1944, and again in Sievers "to the [entire] personal staff of the national leader of the SS [Himmler]" on December 7 (here again broken off, but with *u.ä* = "and similarly").

Work started at the beginning of October—sort of. The authorities at Sachsenhausen had not yet really started any preparations. Also, Boseck had some doubts about the half-barracks to be devoted to his project: while adequate for an envisioned forty workers, it was proximate to the camp's tuberculosis isola-

[466] BDC, Boseck file, Boseck to Wolf, Oct. 5, 1944.

[467] NO-640, Sievers to Gerlach, Aug. 21, 1944; Gerlach to Sievers, Aug. 29, 1944.

[468] NO-640, Süss to Sievers, Sept. 7, 1944. For Mohr, see above, chapter 4, "Hasse's Appointment at Göttingen."

[469] Above, chapter 5, "The Wartime Drafting of Scientists."

[470] NO-1056 (p. 4); NO-640, Sievers to Gerlach, Aug. 21, 1944.

tion ward. By mid-October, however, the rebuilding of this area was completed. Reference books, equipment, and paper were also acquired. As to personnel, however, of the eleven Dachau inmates chosen by Fischer, only five had arrived in Sachsenhausen by the end of October,[471] and of these, Boseck found only one usable. Consequently, he made trips to Buchenwald, and also for "gleanings" (*Nachlese*) again to Dachau (where Fischer had not had a selection from all inmates). Eventually, fourteen more inmates were selected from Buchenwald, but only nine of these were delivered. Finally, on November 14, actual work began with the nine from Sachsenhausen and one from Dachau. These shrunk to seven, to whom eventually the nine from Buchenwald, and presumably more from Dachau, were added in December.[472]

The detailed small numbers are mentioned because they seem to indicate, in part, how little help was to be had in the camps for the mathematics project, though it came from the highest authority. Indeed, Gerhard Maurer, the head of the branch of the WVHA that negotiated contracts for the use of concentration camp prisoners, seemed annoyed that because of Boseck's failure to arrive on August 15, the inmates were not working.[473] While Sievers praised Boseck as having shown himself "very circumspect, capable, and clever. Good cooperation and successful results are therefore guaranteed,"[474] it is also clear that he had his own reasonable doubts about the complete efficacy of employing concentration-camp labor to do intellectual work for the Third Reich. With respect to the Kraków-Plaszów group, he wrote (and sent as a memorandum to Boseck):[475]

> *In Re*: The employment of prisoners for scientific work.
> The reliability of scientific work is conditional. The results are frequently "cooked" (*frisiert*) in order to achieve relief. It was established that then, if certain relief were granted, e.g., permission to be able to work in ordinary clothing, the investigations were immediately more reliable. During the employment [of inmates] for scientific activity, in trial cases (*Bewährungsfalle*), one should go up to release from imprisonment, in order to achieve the impression that those in question are really active as scientists.

[471] Boseck's somewhat ominous phrase in the circumstances, "since in the meantime the remainder has been disposed of otherwise." See Boseck to *Ahnenerbe*, Oct. 26, 1944, in BDC file on Boseck.

[472] BDC, Boseck file, Sievers's report to Himmler, Dec. 1, 1944 (also in NO-640). For other material, see ibid., Boseck to Sievers, Oct. 5, 1944; Oct. 16, 1944; Oct. 19, 1944; Oct. 28, 1944. Sievers's report contains explicit mention that a professor of mathematics at Prague, who was interned at Buchenwald, was not among those transferred to Sachsenhausen. This was presumably Ernst Mohr. However, F. Litten (1996) claims that Mohr was never in Buchenwald, and that in fact the person meant was Prof. Erwin Lohr, of the technical university in Brünn (Brno) (not Prague), who had also been arrested.

[473] BDC, Boseck file, Maurer to Sievers, Aug. 16, 1944.

[474] BDC, Boseck file, Sievers's report to Himmler, Dec. 1, 1944 (also in No-640).

[475] NO-640, memorandum by Sievers dated Aug. 4, 1944, presumably sent to Boseck with the memorandum mentioned in note 448.

Boseck's institute needed a name so that it could be addressed by contracting parties and deal with procurement problems. This was established as "Mathematical division of the institute for military scientific goal-oriented research of the Waffen-SS and police in the office *Ahnenerbe* of the personal staff of the national leader of the SS," or "Short form: 'Institute for military scientific goal-oriented research, mathematical division, Oranienburg, P.O. Box 63.'"[476] This organization, whose "long-form" title is almost a self-caricature of German, particularly Nazi, hierarchical language, naturally engaged in much correspondence of the sort mentioned, but this could be effective, especially when ordering items, only if it were clear that a service division of the military were involved. Thus,[477]

> The utilization of a service stamp (*Dienstsiegel*) is urgently necessary. Therefore an application is made to make available a service stamp (metal stamp) for the Institute of Military Scientific Goal-Oriented Research, Mathematics Division, with the inscription "Waffen-SS, Institute for Military Scientific Goal-Oriented Research"; in order to avoid mistakes and for a distinguishing mark, it is necessary, in contrast to those already available, that this seal should carry the letter "M" under the national eagle (M = Mathematics Division).

The most mundane bureaucracy therefore ground on at a time when Strassbourg and Aachen had both fallen to the Allies, the American General Patton was establishing a bridgehead over the Saar, and the bloody battle of the Hürtgen Forest had just ended with German retreat, though also with many Allied casualties. It is true that the German resistance to further Allied incursion on Germany itself was very stiff and that the last German offensive (the Ardennes offensive, or "Battle of the Bulge") was about to begin, but one would have thought a Waffen-SS bureaucracy would not at the time be worrying about the nature of new marking stamps.

In what is apparently Boseck's last extant official report, on December 28, he described the workers as three German, six French, three Czechs, three Belgians, one Dane, one Portuguese, and one Jew. Of these, the Portuguese and the Frenchman (the noted physicist Georges Bruhat) were ill—Bruhat would die within a year. By the time of this report (intended for the report due to Himmler at the New Year), various airflow problems, tables of functions defined by integrals, and calculation of altitude maps for characteristic surfaces had been completed, or nearly so. In addition to the tables and machines obtained for the work, the concentration-camp command provided the institute "from liquidated effects" (*aus aufgelösten Effekten*) "a series of good magnifying glasses and lead and drawing pencils." French textbooks were borrowed from the University of Berlin.[478]

The "mathematics division," however, went on until early April. In January 1945, there were complaints about the Italian computing machines requiring

[476] BDC, Boseck file, Sievers' report to Himmler, Dec. 1, 1944, item 7 (also in NO-640).

[477] NO-640, Sievers to "Personal Staff of the Reichsführer SS [Himmler]," Dec. 7, 1944.

[478] NO-640 or BDC file on Boseck, Boseck to Sievers, Dec. 28, 1944.

repair and the need to buy a detailed list of books. By early February, evacuation to some other location was being discussed. Boseck suggested moving near Jena (Buchenwald) or Göttingen. Nevertheless, on February 15, there was talk of Boseck's need for dictionaries, as well as important computations to be done for "long-distance weapons" (*Fernwaffen*). In fact, the V-1 bomb had been introduced about a year previously, and the V-2 the preceding September. Somewhat later, still in 1944, the Germans had introduced a rocket-powered airplane—which had the unfortunate characteristic of spontaneously exploding. Also, Boseck apparently needed assistance, since an ever greater range of problems was being handled, and Pohl sent him as *Unterführer* a certain Franz Lippa, who was a war-wounded SS man. Boseck had been told that he, together with the "mathematics division," must leave Oranienburg by February 15, since after February 16 no army transport would be available. Possible places of removal, such as Dachau and Waischenfeld (where the offices of the *Ahnenerbe* were located), were considered and rejected, the latter because of insufficient space and insufficient guards. On February 23, Pohl believed that the mathematics division was on the march to new quarters in Dachau; but on April 4, Boseck was still in Oranienburg.

After mulling through a number of possible places to relocate, Wolfram Sievers had favored moving to the concentration camp at Flossenbürg. Boseck preferred his own independent establishment. Among other places discussed was Boseck's suggestion of Hostischau (modern Hostisov in the Czech Republic). For Boseck, this recommended itself because there were no air raids, there were easy connections to Prague, it was reasonably far from both fronts, and not least, the possibility of keeping the work secret was enhanced. Boseck even described a building he had in mind and the necessary renovations to it—renovations that Sievers found impossible under German circumstances in March 1945. Indeed, the whole idea of moving from near Berlin to Hostischau was impossible given the situation, as Sievers remarked with sarcasm. In fact, some of the difficulties of going to Hostischau, in addition to distance, were the need for extra guards, the suspicion that the Czech prisoners would attempt to escape, and the certainty that the Nazi National Security Office (RSHA) would not allow Czech prisoners in the "Protectorate of Bohemia." Throughout this correspondence, Boseck was most concerned for the secrecy of his group's work. Whether this was self-importance on his part (all his letters are signed "Dipl[om]-Math. K. H. Boseck, SS Untersturmführer [F]"), or simply willful failure to realize that the German military situation was unsalvageable, is unclear—and hardly matters. Boseck certainly behaved as though his mathematics division were essential to the eventual German victory. On March 5, 1945, he heard of Italian calculating machines finally located in Verona, and on March 23 was complaining that their price was exorbitant, and furthermore they needed repair—talk about business as usual. Boseck even discussed amortization of the calculator price over two years. On March 12, Boseck's financial support was increased. On March 23, he was also complaining that machines sent from Holland for the mathematics division had been mistakenly sent to Dachau.

On March 3, 1945, the concentration camp at Ravensbrück was evacuated into Sachsenhausen.[479] With it came Professor K. Walter's computational group of eight people. This led to the kind of bickering two months before Germany's unconditional surrender that not only illustrates Boseck's primary motivations of personal aggrandizement, but, were it not documented, would be scarcely believable.

As Boseck reported to Sievers on March 5, he interviewed five of this group, all Polish Jews, and found only one of them, a former *Assistent* in the university at Lemberg (modern Łwow), usable. The other four were only "average" in ability, and such men "are always obtainable." Also, three of these, Boseck said, were "already no longer in Sachsenhausen." Since mass gassings at Sachsenhausen began in February, this was presumably their fate.[480] Boseck was also amazed that Walter's group worked without direct oversight—he would visit Ravensbrück every two weeks or so to pick up completed work, and leave new tasks to be done. They also apparently had very little in the way of books or other sources of known results, which, said Boseck, had to be discovered anew.

On March 8, Walter visited Sachsenhausen himself, as it turned out, while Boseck was away at the Army Weapons Office. He attempted to visit the mathematics division, but Boseck's assistant there (presumably Franz Lippa) told him to get out immediately and return when Boseck was present. On the twelfth, Boseck discovered that the camp commandant's adjutant thought the Ravensbrück mathematics people would fit in quite well with the local mathematics division. A meeting of Walter and Boseck was set up for the next day.

Already on March 5, Boseck had told Sievers that it was not his job to worry about Walter's group. Furthermore, he would not even trouble to communicate to Walter the arrival of his men and machines at Sachsenhausen. When the two mathematical entrepreneurs met on the thirteenth, neither wished to blend their programs. In telling Sievers of the meeting, Boseck was sarcastic about Walter's interest in "busying" (*beschäftigen*) the prisoners. In Boseck's view, the task was to give them committed employment (*einsetzen*). Furthermore, he refused to allow Walter to see his stock of machines and material. Walter told Boseck he had originally attempted to use Polish and Russian civilian personnel under the auspices of the Institute for German Work in the East for his mathematical problems, but this proved unsuccessful—"especially the Poles were obstinate." The eight Jewish prisoners from Sachsenhausen worked "diligently and inde-

[479] Sachsenhausen was about twenty miles north of central Berlin and Ravensbrück about thirty miles further north. Though Ravensbrück is known as a female concentration camp, in spring 1941 a small men's camp was adjoined, though officially it was a Sachsenhausen satellite. Presumably it is this which was first evacuated suddenly on March 3 into the main Sachsenhausen camp. Ravensbrück proper was evacuated in late March 1945. See *Encyclopedia of the Holocaust*, vol. 3 (New York: MacMillan, 1990), 1226–1227.

[480] A gas chamber was installed at Sachsenhausen in 1943, but was not used regularly. In February 1945, several thousand prisoners too ill to march were killed in it. Ibid.: vol. 4, pp. 1321–1322. It seems unlikely that the prisoners would have escaped.

pendently," although their work was of uneven quality.[481] It turned out that Walter had obtained through Werner Osenberg's office, as a consequence of his computational activity with the Jewish prisoners, the coveted "indispensable" (*uk*) ranking that prevented him from being drafted into the military. Boseck suspected, probably rightly, that it was to preserve this that Walter was insistent on having his own working group of prisoners (after all, Boseck himself had parlayed the leadership of the mathematics division into an SS rank he otherwise would not have received). The adjutant's plan to put Boseck's and Walter's groups side-by-side was a priori impossible in Boseck's eyes. He stressed the essential secrecy of the work in the mathematics division and that he had prisoners capable of being "heavily employed for theoretical considerations," which required large effort on his part, as opposed to Walter's more nonchalant operation. In the end, Walter's remaining five prisoners received another small corner of Sachsenhausen, since he would not release them to Boseck's supervision.[482]

On March 5, Boseck had pleaded with Sievers about the necessity of a "speedy and orderly" removal of the mathematics division from Sachsenhausen prior to any sudden evacuation of the whole camp. On March 28, Sievers felt Flossenbürg would be best. But on April 4, never flitting, Boseck still was sitting in Sachsenhausen, even though its mass evacuation began at the end of March. When the Russian army arrived on April 21, they found only 3,000 ill prisoners remaining in Sachsenhausen.[483] Boseck's fate is unknown. Thus ended the experiment with mathematical slave labor.[484]

Even at the outset of the program, though, dating it as early as possible, say from Himmler's May order, it is hard to see what might have been expected of it. It is certainly true that calculations were done for Walther's Darmstadt institute and, consequently, presumably for the rocket installation at Peenemünde. It is also true that Himmler himself suggested it be used for meteorological calculations with an aim at long-term weather prediction, a program initiated by a certain Dr. Hans Robert Scultetus (also an SS man) in Königsberg, and broken off because of the war.[485] But Sievers did not get around to apprising Boseck of this until about six months later, when "at present all our Atlantic and many other important weather stations have been lost and we are more than ever thrown back upon ourselves."[486] No doubt this was in Himmler's mind six months earlier as well. Sievers's ruminations cited above and based in part on experience seem nearer the mark—expecting concentration-camp inmates to

[481] Presumably these were what was left of those transferred from Plaszow.

[482] Had that occurred, Boseck planned to "separate out" the "useless prisoners."

[483] *Encyclopedia of the Holocaust*, vol. 4, p. 1322.

[484] The story of the end of Boseck's institute is in U.S. National Archives, Captured German Documents, T-580/125/39.

[485] NO-640, Himmler to Pohl, May 25, 1944. Scultetus was also involved in Himmler-initiated researches to justify Alfred Hörbiger's crazy *Welteislehre*, or "World Ice Theory." See Michael Kater, *Das Ahnenerbe der SS, 1935–1945* (1974), under Scultetus, and Martin Gardner, *Fads and Fallacies in the Name of Science* (1957). In chapter 7, below, student work on this theory is discussed.

[486] BDC file on Boseck, Sievers to Boseck, Dec. 7, 1944.

work intellectually, and accurately (and so willingly), for the Third Reich was, at best, a vexed business. If one has a certain mildly Marxist cast of mind, Boseck's effort and the one at Kraków-Plaszów can be seen as instances of a "new war-determined rationality," using people according to their ordinary qualifications in bourgeois life.[487] However, even at the moment, second thoughts of Sievers's sort must have seemed more "rational." Rather than rational, Himmler's order seems desperate. By May 1944, it was probably clear to many Germans that they would not win the war; the question was, how badly would they lose it; might some sort of agreement be reached with the enemy powers? These were issues, however, that none dared speak of openly for fear of arrest and execution. Himmler's May order was a literally last-ditch attempt to use every conceivable sort of scientific resource left in Germany to the benefit of the Third Reich. Four future concentration-camp divisions were even envisioned (in December 1944) for chemistry, physics, electrotechnics, and construction.[488]

This is not to deny strong elements of rationality and rationalization in the concentration-camp mathematical effort. For one thing, it acknowledged the evolved status of the prisoners. Originally, the declared purpose of the camps was to arrest people for purposes of "re-education," or for prophylactic security reasons. Thus, for example, the mathematician Ernst Hellinger and the psychoanalyst Bruno Bettelheim both spent stints in concentration camps and were then released. The first camp opened, Dachau, was publicly celebrated.[489] Conditions were never particularly good, however, and soon terror took over, as did the realization that the camps were an excellent source of slave labor. As slaves, the inmates could be worked to death and, in general, treated as human material, as, for example, for the various medical experiments carried out in the camps. Thus the original purposes of indefinite "preventive detention" and "reeducation" became quickly subsumed by the better-known and more permanent ones of terror and economic/scientific "rationality." The war only accelerated the use of such human material.[490] Terror and sadism were part of this structure, as they were of the extermination camps in Poland. Himmler's May

[487] Ebert n.d.: 5.

[488] NO-640, Sievers to Pohl, Dec. 7, 1944.

[489] The *Völkischer Beobachter* for Tuesday, March 21, 1933, cited Himmler at a press conference: "On Wednesday in the neighborhood of Dachau, the first concentration camp was opened." Here, he said, various functionaries who threatened state security would be kept since it was impossible to keep them for any length of time in the usual prisons and equally impossible to let them go free. "We have taken this measure without any heed for petty scruples in the conviction of thereby acting for the reassurance of the general population and in their spirit." This report also said Himmler stated that this custody would not be maintained longer than necessary and rejected rumors of bad treatment of such prisoners. On March 23, the first shipment of about sixty prisoners arrived. See, e.g., Paul Berben, *Dachau, 1935–1945, The Official History* (1975), for the original intent of the camps. The opening of the *Völkischer Beobachter* article is reproduced in Berben on the page preceding page 1.

[490] See Pohl to Himmler, Apr. 30, 1942, as cited and discussed in Ebert, n.d.: 5–6.

order was an attempt to utilize the concentration-camp inmates other than in a purely physical way.

The date of Himmler's order is also of some interest. After the German overturn of Admiral Horthy's government on March 19, 1944, because he attempted to make peace with the Russians, Hungarian Jews began to be ghettoized in mid-April, and deportations began May 15. Thence came the language of Himmler's May 25 order reflecting the new supply of Hungarian Jews. In fact, however, no effort was made to use the Hungarian Jews in any such capacity—the transports rolled to Auschwitz-Birkenau, where virtually all aboard were gassed upon arrival. Himmler's May 25 directive to Oswald Pohl may have been more "rational," but the logic of the Third Reich intervened. An added irony is that the Nuremberg Trial Document NO-640 dealing with these matters is labeled *Judeneinsatz* (utilization of Jews), when at most one Jew seems to have been employed in the Sachsenhausen institute.

A second way in which the concentration-camp efforts were rational was the concentration on computation rather than on uncheckable intellectual production. As H. J. Fischer, who selected the original batch of prisoners, and who claimed credit for first thinking of a computational institute, remarked:[491]

> The tasks [set the inmates] were themselves auxiliary tasks for other projects, and the participating computational personnel did not learn for what the results of their work were used. Also, the checking of the computational work was not difficult. It was only necessary to busy two separated, working groups with the same task and then to compare both results with one another.

Whether, in fact, this was done is unclear.

Finally, an attempt was made to combine rationally those prisoners removed from Kraków-Plaszów in Poland and working on astronomical calculations in Alt-Thymen (connected to the concentration camp at Ravensbrück), those other prisoners removed from the same Polish camp to the concentration camp at Flossenbürg, and the group at Sachsenhausen (assembled from there, Buchenwald, and Dachau). This, like the December planning for future scientific institutes, though, seems to have been more in the interest of the self-aggrandizement of the participating supervisors than for any actual achievement of efficiency. However, despite his later disagreement with Walter, Boseck did apparently travel to Ravensbrück to see if he could use any prisoners there.[492]

This chapter has examined some mathematical institutions during the Third Reich, both ordinary ones, like societies and journals, and extraordinary ones, ranging from the institute at Oberwolfach, actuated by the finest scholarly and academic activity, to the gruesome debasement of slave labor under concentration-camp conditions.[493] However, no institution, even the last, was perhaps as

[491] Fischer 1985, 2:100.

[492] For a discussion of such motivations, see Ebert n.d. For Boseck's trip, see NO-640, Sievers to Sommer, Dec. 7, 1944.

[493] Curiously, in 1962, there appeared in English a short story "Maxwell's Equations" by the

peculiar as the movement toward discerning an ethnically German mathematics, or "Deutsche Mathematik," as distinct from other kinds. The next chapter is devoted to it and its principal protagonist, the distinguished mathematician Ludwig Bieberbach.

Russian physicist Anatoly Dnieprov, in which mathematics is done by slave labor under the aegis of a former German concentration-camp commander. Whether Dnieprov knew of the Plaszów or Sachsenhausen efforts is unknown to me, but having been born in 1919, he was old enough to have been a prisoner in Plaszów, where the mathematical activity explicitly took place with Russian prisoners. This story appears in at least three anthologies, the earliest of which (from which the others borrow) is *Beyond Amaltheia* (Moscow: Foreign Language Publishers, 1962).

CHAPTER SEVEN

Ludwig Bieberbach and "Deutsche Mathematik"

THE figure of Ludwig Bieberbach has already appeared frequently in the preceding pages.[1] He was a mathematician of high repute who, in 1915, when he was twenty-eight, was described by Georg Frobenius, one of the leading figures of the preceding generation of mathematicians as someone who attacked with his unusual mathematical acuity always the deepest and most difficult problems, and might be the most sharp-witted and penetrating thinker of his generation.[2] He was also, among mathematicians, a leading proponent of Nazi ideology. Yet, somewhat earlier, he had had a reputation as an academic who was politically of a relatively liberal cast, and during and after the First World War was a member of the faculty of one of the reputedly "politically more liberal" universities (and one with a high percentage of Jewish faculty).

Ludwig Bieberbach was born on December 4, 1886, in Goddelau, a town near Frankfurt-am-Main. In secondary school he was already interested in mathematics, being particularly influenced by a teacher who "knew how to lecture very interestingly on his topics."[3] In 1905 in military service in Heidelberg, on the side he heard lectures by Leo Königsberger, "a completely excellent teacher," then near the end of a long career. He reviewed the lecture announcements of the various universities as published in the *JDMV* and noticed that Hermann Minkowski had announced lectures on invariant theory.[4] Unaware of Göttingen's general mathematical reputation, but having progressed far enough in his studies to be able to listen to Minkowski, whose announcement sounded attractive, he decided to attend. Arriving in Göttingen, he became "fascinated" with Felix Klein—the way he lectured and the way he interested students in mathematical matters. Already prepared by material he had heard from Königsberger, he listened to Klein's lectures on elliptic functions. Bieberbach had been attracted to Göttingen by his interest in algebra, and Minkowski's announcement; however, Klein influenced him in an analytic direction. Four years older than Bieberbach, and already "habilitated" in 1907 at Göttingen, was Paul

[1] The title of this chapter is also the title of an article by Herbert Mehrtens (Mehrtens 1987).

[2] See Edgar Bonjour, *Die Universität Basel von den Anfängen bis zur Gegenwart, 1460–1960* (1960), 753–754. The writer was F. G. Frobenius on February 6, 1913 (ibid.: 765 n. 112). Bieberbach was succeeded by Erich Hecke, again with Frobenius's recommendation (ibid.: 754 and 765 n. 116).

[3] On September 21, 1981, Bieberbach (then nearly ninety-five) was interviewed by Herbert Mehrtens. The interview was tape-recorded and partially transcribed. I have a copy of that partial transcription thanks to Prof. Mehrtens. This memory, including the direct quote, comes from that interview, hereafter cited as BI.

[4] BI. Invariant theory was one of the most actively pursued research areas of the day. While its death was once presumed as a result of new interpretations of its problems, and even analyzed by historians of science (see above, chapter 2), it has apparently been reborn in recent years.

Koebe. Koebe would become famous as an analyst who did fundamental work in complex function theory, and infamous as one who was vain and whose papers were not models of clear exposition. The young Koebe also influenced Bieberbach's interest in analysis. In his own words, Bieberbach had, "so to speak, two souls in one breast":[5] Klein's automorphic functions on one side and more algebraic things on the other. Bieberbach satisfied the first side by writing a dissertation under Klein on automorphic functions.

Ernst Zermelo had been a "habilitated" *Privatdozent* rather longer than usual at Göttingen, and was chosen as Erhard Schmidt's successor in Zürich, when Schmidt left for Erlangen in 1910. Zermelo wanted some new doctorand to go with him, and he chose Bieberbach. In 1910 also Bieberbach announced a result from his "algebraic soul" that would initially make him famous. In 1900 David Hilbert had given a well-known lecture in Paris in which he mentioned twenty-three mathematical problems that he thought important for the future. Some of the problems had several parts, and some were not precisely formulated, but by and large they have indeed indicated the directions of twentieth-century mathematics. The eighteenth of these problems had three independent parts, all of which dealt with geometrically formulated problems, though their solution might involve other ideas as well. The first part was decidedly algebraic in character and asked for a generalization to *n* dimensions of a result already proved by Arthur Schoenflies in two and three dimensions. In late 1908, at Hilbert's instigation, Bieberbach gave a seminar lecture on Schoenflies's work and took up the question of its generalization.[6] He announced and sketched his successful solution in 1910, with full publication occurring in two parts in 1910 and 1912. Georg Frobenius had already occupied himself with related questions, and in 1911, before the second part of Bieberbach's proof had appeared, he had simplified the first part. This paper also gave Bieberbach an idea that he exploited in another related paper.[7] Thus, while Frobenius was no doubt honest and accurate in calling Bieberbach a shaarp-witted and penetrating thinker, he was also not exactly a mathematically unbiased observer. Furthermore, Frobenius's comment was in the context of a letter of recommendation. Bieberbach would later return to the theme of this early success in several papers, including one coauthored with Issai Schur.[8]

[5] "Two souls alas dwell in my breast" is a famous line of Goethe's (*Faust*, part I, line 1112).

[6] Ludwig Bieberbach, "Über die Bewegungsgruppen Euklidische Räume I," *Mathematische Annalen* 70 (1911): 297–336. For the statement that Hilbert instigated Bieberbach's work, see ibid.: 298, and Helmut Grunsky's obituary of Bieberbach in *JDMV* 88 (1986): 191. The result was Bieberbach's *Habilitationsschrift*. Thus Mehrtens (1987: 197) is incorrect in implying it was Klein who stimulated this work.

[7] Ferdinand Georg Frobenius, "Über die unzerlegbaren diskreten Bewegungsgruppen," *Sitzungsberichte der königlich Preussischen Akademie der Wissenschaften zu Berlin* (1911): 654–665 (pp. 507–518 in Frobenius, *Gesammelte Abhandlungen*, ed. J. P. Serre [Berlin, Heidelberg, and New York: Springer, 1968], vol. 3). Bieberbach's second part appeared in *Mathematische Annalen* 72 (1912): 400–412 and acknowledged Frobenius's work. The other paper mentioned is in *Göttinger Nachrichten* (1912): 207–216.

[8] *Mathematische Zeitschrift* 9 (1921): 161–162; *Sitzungsberichte Preussische Akademie der Wissenschaften* (1928): 510–535 (this is the paper coauthored with Schur); ibid. (1929): 612–619.

Not only Frobenius was impressed. No sooner was Bieberbach in Zürich than Arthur Schoenflies, whose work Bieberbach had generalized, arranged a teaching position for him at Königsberg, where Schoenflies was Ordinarius. Three years later, Bieberbach left this "very modestly paid" position to become Ordinarius at Basel, but he stayed there only two years.[9] In 1914, a new university was opened in Frankfurt-am-Main on novel principles: it was initially financed directly by the city, a seemingly unique event in the history of European universities.[10]

Schoenflies became the first Ordinarius in mathematics at the new university, and in 1915 he was no doubt active in arranging for Bieberbach's call to a second such position there. Frankfurt developed a reputation both as a relative hotbed of liberalism among the generally conservative German university faculty, as well as the location of numerous Jewish scholars. As many as a third of the Frankfurt faculty apparently had Jewish antecedents.[11] At that time, Bieberbach himself was also accounted something of a liberal.[12] Bieberbach's mathematical work proceeded apace, though mostly on the analytic side, enhancing his reputation—and he also found time to write textbooks in differential and integral calculus, and a brief book on conformal mapping. As if to demonstrate the breadth of his interests, his inaugural address on assuming the chair in Basel dealt with issues in the foundations of mathematics.[13]

On August 3, 1917, Georg Frobenius, who had been so impressed with Bieberbach's algebraic work, died in Berlin. Shortly before, Hermann Amandus Schwarz had been emerited, and his replacement was Erhard Schmidt, who assumed his position on October 1. Largely through his influence, Constantin Carathéodory and Issai Schur were both placed first on the list of three that the Berlin faculty submitted to the ministry (instead of Schur alone in that position); the ministry chose Carathéodory, and he accepted the position effective October 1.[14]

However, Carathéodory left again in 1919 to follow the call of his native Greece to help establish a university in Smyrna (present-day Izmir). In 1922, Smyrna, which is on the Ionian coast, fell to the Turks, and Carathéodory returned to Germany (to Munich) via Athens. Nevertheless, in his two-and-

[9] "Very modestly paid" is the phrase used by Helmut Grunsky in his obituary of Bieberbach, *JDMV* 88 (1986): 191.

[10] Scharlau et al. 1990: 97.

[11] Ibid.

[12] For example, Hans Freudenthal in a letter to me (Dec. 7, 1976) wrote "Bieberbach was known in Berlin as a moderate leftist, a rare phenomenon in German Academe." This is supported also by a conversation with Andreas Defant (Feb. 4, 1988), whose grandfather was an oceanographer who had known Bieberbach in Berlin in those days and later talked about him to his family. See also citations in Mehrtens 1987: 217–218, though Mehrtens's remark about communists is an exaggeration from what Freudenthal wrote me.

[13] "Über die Grundlagen der modernen Mathematik," *Die Geisteswissenschaften* 1 (1914): 896–901.

[14] Biermann 1988: 153–154.

a-half semesters at Berlin, he had supervised one doctorand: Erich Bessel-Hagen.[15] Thus in 1919, it became again necessary to fill the position. Procedures started in December, but 1920 was not a good year in Berlin; for example, Wolfgang Kapp's attempted *putsch* took place there in March and was only prevented by a general strike. All three of those initially suggested for the Berlin position (L.E.J. Brouwer, Hermann Weyl, and Gustav Herglotz) turned the post down, as did Erich Hecke. So a year later, the faculty was still trying; this time focusing on the need for a geometer. First place for the suggested position quite naturally went to Wilhelm Blaschke. In second place was Ludwig Bieberbach, of whom the faculty said, "If also his exposition now and then shows lack of the desirable care, this is by far outweighed by the liveliness of his scientific initiative and the large-scale layout (*Anlage*) of his investigations."[16]

On January 2, 1921, Bieberbach wrote that he would gladly come to a position held by such "eminent men," "longest, presumably, by his fatherly patron, Frobenius." On January 31, the ministry named Bieberbach to the position, effective April 1.[17]

Thus, by age thirty-four, Bieberbach had reached a pinnacle, Ordinarius in one of the two great centers of German mathematics. The other center was Göttingen, where he had done his first significant work. There had long been a natural competition between these two centers, which had early acquired a somewhat personal tinge as a result of animosity between Felix Klein and Karl Weierstrass. This was carried on with even greater acerbity by Georg Frobenius, who seems to have been a man of sharp and sharply expressed opinions, and who had succeeded Leopold Kronecker at Weierstrass's instigation.[18] Stimulated by the American example, Klein had encouraged the involvement of German industrialists in the support of mathematics,[19] and reputedly had the ear of the national educational authorities in Berlin who made appointments. Indeed, in 1917, when the ministry suggested to the university at Berlin that Edmund Landau, a Berlin graduate, might be considered for the position eventually filled by Erhard Schmidt, Frobenius, suspecting the hand of Klein in the suggestion, and though already severely suffering from a heart condition, arose from what would be his deathbed six months later to word an acid rejection of the suggestion.[20] Curiously enough, though Frobenius had been (with Friedrich Schottky) the approver of both Landau's dissertation and his *Habilitationsschrift*, he wrote

[15] Two well-known mathematicians also "habilitated" with Carathéodory during this brief period: Hans Hamburger and Hans Rademacher (Biermann 1988: 185–186.)

[16] Biermann 1988: 193–194. *Anlage* can mean "layout" but also "talent."

[17] Ibid.: 192–194.

[18] Ibid.: 150–152.

[19] Schappacher and Kneser 1990: section 2.2 (pp. 12–15); David Rowe, *Historia Mathematica* 12 (1985): 278–291 (this is an essay review of three historical works); Reinhard Siegmund-Schultze, "Felix Kleins Beziechungen zu den Vereinigten Staaten, die Antfänge deutscher auswärtiger Wissenschattspolitik und die Reform um 1900," *Sudhoffs Archiv* 81 (1997): 21–38.

[20] Biermann 1988: 182–183, 328–330.

disparaging remarks about his student.[21] More curious still, Bieberbach, from his Berlin vantage point beginning in 1925, would launch an apotheosis of Klein's mathematical attitudes, and a denigration of David Hilbert's, culminating in 1934 with praise of the dismissal of the Jewish Landau as an "un-German type" unsuited to teach German students.[22]

The Klein praised by Bieberbach was the intuitive genius who had a natural feeling for the geometric-physical basis of mathematical results. In fact, though Klein and Hilbert had sharply differing philosophical views on the nature of mathematics, in academic politics they worked well together. They collaborated in helping bring Jews like Hermann Minkowski, the famous astronomer Karl Schwarzschild, and Edmund Landau to Göttingen. Indeed, Minkowski (who had taught Albert Einstein mathematics at Zürich) and Schwarzschild, both of whom would die at tragically young ages,[23] were two of the earliest developers of (the Berlin professor) Einstein's relativity theories. Had Hilbert and Klein had serious differences, the former could certainly have accepted one of two offers of a chair in Berlin made to him in 1902 and 1914. A lecture Bieberbach gave in 1926, which will be taken up shortly, presents the spectacle of Bieberbach, established in Berlin, elevating Klein, who had been the bête noire of Berlin mathematics, and attacking Hilbert, whom Berlin had multiply tried to attract.[24] A further irony was that Bieberbach's first prominent mathematical accomplishment had been stimulated by Hilbert. Though the Nazi physicist and scientific functionary (as well as Nobel laureate) Johannes Stark later spoke of the "business concern (*Konzern*) of Göttingen mathematical Jews led for a long time by Klein and Hilbert," Bieberbach not only (more accurately) distinguished the two as to "style," but would later defend both as (different kinds of) German thinkers.

The years 1920–21 were good ones in other ways for Bieberbach. He became secretary of the German Mathematical Society. His article for the German encyclopedia of mathematical sciences, "New Investigations Concerning Functions of Complex Variables," which intended to update results in the field to 1920, appeared.[25]

[21] Ibid.: 163. In fairness, Frobenius did attempt twice in vain to promote Landau to Extraordinarius at Berlin, in 1904 and 1908 (ibid.: 175–177), though the document cited in note 20, even after being perhaps toned down, was still somewhat disparaging about him (ibid.: 328).

[22] Nevertheless, it is quite clear that anti-Semitism had nothing to do with Frobenius's opinions of Landau, as he was an active and ardent promoter of another of his Jewish students, Issai Schur.

[23] Karl Schwarzschild (1873–1916), after leaving Göttingen, became director of the observatory at Potsdam, despite a refusal to be baptized. He apparently died of a skin disease while serving in World War I. Hermann Minkowski (1864–1909), who was perhaps Hilbert's closest mathematical friend, died of appendicitis. See Fraenkel 1967: 86–88.

[24] Although in both 1902 and 1914 the letter recommending Hilbert (which was composed by Frobenius) contained small barbs against him (Biermann 1988: 165–167, 182–183, 310–312, 324–326).

[25] On September 20, 1915, Robert Fricke wrote Bieberbach asking him to undertake this article because the editorial board of the *Enzyklopädie* had decided "to break off all connections to France," and the article to be updated had originally appeared under the editorship of Emile Borel. Klein

Bieberbach and Landau

In 1921, Bieberbach also received a letter from Edmund Landau. In this, Landau criticized a statement Bieberbach had made in a survey article for a Yugoslavian journal (published in Zagreb) indicating how a much better result was almost trivial; gave a simpler proof of a different research result of Bieberbach; and corrected some misprints.

Landau would prove to be of further annoyance to Bieberbach. One of Bieberbach's theorems had to do with bounds on the amount a certain class of complex-analytic maps rotates geometric figures. Paul Koebe had proved an earlier theorem about bounds on the distortions caused by such maps, and Bieberbach's introduction to his paper in volume 4 of the *Mathematische Zeitschrift* (1919) explicitly said that Koebe's "distortion theorem" contributed nothing to his "rotation theorem." There are two questions here: the existence of bounds of a certain type (the qualitative question), and obtaining explicit, perhaps best possible, bounds (the quantitative question). In 1920, in volume 6 of the same journal, Koebe, Bieberbach's former mentor at Göttingen, said that, on the contrary, the qualitative rotation theorem was an immediate corollary of his distortion theorem, though the quantitative one was not. In 1921 (same journal, volume 9) Bieberbach publicly replied, sort of admitting Koebe was right, but saying that quantitative results were his aim, and anyway, both Koebe's theorem and his rotation theorem flowed directly from another theorem of his: "My conjecture that my 'surface theorem' is the true root of all results known up until now about the behavior of univalent mappings has thus found complete confirmation."[26]

In 1922 Landau took up the matter in his advanced seminar and wrote Koebe and Bieberbach a joint letter. Landau said that, as a consequence of this further study, he had come to the conclusion that Koebe was more correct than Bieberbach in their public exchange, but not correct enough! For the "qualitative theorem" followed directly from a well-known inequality (due to others, but appearing in a book by Landau cited by Koebe). Landau indicated that the consequence was so immediate he would never think of appropriating it as an independent theorem. As to the "quantitative theorem," all agreed that this was a new contribution by Bieberbach, but his results were far from best possible.[27] Also in 1922, Landau simplified the proof of and improved another result of Bieberbach.[28]

(Fricke's teacher and coauthor) agreed with him that Bieberbach was the man for the job. See Bieberbach correspondence originally in the possession of Niels Jacob, now deposited with the Niedersächsische Staats und Universitätsbibliothek in Göttingen (hereafter cited as BL). All Bieberbach correspondence cited below and not otherwise annotated is from this collection.

[26] *Mathematische Zeitschrift* 9 (1921): 162. The theorem is known today in the English language literature as the "Area Theorem."

[27] In fact, the best possible result was not obtained until 1936 by the Russian mathematician Gennadii Goluzin (*Mat. Sbornik* 1, no. 43: 127–135).

[28] See numbers 167 and 171 in Landau 1984: vol. 7.

Both Koebe and Bieberbach seem to have been somewhat self-important people who ill-brooked competition. Koebe was notorious as well for appropriating the incomplete ideas of younger mathematicians and finishing them up as his own. It will be recalled that dislike of Koebe was one thing Hilbert and Brouwer could agree upon.[29] When Bieberbach joined the faculty at Frankfurt, he was asked to provide a brief biography, as were all new faculty. When the physicist Max Born arrived at Frankfurt in 1919, Schoenflies, as Dekan, provided him with a copy of this record of his colleagues, and he read some of them, as did Hedwig Born, his wife. Presumably Max Born had known Bieberbach, since they were both in Göttingen at the same time, though by that time, Born had converted from mathematics to physics. In any case, Hedwig Born found Bieberbach's brief autobiography amusing in its vanity, and copied out choice passages for Albert Einstein. Einstein's reply was:

> Mr. Bieberbach's love and veneration for himself and his Muse is quite delicious. May God preserve him, for it is the best way to be. Years ago, when people lived their lives in greater isolation, eccentrics like him were quite the rule amongst university professors, because they did not come into personal contact with anyone of their own stature in their subject, and apart from their subject nothing existed for them.[30]

The picture of Bieberbach in the Pólya picture album[31] around 1921, with all allowance for camera angles, is true to this view of Bieberbach at that time: vain and superficially cocksure.

Both Koebe and Bieberbach were undeniably important mathematicians, yet Bieberbach had the reputation of a careless expositor, while Koebe's exposition was often reputed to be impenetrable. Landau was a mathematician of encyclopedic knowledge of the literature in his special areas of expertise, meticulous to a fault, and always devoted to finding the simplest proof possible of a result. He could not resist sticking pins in people he considered self-inflated, especially when the mathematics was in an area in which he was acknowledged as a leading contributor and expert.

Many of Landau's papers are explicit commentary on the papers of others, refining or improving them. Not infrequently, they deal with papers of friends or acquaintances. However, they usually have the tone of simply advancing the subject, rather than the somewhat sharper edge of the private letter to Koebe and Bieberbach or his private comment on Bieberbach's Yugoslavian publication. Landau could be devastating and could do it publicly in print when he thought he was dealing with a total incompetent whose significant errors in print had not been discovered by others.[32]

[29] Reid 1976: 32–33; and above, chapter 2.

[30] *The Born-Einstein Letters*, trans. Irene Born (New York: Walker, 1971), 12–14. Cf. the portrait of the philosopher Dilthey in chapter 3, above.

[31] G. L. Alexanderson, ed., *The Pólya Picture Album: Encounters of a Mathematician* (1987), 58.

[32] E.g., Landau 1984: vol. 9, no. 244, comments on a paper of Maria-Pia Geppert. This was Landau's last publication prior to Hitler's accession. M-P. Geppert's brother, Harald, became a mathematician who actively promoted the Nazis.

Another somewhat self-important young mathematician whom Landau took to task in print was the young Wilhelm Blaschke.[33] Blaschke was primarily a geometer, arguably the foremost geometer of the first half of the twentieth century, and mostly worked in areas of mathematics completely foreign to Landau's interests. However, in 1915, he wrote a small paper in the *Reports of the Academy of Sciences of Saxony* (Leipzig) that dealt with the kind of analytic question with which Landau was familiar. Landau's follow-up paper appeared in the same *Reports* in 1918 and contained the paragraph:

> However, the proofs that Herr Blaschke gives for each of the two parts of this theorem are not only unnecessarily complicated; also the first part should be regarded as well known, the second part as the immediate consequence of a known product construction of Picard and Mittag-Leffler.

The remaining three pages of this note not only demonstrated this statement, but, in typical Landau fashion, were dense with references to the literature.[34]

Small wonder, then, that a letter from Blaschke to Bieberbach in 1921 ended with the sarcastic query: "Wouldn't you like to free Göttingen from Landau?"[35]

The Frankfurt Succession

The context of Blaschke's 1921 letter is the problem of who would be Bieberbach's successor at Frankfurt. Since Bieberbach responded positively to the Berlin inquiry on January 2, 1921, and was not officially named until the thirty-first, Blaschke, writing on January 27, was apparently among those friends of Bieberbach informed early of his impending move. This discussion is of more than a little interest because it reveals the attitudes of Blaschke, Bieberbach, and the general 1920s mathematics establishment toward Jews and anti-Semitism.

Wilhelm Blaschke was a year older than Bieberbach, born in Graz, Austria, on September 13, 1885. His father was a secondary-school teacher of mathematics who apparently gave his son his first inclination toward the subject. Graz had been Kepler's city for six years, and the house where he had lived was still extant in Blaschke's youth. Graz was also the capital of the Austrian province of Styria, a "Germanic outpost" facing the East. It was the Styrian governor Anton von Rintelen who conspired with the Nazis to become chancellor had the abortive July 1934 coup against the Dollfuss government been successful. In 1938, near civil war in Graz preceded the capitulation of Vienna and *Anschluss*. While

[33] That it was important to Blaschke to be important is my inference from conversations with Natascha Artin Brunswick, Werner Burau, Erich Kähler, Christoph Maass, and Hans Zassenhaus. He was also a well-traveled and cosmopolitan man with a passion for photography.

[34] Nevertheless, Blaschke's paper introduced the important idea that came to be known as "Blaschke Products." Landau's paper (1984: vol. 7, no. 134) is dated June 8, 1918. It may be that the war delayed Landau's seeing Blaschke's paper.

[35] BL, Blaschke to Bieberbach, Jan. 27, 1921. This letter antedates the letter of Landau to Bieberbach cited earlier. Since Bieberbach and Blaschke had been correspondents for some years, they may have had earlier discussion of Landau.

such events took place long after Blaschke's childhood, it is safe to say that Graz and Styria were earlier also outposts of pan-German nationalism.

Blaschke finished secondary school at age eighteen, studied architectural engineering locally for two years, then moved to Vienna, where he obtained his doctorate in mathematics under Wilhelm Wirtinger in 1908. Attracted by the presence of the well-known geometer Eduard Study at Bonn, Blaschke traveled there in 1908, "habilitating" under Study in 1910, shortly after his twenty-fifth birthday; however, between his first arrival at Bonn and his *Habilitation* there, he managed to spend a semester in Pisa and one in Göttingen. As his reputation was steadily growing, and he was not disinclined to travel, he was offered and assumed a succession of positions, staying nowhere more than two years: Greifswald, the German Technical University in Prague, Leipzig, Königsberg,[36] and Tübingen. At this last, he stayed only a semester, being appointed the first professor of mathematics at the newly established university in Hamburg. He remained at Hamburg for the rest of his career and, together with Erich Hecke, also one of the first three *Ordinarien* in mathematics, and later Emil Artin, built it into a mathematics department that rivaled those at Göttingen and Berlin. Artin's two predecessors, successively Johann Radon and Hans Rademacher, also became very distinguished mathematicians.[37]

Blaschke was also an anti-Semite. As Werner Burau put it delicately, mathematics at Hamburg did not have much trouble in the Nazi period because Blaschke had taken care that there were not too many Jews there.[38] In fact, in 1933, apparently the only person teaching mathematics at Hamburg who was Jewish or "Jewish-related" was Emil Artin (whose wife was "partly Jewish").

Bieberbach and Blaschke likely met during Blaschke's semester at Göttingen. In any case, by 1917 there had already been some correspondence, as is clear from a postcard Blaschke sent Bieberbach in February 1917. In this, he "sincerely marveled" at the "many-sidedness" of Bieberbach's "scientific activity"—Bieberbach had shortly before published his ground-breaking paper in analysis on coefficients of univalent functions.[39]

Another card to Bieberbach thoroughly reveals Blaschke's attitude toward Jews in mathematics.[40]

> The idea of ranking [Leon] Licht[enstein] and [Georg] Po[lya] before Wirtinger [Blaschke's *Doktorvater*] is in any case a joke, even if a not very happy one. W[irtinger]

[36] Scharlau et al. (1990: 206) erroneously have Blaschke going to Jena rather than Königsberg from Leipzig.

[37] Blaschke's academic peregrinations are as in Hans Reichardt's obituary of him, *JDMV* 69 (1966): 1–8 (reprinted in Wilhelm Blaschke, *Gesammelte Werke* [Essen: Thales Verlag, 1985], vol. 3).

[38] Interview, Jan. 31, 1988.

[39] BL, Blaschke to Bieberbach, Feb. 28, 1917.

[40] BL, Blaschke to Bieberbach, Mar. 6 [1921]. This letter also is sarcastic about Tübingen. In the end, Bieberbach was succeeded by Max Dehn and Schoenflies by Carl Ludwig Siegel (Scharlau et al. 1990: 98–99). Both Schoenflies and Dehn were assimilated Jews. Compare the letter of Weyl cited in the next note.

> does not have the advantage of being a Jew, however, instead of it, he is indeed by far the better mathematician. . . . Wirt[inger] would with his Alpine primitivity (*alpenländische Ursprünglichkeit*) bring a fresh breeze into the varnished politeness of Frankfurt.

Pólya and Lichtenstein are presumably mentioned in connection with the filling of appointments at Frankfurt: in addition to Bieberbach's departure for Berlin, Schoenflies retired that year, and the same names appear in a letter of Hermann Weyl to Bieberbach on February 16, 1921.

Weyl found Pólya's sort of mathematical activity foreign to him and evaluated Pólya as more a problem-solver than one who developed fundamental theories. Nevertheless, he found Pólya's papers "bold thrusts," his questions "original," his style "clear." He had "the greatest respect" for Pólya, "one of the people for whom he had the most regard," even though Weyl was not close to him. Pólya, while "capable of great and inwardly felt loyalty," was "in no way" "lovable." Pólya was "fabulously sincere," "witty," and given to outbursts of temper. This last, as well as his pacifism, had hurt him.

Lichtenstein, found Weyl, was "from a scientific perspective a serious competitor" for Pólya; for though he was not as "full of ideas, original, and sharp" as Pólya, he "impresses by his working power." Weyl also ranked Johann Radon and Arthur Rosenthal behind Pólya and Lichtenstein and thought that Robert König was better than either of these but unfairly overlooked because of some early difficulties: "A man's reputation has a very considerable coefficient of inertia."[41]

Blaschke's already cited letter of January 27 also emphasized his attitude toward Jews: "That you designate Pólya and [Otto] Blumenthal as non-Jews is certainly only meant jokingly." Blaschke thought Pólya far better than Blumenthal, and mentioned that Blumenthal's "fame" stemmed primarily from von Kármán's desire to get rid of him at Aachen. Earlier, von Kármán had spoken very negatively of Blumenthal. He also mentioned Radon as Pólya's equal (in contrast to Weyl—Radon was in fact at the time Blaschke's colleague at Hamburg, and Weyl admitted to scarcely knowing him), though Blaschke would not wish to lose him from Hamburg. Blaschke also returned to the Jewish theme: "R[adon] is a born German-Bohemian and non-Jew." Shortly after followed the last sarcastic sentence about Landau mentioned above—"Wouldn't you like to free Göttingen from Landau?" Presumably Blaschke meant why not this annoying Jew at this annoyingly Jewish university, as well as the insult about a senior colleague leaving a premier position.

Blaschke will be further discussed in chapter 8, but in the present context, his free expression of his feelings about Jews to Bieberbach raises questions about Bieberbach's own attitudes. Herbert Mehrtens claims that Bieberbach had at this time no bias against Jews, only against foreigners.[42] Unfortunately, Bieberbach's letters to Blaschke do not seem to be extant; however, Blaschke's freedom with his anti-Semitic remarks would seem to lend credence to the idea

[41] BL, Weyl to Bieberbach, Feb. 16, 1921.
[42] Mehrtens 1987: 200.

that Bieberbach shared them—furthermore, for anti-Semites, Jews were and would be typified as foreigners in the German body politic.

On the other side, it must be admitted that Georg Pólya (a year younger than Bieberbach), whose temper, pacifism, and Jewishness were all strikes against him, sensed no animosity on Bieberbach's part and, in responses to queries from him, wrote to him quite openly about these matters.[43]

The "temper issue" arose from Pólya's bad-tempered altercation with someone in a railway car in 1913 whom he struck; the victim turned out to be both the son of an important person and a student and consequently caused Pólya some trouble in Göttingen. The "pacifism" issue—more accurately, rumors that Pólya had evaded service in World War I—arose as follows. Pólya was a native Hungarian (so Austro-Hungarian). At the beginning of 1917, he wrote a letter to the Austrian consulate (in Zürich) in which he

> declared (in somewhat different fashion) that I do not wish to participate in such a plainly senseless, unjust, and hopeless undertaking as this war. . . . (No decision has cost me a similarly long and painful consideration as this did, however, also with no other decision am I more satisfied, and will remain satisfied, even if, as it seems, it should bring me a life-long neglect [for a position].)

As to the Jewish issue, Pólya's father was a Jew, who was baptized a few years prior to his son's birth, and Pólya was raised a Roman Catholic, as appeared in all his papers. To this fact, Pólya appends a sarcastic remark reflecting on the anti-Semitic atmosphere: "What I am unofficially, I certainly do not need to explain."

After these explanatory remarks, Pólya (a naturalized Swiss citizen since 1918) reflected on the Weimar academic atmosphere: "You see therefore [from these facts] that for me all possibilities are closed off in advance. Several months ago, besides, that was completely clear to me."

As though the above were not enough, there were also rumors that Pólya was a "Bolshevist," a problem not mentioned by Weyl, which Pólya also refutes in the letter to Bieberbach:

> I must make this somewhat surprising statement [of refutation], because in [Bad] Nauheim [where there had been a recent meeting of the German Mathematical Society], a new small (very small) Ordinarius had the friendliness also to suspect[44] me of that. (I wish between us to remark that I had strong sympathies for the Socialists; however, since the events of the past two years in particular; since in Hungary they make economic policy as foolishly as vulgarly, I detest them with my whole heart, as say some majesties whom I do not wish to describe more closely.)

In any case, Pólya's economic situation was poor, and his outlook for economic improvement poor, despite his acknowledged brilliance and his breadth of learning (he had studied classical philosophy, law, and physics before turning to mathematics). He asked Bieberbach for mathematical correspondence,

[43] BL; Bieberbach wrote Pólya on Jan. 4, 1921, and Pólya answered on Jan. 18.

[44] In the manuscript is a caught "Freudian slip": *verteidigen* ("defend") appears, but is crossed out and replaced by *verdächtigen* ("suspect").

and pleaded with a certain humor at his own situation, "If you should have the opportunity to recommend an associate professor (Extraordinarius) to a Hottentot Ordinarius, so please think also of me."

If Bieberbach were primarily opposed to foreigners, as Mehrtens claims, it seems strange that he should think almost first off of Pólya as a possible successor, since Pólya, for all his brilliance, was a native Hungarian, married to a Swiss woman (and a naturalized Swiss as well), who had been plagued by rumors of an uncivil personality and draft-dodging.

One other mathematician whom Bieberbach wrote about the formation of the list for his successor at Frankfurt was Erich Hecke. Hecke was the other senior Ordinarius (besides Blaschke) at Hamburg. Hecke, also responding on January 27, recommended "above all" Max Dehn; as to the others whose names were apparently raised by Bieberbach, Radon was "very good," but Hecke did not indulge in the sort of panegyric Blaschke had. As to Pólya, he shared Weyl's opinion that Pólya was indeed very clever, though perhaps a bit too "artificial." As if confirming Pólya's own beliefs, Hecke remarked, "Above all, I believe practically, [to put] Pólya [on the list] will only be a beautiful gesture; I cannot imagine that he would actually be called [by the faculty]." On Blumenthal, Hecke in fact shared Blaschke's view: "Kármán recommends him very warmly, that is already not a good sign." Hecke also suggested Ernst Steinitz and Issai Schur[45] as possibilities.[46] In fact, Dehn (who was an assimilated Jew) would become Bieberbach's successor at Frankfurt.

Bieberbach's Conversion to Intuitionism

Within a few years of moving to Berlin, Bieberbach gave a remarkable address apotheosizing Felix Klein and castigating David Hibert in tones almost as severe as those used by Max Steck twenty years later.[47] This lecture was given to an association dedicated to the objectives of furthering mathematics education,[48] and thus quite appropriately took as its starting point the recent death of Felix Klein.[49] His text was never published; however, a copy was retained by Bieberbach.[50]

Entitled "Concerning the Scientific Ideal of the Mathematician," the lecture began with an attempt to understand the bases for the diametrically opposed and strong views mathematicians had about Klein, but ended with the total rejection of Weierstrassian and Hilbert-style formalism as a transitory period between Klein's view of mathematics and the coming (in Bieberbach's view)

[45] Hecke writes "F. Schur," but it is clear he means Issai Schur (whom he mentions as a "personal Ordinarius" at Berlin). Friedrich Schur, a mathematician at Breslau, was then sixty-five.

[46] BL, Hecke to Bieberbach, Jan. 27, 1921.

[47] Above, chapter 6, "Max Steck and the 'Lambert Project.'"

[48] BI.

[49] The lecture was delivered on Feb. 15, 1926. Klein died on June 22, 1925.

[50] I am indebted to Prof. Herbert Mehrtens for a copy of this talk, from which the material below is taken. Page numbers given are to this copy.

ascendancy of Brouwer's intuitionism. This is such a remarkable view of the history of mathematics, and was so even in 1926, that his peroration deserves quotation:

> So Formalism appears as a period that conveys the mathematician from the naive romantic intuitionism of Klein to the modern intuitionism of Brouwer and Weyl. Should I see this development correctly, the time will not be far off when the sole remaining importance of Formalism will be this historical role, where the overvaluation again fades away and the catastrophic consequences that it entails, namely the turning away from problems of concrete reality, will belong to the past.

Bieberbach discerned two mathematical ideals. One, which he associated with Klein, devolves in an interpenetration of physical and mathematical thought. Klein becomes represented as the ideal *anschaulich*[51] mathematician both in research and in exposition. In Klein's mathematical exposition, "Every conclusion that appears has an immediately visible, concretely serviceable (*sachdienlich konkrete*) meaning. Never does the route of thought ramble from the theme, in order to find its way back to the topic only by first going through the underbrush of many lemmas and auxiliary calculations."

As this passage already hints, the polemics also started early in this address. Klein's suggestive, intuitive, physically aware style as a mathematician, which might, but did not always, lead "to complete certainty in the strictest mathematical sense," had its opposite in "the great school of Weierstrass, which was on its way to conquer the world."

> First it [Weierstrass's school] shattered . . . the trust in the certainty of intuitively based conclusions and simultaneously, through a form of representation operating suggestively in its pedantry, secured the impression that by sharpening the power of drawing conclusions used and recommended by it, it also possessed the means to assist where it had just previously pulled the rug out from under intuition.

Bieberbach went on to speak of the "often seemingly pedantic exactitude of mode of expression in the writings of Weierstrass and his greatest pupil [Hermann Amandus] Schwarz. Therefore also the occasional utterance of Schwarz . . . that indeed Klein did not come under consideration for real instruction."

Klein and Weierstrass did not like one another. The antipathy felt for Klein by Frobenius, Bieberbach's predecessor in his chair at Berlin and his earlier promoter, seemed to know no bounds.[52] Yet here was Bieberbach, speaking as a Berlin Ordinarius, apotheosizing Felix Klein, while condemning Weierstrass, arguably the greatest of Berlin mathematicians, and his student Schwarz as pedantic. While disliking Klein intensely, pre-Bieberbach Berlin had tried twice in vain to lure David Hilbert from Göttingen. Bieberbach's praise of the recently deceased Klein was at the expense of the living Hilbert. As he said, "Thus we have the psychological basis for the tendency that is worked out at the extreme in Hilbert's axiomatics

[51] That is, Klein's intuitions were apprehensions of physical reality.

[52] Biermann 1988: 151–52, 166–167, 182–83, 305–307, 312–313.

and that doesn't wish to know anything of the intuitively objective."[53] According to Bieberbach, for Klein, the "intuitively objective" provided some certainty to the content of mathematical structures; for Hilbert, this was no longer true. It is not surprising to find Bieberbach taking a swipe at the French as well:[54] "The exactitude of foundation toward which Newton still strove [and] of which Leibniz had never lost sight, gave way, namely in Euler's hands, and in the hands of the French encyclopedists, to an opportunism." Furthermore,

> Hilbert's scientific ideal is directly inimical to the needs of applications. Under the aegis of Formalism,[55] applied mathematicians have, so to say, died out, and this shortly after Klein's initiative had inaugurated a new blooming of applied mathematics, shortly after Klein succeeded in rescuing it from the assaults of the Weierstrass school.

Bieberbach did have a valid point about the importance of geometric intuition in mathematics, and that mathematics should be more than the surety of proof. But his polemical exaggeration of the cleft between such intuition and the importance of rigorous proof, of the difference between such intuition and the role of axiomatics in declaring clearly what is being assumed, was both historically invalid and untrue to the everyday practices of mathematicians. Furthermore, the demonizing of Hilbert and the sanctifying of Klein was completely untrue to the relations between the men. For while Klein was skeptical of set-theoretic and axiomatic foundations of mathematics, his "Erlanger Programm" might be said to have been an early precursor of the axiomatic trend.[56] Also, Klein was instrumental in bringing Hilbert to Göttingen. Truer to Klein's view is that "despite his reserve with respect to axiomatically based mathematics, he personally took pains to have the most learned people in this area, in particular Hilbert."[57] In fact, Klein took regular weekly walks with Hilbert and Minkowski or Runge and, even after his retirement, worked together with Hilbert to bring the best people in mathematics and physics to Göttingen.[58] Klein was devoted to elevating the status of applied mathematics,[59] but his attitude is perhaps best expressed in his own words:[60]

[53] Klein is touted as *anschaulich*. On the other hand, Weierstrass showed on more than one occasion, perhaps most famously in connection with the original proof of the so-called Riemann Mapping Theorem, that intuitive ideas could lead to error.

[54] Klein and the great French mathematician Henri Poincaré had been rivals.

[55] In the original manuscript, "axiomatics" originally appeared and was struck out and replaced by "Formalism."

[56] As acknowledged by Bieberbach on p. 15. For Klein's "Erlanger Programm," see, e.g., his *Gesammelte Abhandlungen* (Berlin: Springer, 1921), vol. 1, 460–497. A hundred years after Klein's address, it was still sufficiently important to be reissued as an annotated separatum (Leipzig: Akademische Verlagsgesellschaft, 1974).

[57] Renate Tobies, *Felix Klein*, Biographien hervorragender Naturwissenschaftler, Techniker, und Mediziner, Band 50 (1981), 66.

[58] Ibid.: 67, 86.

[59] For example, the famous Berlin mathematician Ernst Eduard Kummer (1810–93) called applied mathematics "dirty mathematics" (Tobies 1981: 67), and Edmund Landau connected applied mathematics with *Schmieröl* ("grease") (Reid 1976: 26).

[60] Felix Klein at the International Mathematical Congress in Heidelberg (1904), as cited by Friedrich Hirzebruch, *Sonderheft der Mitteilungen der DMV* (Dec. 1990), 24.

> Without doubt, necessary for the prospering of science is the free development of all its parts. Applied mathematics thereby undertakes the double task of directing to the central parts again and again new stimulation from the outside and conversely making operant the products of the central research on the outside.

Some time has been spent on Bieberbach's address (though far from exhausting its varied remarks) and its tone, because Felix Klein became a posthumous divinity for the "Deutsche Mathematiker," those who attempted to discern a specifically German mathematics as distinct from other ethnic sorts, and Ludwig Bieberbach was their leader.

The questions arise as to why Bieberbach so distorted the true situation, and why the praise of Brouwer's "intuitionism," the condemnation of Hilbert and Weierstrass, the emphasis on the *anschaulich*, the contempt for set theory, and the elevation of "applied mathematics." This last is especially curious, since Bieberbach's own mathematical work was far from immediate application. The address becomes even more curious if it is realized that only twelve years earlier, Bieberbach had praised Hilbert and basically assumed a formalist standpoint. In 1913, Bieberbach became Ordinarius at Basel. A year later he gave an inaugural address entitled "Concerning the Foundations of Modern Mathematics."[61] While acknowledging that the formalists had difficulties carrying out a program that might be impossible,[62] and calling Brouwer's address on intuitionism and formalism "brilliant,"[63] Bieberbach nevertheless had difficulties with intuitionism because its adherents of necessity "denied broadly fruitful areas of modern mathematical research." For Bieberbach in 1914, the intuitionist could not meet the demands occasioned by scientific activity.

Bieberbach suggested his own way out of this dilemma, which might be called "contingent formalism." Mathematical objects should be "all objects of thought for which the axioms of analysis, i.e., of transcendental set theory, hold without contradiction." This raised the question of whether there are any such objects at all, whether therefore mathematics in fact exists (in a meaningful, nonself-contradictory fashion). Bieberbach's reply was that this question may be ultimately unanswerable, but that one can proceed in mathematics with a contingent notion of truth that seems not dissimilar to the notion of contingent scientific truth: "An object or a concept only has 'mathematical citizenship rights' so long as its use does not result in any sorts of contradictions." The last phrase of Bieberbach's address, "The truth of mathematics rests solely in its logical correctness and consistency," is thoroughly formalist in tone.[64]

Clearly, profound changes had taken place in Bieberbach's attitudes between

[61] Bieberbach, "Überdie Grundlagen der Moderne Mathematik," *Die Geisteswissenschaften* 1, no. 33 (1914): 896–901 (journal date is May 14, 1914). Citations are from this publication.

[62] Kurt Gödel's proof that the formalist program could not be completely carried out was still sixteen years in the future.

[63] Brouwer gave an address "Intuitionisme en formalisme" in Amsterdam in 1912. This appeared in *Wiskundigtijdschrift* (1913): 180–211, and an English translation by Arnold Dresden is in *Bulletin of the American Mathematical Society* 20 (1913): 81–96.

[64] Bieberbach 1914: 901.

1914 and 1926. In this light, the 1926 lecture seems a way station to the 1934 lecture on the structure of personality and mathematical creativity, the consequent conflict within the German Mathematical Society,[65] and the creation of the journal *Deutsche Mathematik* in 1936.

The Bologna Congress

One more such putative "way station" needs brief mention. An international mathematical congress was scheduled for Bologna in 1928 (two years after Bieberbach's address). Since the end of World War I, Germans had been barred from such congresses, as from many such international meetings, largely through French influence.[66] However, by the time of the 1928 congress, some saner heads on the international scene had prevailed, and a German delegation was invited to attend. In fact, at Toronto in 1924, the American delegation (seconded by Italy, Denmark, Holland, Sweden, Norway, and Great Britain) had proposed a motion to lift national restrictions (in addition to Germans, Austrians, Hungarians, and Bulgarians were excluded). In June 1926, this was adopted, and the previously excluded were invited.[67] However, not all Germans were happy with that result. Mathematicians of nationalist bent, among them Hellmuth Kneser, Erhard Schmidt, and Ludwig Bieberbach, outraged at their previous exclusion in 1920 and 1924, proposed a counterboycott. Schmidt was at the time president of the DMV, which rejected the offer to send an official delegation. There were at least two ostensible reasons for this rejection, which are laid out in a letter from Schmidt to Kneser.[68] The official reason was that the congress was sponsored by the "Union Mathématique Internationale," an organization from which Germans had been excluded. A further provocation from the German nationalist point of view was the plan for the official congress excursion to be to the electrical plant in Ledrosee. Ledrosee was in the South Tirol (or, as the Italians called it, Alto Adige): Austro-Hungarian until 1914, Italian after 1918. The irredentist issue of to whom this territory (with many German-speaking inhabitants) should belong still echoed in Austrian politics fifty years later, long after World War II. A mere ten years after the fact, Schmidt imagined chauvinistic French and Italian mathematicians exchanging remarks about the "liberated areas."

The organizers of the Bologna congress also had sent invitations to various corporate organizations of mathematicians. Schmidt found this contrary to

[65] Above, chapter 6, "The Bieberbach-Bohr Exchange and the 1934 Meeting of the DMV."

[66] E.g., Brigitte Schröder-Gudehus, "Challenge to Transnational Loyalties: International Scientific Organizations after the First World War," *Science Studies* 3 (1973): 93–118.

[67] *Proceedings of the International Mathematical Congress* (Toronto, 1924), vol. 1, p. 66; and *Atti del Congress Internazionale del Matematici, Bologna* (1928) Tomo 1, p. 5 (abbreviated *Atti* below). The statement of this decision, recounted here in Italian, is also reproduced in French translation in a footnote.

[68] HK, Schmidt to Kneser, n.d. The letter itself is also undated and is a response to an inquiry to Heidelberg to send a delegation.

usual practice, and hoped they would equally reject the offer to send official representatives. The Prussian Academy of Sciences would, and he believed the University of Berlin would as well (it did). Of course, no one would or could prevent individual attendance in Bologna; however, Schmidt thought individuals should attend only when there was a "pressing factual scientific task." In general, Schmidt wrote: "I would deeply deplore numerous attendance at the congress by German mathematicians." He hoped that the congress would prove a disaster because of the failure of German participation, for only then, through a complete reconstitution of a new international organization, could a "truly international" congress take place.

Another figure who felt strongly that the invitation should be rejected was L.E.J. Brouwer. Though Dutch, Brouwer was sufficiently a pro-German nationalist to see an affront in the present invitation by an organization that termed itself international in name but was purposely not international in fact. He worked behind the scenes for the German cause of a reconstituted, truly international body, and managed at least that the "third preliminary announcement" of the Bologna congress no longer mentioned the despised "union." Brouwer himself considered this concession worthless, and did not plan to attend the congress. Bieberbach felt similarly. On June 18, 1928, Bieberbach answered an inquiry of the Rektor at Halle. This was relayed to him by the Office for Academic Information as the then-Dekan of the Berlin philosophical faculty (which included mathematics and natural science). Bieberbach attacked the forthcoming Bologna congress in language similiar to Schmidt's—in fact, well over half of it is almost word for word the same.[69] However, in what seems an attitude and turn of phrase particularly Bieberbach's, Brouwer's achievement in getting the Italians to stop mentioning the "Union Internationale" in preliminary announcements of the congress became evidence that "the Italians have always greatly concerned themselves with covering up [the role of the 'Union'] with great adroitness."[70] Bieberbach's letter became widely distributed, and Hilbert learned of it through his Rektor. As a consequence, on June 29, Hilbert wrote a letter to all Rektors and leaders of mathematical seminars, expressing on behalf of himself, his Göttingen colleagues, and many others the "diametrically opposed conviction" to Bieberbach's. The previous autumn, German reservations about a congress connected to the "union" had been expressed.[71] In May, said

[69] The university archive in Greifswald kindly provided me with copies of Bieberbach's two letters and Hilbert's letter cited below. Constance Reid claims that Bieberbach's first letter was sent to all German secondary schools and universities (1970: 188). In contrast, Herbert Mehrtens says that Bieberbach's first letter "was very likely never intended to be widely publicized, but was circulated in German universities" (1987: 214). The close identity between Schmidt's letter and Bieberbach's suggests that Bieberbach (as secretary of the DMV) may have drafted Schmidt's letter as well (which was an "official" response). In any case, Hilbert certainly thought Bieberbach's letter was likely widely circulated and directed his remarks accordingly. On balance, Reid seems nearer right than Mehrtens on this matter.

[70] Bieberbach to Ziehen (Rektor at Halle), June 18, 1928 (as in copy at Greifswald, received there July 5, 1928).

[71] See note 68.

Hilbert, the Italians dissolved the relationship to the "union." From his point of view, "Italian colleagues have with the greatest idealism and application of time and effort troubled themselves for some time (*seit Jahr und Tag*) to bring into being a truly international congress." Not only did Hilbert's view of the Italians contrast sharply with Bieberbach's, but he found it in "the interest of German science and German respect" that everyone invited should accept the Italian invitation. Bieberbach replied with an equally widespread letter that he wrote "greatly against my inclination" since his "much-honored teacher" had "sharply attacked" him. In this he suggested that Hilbert had inadequate national feeling.[72]

Actually, the truth of the matter exists in published documents, though many writers somehow fail to allude to them. As already noted, the move to reinclude the formerly excluded had started no later than August 15, 1924, with an American motion presented at the congress in Toronto, and was positively affirmed in June 1926. However, the rules for the "union" did not allow "international congresses" to invite countries other than those belonging to the "Conseil International des Recherches," an organization that promoted the boycott of German scientists and consequently was detested by them. Actually, these rules had already been honored in the breach by mathematicians at the 1924 meeting in Toronto, to which representatives from Russia, Spain, India, and Georgia had all been invited, though none were members of the "Conseil." The decision of 1926, however, had been to invite Germany and the other purposely excluded nations to join the "Conseil." The Germans, at least, hesitated to do this. Nevertheless, the organizers of the Bologna congress invited Germans to attend, and on April 26, officially informed the "Union" of this.

After conferring with the president of the "Conseil," who at the time was the famous mathematician Emile Picard, the secretary-general of the "Union" replied a month later that the Bologna congress was consequently "illegal" and could not be represented as associated with the "Union." As a result, Salvatore Pincherle, the Italian mathematician who was president of the organizing committee, wrote a long and detailed letter (in French) to Picard, pleading diplomatically but forcefully the cause of true internationalism.[73] Picard's reply was, however, negative, and in the end, he refused to attend the congress. Ironically, Pincherle at the time was also president of the "Union"—thus, in the interests of international amity, he was running a congress that his own organization had said he should not run. This aim is explicit: "The Organizing Committee pursued its work intended to bring peace to people's spirits, reconcile countries that had been divided by the war, and reestablish the collegial relationships that had characterized mathematicians in prewar congresses."[74]

Thus Picard and Bieberbach at about the same time were both trying from ultranationalist but opposite viewpoints to prevent German participation. In the

[72] Bieberbach on July 3, 1928, as in note 70.

[73] *Atti* 1928: Tomo 1, pp. 5–10. I am indebted to Raffaella Borasi for an accurate translation of the Italian on these pages.

[74] Ibid.: 9.

true light of the Italian attitude, one cannot but see Bieberbach's opinion, admittedly shared by some colleagues, especially in Berlin, as a sort of revanchism that was more interested in smashing the "Union" than in international mathematical cooperation. Since German mathematics had flourished despite its exclusion from international circles, perhaps there was also an air of imperialism about Bieberbach's attitude: there should be a new organization, and Germany would play a leading role in constructing it.

Brouwer's attitude was in the same thoroughly German nationalist tone. He presumed to act as a sort of German agent in dealing with the Italians and the international organization. Karl Menger, who had just been Brouwer's *Assistent* in 1925–27, spoke of "Brouwer's precarious relations with the French officials of the Union Mathématique Internationale"; how "Brouwer's hatred in those years was concentrated on the French; and these feelings, which greatly bothered me, also tainted his mathematical judgment"; and that this "aversion against anything French" was what kept his "endless and very unpleasant correspondence" going with the international organization.[75] In 1928, of course, this crisis reached its culmination with the Bologna congress. Brouwer also circulated a letter urging a German counterboycott of the Bologna congress, about two weeks prior to its opening.[76]

Thus, not only was the Bieberbach of 1914, who was then a modified formalist, by 1926 enthusiastically and aggressively in the opposing mathematical-philosophical camp led by L.E.J. Brouwer, and viewing intuitionism as the coming future of mathematics, but in 1928 he was making common mathematical-political cause with Brouwer in advocacy of an extreme German nationalist position. Furthermore, he was supported in this by two of the other three Berlin *Ordinarien*: Erhard Schmidt and Richard von Mises.

In this way, by 1928, corresponding lines of intuitionist versus formalist, nationalist versus internationalist, Berlin versus Göttingen were drawn, and Bieberbach was on the Berlin, extreme nationalist, intuitionist side. Lest one read more into these juxtapositions than is really there, it should be remarked that Hermann Weyl, arguably Hilbert's most distinguished student, shortly after World War I was also attracted to intuitionism, and made forceful contributions to it. Weyl, in addition to his extraordinary mathematical ability, was a man of great literary and philosophical interests with a cosmopolitan outlook. By 1927, his "enthusiasm for Brouwer's ideas had abated," and he did attend the Bologna congress.[77]

However, as Weyl's interest abated, Bieberbach's apparently increased. Two other items need to be mentioned here in the attempt to understand Bieberbach's transitions. One is a preliminary nationalist mathematical skirmish in

[75] Karl Menger, "My Memories of L.E.J. Brouwer," in *Selected Papers in Logic and Foundations, Didactics, Economics* (1979), 242–243, 248–249.

[76] Ibid.: 249. This letter is explicitly (but anonymously) referred to in *Atti* 1928: Tomo 1, pp. 9–10 (and cited in German).

[77] Reid 1970: 148–157 and 186–187. The phrase about Weyl's change of attitude is Reid's (ibid.: 186). *Atti* 1928: Tomo 1, p. 61.

1925 over whether French authors, and in particular Paul Painlevé, should be invited to contribute to a volume of the *Mathematische Annalen* memorializing Riemann on the 100th anniversary of his birth (September 17, 1826). Here again we find Otto Blumenthal (as effective managing editor of the *Annalen*) on one side, suggesting French participation, and L.E.J. Brouwer on the other, arguing against it in the name of German nationalism. Here Bieberbach, though mostly on Brouwer's side, proposed a compromise: selected Frenchmen other than Painlevé.[78] Approach to the French was to be via Albert Einstein, then one of the *Annalen*'s collaborating editors.[79] In the end, however, no French author appeared in the volume, though in addition to articles of German authorship, there were articles by a Russian (Serge Bernstein), a Dutchman (Brouwer), two British (G. H. Hardy and J. E. Littlewood), an Italian (Tullio Levi-Civita), and a Pole (W. Sierpinski), as well as by representatives of former members of the Central Powers: Hungary (Léopold Fejér and Alfred Haar) and Austria (Wilhelm Wirtinger). In fact, Bernstein, Levi-Civita, and Sierpinski wrote in French.[80]

Another "event" was the general attitude toward Brouwer's ideas in Berlin. Hans Freudenthal remarked that soon after his arrival as a student in Berlin in 1923, he discovered that Brouwer's intuitionism was "all the rage" (*Tagessprache*) there. Indeed, the young Karl Löwner, about to become famous as a complex analyst, gave a Berlin course in 1923 in differential and integral calculus on an intuitionistic basis.[81] In 1926–27, Brouwer gave a series of lectures on intuitionism in Berlin that excited great attention; around the same time he also lectured in Göttingen and was received much less favorably; in March 1928, he lectured in Vienna, where he spent much time calumniating the upcoming congress in Bologna. Hans Hahn, among others, tried to calm him down, arguing for the virtues of forgetfulness, but to no avail—Hahn would be one of those formerly excluded who attended the Bologna congress.[82] Berlin seems to have become Brouwer's bastion, both politically and mathematically.

Max Born, a distant but concerned observer of the multifarious conflict, and one of Hilbert's students, wrote from Göttingen to his friend Albert Einstein in Berlin:[83]

[78] Mehrtens 1987: 213; Schappacher and Kneser 1990: 55. Paul Painlevé was prominent as both a mathematician and a politician. As a mathematician, he is still remembered for "Painlevé transcendants" in the theory of differential equations. As a politician, he founded a military-scientific institute in 1914 and by 1917 had become the French minister for war. In this capacity, he was responsible both for the negotiations with Woodrow Wilson for American entry into the war and for the appointment of Maréchal Ferdinand Foch to head the Allied armies. In the event, however, Painlevé worked hard to cancel the German exclusion, and was active in bringing about the June 1926 repeal of the exclusion clause. See Schröder-Gudehus 1973: 110–111.

[79] Mehrtens 1987: 213.

[80] *Mathematische Annalen* 97 (1927).

[81] Freudenthal 1987: 7, 10.

[82] Ibid.; Reid 1970: 184–185; Menger 1979: 249; *Atti* 1928: Tomo 1, p. 45.

[83] Born 1971: 98. Despite the difference in their politics, Born and Schmidt were old friends who shared a mutual respect (ibid.: 100). The date given for this letter (no. 58) on p. 96 is clearly wrong; presumably it should be December 20, 1928.

> But the worst of it all was that the Berlin mathematicians were completely taken in by Brouwer's nonsense. . . . I can understand this in Erhard Schmidt's case, for he always did lean to the right in politics, as a result of his basic emotions. For Mises and Bieberbach, however, it is a rather deplorable symptom.

Thus, even Bieberbach, by now committed to both intuitionism and an aggressive hypernationalism, was still seen by some as just falling away from his formerly more liberal position.

In the end, Hilbert led a delegation of seventy-six German mathematicians to the Bologna congress, where they were the largest number of foreigners. Also, nine Austrians, five Bulgarians and twenty-two Hungarian mathematicians attended. Apart from the 336 Italians, the formerly excluded were more than 22 percent of the other attendees. Moreover, the presidents at section meetings included not only the Austrian Hans Hahn but Germans such as Edmund Landau, Paul Koebe, Richard Courant, Leon Lichtenstein, Emil Gumbel, and Wilhelm Blaschke and the Hungarians Frigyes Riesz and Alfred Haar. Plenary addresses were given by Hilbert, Hermann Weyl, and Theodor von Kármán. All these (except Hahn) were also "official delegates" representing some organization or other. All, and many other well-known attendees, had been banned from "international" mathematical gatherings for the preceeding ten years.[84] Indeed, a German mathematician who failed to attend the congress because he felt constrained by an official position nevertheless wrote to it, that by its actions it had taken the first giant step toward the healing of relationships and that they would retain the fame of being the brave path-breakers—there would be no such political problems at the next congress four years thence.[85] Furthermore, the "Union," whatever the opinion of its general secretary (who apparently did not attend at Bologna), at its congress meeting unanimously acclaimed Pincherle's behavior in organizing the congress.[86]

Throughout the affair, Hilbert viewed Bieberbach as Brouwer's German cat's-paw:[87]

> In Germany, a political blackmail of the worst sort has come into being: you are no German, unworthy of German birth, if you do not speak and behave as I now prescribe for you. It is very easy to be free of these blackmailers. One only needs to ask them how long they have lain in German trenches. Unfortunately, however, German mathematicians have fallen victim to this blackmailing; for example, Bieberbach. Brouwer has understood how to make use of this condition of the Germans without himself being active in the German trenches, all the more to have a care for the incitement and the division of the Germans in order to set himself up as master over German mathematicians. With complete success. He will not succeed a second time.

[84] *Atti* 1928: Tomo 1, pp. 63, 25, 67, 26, 28, 34. Reid (1970: 188) is incorrect in giving the number of German mathematicians attending as sixty-seven.

[85] *Atti* 1928: 10. Possibly this was Georg Faber—the author is again anonymous.

[86] *Atti* 83.

[87] Schappacher and Kneser 1990: 57. Mehrtens 1987: 214–215.

The above note was Hilbert's private comment in late June 1928 about the agitation concerning the Bologna congress. The "first time" implied is presumably the capitulation, from Hilbert's point of view, of Bieberbach (and others) to Brouwer mathematically. Neither Bieberbach nor Brouwer served in World War I, and Hilbert's comments about trenches, especially given that Brouwer was Dutch, are pointed.

It was shortly after returning from the Bologna meeting (September 3–September 10, 1928) that Hilbert instigated the "*Annalen* crisis" mentioned earlier by his eventually successful attempt to remove Brouwer from its board.[88] Here again we find Bieberbach collaborating with Brouwer. They paid a joint visit to Ferdinand Springer, the *Annalen* publisher, and threatened him with attacks as a publisher lacking in German national feeling if he allowed the removal of Brouwer (who was Dutch) from the *Annalen's* editorial board. Springer apparently informed Harald Bohr, Richard Courant, and Albert Einstein, among others, of this visit.[89]

Also, in 1931, against Bieberbach's (and Brouwer's) wishes, the abstracts journal *Zentralblatt der Mathematik* was started by Springer with Otto Neugebauer as editor. The stimulus from mathematicians for such a new journal was that the current one, *Jahrbuch für die Fortschritte der Mathematik*, was several years behind in abstracting papers; from Springer's point of view, this was no doubt an opportunity to fill a publication vacuum to his own advantage. Through Bieberbach's influence in 1928, the Berlin Academy of Sciences had taken responsibility for issuance of the *Jahrbuch*, with exactly this problem in mind, and put Bieberbach in charge of the publication. But success was slight.[90] Here again we see Göttingen (the *Zentralblatt* and Neugebauer) versus Berlin (the *Jahrbuch* and Bieberbach). Actually, work on the *Jahrbuch* continued consistently throughout World War II; in 1940, its offices became shared with those of the *Zentralblatt*, and the German Mathematical Society became its copublisher. The Berlin mathematician Georg Feigl, who had been editor, took a professorship in Breslau (modern Wrocław) in 1935 and was succeeded by Bieberbach's student, Helmut Grunsky (who "habilitated" in 1938). In 1939, Grunsky was called to the German Foreign Office and was succeeded by the ardent pro-Nazi Harald Geppert, who committed suicide in 1945 at the end of the war. In 1934, the *Jahrbuch* was still working on publications for 1926; by April 1945, it was almost, but not quite, caught up. The academy, despite (or because of) the time delay, had called the *Jahrbuch* an activity "important for the war effort," and this had allowed the continuing work on it. In 1945, it ceased to exist.[91]

[88] Above, chapter 6, "The Case of Otto Blumenthal."

[89] Born 1971: 98; Mehrtens 1987: 214.

[90] Mehrtens 1987: 216; Siegmund-Schultze 1984a: 92–93; *Die Berliner Akademie der Wissenschaften in der Zeit des Imperialismus*, vol. 2, 1917–1933 (1975), 181. In ibid.: 341, one reads, "[For 1933] a shortening of the temporal discrepancy between mathematical research and information was to be hoped for."

[91] *Die Berliner Akademie* 1975, 3:384–386. Grunsky also "habilitated" in Berlin. Siegmund-Schultze 1984a: 95 and passim. Geppert was the brother of Maria-Pia Geppert (above, note 32).

The Question of Bieberbach's Motivations

What can account for Bieberbach's change from formalism to intuitionism, and for his development into an aggressive superpatriot, which would lead him into active promotion of the Nazi agenda, in both word and deed? For Bieberbach not only paid lip service to Nazi ideology through addresses like "Styles of Mathematical Creativity," but was among the most prominent agitators for the Nazi cause within mathematics.

Everyone who writes about Bieberbach is puzzled by this transformation.[92] The sole reasonable explanation to me seems to be that Bieberbach was originally a person of no truly fixed or well-thought-out philosophical or political ideas. He was the son and grandson of state employees,[93] and he grew up in a thoroughly Wilhelminian atmosphere—he was four when Wilhelm II dismissed Otto von Bismarck, and the idea of the "apolitical" civil servant came naturally to him. At the same time, he was someone not a little given to pompous self-inflation. As a young man, he did significant work on one of the "Hilbert Problems," attracting the admiration of famous figures like Schoenflies and Frobenius. At the age of just thirty-five, he was offered and accepted an Ordinarius position—indeed, what had been Frobenius's position—in Berlin, then the only rival to Göttingen in Germany.

He may never have known, and it certainly would have mattered little to him, that Brouwer, Weyl, Herglotz, Hecke, and Blaschke had all been preferred by the faculty to him.[94] Whether this desire on the part of the Berlin faculty to have Brouwer or Weyl indicated some leaning toward intuitionism already is unclear, since both were famous for their more conventional and considerable mathematical accomplishments. It may possibly have been an attempt at an anti-Göttingen emphasis. As already noted, Hilbert had twice turned down an offer from Berlin, in 1902 and 1914, and Klein was disliked in Berlin. Weyl, on the other hand, was Hilbert's distinguished pupil, who had just taken up the exposition and defense of the mathematical-philosophical ideas of Brouwer, ideas that were anathema to his teacher.[95] In any case, Bieberbach went to Berlin and became a passionate (if unpublished) supporter of Brouwer.

Aside from his undeniable mathematical ability, Bieberbach seems to have had a gift for seeking out or creating niches where he could shine. He filled those roles with great ability and efficiency, such as his effective, even though not statutory, running of the mathematical society's journal for a long time. Yet, without any personal philosophical or political compass directions to steer by, he seems to have simply sought the main chance for himself. He may have

[92] E.g., Mehrtens 1987: 217–218. Biermann 1988: 198.

[93] Mehrtens 1987: 197.

[94] Biermann 1988: 192–193.

[95] For Hilbert's rejection of the Berlin offers, see ibid.: 165–167, 187. For Weyl and Brouwer's ideas, see Reid 1970: 151.

become an intuitionist because part of being a "Berliner" for him was Berlin *contra* Göttingen, and there was no profit in being the Hilbert epigone he had been. At the same time, there was then no explicit champion of Brouwer among the Berlin *Ordinarien*. As to politics, aside from a genuine love of country, he seems, more even than most of his colleagues, to have been blown where the wind listed, thus obtaining self-aggrandizing advancement. During the Nazi period, he may also have sought to advance the station of mathematics thereby. It is not as though, objectively seen, Bieberbach had been denied due honors by that time. When Hitler came to power, Bieberbach was forty-six years old, had been an Ordinarius in Berlin for nearly twelve years, was well known for significant contributions to two distinctly different areas of mathematics, had written encyclopedia articles and well-received texts, and had been selected by a broader spectrum of Berlin colleagues for a term as Dekan. Yet it may be that he aspired to no less than becoming a czar (or *Führer*) of mathematics who would obtain for his subject matter (and for himself) its rightful station.

Whether Bieberbach was tacitly anti-Semitic prior to 1933 seems unclear. In mid-1932, Bieberbach apparently refused to lodge in a vacation hotel decorated with a swastika. Even in early 1933 he told Irmgard Süss that in Spain the Jews had been expelled, only to be soon recalled because one saw that without them decline set in. However, she says that, nevertheless, he soon became a party member.[96] These tales may be further evidence of Bieberbach's attempts to discern the "winning side" and join it as, on April 13, 1932, the SA and SS had been prohibited, and Hitler's first cabinet contained only three National Socialists (including himself).

Ironically, there were apparently occasional rumors that Bieberbach had had Jewish antecedents.[97] These may have stemmed from the fact that his full christening name was Ludwig Elias Georg Moses Bieberbach, from his "moderate liberalism" prior to 1933, and from the fact that there was apparently no baptismal church known for his paternal grandfather Elias Bieberbach (his maternal grandfather was Georg Ludwig, which explains two other of his given names).[98]

[96] Irmgard Süss, "Erinnerungen," an unpublished, undated typescript consisting of three pages labeled "Vor der Machtergreifung" and twenty-three labeled "Unmittelbar nach der Machtergreifung" (i.e., before and after Hitler's seizure of power). I am indebted to Prof. Martin Barner for obtaining a copy of this for me as well as for arranging an interview with Frau Süss on March 25, 1988: citations above are from "Vor," p. 2, and "Nach," p. 6. The story of the recall of Spanish Jews is, of course, false. Also, Bieberbach did not become a party member at this time, though he did join the SA in November 1933.

[97] For example, *The Brown Book of the Hitler Terror*, published September 28, 1933 (New York: Knopf) under the editorship of Lord Marley, lists Bieberbach among the dismissed Berlin professors (p. 155). Given its early publication date, despite assertions to the contrary in its preface, clearly not every purported fact could be given the necessary scrutiny, and this is well known concerning this publication. For example, the names of Courant and von Mises are misspelled. However, the rumor appears elsewhere as well; see Biermann 1988: 200 n. 2 and Littlewood 1986: 157–158.

[98] See the genealogical formulary Bieberbach filled out in 1933(?) to be a member of the *Reichsschriftumskammer*, which is in the BDC file on Bieberbach. He gives no baptismal church for his

His behavior under the Nazis seems to have been a singular mixture of naiveté and thoughtless aggression (as required of anyone aspiring to the political forefront). It may be that a sympathy with the nationalist resentment evinced by pro-Nazi agitating students also influenced him. Horst Tietz, who was a secret (half-) Jewish auditor of Hecke's classes at Hamburg (with Hecke's knowledge) and who spent time in a concentration camp, has said that in the mid-1950s, Bieberbach came to him in tears, saying he had never before known about the true conditions in the concentration camps, or about the Polish death camps.[99] Given that Karl-Heinz Boseck had been a student at Berlin and was an active pro-Nazi *Assistent* there, he certainly could have known. It seems, therefore, assuming the truth of his declaration to Tietz, that he did not want to know earlier.[100]

This explanation of personal self-aggrandizement as a governing theme in Bieberbach's professional life and his lack of deeply held beliefs may seem unfair, but Bieberbach might be contrasted with his fellow Berlin Ordinarius, Erhard Schmidt. Schmidt was a thoroughly decent man who, when Issai Schur was dismissed as a Jew, opened his lecture with a protest against what had been done to his friend and fellow Ordinarius. Schmidt worked successfully to get him reinstated (until autumn 1935)[101] under the exceptions clause for civil servants in office prior to 1914. Schmidt was among the few people with the courage to visit Schur after *Kristallnacht* (at which Schur went into hiding for a few days). But Schmidt was also a thoroughly conservative and nationalist man who could tell his friend Schur in late 1938: "Suppose we had to fight a war to rearm Germany, unite with Austria, liberate the Saar and the German part of Czechoslovakia. Such a war would have cost us half a million young men. But everybody would have admired our victorious leader. Now, Hitler has sacrificed half a million Jews and has achieved great things for Germany. I hope some day you will be recompensed but I am still grateful to Hitler."[102]

Yet, while from our present perspective Schmidt may have been momentarily callous and thoughtless, he was, as Menahem (Max) Schiffer said immediately after relating this anecdote, "a great scientist, a decent man, and a loyal friend." As good a witness as Werner Fenchel judged him as "conservative but free of prejudice. . . . During the Nazi period, he was *persona non grata*."[103] Erhard

grandfather Elias. It was common baptismal practice to preserve familial, especially grandparental, names, though only one of these many would generally be used. The date of 1933 for this formulary is suggested by the attached correspondence.

[99] Interview, Apr. 5, 1988.

[100] For Boseck, see above, chapter 6, "Mathematics in the Concentration Camps."

[101] Pinl 1969: 191.

[102] M. Schiffer, "Issai Schur, Some Personal Reminiscences"; this was to appear in a publication called *Mathematik in Berlin*, edited by H. Begehr (1998), but it does not seem to have done so. It was originally presented at an international colloquium in honor of Schur at Tel Aviv University in 1986. I am indebted to Prof. Begehr for a copy.

[103] Werner Fenchel "Erinnerungen aus der studienzeit," *Überblicke Mathematik* (1980): 161; cf. Hans Rohrbach, "Erhard Schmidt, Ein Lebensbild," *JDMV* 69 (1968): 209–224.

Schmidt, like Hellmuth Kneser,[104] shows that one could have been originally conservative nationalist in politics, have initially welcomed Hitler's accession to power, and yet have been possessed of a sufficient sense of great personal integrity gradually (if necessarily tacitly) to have changed one's mind while maintaining academic standing and reputation. There are many similar examples of nationalist conservatives who originally supported Hitler, but came to oppose him. Among others more outspoken and generally better known were the famous pastor Martin Niemöller (an ex-submarine commander and ex-*Freikorps* fighter), who stated accurately at his trial in 1938 that he had voted Nazi since 1924 and had no love for republics, and the economist Jens Jessen. Jessen had a spectacular career as an enthusiastic Nazi academic, using his political service to further his academic advancement. Jessen was no more a democrat than Niemöller, yet he gradually became involved with Carl Goerdeler and General Ludwig Beck and a conspiracy against Hitler's life. After this failed, he was executed in November 1944. These examples are mentioned to emphasize that there could and did develop a conservative right-wing opposition to Hitler within Germany, and some conservative academics tacitly or more explicitly took part in it.[105]

Kneser's good friend, Wilhelm Süss, was somewhat similar to him, though his position as leader of the German mathematical community required continual interaction with Nazi officialdom, and prevented the "inner emigration" available to Kneser and Schmidt. In any case, the sort of passionate wholehearted enthusiasm for Naziism and its consequences exhibited by Bieberbach without pause throughout the duration of the Third Reich was in no way predicated or necessitated by a conservative-nationalistic viewpoint and initial enthusiasm at the outset in 1933. The zeal of the convert is banal; but neither was Bieberbach any sudden convert to a conservative-nationalist point of view in 1933. Thus the conclusion that Bieberbach had no firm unshakable political or philosophical values. The contrast between Schmidt and Bieberbach earlier in 1928 (when they were drawing the same extreme political consequences for mathematics) is unequivocally shown later by an incident in which Schmidt prevented the *Habilitation* of the *Assistent* Karl Molsen on the basis that his *Habilitationsschrift* was inadequate, though Bieberbach and Werner Weber had already approved it, because Molsen was a deserving National Socialist.[106]

[104] Coincidentally, both came from Dorpat (modern Tartu in Estonia). For Kneser, see HK, passim; also see his obituary of Erich Kamke (*JDMV* 69 [1968]: 206–208).

[105] For Niemöller, see Hans Buchheim, "Ein NS-Funktionär zum Niemöller Prozess," *Vierteljahrshefte für Zeitgeschichte* 4 (1956): Dokumentation, pp. 307–315. For Jessen, Matthias Gross, "Die nationalsozialistische 'Umwandlung' der ökonomischen Institute," in Becker et al. 1987: 147–148, as well as several other citations in that volume. The two best-known early strong supporters of Hitler who eventually fled Germany are perhaps Ernst ("Putzi") Hanfstaengel and Hermann Rauschning.

[106] Schappacher and Kneser 1990: 36. This *Habilitationsschrift* is presumably essentially the article appearing in *Deutsche Mathematik* 2 (1937): 117–126. Dealing with algebraic irreducibility criteria for some very special polynomials in a limited and computational way, it is an adequate publication, but too insubstantial to be a *Habilitationsschrift* (at least in Schmidt's view). After his necessary

Curiously enough, Molsen's doctoral dissertation was approved in 1935 with Issai Schur and Bieberbach as approvers. It seems a creditable piece of work and is in an area in which Schur was an expert.[107] The rejected *Habilitationsschrift* was in the same area, but seems slighter. If Schappacher and Kneser's political characterization of Molsen is accurate, this is yet another example of the active Nazi with a Jewish supervisor.

Mathematics and Typological Psychology

The general lay view of mathematics in 1933 Germany was not much different than in the United States of today—many thought mathematics necessarily inaccessible and dull, only of interest to "geniuses" who were less than humanly interesting in other ways.[108] It was also, of course, *the* rationalist subject. The *Mathematische Reichsverband* in fact had been originally founded to counter mathematics' bad reputation, and to give it some cachet among the general educated public. With the sudden Nazi accession to power came a sudden *political* emphasis on the value of feeling as opposed to intellect, an emphasis even promoted by some academics.[109] Rather than the union of *Geist* and *Macht*—Intellect and Power, long dreamt of by German academics[110] now *Geist* was seen *als Widersacher der Seele*—"Intellect as the Adversary of the Soul." "Soul" was all-important, and irrationalism was promoted as fiercely within the academy as without it. Erich Jaensch in his worries about this irrationalism mentioned particularly the "confusion-creating propoganda of a sectarian discipledom of Ludwig Klages."[111] In such an atmosphere, there may be some excuse for Bieberbach having a desire to save respect for mathematics. Indeed, the addresses of Robert König and Gerhard Thomsen discussed in detail in chapter 5 can easily be read in this light. However, Bieberbach seemed also motivated by self-aggrandizement, and a desire to become the effective *Führer* of mathe-

departure from Göttingen (above, chapter 4, "Hasse's Appointment at Göttingen"), Werner Weber spent a summer semester interlude as replacement at Frankfurt for Carl Ludwig Siegel (on leave at Princeton) before moving on to Berlin at Bieberbach's behest.

[107] See Karl Molsen, "Über spezielle Klassen irreduzibiler Polynome," *Schriften des Mathematischen Seminars* (Berlin, 1935–37): 35–48.

[108] See chapter 5, "The Value of Mathematics in the Nazi State."

[109] E.g., Martin Heidegger's infamous address: "Die Selbstbehauptung der deutschen Universität," which was originally delivered May 27, 1933, and was reissued in 1983 (Vittorio Klostuman, Frankfurt) by Heidegger's son, Hermann Heidegger, together with previously unpublished reflections by Heidegger on his year as Rektor at Freiburg. The tone of this publication is exculpatory, and this is certainly not the place to enter the massive ongoing Heidegger debate.

[110] Ringer 1969: passim.

[111] "Geist als Widersacher der Seele" is the title of the philosopher Ludwig Klages's famous book. For Klages's personal positive relationship to Naziism (though he remained in self-imposed Swiss exile), see, e.g., correspondence in U.S. National Archives, Captured German Documents, roll T-580/125/38. The citation from Jaensch is in Jaensch and Althoff 1939: vii.

matics. He almost immediately used his already considerable position within the mathematical community to promote himself as its leading Nazi ideologue. His first public attempt to link *anschauliche* mathematics with racial type occurred on July 13, 1933.[112]

This was followed by the Easter Tuesday 1934 address later published as "Personality Structure and Mathematical Creativity" (*Persönlichkeitsstruktur und mathematisches Schaffen*)—the paper that called forth the international rejoinders from G. H. Hardy and Harald Bohr and led, because of Bieberbach's desire to appear (illegitimately) as spokesman of the German Mathematical Society in responding to Bohr, to the crisis at Bad Pyrmont described earlier. Also, in the same year, Bieberbach published a similar article on "styles of mathematical creativity."[113] The content of these two articles, which overlapped significantly, pretended a serious intellectual foundation for its anti-Jewish and anti-French remarks. There *are* differences in mathematical style and presentation, in mathematical interests, and in attitudes toward those interests, which vary among individuals. What Bieberbach did in these lectures was attempt to connect these undeniable individual differences to psychological types that were racially or nationally determined. Thus the articles move in an unclear mixture of "racial science" and typological psychology. Nevertheless, they are worth examining in some detail because they reveal the intellectual links forged by Bieberbach between the various expressions of his political, philosophical, and intellectual ideas.

The psychological typology used by Bieberbach was that of Erich Rudolf Jaensch, who himself in 1933 had made a connection between the kind of psychological anthropology he had been developing since the 1920s and Hans F. K. Günther's "racial science."[114] Jaensch was perhaps the academic psychologist who, after 1933, most thoroughly accommodated his theories to National Socialism. Yet earlier, though a nationalist who believed that Hitler's movement represented a cultural renewal for Germany, anti-Jewish statements were apparently absent from his work.[115] However, after 1933, such remarks were frequent. Chapter 3 of his apparently instantly famous 1938 book *Der Gegentypus* (The anti-type) is 100 pages entitled "The Anthropological Goal of the German Movement and the German National-Becoming (*Volkwerdung*) in the Light of the Basic Organic Human Types." Whether section 14 of the same, entitled "The Anthropological Way of Thought of the German Movement Furthers Not Opposition, But Rather Understanding between Nations," involves a reversion to an earlier pacifist inclination toward the mutual assistance of nations in the interest of their self-development, or is just more of the Nazi-inspired "peace in

[112] *Sitzungsberichte Akademie Berlin* (1933): 643. The importance of reestablishing a German style of mathematics is already stressed here.

[113] "Stilarten mathematischen Schaffens," *Sitzungsberichte Akademie Berlin* (1934): 351–360.

[114] Ulrich Geuter, "Nationalsozialistische Ideologie und Psychologie," in *Geschichte der deutschen Psychologie im 20. Jahrhundert* (1985), 172–200.

[115] Ibid.: 183–185, 190.

our time" talk popular in 1938, is unclear.[116] What is clear is that it was easy for the nationalist Jaensch, who believed in a biologically based psychological anthropology and had formulated an elaborate such system, to accommodate to German National Socialism.

Bieberbach based his own ideas on the exposition in Jaensch's 1931 opus *On the Foundations of Human Cognition*,[117] applying Jaenschian categories to mathematicians and their style of exposition. The lecture on personality structure exists in two published forms: a condensed version in *Forschungen und Fortschritte* for 1934, which is what caused Hardy's letter of response in *Nature*, and a fuller version published in a pedagogical journal.[118] Harald Bohr's article against Bieberbach was occasioned not by the lecture itself, but by the report of it in *Deutsche Zukunft*.[119] The "styles" paper contained some of the same material and was delivered to the Berlin Academy of Sciences the same year.[120] "Personality Structure and Mathematical Creativity" began with a justification of the boycott of Edmund Landau's classes.[121] Landau's (in)famous definition of π in his differential and integral calculus as twice the smallest positive root of the cosine function (as defined by power series) was taken to exemplify his "inorganic" manner that was "foreign to reality" and "inimical to life." Landau was contrasted with the true German (and Berliner) Erhard Schmidt. For a definition of a "Jewish thought type," Bieberbach cited a certain Paul Ernst: "Jewish thought always commences from something already mental (*etwas schon Gedachtem*), it never comes from Nature and human experience."

While Bieberbach praised the rejection of Landau by the Göttingen students as a recognition of an inappropriate foreign style that demanded rejection, nevertheless he explicitly said that his remarks had nothing to do with the importance and degree of scientific ability. "Thus the preceding exposition also does not treat the unarguable service of Landau in the discovery of new scientific facts."

Jaensch had originally introduced his negative S-type as that of the French, more particularly of the cosmopolitan Parisian,[122] and Bieberbach now introduced Jaenschian typology in connection with a remark of Henri Poincaré about how difficult it was for Frenchmen to read James Clerk Maxwell. (Of course, many other people also found Maxwell difficult.) The S-type, who, in Bieberbach's words, "only values those things in Reality which his intellect infers (*hineinsieht*) in it," was contrasted with the I-type, who "is wide-open to Reality"

[116] Geuter (ibid.: 184) inclines to the former view. Erich Rudolf Jaensch, *Der Gegentypus. Beiheft* to *Zeitschrift für angewandte Psychologie und Charakterkunde* (1938), 194.

[117] Erich Rudolf Jaensch, *Über die Grundlagen der Menschlichen Erkenntnis* (1931).

[118] *Unterrichtsblätter für Mathematik und Naturwissenschaften* 40 (1934): 236–243.

[119] Above, chapter 6, "The Bieberbach-Bohr Exchange and the 1934 Meeting of the DMV."

[120] *Sitzungsberichte der Preussischen Akademie der Wissenschaften* (1934): 351–360.

[121] For details of this, see above, chapter 4, "Hasse's Appointment at Göttingen." All citations below are from the lecture published in *Unterrichtsblätter für Mathematik und Naturwissenschaften* 40 (1934): 236–243.

[122] Geuter 1985: 191; Jaensch 1938: 194–197. Of course, Jaensch's condemnation of sophisticated Parisian cosmopolitan culture overlooks the culture of Weimar Berlin.

and who "lets the influence of experience stream into him." "S-type" stands for *Strahltypus* or "ray (or 'radiating') type" while "I-type" stands for *Integrationstypus* or "integration type."[123] In brief, the S-type constructs the world to match his intellectual preconceptions, while the I-type infers the world as it is. Landau (the Jew) was construed as an S-type, though Bieberbach said that there are, of course, different sorts of S-types, and Landau and the noted contemporary French mathematician Edouard Goursat would not appreciate one another's style. Bieberbach contrasted the definition of complex numbers by Goursat or his predecessor, Augustin Louis Cauchy, with their introduction by Carl Friedrich Gauss. Naturally, while Gauss (and the Englishman William Rowan Hamilton) was "organic" and "concrete," the French were inorganic symbolists. Gauss was also contrasted with Carl Gustav Jacobi (a Jew). Jacobi was "oriental" and had a "heedless will to push through his own personality." Gauss was characterized as "nordisch-falisch," a term borrowed from H.F.K. Günther's racial theories; similarly, Euler was "ostisch-dinarisch," another similar term. Gauss and Euler exemplified different positive types as contrasted with the S-type Jacobi. Gauss, Riemann, and Klein were all praised for their closeness to applications, while Jacobi was also characterized by his striving "to form the human intellect (*Geist*) abstractly."

Bieberbach did face up to the contrast between Klein and Weierstrass, both *echt Deutsch*, whose "mathematical style as well as outer appearance" seemed completely different. Bieberbach validated both Klein and Weierstrass as truly German mathematicians who took proper concern for the intuitive as derived from inspection of nature, and attached different Güntherian race-theoretic adjectives to them—despite the fact that they were enemies both personally and in terms of their mathematical emphases and styles. However, Bieberbach's anti-formalist stance also found expression here:

> The unity, which, however, at least still existed in his [Weierstrass'] inner person first vanished among his pupils. That came about since his influence (*Wirken*) fell in a time when in other places the compulsion toward abstraction and formalization seemed to exist. For, remarkably, the paradoxes of set theory did not have the result that one turned one's back on a use of formal understanding divorced from reality, but, on the contrary, corresponding to the assessment of those who concerned themselves above all with set theory, the development went in the direction that banned intuition. For the errors and paradoxes were supposed to come from human intuition and human understanding. They were to be driven out by an absolutizing and dehumanizing of mathematical science. The Cartesian anxiety over error, so-called by Jaensch, had influenced our scientific and instructional activity lastingly and negatively.

Thus certainty of being error-free became more important than traffic in mathematical things. This is again blamed on the "unhealthy" influence of Ja-

[123] Originally the S-type was supposed to suffer from synaesthesia, and so the letter indicated this as well. In Jaensch's work, the letter *J* is used for the uppercase *I*, so actually the "integration type" is the "J-typus." Writers in English have used both "J-type" (as G. H. Hardy) and "I-type" (as above).

cobi's school.[124] This continuing diatribe against formalism as a mathematical philosophy set in the context of a diatribe involving the categories of racial science and Jaenschian typology represented for Bieberbach an intellectual connection with his previous antiformalist attitudes. What he did was tie those attitudes he disagreed with to Jewish and French influence, thus attempting to gain positive political recognition of his own views. Nor did he miss an opportunity to make the positive political point. Immediately following the attack on formalism partially cited above, he remarked:

> An SA comrade recently displayed to me during a service break how such misappropriately regimented education (*Verschulung*) operates when he laid the same question before some mathematicians and some unlearned men. The unlearned men answered it correctly instinctively, the mathematicians began to think it over.[125]

Bieberbach maintained that it was neither necessary nor welcome to hand mathematics over to the "anti-type"; a recent address on these issues by Hans Hahn[126] made him a speaker for these anti-types. Gödel's work suggesting the impossibility of carrying out Hilbert's program in the philosophy of mathematics was cited and given simultaneously intuitionist and racist consequences: "There can be no self-sufficient mathematical kingdom independent of human activity and intuition, therefore also none independent of the styles in which human racial membership expresses itself." The first half of this sentiment would have been applauded by Brouwer; how he may have felt about the second half is less clear.

However, although Bieberbach wanted to attack formalism, he could not completely dismiss axiomatics as a result of pernicious influence. After all, two of the principal heroes of the axiomatic method were David Hilbert and Richard Dedekind, and no one could have had more thorough German lineage than either of these.[127] Although the Bieberbach of 1934 apparently despised Hilbert's formalism, he found it necessary to validate Hilbert as a true German mathematician. Faced with a problem rather like his Klein-Weierstrass problem, Bieberbach solved it somewhat similarly. Dedekind and Hilbert were indeed "integration types," but of the "idealist form" of the type. This allows a "bridge" from their thought to that of S-types (like, presumably, Adolf Hurwitz and Emmy Noether, both Jews active in axiomatic work, though Bieberbach did not explic-

[124] It should be remembered that Bieberbach was well known, despite his brilliance, for a certain carelessness, and so this also amounted to a sort of self-justification against the likes of Landau.

[125] The problem: Two runners run along a track 100 meters long and one meter wide. One runs always in the middle of the track, the other runs diagonally across the track, so that at the fifty-meter point he has reached the other side, and then he runs diagonally across the track a second time [to its end]. Approximately how much is the difference in path length for the two runners? The answer is approximately 1/50 m. The precise difference in path length is $2(\sqrt{2501} - 50)$.

[126] It should be noted that Hans Hahn had also opposed Bieberbach in the matter of the Bologna congress.

[127] Nevertheless, David Hilbert appeared (erroneously) on page 1129 of volume 3 of the anti-Semitic encyclopedia *Sigilla Veri*.

itly mention them). Mathematicians like Dedekind and Hilbert, according to Bieberbach, do show a certain preference for thinking over intuition, but this is distinct from the S-type, who "denies the connection to an outer reality that is not mentally constructed." Naturally enough, German founders of axiomatics like Dedekind were still closer to reality than Frenchmen like Poincaré. But for all the hedging about typological variation, Bieberbach was plainspoken:

> Generally, I am of the opinion that the whole dispute over the foundations of mathematics is a dispute of contrary psychological types, therefore in the first place, a dispute between races. The rise of intuitionism seems to me only a corroboration of this interpretation.

Thus was a philosophical doctrine linked to the Nazi point of view.

The allowance of variations among integration types also permitted variations among S-types. Thus, for example, the Jew Hermann Minkowski showed some traits of the integration type, as did P.G.J. Lejeune-Dirichlet (whom Bieberbach characterized as French).[128] Similarly, while Jacobi's papers show evidence of occasional occupation with mathematical applications, he almost never lectured on such topics.

One intellectual problem that remained for Bieberbach was that Felix Klein, once again apotheosized in this article as exhibiting the ideal type of German style in mathematics, was often considered a forerunner of formalism. This was because Klein was an early proponent of the isomorphism of superficially different mathematical structures. However, said Bieberbach, for Klein the central idea was not the identical logical structure of different areas of mathematics, but that the same logical structure could be filled with different intuitive content. Other than rhetorically, and the thought is set about with rhetoric, it is difficult to understand the distinction Bieberbach is making. After all, to consider "filling" an "empty" logical structure means that it must have been stripped of some other content and considered in isolation.

Actually, Bieberbach went so far as to bring skin complexion into his discussion of race:

> If, namely, one marks the provinces in which our great German mathematicians are rooted through generations on a race-information map of the areas of diffusion of blonde and swarthy (*der hellen und der dunklen*) races in Germany, there comes the remarkable discovery that they almost all come from the diffusion area of the blonde races, reaching partly to the boundary of the swarthy areas, and seem to fall completely in the swarthy area only in the case of Euler.[129]

[128] Peter Gustav Lejeune-Dirichlet was born in Düren (about halfway between Aachen and Cologne) and attended school in Bonn and Cologne, completing the leaving examination at the early age of sixteen. He then went to study at the Collège de France in Paris because the level of pure mathematics in the German states was so low at the time (1822). See, e.g., *Dictionary of Scientific Biography* 1970, under Dirichlet (article by Oystein Øre).

[129] Leonhard Euler was born in Basel, which had been his family's home since the end of the

From this "strong Nordic dash" in all the great German mathematicians save Euler, Bieberbach draws the consequence that German mathematical education ought to be directed toward suitability to the Nordic racial type.[130]

The concluding item in Bieberbach's talk dealt with the fact that applications of mathematics are a reason given for its "cultural necessity" (*Volksnotwendigkeit*), to which he added its necessity for defensive purposes. He also argued that practical utility is not the sole justification of mathematics—it is also connected with "influences of blood and race" and, as an activity of an idiosyncratically German nature, needs no further justification.

This last point may indeed indicate a fear that mathematics would suffer as a result of the emphasis on proper irrational feeling popular in Germany. However, it is important to take this article seriously. Not only is it a justification of mathematics, but it creates intellectual links between political nationalism, racism, psychological typology, intuitionism, and the teaching of mathematics. It was seriously meant, and not just a pacification of the powers that were in the interest of mathematics. Furthermore, Bieberbach, in footnotes, specifically linked these ideas to those of his unpublished 1926 paper discussed previously and to his other paper appearing in 1934 on this subject. Thus he explicitly made the case that his thought had been in one consistent pattern (at least since 1926), and through the mediation of Jaenschian typology and now thoroughly respectable racial theories had found adequate expression.

The second 1934 paper by Bieberbach on this subject, entitled "Styles of Mathematical Creativity" (*Stilarten mathematischen Schaffens*), is much the same in content. However, it appears in the *Proceedings of the Berlin Academy*, and thus is addressed to a more elite audience. Not only is the content similar, but occasionally sequences of sentences are identical.[131] The paper begins, however, with a lengthy "scholarly" comparison of Cauchy's introduction and development of complex numbers (1821) and Gauss's (1825). Nevertheless, the material of the "personality structure" paper is recycled, only with different emphases. Many of the examples are the same, the evaluation of Weierstrass is the same. Klein, as the ideal German type, receives, if possible, even more emphasis. The praise of intuitionism is the same, as is a remark about the "blonde races." Hilbert is again an I-type of the sort open to S-type influences. In short, "Styles" is essentially the same paper as "Personality Structure" with an argumentation especially suited to Bieberbach's elite audience of scholars.

Three small differences are perhaps worth noting. One is Bieberbach's de-

sixteenth century. At that time, his great-great-grandfather had moved there from a town on Lake Constance. Thus he was thoroughly Swiss. See *Dictionary of Scientific Biography* 1970, under Leonhard Euler (article by A. P. Youshkewitch).

[130] Bieberbach's own ancestors came from the area just south of Frankfurt-am-Main. See his *Abstammungs-Nachweis* in Berlin Document Center, under Bieberbach.

[131] E.g., among others, the glossing of Poincaré's remarks about Maxwell, or some of the remarks about Felix Klein's relationship to formalism, and comparing Weierstrass and Klein.

fense of the intuitionist rejection of the *tertium non datur*, Aristotle's "law of the excluded middle":

> I might attempt to make clear with a somewhat drastic example where the feeling of the intuitionists strives against the unhesitating use of the *tertium non datur*. If I say, no fact known to me contradicts that in the records of the Academy [to which he was speaking] there is a letter in Gauss' own hand, so no one will draw the conclusion therefrom, that therefore such a letter may be found among the records. The formalist, however, demands of his follower that he recognize such existence proofs.

Bieberbach must have known, and many of his intellectually elite audience must have known, that this example is fallacious and thoroughly specious. Only if the nonexistence of Gauss's letter led in logical thought to some counterfactual conclusion (like the nonexistence of the Academy) could the formalist conclusion that the letter must therefore exist be drawn.

A second difference is the failure to praise the boycott of Edmund Landau's classes in Göttingen. Landau had been a corresponding member of the Academy since 1924 and apparently would remain so at least until 1937.[132]

The third matter perhaps worth mention is that in the printed version of Bieberbach's "Personality Structure" paper, Klein's informal remarks about racial typology in mathematics made in English at Northwestern University[133] were in the original language, while in "Styles" they were translated into German. This is curious, though its significance is somewhat unclear, as in both 1934 papers Bieberbach touted Maxwell as a German-type thinker in contrast to Poincaré (with whom Klein was a bitter competitor).

These two lectures certainly established Bieberbach as the most "progressive" (in a Nazi sense) of prominent German mathematicians as well as a mathematician who could claim that his *Weltanschauung* had long been congruent with the Nazi one. They provided him with the political credentials that might possibly lead to the leadership (in the Nazi sense) of the German mathematical community. Many years earlier, Hedwig Born and Albert Einstein had noted Bieberbach's vanity with amusement. In the circumstances of 1934, it was not as harmless as it had once seemed. The international storm raised by Bieberbach's articles and the consequent stressful meeting of the DMV in Bad Pyrmont have already been discussed. Bieberbach's desire that there be a *Führer* of the DMV seems heartfelt, a mode of political accommodation for mathematics, whose reputation for rationality might otherwise endanger it, and a fulfillment of personal ambition. Already an ardent nationalist, seeing the opportunity, he seized it, even if that meant abandoning quondam friends like Otto Blumenthal and Issai Schur. Bieberbach's suggestion at Bad Pyrmont of Erhard Tornier as *Führer* was perhaps simply "window-dressing" for his own ambitions, since Tornier was already probably known as a somewhat erratic sort and, in any case, did

[132] See *Die Berliner Akademie* 1979, 2:253 and 3:63–64 and n. 267.

[133] See, for these, chapter 6, "The Bieberbach-Bohr Exchange and the 1934 Meeting of the DMV."

not have the mathematical *bona fides* Bieberbach did, something presumably important to the mathematical community.[134] Perhaps Tornier and Bieberbach even arranged their complementary nominations to be *Führer* ahead of time.

Efforts to Ideologize Mathematics

One offshoot of the several months' turmoil within the German Mathematical Society following the Bad Pyrmont meeting was Bieberbach's decision to found the journal *Deutsche Mathematik*. Perhaps he expected that through political support of this sort of venture, he could become the even more important figure within the German mathematical community, if not within its present organization, that he hoped to be.

On October 14, 1934, a month before Bieberbach finally sent the proposed changes in the mathematical society's by-laws to Leipzig for approval, and the turmoil in the society that ensued,[135] he wrote the *Notgemeinschaft der Deutschen Wissenschaft* a letter about founding a new "German journal for mathematics."[136] The *Notgemeinschaft* was a state funding organization for science and scientists that was reorganized as the DFG and whose president at the time was Johannes Stark, a Nobel laureate physicist and an ardent supporter of Hitler. Several things about the letter are interesting. For one, Bieberbach wrote that it was "impossible to bring in the German Mathematical Society," "since it has distinctly proved at its Pyrmont meeting that, for the present, it did not wish to be the bearer of national interests." Thus the failure of the DMV to adopt a conventional *Führerprinzip* was the direct impetus for the journal *Deutsche Mathematik*. Furthermore, Bieberbach had been in contact with the national association of mathematical students (presumably Fritz Kubach) about such a journal, and he foresaw a substantial section of the journal directed toward students. Students and professors would work together on it. Also, "especially student" subscribers were foreseen as well as the usual scholars, teachers, and libraries, and the journal would explicitly depend on the support of the various divisions of mathematics students at the universities. Curiously, while France, the United States, and Great Britain at the time all had mathematical journals with material accessible to interested but inexperienced students, Germany does not seem to have had such a publication.[137] A journal of the sort Bieberbach proposed would consequently seem to plug a pedagogical hole irrespective of its avowed Nazi intention of filling the "long-felt need" of bringing "men and science, national

[134] Georg Hamel did have such *bona fides* and established the *Führerprinzip* in the *Mathematische Reichsverband* with himself as *Führer*. He was thus an obvious mathematically and politically correct person to succeed Blaschke (if only for a year). See above, chapter 6, "The Bieberbach-Bohr Exchange and the 1934 Meeting of the DMV."

[135] See chapter 6, "The Bieberbach-Bohr Exchange and the 1934 Meeting of the DMV."

[136] BAK R73 15934. The citation of this letter below is from this source.

[137] *The Mathematical Gazette*, the *American Mathematical Monthly*, and *L'Enseignement mathematique* (actually Swiss, but French-language) were founded respectively in 1899, 1894, and 1894.

(*völkische*) membership and scientific accomplishment" closer to one another. Such at least was the avowed purpose, and a reduced price (compared to other mathematical journals) was envisioned in order to encourage student subscription. As with Weiss' mathematical camps, while the motivation and intention of Bieberbach's letter seems reprehensible, its avowed purpose, when divorced from ideology, seems to have recognizable pedagogical value. By July 1935, the publisher S. Hirzel in Leipzig was calculating necessary costs, price, size, and format, and by November, the DFG had approved a subvention to the proposed journal, requested the necessary publication permission from the *Reichspressekammer*, and awarded Bieberbach an annual honorarium of 2,000 RM for his present and future efforts.[138]

Before pursuing the future of the journal *Deutsche Mathematik*, it is useful to realize that Bieberbach was far from alone in promoting the idea of a peculiar "Nordic" style in doing mathematics, as distinct from other styles, nor was its only basis Jaenschian psychology. It is also worth looking at some of Bieberbach's other efforts to place himself in the ideological forefront.

Already noted is that Udo Wegner had ideas (if unclearly formulated) along these lines, as did Oswald Teichmüller.[139] As early as 1923, Theodor Vahlen in his inaugural address at Greifswald had expanded on Klein's well-known expressions about the gifts of various "races" for different kinds of mathematics.[140] Following Hitler's failed *putsch* attempt in November 1923, Vahlen would become an early Nazi, and when Hitler came to power, he would rise to a reasonably high position in the education ministry. While Vahlen's address did cite Houston Stewart Chamberlain (on Euclid), it also made favorable remarks about Jewish mathematicians like Alfred Pringsheim, Carl Gustav Jacob Jacobi, Leopold Kronecker, Hermann Minkowski, and Felix Hausdorff. It had a good word to say about relativity theory. It also found Cantor's statement about the essence of mathematics being in its freedom a quite positive one properly interpreted, in contrast to Bieberbach's negative view of it as typically Jewish. Vahlen did say: "Thus mathematics is a mirror of the races and proves the presence of racial qualities in the intellectual domain with mathematical, thus incontrovertible, certainty." This is a startling anticipation of Jaensch and Bieberbach. But his attitude in 1923 was more in line with Klein's, that different peoples had different contributions to make, than with the one he would adopt ten years later as a Nazi official.[141]

Another expression of Nazi ethnic particularism in mathematics is the mono-

[138] BAK R73 15934, Heinrich Höter to Griewank, July 2, 1935; E. Wildhagen to *Reichspressekammer*, Nov. 26, 1935 (letter drafted by Griewank); Stark to Bieberbach, Nov. 28, 1935.

[139] Above, chapter 4, "Hasse's Appointment at Göttingen." See also chapter 8, "Oswald Teichmüller."

[140] See chapter 6, "The Bieberbach-Bohr Exchange and the 1934 Meeting of the DMV." Klein was not the only mathematician to make such remarks earlier. As noted above, there were well-known expressions of such sentiments by Poincaré and Weierstrass, among others.

[141] Theodor Vahlen, Greifswald Universitätsreden, 1923.

graph *Raum oder Zahl* (Space or number) by Cl. H. Tietjen.[142] This is directed at the elementary school and has a sort of imprimatur by Vahlen. It was published in 1936, the same year as the appearance of *Deutsche Mathematik*'s first issue. Tietjen says in his foreword that the question of space or number amounts to the question of German (Space) or Jewish (Number). While he seems to have developed his ideas himself, his exposition leans on already familiar sources (Klein's Northwestern address, Bieberbach's Prussian Academy lecture, Poincaré on his difficulties reading Maxwell, Vahlen's 1923 address [misdated as 1933]), with already familiar rhetoric. It deserves mention for two reasons. One is that "Number" is obviously not rejected—schoolchildren do need arithmetic. However, the concepts of elementary-school mathematics are to be built upon the fundamentals of Observation, Direction,[143] and Space (*Schau*, *Richtung*, *Raum*). "Time and Number will then be gained observationally."[144] The other is that Tietjen expressed the same genuine fear as Jaensch that mathematics as a subject matter was threatened by misunderstandings of the new *Weltanschauung*, and so there arose the necessity of building up a "new" Germanic mathematics.

As to the older "Germanic" mathematics, in 1936, a collection of articles entitled *German Seed in Foreign Soil*[145] appeared. The preface by one Karl Bömer, apparently also the publisher, explicitly eschewed any sort of German chauvinism; rather, "out of the deep love of our own people, which is the foundation of the Third Reich, indeed, also grows respect when faced with foreign idiosyncrasy (*Eigenart*) and foreign achievement." Bieberbach made a contribution to this volume concerning German mathematical achievement. True to the book's preface, these three pages reiterate Bieberbach's earlier ideas in softer international clothing. Here we find the names dropped of many famous German mathematicians, and

> thus more than one name comes to mind, whose bearer is himself not conscious how exactly by his accomplishment he embodies German style, more than one who in his modesty believes himself only a leaf on the international tree of science, more than one who even thinks his place at the right hand of Apollo will vanish, if he seeks the roots of his power in his nation (*Volk*). That, however, does not alter the circumstance that thus more than one named and unnamed belongs by the style of his creativity to the German nation and could not at all thrive in foreign soil.

Bieberbach emphasized the "weight of personalities" and how times of national agitation are times of extraordinary scientific as well as political accomplishment, giving German and French examples. Furthermore, "It is also well known that the entry of the Jewish nation into science begins in that moment that the emancipation commences to unburden them from the feeling of inferiority."

The kind of soft-spoken advocacy we find here of the value of international

[142] Cl. H. Tietjen, *Raum oder Zahl* (1936). This is apparently a selection from a larger work.

[143] Or "Line."

[144] Tietjen 1936: 18.

[145] Bömer, ed., *Deutsche Saat in fremder Erde* (1936).

apartheid in mathematics was also used in defense of Bieberbach's 1934 articles by his student Eva Manger. Thus, in her article "Felix Klein im Semi-Kürschner!"[146] which defended Klein against the accusation of being a Jew, she also defended Bieberbach's article, "Personality Structure and Mathematical Creativity." In fact, her article directly followed Bieberbach's "open reply" to Harald Bohr discussed earlier. Here again, the tone of international separateness is maintained:

> We are proud of our German mathematicians and their accomplishments, however, avoid decorating ourselves with foreign fame, on the contrary, are only too ready to recognize foreign accomplishments as such, and to value them. Indeed, it is from time immemorial a German quality—one could already say weakness—so to treasure the foreign that we forget what is ours (*das Eigene*).

She said the reason for all the emphasis in Germany on the Jewish question was the huge influence of the "foreign race" of Jews in economic, political, and cultural areas. She cited statistics, such as that Jews made up one and one half percent of the German population but 30 percent of the academic faculties. Indeed, the whole argumentation followed a party line laid down slightly earlier, as is evident from her citation of an article by Wilhelm Frick[147] in the *Völkischer Beobachter*: "To put a stop to this foreign infiltration (*Überfremdung*) had become a life-and-death question for the German people, so that its promulgation of racial laws represents only an act of self-defense (*Notwehr*) and not one of hate." Her article reviewed Bieberbach's with the aim of emphasizing its "nonracial" aspect, which does not involve the "subjective valuation or even devaluation of the styles of foreign races," justifying in this way statements like Bieberbach's that "the entire foundational struggle in mathematics is to be explained as a struggle between opposing psychological types, therefore in the first instance as a racial struggle."[148] Thus, for Bieberbach, she said, a "racial struggle" does not imply a valuation of races.

There is much more in this vein, including an excoriation of "P.S.," the reviewer of Bieberbach's lecture in *Deutsche Zukunft*—the review, with its lightly ironic title "New Mathematics," that led to controversy with Harald Bohr and at Bad Pyrmont. Bieberbach may have emphasized style in thinking, pedagogy, and perhaps even the selection of problems, as showing the influence of "blood and race on the type of mathematical activity"; whereas "P.S." in his *Deutsche Zukunft* article emphasized (naturally) conclusions from it about the work of

[146] *JDMV* (1934): Abteilung 2, pp. 4–11. "Semi-Kürschner" was the familiar term for the anti-Semitic encyclopedia *Sigilla Veri*.

[147] Wilhelm Frick was an early National Socialist (he took part in the Hitler *putsch* attempt of 1923) who in Thuringia on January 23, 1930, became the first Nazi in a state government. It was as Thuringian education and interior minister that he prevented the *Habilitation* of Max Herzberger at Jena. He was Hitler's first Minister of the Interior and as such was instrumental in the formulation of both the April 7, 1933, Law for the Restoration of the Civil Service and the Nuremberg Laws of 1935.

[148] *JDMV* (1934): Abteilung 2, p. 7; and Bieberbach 1934a: 241.

persons. However, those conclusions are implicit in the lecture, despite all attempts by Bieberbach and Manger to deny them. As Bieberbach before, Manger justified the "rejection" of Landau on these principles, putatively dealing only with matters of style, each of which was validated for its own "people." The point for Bieberbach and Manger was the separation of the Jewish "people" from the German people; this consequently meant the expulsion of the Jewish people from the German body pedagogical—whither was of no concern. Manger explained, "One may no longer count as Germans many researchers previously so counted." She explicitly (as Bieberbach) did not deny the value of their work, and disingenuously said such separation "should really be in the interests of each people." That the labeling and discernment of style was, in effect, done a posteriori on the basis of the "racial origins" of the creators was unremarked, though, as already noted, casuistical efforts were made to include, for example, David Hilbert and Richard Dedekind among the "Germanic types" while excluding Emmy Noether and Adolf Hurwitz therefrom. Manger's conclusion: "Thus German science relinquishes Jacobi and leaves it to Judaism to see in him one of its greatest sons. . . . Thus we defend ourselves against every attempt at contesting that Felix Klein is one of our great Germans."[149] This sounds reasonable, but, of course, it was written in the context of the effects of the segregation she advocated: the forcible expulsion of "Jews" from academic life that had been going on for over a year. "Jews" is given in quotation marks because the definition of non-Aryan in the April 7 law was given in a mixture of "racial" and religious categories.[150]

Eva Manger also wrote to *Deutsche Zukunft* protesting in similar but briefer tones the interpretation "P.S." made of Bieberbach's lecture.[151] The rejection of foreign style does not mean no more reading of foreign authors, she said, but being conscious of one's own style, and remaining faithful to it while absorbing foreign material. Also, historical examples show that it is not a question of "new mathematics" but of what "has always been manifest in all great mathematicians and their works—their national provenance (*Volkstum*)." Manger's articles clearly have Bieberbach as guide as well as inspiration (for example, she cited in both places Bieberbach's unpublished 1926 address).

What is striking about both Bieberbach's open letter to Bohr and Bieberbach's piece in the "German Seed" volume is their defensiveness. Observers as varied

[149] Later editions of *Sigilla Veri* had a paste-in insert correcting the mistaken attribution of Jewishness to Klein. A genealogy (*Ahnentafel*) for Klein follows Manger's article. The original mistake of some Nazis about Klein ironically seems to be based upon a Jewish mistake. See *The Jewish Encyclopedia*, vol. 7 New York (Funk and Wagnalls, 1912), 521. The ascription is by a Brooklyn M.D. named Haneman.

[150] From the April 11, 1933, elaboration and executive order for the April 7 law: "As non-Aryan counts a person who is descended from non-Aryan, in particular Jewish, parents or grandparents. It suffices if one parent or one grandparent is non-Aryan. In particular, this is assumed if one parent or one grandparent has belonged to the Jewish religion." Cited from Bruno Blau, *Das Ausnahmerecht für die Juden in den europäischen Ländern, 1933–1945* (New York, 1952), I. Teil, Deutschland: 19.

[151] *Deutsche Zukunft* (May 13, 1934): 15.

in inclination and ideology as "P.S." and G. H. Hardy drew similar obvious conclusions from Bieberbach's lecture/article; why, then, did he and Manger attempt an intellectually convoluted defense of his ideas? This is a defense that actually extends to mathematics itself. Thus Manger said in her *Deutsche Zukunft* piece: "What is of moment here is that, once and for all, it is stated that mathematical creativity also does not hover in airless space, that also mathematicians as mathematicians are rooted in the people to which they belong." Bieberbach's "German Seed" article of 1936 even seems to extend this defensiveness to German science, commencing: "The words 'German Science' arouse in many scholars the reaction that this concept demands or foresees a dismemberment of an organic whole." This last is easily explained, since Bieberbach went on to distinguish German science from science done in Germany, just as for him, German mathematics was concrete, organic, and systematic.

Remarks like Manger's, which seem to defend not only Bieberbach, but also mathematics, perhaps indicate that one motivation for Bieberbach's articles was to assure rational mathematics a place in an academe governed according to the Nazi *Weltanschauung*. Other examples of this are the remarks by Jaensch or the speeches of König and Thomsen cited earlier. In fact, the somewhat ironic tone used by "P.S." may derive from suspicion that this was *the* motivation of Bieberbach's lecture, and "P.S." found this apparent eagerness to make mathematics *gleichgeschaltet* ironically amusing.

But why, then, be so casuistically defensive about the manifest content of Bieberbach's ideas—a content that would certainly find favor with the powers that were, should they come across it? From the point of view of April 1934, or even 1936, in Germany, it was probably not clear to most what the developing international political relationship would be, let alone the intellectual one. After all, even by 1936, the *putsch* in Austria had failed, though Dollfuss was successfully assassinated; Poland and Germany had signed a nonaggression treaty; the Sudetenland was still Czech. On the other hand, the Saar had voted overwhelmingly to reunite with Germany, and, in March 1936, Hitler had made his bold move reoccupying the previously demilitarized Rhineland, though militarily he was too weak to fight if he met French or British resistance—which he did not. German academics sympathetic to the Nazis were nonetheless interested in preserving good international relations, especially in those disciplines in which Germany had been an acknowledged leader (even during the intellectual boycott), like mathematics, physics, and chemistry. Thus, there would have been a tendency among German academics like Bieberbach to set themselves up internally as devotees of the new ideological dispensation *à l'outrance*, while at the same time externally minimizing those consequences of that devotion that might give offense in other countries.

Two other aspects of Bieberbach's brief "German Seed" article deserve attention. One that may seem at first strange in a nationalist article is the mention of "a foreign guest on German soil, the Greek Carathéodory, whose activity can scarcely be thought of apart from German mathematics." However, not only does this reflect the oft-remarked German fascination with Greece, but no one less than Adolf Hitler had remarked: "A culture combining millenniums and embracing Hellenism and Germanism is fighting for its existence."[152]

The second is the elevation of Theodor Vahlen by mentioning him in the same breath as Gauss and Klein, and concluding his article with:

> As a young man [Vahlen was] richly decorated with laurel in the area of pure mathematics;[153] he has in a second flowering of his creativity accomplished fruitful results in various applied areas. What he creates as leader of the university division of the National Education Ministry appears as though it should be completely appreciated in its importance only by a coming generation.

This is the purest obsequious flattery. Bieberbach was certainly well aware of his own rather high international mathematical standing, and Vahlen's considerably lower standing, despite the belief at the time that Vahlen had solved an unsolved problem of Kummer. It seems clear, especially considering the role Vahlen played in the 1934–35 contretemps in the German Mathematical Society discussed in chapter 6, that Bieberbach had aimed and perhaps was still aiming at being the effective *Führer* of German mathematics, either in fact or as *éminence grise*.

There is no denying, however, that around this time Bieberbach seemed to truly believe in Nazi racial ideas. In late 1935, a mathematics student named Otto Richter at Berlin was apparently considering the racial background of German Nobel laureates and sent out detailed questionnaires. The great physical chemist (the "third law of thermodynamics") Walther Nernst refused to fill it out. Nernst at the time was seventy-one, and had retired the previous year. In fact, Nernst was born in and his early education was in West Prussia (now part of Poland), and his parents were eminently "Aryan." However, Nernst thought the form was ridiculous and returned it with a note saying he had more important things to do. It should be noted that two of Nernst's daughters had married "non-Aryans" and that he himself had made considerable argument against the dismissal of the (Jewish) spectroscopist Peter Pringsheim in the spring of 1933. These facts were probably known to Richter, who complained to the propaganda ministry. The ministry referred him to Bieberbach, then Dekan at Berlin. Bieberbach provided him with Nernst's birthplace and birth date and suggested a check of church registries; indicating his personal interest in the result. However, in 1935 Nernst's birthplace was in the "Polish Corridor." For whatever reason, Bieberbach took the matter in his own hands and sent Education Minister Rust a note complaining about Nernst's failure to appreciate "the basic tenets of the new Reich," and suggesting that Nernst be forced to fill out a form designed for professors suspected of non-Aryan descent. By February 29, 1936,

[152] Adolf Hitler, *Mein Kampf* (1943 ed.): 423.

[153] In 1891, Vahlen published an example that purported to settle an open question about curves in ordinary three-dimensional Euclidean space. In 1941 Oskar Perron showed that Vahlen's example was fallacious. It is thought that Perron had known for several years that Vahlen's example was wrong, but waited until its fiftieth anniversary to publish his disproof as a way of indirectly showing his contempt for Vahlen's politics. See Oskar Perron, *Mathematische Zeitschrift* 47 (1941): 318–324.

Bieberbach had sent Richter Nernst's completed form—which ironically proved his Aryan status.[154]

It would be a mistake to think all people associated with mathematics and also sympathetic to, or even actively involved in, the Nazi movement adhered to Bieberbach's brand of discerning a *völkisch* mathematics. On the one hand, long-time supporters of the Nazi cause, like the Nobel laureate physicist Phillip Lenard, thought that much of contemporary mathematics should simply be trashed. In 1936, he published a book (in four volumes) entitled *Deutsche Physik* that attempted to discern a *völkisch* German physics, much as Bieberbach wanted to do for mathematics. Concerning the nature of mathematics, Lenard, an experimental physicist and an opponent of relativity theory, said:[155]

> Because of its [mathematics'] fixed and clear inner construction, which gives certainty every time that it operates only with the necessary thought processes of the Aryan spirit, and in correct and honorable application eliminates every arbitrariness, so that complicated conclusions or every endangerment of certainty is able to be overcome, so was it also rightly called the "royal aid" to natural research. Aryans have developed it to such high accomplishments from Pythagoras forward to Newton, Leibniz, and Gauss. Gradually, presumably from approximately Gauss' time on, and in connection with the penetration of Jews into authoritative scientific positions, however, mathematics has in continually increasing measure lost its feeling for natural research to the benefit of a development separated from the external world and playing itself out only in the heads of mathematicians, and so is this science of the quantitative become completely a humanities subject (*Geisteswissenschaft*). Since the role of the quantitative in the world of the spirit is, however, only a subordinate one, so this mathematics is presumably to be designated as the most subordinate humanities subject. It works with that part of the human spirit which is left over when all the higher and highest capacities, standing completely apart from the quantitative, are disregarded. The capacity to count, and what is related to it, then remains left over. It is certainly not good to allow this humanities subject with all its newest branches any large space in the school curriculum.

Indeed, Lenard thought that contemporary physics teachers were "overfilled with too much new mathematics."[156] For Lenard, mathematics that Bieberbach was at pains to defend as truly German—the mathematics of Dedekind and Hilbert, as well as Riemann, Klein, and Weierstrass—was, in fact, suspect. Lenard shared Bieberbach's ideas about *völkisch* science, but drew negative consequences for mathematics from them. For Lenard, German physics was "classi-

[154] Mendelssohn 1973: 152. The quotation is as cited there. There are no scholarly annotations in this book, but I presume the story comes from Nernst's Berlin *Nachlass*, given the details and dates in Mendelssohn's presentation. However, Mendelssohn's gratuitous mention of Nobel laureates Otto Wallach and Richard Willstätter in this connection is erroneous, since Wallach had died four years previously and Willstätter was still living in Germany.

[155] Lenard 1936 1: 7.

[156] Ibid.: xi.

cal experimental physics," and the only pertinent mathematics was the mathematics pertaining to it.[157]

Lenard had been the only *Assistent* to the famous physicist Heinrich Hertz (1857–94), whose father came from a Jewish family, though his mother was "Aryan" and the family was a thoroughly assimilated practicing Lutheran one. In Lenard's book, *Great Men of Science*, Hertz was given generally laudatory treatment. However, Lenard showed himself somewhat sympathetic to ideas about racial style: "[In his book (posthumously published by Lenard), *Principles of Mechanics*] suddenly—deceptive, given the then lack of racial science—a Jewish spirit (*Geist*) strongly broke out which in Hertz' earlier fruitful works remained hidden."[158] Thus Lenard did not necessarily argue with Bieberbach's ideas; he did argue with the attempt to give then-modern mathematics, especially as applied to physics, a "truly German" character.

If Lenard was a Nazi physicist who opposed contemporary mathematics, there were also Nazi sympathizers among physicists who used it, supported it, but would have nothing to do with Bieberbach's *völkisch* ideas. One such was (Ernst) Pascual Jordan. Jordan was a student of the well-known theoretical physicist Max Born, and working together they published the first thoroughgoing description of Werner Heisenberg's "matrix mechanics." Born was a thoroughly assimilated Jew who resigned in 1933 and then emigrated. As for Jordan, "in spite of his sympathies for the National Socialist movement, Jordan never broke with the tenets of modern theoretical physics."[159] Indeed, Jordan made a not-so-veiled attack on Bieberbach in a small book, *Physical Thought in Modern Times*.[160] This shows the lack of unity on ideological consequences by distinguished scientists, each purporting to derive such consequences for mathematics from the Nazi *Weltanschauung*.

Jordan was stimulated by an attack on him by the philosopher Kurt Hildebrandt at Kiel,[161] who "undertakes moreover the attempt at a political defamation of *all of mathematical-physical research*, which according to his opinion 'leads to (*darauf aus ist*) the burying of the people-nation (*Volkstum*).'"[162]

[157] See, for example, the preface to his *Wissenschaftliche Abhandlungen* (Leipzig: S. Hirzel, 1942–44). Originally four volumes were envisioned, but only three ever appeared.

[158] Lenard, *Grosse Naturforscher* (1943): 330. The cited passage appears in the editions of 1929, 1933, 1936, 1940, 1943.

[159] The quotation is from Karl von Meyenn's article on Jordan in the *Dictionary of Scientific Biography* 1970 (vol. 17 [suppl. 2], 448–454, p. 451). Lenard and Jordan are further examples among physicists of the unexplained phenomenon noted earlier among mathematicians: ardent Nazis or Nazi sympathizers suprisingly often did doctoral work with or collaborated with those whom the Nazis called Jews.

[160] Jordan, *Physikalisches Denken in der neuen Zeit* (1935).

[161] *Zeitschrift für die gesamte Naturwissenschaft* 1 (1935): 1–22.

[162] Jordan 1935: 9. Emphases in original. The argument between Hildebrandt and Jordan is really about the philosophical stance known as positivism. Interestingly, both opponents couched their arguments in terms appealing to the revolution, in ideas brought about by the Nazis: Hildebrandt appealed to *völkisch*, traditional German and somewhat romantic-mystical ideas, Jordan to ideas about the importance of modern, especially military, technology, and taking one's rightful place

He added, in words that might also have been applied to Lenard the following year:[163]

> It seems opportune to give a brief answer to the attempt to represent precisely as a supposed consequence of National Socialist engagement a pleasure in defamation of mathematical-physical research. *We live in the era of technological war*: An attempt to sabotage Germany's leading position in the area of mathematical-physical-chemical research must therefore be judged according to the same principles as are standard for the judgment of every other work of disintegration aimed against the defensive capacity of the National Socialist state.

At the end of the book, Jordan explicitly took up Bieberbach's concerns and attacked them, though without mentioning him. The tone of this 1935 statement is so remarkable that I hope the reader will bear with its reproduction.[164]

> First of all, the *stylistic differences* between Greek and Western mathematics emphasized by Spengler[165] should not become overvalued in their importance: the *correctness* of mathematical theorems is completely *independent* of them. . . .
>
> Or are there real differences, say, between *German* and *French* mathematics? Recently that has actually been asserted: the stylistic differences between German and French mathematics are immensely large and it could be asserted that an occupation with *German* mathematics—and careful avoidance of *French* mathematics would uncommonly strengthen the schoolchild or student in their German consciousness. These theses probably arose from the worry that from a widespread aversion to "objective science" must arise a negative valuation of mathematics—and the conviction that it may be easier and richer in prospects to recommend mathematics through veiling its objective character, than to limit the objections against objective science to their legitimate amount. However, one renders National Socialism no service if one offers as bases for the detail of its decisions points of view that are selected only according to convenience, without regard to their truth content.
>
> *The distinctions between German and French mathematics are not more real than the distinctions between German and French machine guns*. Therewith is recognized that

among nations. Though the debate itself could have taken place in many another time, its sociopolitical context at this time gives both sides peculiarly Nazi flavors.

[163] Jordan 1935: 9. Emphases in original. In March 1935, Hitler began openly to build up the German military in repudiation of the Versailles Treaty. Not only did France, Great Britain, and Italy limit their protests to verbal ones, but Great Britain even signed a naval agreement with Germany. Interestingly enough, Jordan hints at the potentiality of atomic weapons (p. 49). He also appropriately directs his attention toward the importance of contemporary physics for biology, the fundamental science in the Nazi context.

[164] Ibid.: 56–59. Emphases in original. Readers familiar with Jeffrey Herf's book *Reactionary Modernism* (1984) or the writing of Ernst Jünger will recognize themes in this passage.

[165] Chapter 2 of volume 1 of Oswald Spengler's famous book *The Decline of the West* (*Der Untergang des Abendlandes*) is entitled "The Meaning of Numbers" and distinguishes sharply between "number as pure magnitude" (Classical, "Apollonian") and "numbers as pure relation" leading to "the idea of Function" (Western). For the possible influence of Spengler's book on mathematicians and physical scientists, see Forman 1971.

there actually are also in the mathematical sciences certain very fine differences of style of a national sort. If one (and the opportunity occurs now and then in the cinema) compares the appearance of Japanese warships with European ones, one recognizes distinctly that even in such an instrument of technical precision, Japanese *feeling for style* is able to assert itself: somehow also the shape of such a warship shows the characteristically un-European features that represent Japanese *art* to us. Perhaps a very sensitive analysis could reveal indeed a rationally determined difference in style between a German and a French machine gun. *However, the value of a weapon rests directly not on this*: what matters is solely the *effectiveness* of the machine gun, and for this question there prove to be standards *from military experience* of "objective" validity going beyond the differences in taste and style of the different nations.

Therefore it completely misses the nub of the matter if one wishes to recommend mathematical-school instruction by the assertion that the students may gain from *German* mathematics a strengthened German consciousness. If *therein* lay the actual task and value of mathematical instruction, then it were high time to *completely abolish* this torment, since for this end there are *better* means. However, as is well known, our youth capable of defense will not be instructed in the use of a machine gun *for the reason* that they experience in their association with German weapon factories a strengthening of their Germanhood (while through the use of French factories they must become Frenchified . . . [ellipsis in original]). On the contrary, the education in a machine gun occurs because of the importance of this instrument for *international* intercourse, and nations who must buy their weapons in foreign countries pay not for the finest traces of national peculiarities of style contained therein, but for *objective effectiveness*.

These considerations suggest that also the *concept of scientific objectivity* is a *politically definable* concept. Objective standards, i.e., standards of supranational validity, exist for all things that possess a *connection to war*. War is the most distinguished means for creation of *objective historical facts*—i.e., such facts whose *factuality* must also be recognized by the conflicting nations. And war represents the *objective test* for the relation of the forces and weapons on both sides.

It reminds us—compared with the grotesque misunderstandings with which we must occupy ourselves—that the computation of bullet trajectories, of airplanes and armored ships, *depend upon nothing else as solely and exclusively as the objective correctness of the computational results*. Therefore, that the mathematical-physical sciences perhaps present in the most refined secondary traces turbidity of their objective content brought about by national peculiarities of style must not be *cultivated* but *overcome*.

If Bieberbach had not earlier seen this attack from a significant scientist also sympathetic to Nazi aims, it was called to his attention by Hildebrandt's colleague at Kiel, L. Wolf, in the physical chemistry department.[166] Bieberbach used the pages of *Deutsche Mathematik* for a brief but sharp reply. In the format of a

[166] BL, L. Wolf to Bieberbach (postcard) (1935?).

book review, but specially headed "Criticism," he cited the last sentence quoted above (in fact, the last sentence of Jordan's book) and then added:[167]

> To see more in mathematical science than a collection of facts and to recognize in the national peculiarity (*völkischen Eigenart*) of their treatment something other than a "secondary turbidity of their objective content" presupposes a certain maturity of activity that also peoples (*Völker*) only achieve after a certain period of occupation with mathematics, and achieve all the easier, the prouder they are in general to be careful to cultivate their national manner (*völkische Art*), and thereby increase its accomplishments. This short indication may suffice here, the more so since the arguments of the author [Jordan] before their appearance were already contradicted by diverse lectures. These have plainly remained unknown to the author.

Bieberbach appended a footnote referring to his two 1934 papers, and his already analyzed unpublished 1926 lecture, as well as a 1935 paper entitled "Two Hundred and Fifty Years of Differential Calculus." Not only did Bieberbach adopt the disingenuous stance that Jordan would not write the way did if he knew Bieberbach's papers (when, in fact, they contain the very ideas Jordan was addressing, though without explicitly naming Bieberbach), but also, curiously enough, he did not adopt the convoluted intellectual defense of his ideas that he and Eva Manger had used two years earlier. Here there was no talk about distinguishing mathematical facts from problem selection or mode of mathematical treatment or presentation. Bieberbach simply said condescendingly (as a mathematician to a physicist?!) that Jordan was wrong. G. H. Hardy and Harald Bohr had attacked Bieberbach's ideas, but they were foreigners. "P.S." had adopted them somewhat ironically. All three in 1934 saw the attempt to separate mathematical style from content as fallacious and recognized its human consequences; for Hardy and Bohr they were evil, for "P.S.," good. Jordan, however, said that the distinction is fallacious and bad for National Socialism. This was a debate that Bieberbach refused to enter. As a mathematician, he, and not some young, however distinguished, physicist, would say what was appropriate mathematics under National Socialism. Bieberbach would later explicitly suggest that there are degrees of "non-Aryan" mathematics, a consequence already seen by supporters and detractors alike. Also, in the two years since the German Mathematical Society contretemps, Bieberbach's racial view of mathematics had received official approval in the form of supporting funds for his journal.

It is perhaps possible to characterize briefly, if not entirely accurately, the various attitudes that have been under discussion. As a supporter of the Nazi *Weltanschauung*, Lenard was a conservative reactionary, Bieberbach a romantic revolutionary, and Jordan a pragmatic nationalist militarist. It is perhaps of some significance that in 1933, Jordan was thirty-one, Bieberbach forty-seven, and Lenard, seventy-one; but not too much should be made of this age differential.

[167] *Deutsche Mathematik* 1 (1936): 109.

Bieberbach's talk on the history of calculus that he called to Jordan's attention had been given to an association for mathematical and scientific instruction in November 1934, and though it contains untendentious observations about the history of mathematics, it also serves as a further vehicle for already familiar ideas.[168] Worth noting additionally, however, is that Joseph Louis Lagrange is credited with "a strong will to a constructively systematic building-up [of mathematics] directed by a sound critical sense," ascribed as consistent with the fact that his mother came from an old Cisalpine family and "his appearance . . . revealed discernible Nordic features."[169] Lagrange is called "dinaric" in positive contrast with the less systematic Euler (who is "eastern").[170] System-building is a "typically German" trait and is connected with the "Nordic dash" in the great German mathematicians (apparently lacking in the Swiss Euler). Not absent either in this talk is the denigration of some French mathematicians as non-systematists, the praise of intuitionism, and a sneer at the "philosophy of rationalism marching under the banners of the Jesuit-reared Descartes" (which had shattered a Leibnizian unity). Set theory is described as uniting within itself "fruitful and disintegrative" modes leading to the "crisis in foundations." The criticism of the formalist use of the *tertium non datur* with false simplistic examples in popular language persisted,[171] but this time Bieberbach's pseudo-examples in favor of intuitionism are striking:[172]

> No moment known to me speaks against a fly having been squashed between both successive pages of this volume. No one will see that as a sufficient proof that a dead fly is found there. Everyone will rather first look; so also the intuitionist. He wishes to construct mathematical objects in order to believe in their existence. Or, another example. If Herr Hopfenstang[173] declares that no moment is known to him from which it follows that he may be of Jewish lineage, so from that it in no way (*noch lange nicht*) follows that he is an Aryan. Our laws handle him as a non-Aryan from that moment on in which a Jewish ancestor will be found.

The apogee of Bieberbach's personal exposition of these ideas of "racial" or ethnic differences in mathematics occurred in 1940, when German troops seemed triumphant everywhere in the early months of World War II. Two items date from this year: a talk given to German troops in conquered Krakow, and

[168] "Zweihundertfünfzig Jahre Differentialrechnung," *Zeitschrift für die Gesamte Naturwissenschaft* (1935): 171–177. The original address was given to the Berliner Verein zur Förderung des mathematischen-naturwissenschaftlichen Unterrichts on November 13, 1934.

[169] Ibid.: 174. Joseph Louis Lagrange was in fact born Giuseppe Lodovicio Lagrangia in Turin in 1736. His father's family was French but had moved to Italy three generations previously. Until he was thirty, Lagrange lived in Turin.

[170] These terms originate with the racial theorist Hans F. K. Günther.

[171] See discussion of Bieberbach's address to the Berlin Academy of Science, above.

[172] "Zweihundertfünfzig Jahre Differentialrechnung" (as in note 168), 177. Bieberbach's caricature of the formalist Aristotelian argument by contradiction seems to demonstrate a low opinion of his audience. He surely knew such argumentation as follows below was philosophically and mathematically fallacious—its only excuse seems to be the anti-Jewish sarcasm that follows.

[173] *Hopfenstang* = lamppost.

the printed form of an address at the University of Heidelberg given almost a year previously on June 19, 1939. Heidelberg was the university at the time of Philip Lenard, who, though then just seventy-seven, was an influential *alter Kämpfer* in university matters; the astronomer Heinrich Vogt (who in 1933 had "already been a National Socialist for a long time" and owed his appointment to Lenard's influence); and the national student leader (and sometime mathematics student) Fritz Kubach.[174] The operational director of the mathematics department was Udo Wegner, who would become Dekan in 1941. Thus Heidelberg provided a receptive atmosphere for Bieberbach's ideas.

The Krakow address on March 2, 1940, was a brief affair stimulated by Krakow as the university of Copernicus, and mostly devoted to the proposition that Copernicus was a German and not a Pole.[175] It also ascribed the "uncritical" medieval assumption of the Ptolemaic system partly to "the spirit of the Old Testament, which contained the idea that solely this earth (which is promised to them) can be of interest for the 'chosen people,' and that satellites like the sun must halt in their course if a Jewish prophet commands them." It closed with the assertion that the "moral meaning of the Copernican discovery" is the destruction of the Old Testament conception and a "modern expression of the old Nordic belief in the Sun." Indeed, "Astronomy must be designated as a pronouncedly favorite discipline of the Nordic race."[176] In this way, the Copernican revolution was fitted into the German effort to destroy pernicious Jewish influences.

The Heidelberg address is simultaneously the most detailed and the last public exposition of Bieberbach's *völkisch* mathematics. Although it emphasizes the importance for mathematical pedagogy of the Jaenschian typology and has a tone of continuing struggle in this regard, it is hard to see what further could be achieved in this direction in June 1939, when the talk was given, let alone in 1940, when it was published. All even "part-Jewish" or "Jewish-related" people had long since been purged from teaching faculties at all educational levels. German schoolchildren did not even have Jewish-German classmates. By mid-1940, there was also no danger of rapprochement with putatively pernicious French ideas.

Nevertheless, because this address is the longest and most reasoned exposition of Bieberbach's ideas, running to over thirty pages, it is worth some consideration. Here none of the polemical phrases of the 1934 articles, like "foreign lust for mastery gnawing on its [the German people's] marrow," appear.

After introductory bows in the direction of Lenard and Vahlen, Bieberbach described his theme: though mathematics has to do with knowledge of "uncon-

[174] Lenard's birthday was June 7. Bieberbach's talk may well have been in celebration of it. The phrase about Vogt is Lenard's as cited by Vezina (1982: 147). For more about Vogt, see above, chapter 4, "Hasse's Appointment at Göttingen."

[175] The text is in BL. There has been much argument about Copernicus's nationality, occasioned by the fact that he was a German-speaking Polish subject. See "Biography of Copernicus" (pp. 313–412), in Edward Rosen, *Three Copernican Treatises*, 3d ed. (1971).

[176] Citations are from, in order, pp. 3, 6, and 1 of text as in BL.

testable truth" and "apodictic certainty," nevertheless it is created by "human beings (*Menschen*), human beings captured (*verhaftet*) by their national identity (*Volkstum*)." Consequently, the ideas, results, methods, value-conceptions, and thought structure of mathematicians are also variable and dependent on national typology. Felix Klein is contrasted with Henri Poincaré—a contrast that dated back to their "competition" in the 1880s. Bieberbach characterized each by a citation. For Felix Klein: "While I fight for the right of intuition in my scientific area, I in no way wish to neglect the importance of logical development. According to the conception I represent, mathematics only finds its complete validity where both sides come to development next to one another." For Henri Poincaré, "the greatest French mathematician of modern times": "All that is not thought (*Gedanke*) is pure nothingness. . . . Thought is only a gleam in the midst of a long night. However, this gleam is everything."[177] Thus intuition plus logical development (Klein) versus ideas alone (Poincaré) is made the fundamental contrast between German mathematics, which has a holistic quality, and one-sided French (and Jewish) mathematics.

While most would acknowledge that there are different types of mathematical ability, said Bieberbach, many would think them of little importance, since with some effort an individual can understand another's thought processes. However, the first duty of the "creative mathematician" is to discover original material, and that "springs in a completely different fashion [from the understanding of another's thought] from the interior of the human being." He continued,

> It will be a certainty to every National Socialist that in everything that we do for ourselves, are we dependent on and influenced by the talents that our descent places in the cradle, and indeed all the more, the more we are ourselves in our achievements.

While this thought could be read as the most bland and self-evident "Mendelism," in the context "descent" (*Abstammung*) had intended ethnic and "racial" connotations. Bieberbach adopted the frequent (among mathematicians at least) comparison between mathematical and artistic creativity. Just as there are many styles of (German) poesy, so are there many styles of (German) mathematical creativity: one would no more confuse the different styles of Euler and Gauss or of Weierstrass and Klein than one would confuse a poem by Mörike with Rilke, or Schiller with Goethe, or Adalbert Stifter with Kolbenheyer.[178] Jaenschian psychological typology, particularly with reference to the Jaensch and Althoff monograph *Mathematical Thinking and the Form of the Soul*, is brought in in this context. Quite modestly, Bieberbach acknowledged that one is only at the beginning of understanding the connection between "the type of mathematical activity and the structure of the creative personality"; however, "racial science and psychology" will show the way.

[177] These are the last lines of *The Value of Science*. The translation is that of George Bruce Halsted, which was reprinted by Dover Publications (1958 et seq.), 142.

[178] Erwin Guido Kolbenheyer, who may be less well-known than the others, was not only a famous German author, but a contemporary biologistic thinker. See chapter 5, "The Value of Mathematics in the Nazi State."

Gauss, Klein, and Weierstrass were delineated as respectively Jaensch's J_1-, J_2-, and J_3-types, all "integration types," while once more Edmund Landau represented the "S-type," whose "solipsistic (*autistisches*) thinking radiates out into reality and at best is concerned to find again in reality that which his thoughts produce, his ideas, but not as a confirmation of his thinking, rather as an *epitheton ornans* [Bieberbach's phrase] of reality." Landau is contrasted with the "German" Gerhard Kowalewski, and Bieberbach took this opportunity to warn against a misunderstanding. It is not necessarily true that every German is one of the J-types and every Jew an S-type. In the first place, not every human being falls into one of the Jaenschian type-classes. Furthermore, there is no firm a priori connection between racial science and Jaenschian typology, though there are starts toward clarifying the relationship between the two. Again, this is a moderate statement, consistent with Nazi ideology, but not polemical, like the earlier justification of the boycott of Landau's classes. Similary "moderate" compared to his earlier disquisitions on these themes was Bieberbach's assertion that "certainly also S-types can achieve useful, perhaps even important, work. They always run the danger, however, of losing the connection with a larger whole."

Indeed, these are the words prefacing his discussion of Landau, with no explicit mention of the boycott. Bieberbach also discussed various other German mathematicians of mixed Jaenschian type, such as Hermann Amandus Schwarz, or David Hilbert, or Richard Dedekind, as well as other unmixed ones, like Johannes Kepler. For Bieberbach's position to be intellectually coherent, he again needed to distinguish "German" mathematicians whose greatest contribution was the development of new axiomatic theories from "Jewish" mathematicians who contributed to or even initiated such developments. This is a harder problem than simply neglecting the enmity, mathematical as well as personal, that existed between Klein and Schwarz or Weierstrass. Hilbert was mentioned only briefly "since he is still alive" and then given a Jaenschian classification that was mostly like Weierstrass's (J_3) with some admixture of J_2. The role of the Jew Stefan Cohen-Vossen in the well-known book *Intuitive Geometry*,[179] which he authored jointly with Hilbert, was reduced to that of an amanuensis. Dedekind's type is similar to Hilbert's. Some inclined to place Dedekind closer to the S-types than Weierstrass, said Bieberbach, but that was in his opinion wrong.

According to Bieberbach, Dedekind's methods of argument have exercised and still do exercise a peculiar attraction for S-types. The distinction Bieberbach perceived was that the "master of the machinery" (Dedekind) had created it to solve a particular circle of problems, and to investigate the machinery freed from its problem-context has no content and is the sort of "building in the air" done by S-types. In fact, for Bieberbach, Dedekind's "inner relationship" to Riemann's style proves he can be no S-type. Similarly, the fact that Gauss,

[179] *Anschauliche Geometrie* (1932). Bieberbach took care to emphasize that his only citations are those ascribable solely to Hilbert instead of Cohn-Vossen, a Jew who, prevented from lecturing, emigrated in 1933 to Moscow, where he died three years later.

Kummer, Dedekind, and Hilbert, the "greatest nineteenth-century number theorists" and all German, could none of them be counted as S-types is taken as self-evident. Dedekind emphasized that his theory of ideals is built up on the inner qualities of ideals as Riemann's function theory bases itself on the inner qualities of functions—on concepts instead of on computations. In contrast, said Bieberbach, the method applied by (the Jew) Adolf Hurwitz to such questions makes use of an external form of representation (similar to Weierstrassian function theory).

It is fruitless to attempt to discern what the "inner qualities of functions" were or to analyze how Bieberbach thought Hurwitz' variant of ideal theory differed from Dedekind's. The reference to Weierstrassian function theory refers to Weierstrass' taking the fact that analytic functions have a power series representation as the fundamental property on which to build a theory. The last parenthetical phrase is footnoted as follows: "To conclude from this that Weierstrass might be a Jew or respectively Hurwitz a German, or that between the two no more essential (*wesensmässiger*) difference might exist, is plainly illogical." This might even be humorous to someone not a convinced Nazi, were it not so deadly serious. Bieberbach believed his arguments; some cases (Dedekind, Weierstrass) might be in some aspects confusing to an unskilled observer, especially in the inchoate state of application of Jaenschian typological psychology and racial science to mathematics, but closer examination, he believed, reveals the truths they have to offer and the necessary distinctions. A similar tone suggesting that further examination would straighten out apparent difficulties underlay Bieberbach's approach to Jaensch's "discovery" of a close connection between "S-type thinking" and "tuberculous processes." Even Jaensch admitted that there are exceptions, and so there was no need to worry about great tubercular mathematicians like Nils Abel (who was Norwegian) or Bernhard Riemann being S-types (let alone Schiller). Riemann is analyzed in more detail as J_2/J_1.

The issue of mathematical correctness was again handled by addressing the different interests, style, and attitudes of mathematicians. This had been met before, but Bieberbach used it in a different and more moderate way, a way that did not involve, for example, explicit justification of the expulsion of Jewish teachers (of course, by 1939 such justification was hardly necessary).

> In the face of such different types of mathematical thought, one notices that the content of mathematics, despite that, largely seems to be independent of the thought-type. In fact, it would be hard to give a correct mathematical theorem that not every mathematician recognized as correct. As soon, however, as the question arises whether the theorem concerned might be important, interesting, or highly relevant, then one will already hear the most various judgments. Opinion about it depends largely upon the [Jaenschian] type of the judge.

Such thoughts led Bieberbach to the issue of formalism versus intuitionism, and here his earlier polemics, dating back at least to 1926, gave way to the mild remark that whether one adopts a formalist or intuitionist position is "conditioned by one's worldview," and that the pure J_1 type would hardly have any interest in questions of logical foundations. Perhaps this reflects the fact that by 1939, Brouwer's program had been, as a practical matter, rapidly losing interest for most mathematicians.

Why would anyone care about studying mathematical types? What possible utility does such a study have? Here Bieberbach adopted the same point of view as Jaensch's student Fritz Althoff.

> Generally the mathematical is only a fractional part of intellectual (*geistigen*) behavior. However, the peculiar character of the mathematical, in contrast to other natural sciences, its large participation in thinking in the interior world, not only in the construction of mathematics, but also in the creation of its often purely mental objects, discloses the types of scientific thinking directly in mathematical behavior in an especially pronounced way, and therefore also gives the best starting points for an education of a suitable kind of thinking (*artgemässen Denkens*).

Thus mathematics, far from being a subject in danger of being rejected in the Nazi school atmosphere, should become the primary vehicle for discerning true German modes of thought and for arming against S-type influences. Indeed, Bieberbach made a plea not to construe such mathematical education too narrowly in order that all the varied German thought-types could be cultivated.

As though to emphasize this point, at another place in his talk Bieberbach cited Jaensch's "experiments" with students in which the various types were revealed. A student of the to-be-rejected S-type is exemplified by the statement, "I find that mathematics is a pure thought-structure (*Gedankengebilde*) aside from what is concerned with the properties of space. However, they actually belong not at all therein. He who is a logician will also be mathematically gifted." On the other hand, the "German types" J_1, J_2, J_3 reveal themselves as follows.

> J_1: My engagement with mathematics can be said with one sentence: I love it, I get on (*vertrage*) with it.

> J_2: The goal of life is the striving for the truth. Science envelops this truth in an ever smaller neighborhood. . . . However, the feeling that says to me that it is our task to strive for the truth, induces me to believe that indeed there is some final knowledge whose bounds a higher power has composed for us.

> J_3: It is a great attraction for me to systematize material in order thereby to come to clarity and to mastery over it. I like mathematics as something organic. I must always know how and whence the single item comes.

In short, said Bieberbach, the distinction between S-types and J_3-types is that the J_3-type, so Jaensch occasionally formulates it, builds from below to above, and the given is the basis upon which the thought (*gedankliche*) raises itself. The S-type, on the contrary, builds from above to below. That is a procedure whose possibility does not illuminate healthy feeling, and that therefore, to us, wherever it occurs in pure form, and wherever it resonates (*anklingt*), always feels particularly foreign.

In this address Bieberbach also reached the natural conclusion of his ideas with respect to mathematics. Some mathematics is less German than others; S-types are characterized by thinking that Cantorian set theory is the basis on which all mathematicians think, whereas Germans are more reserved.[180] Furthermore,

> One only needs to remember that none of the theorems of point set theory important for real analysis could be named after Cantor. They go back to Germans or mathematicians of a related sort, for whom a rigorous construction of material become historic lay close to their hearts. Certainly, German mathematicians are as good as not involved at all in the modern development of the theory of real functions. Here is a playground for S-types.

Bieberbach's ideas need to be considered as something more than just Nazi sloganeering. They involved an elaborate intellectual rationale that Bieberbach and others seem to have genuinely believed. Certainly they were casuistical in part, but the thoughts that there are different kinds of mathematics or science, and that some are preferable to others, did not disappear with the Nazis, as any observer of the contemporary educational scene can verify.

This does not contradict the earlier suggestion that Bieberbach wished to become *primus inter pares* of German mathematicians. In fact, perhaps Bieberbach's more moderate, more "scholarly" tone in this lecture/article reflects that in 1939–40, his battle on the political front was long won, and he was trying to move toward an accommodation with his less ideologically oriented fellow mathematicians. After all, Wilhelm Süss, the de facto president in perpetuity of the DMV, had been his doctoral student, *Deutsche Mathematik* was seemingly successful, and the end of the Nazi hegemony in Germany could hardly be envisioned. It was convenient for Bieberbach that he could make his disciplinary-political aims, his political-ideological beliefs, and his pedagogical interests coincide and support one another. The result may have been a historical-political-psychological mélange from our vantage point sixty years later, but that does not mitigate its serious intellectual purpose for Bieberbach and his supporters. Bieberbach did genuinely have pedagogical aims, as his letter proposing *Deutsche Mathematik* stated; he did want to improve the mathematics involvement of university students and also help build a National Socialist youth thereby. In this respect, his aims were not very different from those of Ernst

[180] Bieberbach certainly knew, though he avoids mentioning it, that David Hilbert was a great supporter of Cantorian ideas, while Cantor's primary opponent in his lifetime was Leopold Kronecker, who was Jewish (by Nazi standards) and came from a banking family. Bieberbach's mention of Cantor in the talk again shows its relatively moderate tone: he cited as undocumented the (false) assertion of Eric Temple Bell that Cantor was "of pure Jewish descent on both sides," and, as documented, the fact that Cantor's father was already a Lutheran in religion by 1845, and his mother was Catholic. Of course, both statements could be true. The documentation stems from A. A. Fraenkel (who was Jewish).

Weiss. In fact, Weiss assisted with the production of *Deutsche Mathematik*. Having seen how Bieberbach's own ideas and their expression varied, it is now time to turn attention to the journal that was to be their concrete embodiment.

Deutsche Mathematik

Serious pedagogical as well as ideological ideas informed *Deutsche Mathematik*. Bieberbach's devotion to students (although he was a sloppy mathematical expositor) and his devotion to the Nazi cause were both real. It would be a mistake simply to think of *Deutsche Mathematik* as some "racist rag." For Bieberbach, not only were there serious intellectual ideas behind his journal, but also he apparently maintained their validity beyond the loss of World War II. In fact, it was sometime in the mid-1950s before he separated the murderous brutality of Nazi acts from the Nazi theory with which he identified—merely depriving someone of occupation and livelihood seemed to be for him a necessary and fitting consequence of the German renascence under Hitler.

Volume 1 of *Deutsche Mathematik* appeared in six issues beginning January 20, 1936. Its 898 pages of text certainly seemed to fulfill admirably Bieberbach's expectations of a journal that would be somewhat student-oriented in content and contain pedagogical articles, research articles, and book reviews, as well as articles exhorting German mathematics. These last appeared under the rubric "Work" (*Arbeit*), as distinct from "Research" (*Forschung*) and "Pedagogy" (*Belehrung*). Among such articles of exhortation are ones by Fritz Kubach, the national leader of mathematics students, and Erhard Tornier.[181] Kubach's article[182] is a call to students to be in the forefront of this delineation of a truly German mathematics and an emphatic repetition of Bieberbach's arguments.[183]

> Decisive . . . therefore are not the formulas or otherwise rationally apprehensible results, but on the contrary, decisive is singly and solely the question of the creative form that leads to these results: the kind of question-setting, the selection of problems, the mode of thought and the way it is carried out.

Students were called to their subversive role, discussed earlier, in the Nazi approach to universities: Kubach complained that *Assistenten*, lecturers, and professors, "especially the latter," still contained only a small number of supporters. Thus, the students had to be in the forefront of expositing mathematics' ideological (*weltanschauliche*) meaning. As with Weiss, the mutual influence of mathematics and "character" was also stressed. Rather strikingly, Kubach declared that not only will "those formative and educational powers which inhere to mathematical work and research" be made manifest by such activity, but also the image of the mathematician as a laughable figure of scorn will be overcome.

[181] For Tornier's career at Göttingen, see above, chapter 4, "Hasse's Appointment at Göttingen," and Schappacher 1987.

[182] "Students in Front!" (*Studenten in Front!*), *Deutsche Mathematik* 1 (1936): 5–8.

[183] Ibid.: 6.

He proposed a three-point program of student investigation. One was "the treatment of more general and more fundamental questions concerning mathematics and worldview, racially connected mathematical creativity, and similar themes." This area led naturally to another: historical investigation, which had been "too strongly neglected and was and is even today unfortunately partially in Jewish hands." More particularly, the "historical development of individual mathematical institutes" was to be investigated, especially the "hugely important question in mathematics of the influence of Jews."[184]

Various communications from student groups published in this first volume responded to Kubach's exhortation. For example, a group of Heidelberg students contrasted the "soul-structure" of Kepler and Newton (styled a "Germanic researcher") with that of Einstein. Einstein was viewed as a materialist who, far from being in the long line beginning with the religiously grounded Kepler and Newton, indeed posed a challenge to their work, "with the goal of its destruction."[185] Another group at Königsberg contrasted Leibniz and Descartes. Descartes was condemned as a materialist, in contrast to the "energistic-vitalistic worldview" of Leibniz.[186] For both the Heidelberg and Königsberg study groups, the concept of "force" was particularly Germanic, and its elimination materialist, whether French or Jewish. For both, religious grounding was Germanic. For both, Germanic attitudes were unitary and non-Germanic attitudes divisive (e.g., the Cartesian mind-body dualism).

Somewhat different were the students at Giessen, who held a mathematics camp in April and again in October 1935. The report of the first of these remarked with a touch of *Wandervogel* romanticism on the need to form a working community such as could never be achieved in lecture halls, but recognized that the camp is no replacement for these, though it affords an opportunity to realize connections between various subjects. However, its main emphasis is on improving the image of the mathematician so as to eliminate the rightly scorned "freely wandering brain-acrobat" from the lecture halls, and on creating a closer relationship between student and teacher. The latter must be a "good fellow and comrade" who is worthy of his responsibility to separate out the best, an effort the students desired and he should support. The second Giessen report mentioned that only the Bonn mathematicians (under Weiss) have established similar camps. For the Bonn students, as has been seen, a result was individual mathematical accomplishment as well as building National Socialist character. For the twenty participating Giessen students, there seems to have been ample mathematical instruction at the camp (though not the stimulation to individual work aimed at by Weiss), but

> we see the existential kernel of our mathematical work camp in the opportunity resting on comradeship within a camp community of young seekers . . . to ever and again

[184] Ibid.: 7–8.

[185] B[runo] Thüring reporting, *Deutsche Mathematik* (*DM*) 1 (1936): 10–11. cf. also ibid.: 705–711.

[186] O. Freytag reporting, ibid.: 11–12.

> situate ourselves with respect to the demands and tasks that are set us as German scientists and in particular as German mathematicians. . . . The young student wishes to have clarity about the task that he has to fulfill, in order to be able to stand as a worker with his head before our people. Only in the possibility of being able to serve the people with his work does he find that satisfaction which can stimulate him to achievements. Once, in a time of decay, we had begun our studies out of love for our discipline.

Also, whereas the first Giessen camp had discussed the importance of mathematics for military education and in defensive sports, the second took up "the question of the importance of mathematics for national-political education and further the question of what mathematics can contribute to the understanding of the racial and population measures of the Third Reich." Fritz Kubach visited this second encampment, and indicative of the spirit of the whole enterprise is that in the report of his visit he is denominated neither by official title nor as "[Nazi] party member" but as "comrade."[187]

Twenty Hamburg students and eight faculty (including Wilhelm Blaschke, Hans Petersson, and Gunther Höwe)[188] held a camp that took up the pedagogical issue of

> giving the schoolchild an insight into the importance of mathematics as an expression of precisely the German-Nordic will to intellect and culture. Unfortunately this was for a long time seriously unrecognized. This answers the schoolchild's so often remaining question, "Why mathematics in school?" by its inner connection to German intellectual life.[189]

This was the concluding event for a student working group. They also discussed Oswald Spengler's treatment of mathematics, praising him for his recognition of its cultural significance, criticizing him for sometimes misunderstanding mathematics, and for his pessimism, which contradicted "our new feeling for life," and for the fact that he talked about "Western" rather than "German" mathematics.[190]

Another group of Heidelberg students took up the study of the historical influence of Jews in the university's mathematics department. While this was purely along Bieberbach's lines, they had some difficulties:[191]

> An especially important, however, also especially difficult task was the determination in an objection-free manner of the Jewish or Aryan descent of individual faculty members. For almost all faculty, clarification of the question of racial membership was

[187] The reporter was H. Gortler; ibid.: 12–13, 117–121, citations are from pp. 13, 118, 120.

[188] For Blaschke, see particularly chapters 6 and 8 as well as in this chapter; for Petersson, chapter 8; for Höwe, chapter 3.

[189] In German, this is a single turgid (longer) sentence.

[190] *DM* 1 (1936): 121–122.

[191] The reporter was H. J. Fischer, the mathematician who became a member of the SD and whose autobiographical memoir was cited earlier; ibid.: 115.

successful. . . . The material collected up till now is still not sufficient though for a completely clear comparison of German and Jewish creativity.

A small group of Berlin students pursued the lines of Bieberbach's 1934 lectures, taking as the best exemplar of the Jewish spirit in mathematics Landau's *Differential and Integral Calculus*. In the same spirit, they contrasted Kepler "and other German physicists" with Einstein.[192] Since Berlin was Bieberbach's university, it is perhaps not surprising that his articles were more definitive texts there than for student groups elsewhere. In fact, in 1936, Bieberbach offered for credit a course called "Great German Mathematicians."[193]

As Karl-Heinz Boseck[194] reported, a much larger group of Berlin students took part in a working group in summer semester 1936 that studied the following subjects: "1. Fundamental works of National Socialism in their relationship to the natural sciences; 2. Mathematics and Biology; 3. The world of ideas of Greek mathematics; 4. The mathematics of insurance and German socialism; 5. German Physics; 6. The influence of Aryans on the formation of the astronomical world picture; 7. Computation of determinants."[195]

The work on mathematics and biology was put into publishable form for *Deutsche Mathematik*.[196] It studied, unsurprisingly, the effects of family planning and compulsory sterilization under the most basic of simplifying assumptions. Its tone is easily captured:

> The diagram offers a visual picture of the extermination of the valuable that must take place in the course of only two generations if the relationships of numbers of progeny and generation length are maintained. . . .
>
> Only a people that, the danger known, allows the number of progeny to increase with the racial value of the parents, can turn aside this danger of the eradication of fitness. [A basic empirically derived hypothesis was that this number decreases rather than increases.]

One even finds tendentious introduction of mathematical constants:

> If a people is strictly separated into castes or classes or if the selection of spouses takes place according to racial viewpoints, then b is large; for people with heterogeneously mixed (*buntgemischten*) marriages (Pan-mixture), the number b is small.

Extracts like those cited show that Bieberbach was correct in anticipating widespread student support for his efforts. Students who wish to overthrow the *status quo ante* are a commonplace in universities at almost every time and place. In the Nazi context, however, the political establishment was on their side, and had provided a revolutionary ideology that had transformed the state. Bieberbach, established and still relatively young (he was forty-seven at the end of 1933), did seem to have genuine mathematical-pedagogical aims, as well as

[192] Ibid.: 116.

[193] Ibid.: 430.

[194] For Boseck, see above, chapter 6, "Mathematics in the Concentration Camp."

[195] *DM* 1 (1936): 423–424.

[196] Ibid.: 424–429. The citations below are from p. 427.

political ones. He saw *Deutsche Mathematik* as a teaching vehicle both for mathematics and for the new National Socialist youth; at the same time, he was among the few "comradely" professors, especially among mathematicians, who were at one with the Nazi movement among students. It would even seem he explicitly wished to use these students to overcome the reluctance of his more conservative colleagues to politicize themselves actively. Since the German Mathematical Society's behavior in 1933 showed that it was more than ready for a traditional sort of passive acquiescence in the political situation, this engagement of student activism by Bieberbach again seems an attempt to manipulate himself into the paramount position among mathematicians, something denied him by his colleagues at Bad Pyrmont in autumn 1934. The relative amounts of sincere Nazi enthusiasm and cynical opportunism in Bieberbach's motivations are impossible to determine.

In any event, *Deutsche Mathematik* was declared an official organ of the German students' organization (*Deutsche Studentenschaft*); consequently, all local organizations of mathematics students were expected to receive at least one copy. Furthermore, all students were exhorted to send in original contributions, or reports of work groups or mathematics camps—work groups seemed often to cap their work with a camp experience. Fritz Kubach in particular also saw the journal as the center of a "new community of German mathematicians," much in the spirit of Bieberbach's letter proposing the journal.[197]

In addition to Kubach and Bieberbach, the editorial board for volume 1 contained, among others, the following who have already been mentioned elsewhere in these pages: Alfred Klose, Heinrich Scholz, Wilhelm Süss, Erhard Tornier, Egon Ullrich, Werner Weber, and Ernst August Weiss.[198] Although Bieberbach was the responsible managing editor, his name did not appear on the title page as "publisher." Rather, Theodor Vahlen's did. This was not as originally foreseen, since still-extant Hirzel Verlag mock-ups of the title page show Ludwig Bieberbach in this position. Presumably, Vahlen's agreement to serve as nominal *Herausgeber* ("publisher") of *Deutsche Mathematik* more firmly anchored it to the powers regnant, a no doubt wise political move on Bieberbach's part. These mock-ups also show that the journal originally had the subtitle *A Monthly for the Protection of the Interests of German Mathematicians*, which was dropped in the published version.[199]

A pendant to Kubach's exhortation of students is a short diatribe by Erhard Tornier attempting to give a necessary condition for "German mathematics" entitled "Mathematicians or Jugglers of Definitions." Tornier saw the "right to existence of a mathematical theory" in its "applicability." Applicability, in Tornier's sense, means ability to solve concrete problems with real objects or to intellectually unite various circles of questions. For "pure mathematics," "real objects" means integers or geometric figures. The rhetoric of this one-page arti-

[197] Ibid.: 122–123; cf. ibid.: 9.
[198] Ibid.: page preceding page 1.
[199] Ibid.: and BAK R73 15934.

cle need not be repeated; suffice that the adjective "Jewish-liberal" appears four times in this short space as associated with "aesthetic beauty [of a theory]," "technique of illusion," "rootless artistic intellect," "solipsism," and "obfuscation," not to mention "juggling with definitions."[200]

A word is in order about Heinrich Scholz's collaboration with *Deutsche Mathematik*. As has been noted, Scholz was a theologian turned mathematician, and originally a conservative nationalist, perhaps not dissimilar in his original political attitude toward the Nazis to Hellmuth Kneser, Wilhelm Süss, or Erhard Schmidt. Whether it was the wartime attitudes toward Polish intellectuals or earlier events that began to alter his opinion is unclear. In any case, as an internationally respected logician, founder of a school, he was certainly interested in continuing to promote his mathematics in the Nazi atmosphere. Yet the *Deutsche Mathematik* of Tornier or Bieberbach seemed inimical to logic above all: recall that Bieberbach remarked that questions of the foundations of mathematics were at bottom racial questions. Thus for Scholz to publish (with his student Hans Hermes, a future distinguished logician) a forty-page article in mathematical logic in the first volume of *Deutsche Mathematik* seems strange at first, even if it dealt with the work of the undeniably German logician Gottlob Frege. I do not know what arrangements Scholz made with Bieberbach, but the article with Hermes appeared also as the first of a new series entitled "Research in Logic and in the Foundation of the Exact Sciences," published as separata by Hirzel and supported by Bieberbach.[201] The series was advertised as "a point of collection for German work in the area of the new mathematical logic and foundational research." Perhaps the "certificate" issued by Griewank on January 12, 1943, requesting an allocation of paper for the series of separata, comes nearest the point.[202]

> [The series "Research in Logic" under the editorial direction of Prof. Dr. Heinrich Scholz] treats a scientific discipline that is only weakly represented in Germany, to which however a certain European importance is attributed. It will also be considered necessary on the part of the national scientific ministry that Germany come forward further with a certain production in this area.

Indeed, perhaps Bieberbach even wished to represent Scholz's school of logic as a "truly German" one, in contrast to work stemming from other ethnic sources. As has been seen, for all his initial conservatism, Scholz was not of this opinion. Indeed, his letter to Griewank in November 1939 says he is coming to Berlin "in order to personally care about the help that must be given to our unhappy friends in Warsaw before it is too late."[203] In any case, in Scholz, Bieberbach obtained an internationally recognized logician as support for his

[200] *DM* 1 (1936): 9–10.

[201] "Forschungen zur Logik und zur Grundlegung der exakten Wissenschaften." See BAK R73 15934, Scholz to Griewank, with Griewank's handwritten notation, Nov. 26, 1939; Bieberbach to Griewank, Feb. 25, 1940; and Scholz to Griewank, Mar. 14, 1942.

[202] BAK R73 15934, Griewank, "Bescheinigung."

[203] Ibid., Scholz to Girewank, Nov. 26, 1939.

journal, and in *Deutsche Mathematik*, Scholz obtained a publication venue that ensured there would be no frivolous ideological interference in publications with so rational a content as logic. Indeed, in June 1943, Scholz published a lengthy pedagogical article in *Deutsche Mathematik* on formalized metamathematical research during which, as described earlier, he strongly and sarcastically attacked Max Steck's "Nordic," antiformalist views that had denigrated Hilbert and, incidentally, attacked Scholz.[204] Steck was not only a sometime collaborator on *Deutsche Mathematik*, but the preceding issue of the journal had contained a review by him that began with a list of mathematicians who had upheld the "genuinely German geometric tradition" sometimes as a burden of considerable weight "in opposition to the so-called formalist and logistical 'successes' in mathematics."[205]

Volume 1 also contains numerous other mathematical articles, many of them short and with reasonable mathematical content. Bieberbach clearly solicited articles from mathematicians he thought might contribute to launching his enterprise, and this probably accounts for the articles by Paul Koebe, Gerhard Kowalewski, and Hellmuth Kneser.[206] Some articles were by young students just starting their careers, like Georg Aumann in Munich or Willi Rinow and Gunther Schulz in Bieberbach's Berlin. All three of these became professional mathematicians, Rinow especially establishing a considerable reputation, and, whatever they may have believed in 1936, a mathematical article in *Deutsche Mathematik* at that time would not hurt their futures. There were four research articles by the brilliant young mathematician and dedicated Nazi Oswald Teichmüller, who would disappear on the Russian front in 1943 at the age of thirty.[207] Among Bieberbach's ideological coreligionists who contributed mathematical articles, but who were of rather minor mathematical moment, were (with parentheses indicating number of articles in the first volume) H. J. Fischer (1), Max Steck (3), Erhard Tornier (1), Werner Weber (5), and Udo Wegner (3). There was also an article by Vahlen. It should be stressed that these articles denominated "Research" had solely mathematical content.

For the most part, this was also true of the so-called pedagogical articles, which included a five-part, partly historical paper by E. A. Weiss. However, there was an article by Friedrich Drenckhahn in Rostock entitled "The Law for Protection of German Blood and German Honor of September 15, 1935 in the Light of Population Statistics."[208] Its very first sentence spoke of the infiltration of foreign blood in the German people. The law referred to is the "Nuremberg Law" forbidding sexual relations between Jews and non-Jews. Drenckhahn (who

[204] *DM* 7 (1943): 206–248.

[205] *DM* 7 (1942): 120.

[206] However, shortly thereafter, Erich Trefftz declined to contribute even the content of his work in 1936, pleading secrecy restrictions placed on his aeronautical work. BL, Trefftz to Bieberbach, Jan. 5, 1937. For Trefftz's attitudes toward the Nazis, see above, chapter 6, "Applied Mathematics in Nazi Germany."

[207] See below, chapter 8, "Oswald Teichmüller."

[208] *DM* 1 (1936): 716–732. The citation below is from p. 716 n. 1.

taught at the teachers college in Rostock) mentioned a more mathematical discussion in a paper by Hans Münzner, the Göttingen statistician who had been a pupil of Felix Bernstein and later reportedly attempted to terrorize his family.[209] Münzner's brief article on the rapidity of racial mixing appeared in a new Nazi journal devoted to mathematical economics and social research. Drenckhahn's article was extracted from lectures at Rostock and was intended to be equally informative to the student readers of *Deutsche Mathematik*; its publication was "to show how contemporary events will be brought into the circle of mathematical lectures." There is no need to further discuss its tendentious and very elementary mathematical content, except to note that few *mathematical* articles, whether research or pedagogy, in *Deutsche Mathematik* were of this sort—though it was probably the only mathematical journal where such an article could find publication. Indeed, the same volume (though in an earlier number) contained a straightforward pedagogical article by Münzner on statistical correlation coefficients that had no hint of ideology.[210]

Three other aspects of volume 1 will round out the view of the journal's original intentions—for volume 1, issued in a time of peace and national success, arguably represented Bieberbach's ideal of a mathematical-political-pedagogical journal that was accessible and of interest to university students. One is the hortatory quality of the journal itself, irrespective of the articles appearing in it. Each of the six issues composing volume 1 was prefaced by a boldface quotation standing alone on a page. When, for financial reasons, elimination of these expensive pages was later suggested, Bieberbach resisted.[211] Volume 1 contained quotations from Hitler (issue no. 1), Paul Ernst (no. 2), and Immanuel Kant (no. 3). The same quotation from Kant prefaced issue no. 4. Goethe was author of the epigraph for issue no. 5 and the Nazi party ideologist Alfred Rosenberg for issue no. 6.

Second are the book reviews. Even when dealing with intrinsically nonideological material, *Weltanschauung* can find its way in. Thus Kubach, whose dissertation was a historical one on Kepler, in reviewing briefly a book about him, managed to speak of "our time, which, after years of crisis and decay, has again gotten solid ground under its feet through the new formation of an ideology (*Weltanschauung*) directed at the whole." Similarly, Bieberbach in a review of a volume of the well-known scientific biographical handbook "Poggendorf" remarked that in the future he hoped that it would provide the ethnic provenance (*Volkszugehörigkeit*) of individual scholars, which is more important than their generally more recognizable national identity (*Staatsangehörigkeit*).[212] One can imagine, then, the reviews by Bieberbach of Bruno Kerst's pamphlet "Breakthrough in Mathematical Instruction"[213] or of Adolf Dorner's "Mathematics in

[209] Above, chapter 4, note 272.

[210] *DM* 1 (1936): 290–307.

[211] BAK R73 15934, Bieberbach to DFG, Feb. 17, 1937.

[212] *DM* 1 (1936): 538.

[213] Ibid.: 110.

the Service of National-Political Education."[214] In the first of these, Bieberbach stressed again how all-encompassing the breakthrough is, including mathematics: mathematicians must think of themselves as the educators of German youth to German citizenship. This stance again protects mathematics from irrationalist attacks. A similar tone colors his review of a new journal in mathematical economics and social research in which mathematical methods are touted as the future, replacing "juggling," and "the long-overtaken eccentric ideas of liberal and Marxist opinions."[215] A mathematical economist of the present might think similarly (though the tone of expression might be different); for Bieberbach, however, the context was establishing the importance of mathematics to the National Socialist future.

Bieberbach's review of the Dorner book reveals his seemingly almost religious passion for National Socialism at this time:[216]

> Unfortunately it must be emphasized that the detailed carrying out [of the book] does not reach the praiseworthy goal. Above all one must become rightly skeptical in judgment of the whole, if one becomes aware of the evil (*üblen*) profanation that is practiced under the heading National Community (*Volksgemeinschaft*) with the symbol of the movement. It must wound most grievously the feeling of each and every fellow member of our people (*Volksgenossen*) if the symbol of the movement is degraded to the object of shallow school exercises.

Similarly, the Kerst review ends, "Just as truly as the life of our people is a whole, just so truly will the movement not make a halt before any artificially jamming door."[217]

Other reviews include Kubach on the first volume of Lenard's *Deutsche Physik*. This was highly laudatory, despite Lenard's negative view of contemporary mathematics discussed earlier. Kubach considered Lenard's book as a contributory effort to the realization that all science is dependent on the ideology, race, and "blood" of its creators. He also averred that anyone who worked through Lenard's views of mathematics as an independent science would not make the superficial mistake of thinking them falsely conceived.[218] There are also straightforward mathematical reviews of mathematical books by *Deutsche Mathematiker* similar to those already described.

The third matter really has to do not with the content of *Deutsche Mathematik*, but with its apparent survival value. In 1966 the Dutch firm Swets and Zeitlinger decided to reprint *Deutsche Mathematik* with the permission of Hirzel (which had moved from Leipzig to the then–West German location of Stuttgart). Their motivation was to make the mathematical content rather more accessible than it had been. With the agreement of Hirzel, says a preface, of the two kinds of articles, "pure-mathematical and ideological," it was decided to

[214] Ibid.: 255.
[215] Ibid.: 699–700. This is the journal in which Münzner's article cited by Drenckhahn appeared.
[216] Ibid.: 256.
[217] Ibid.: 110.
[218] Ibid.: 256–258.

reprint only the former and blank out the latter. For example, Kubach's "Students in Front" or material on the student camps are replaced by blank pages. However, what has remained is often curious, and can perhaps be most charitably explained as great carelessness upon the part of the appointed censor, who did not trouble to read even brief parts of the journal. The less charitable explanation, of course, is that someone wished to preserve and pass on this ideological use of mathematics, while others failed to make themselves aware of it. This is not an argument against an uncensored reproduction of the whole for historical purposes, but merely to say that the 1966 reprinting is ingenuous in its claim to have only reprinted "pure mathematics," and consequently this may give unwanted and unwarranted weight to the ideological material that remains. A few examples (not the only ones) will suffice. This reprint contains Bieberbach's reply to Jordan, as well as his reviews of the Kerst and Dorner books, and "Poggendorf." While omitting the brief description of student summer semester work in 1936 at the University of Berlin, it reprints the racist piece "Mathematics and Biology" that emerged therefrom (and whose very first line contains the phrase "racial hygiene"). Similarly reprinted was Drenckhahn's "pedagogical" article discussed earlier, when its very title (for someone who read German) revealed its racist and pro-Nazi ideological content (and whose last sentence justifies the "Nuremberg law" referred to in the title). These examples come from volume 1; just silly is the reprinting in volume 4 of the full-page portrait of Theodor Vahlen wearing Nazi insignia, but omitting the *laudatio* on Vahlen's seventieth birthday by the well-known mathematician (and apparent Nazi sympathizer) Friedrich Engel.[219] On the other hand, twice-suppressed is the repeated quote from Kant, which is about avoiding foreign expressions whose use, he says, reflects either mental poverty or negligence, and which are discomforting to see. Whether or not this is in line with Nazi propaganda, the suppression of Kant seems curious (Goethe is not suppressed).

Bieberbach's enterprise was apparently initially successful with respect to content, which places added interest on how it fared otherwise. Originally a great deal of the cost of *Deutsche Mathematik* was to be subsidized by the DFG. The journal was sold substantially below cost in order to attract subscribers, and the initial subscription goal was 500. At this time the membership of the German Mathematical Society was about 1,100, and the most highly subscribed mathematical publication was its journal (at about 725), but one should recall that *Deutsche Mathematik* was intended to be attractive to students and others who were mathematically involved but outside of university faculty.[220] All sec-

[219] Friedrich Engel (1861–1941) was a distinguished mathematician both in his own right (e.g., the "Engel condition" in the theory of Lie groups) and as the student, colleague, and interpreter of the famous Norwegian mathematician Sophus Lie, whose work was sometimes difficult for other mathematicians to understand. As an elderly man, he seems to have been attracted by the National Socialist brand of nationalism (he had been a friend of Theodor Vahlen for many years).

[220] BAK R73 15934, Verlag S. Hirzel to Griewank, July 2, 1935; Verlag S. Hirzel to Börsenverein der Deutschen Buchhändler zu Leipzig, Feb. 11, 1936; Verlag S. Hirzel to Bieberbach, Feb. 17, 1936. Cf. Bieberbach to Präsidium der Notgemeinschaft der Deutschen Wissenschaft, Oct. 14, 1934.

ondary-school principals received a copy of volume 1, no. 1, with the suggestion they subscribe on behalf of their mathematics teachers (especially given its very low price).[221] In the event, volume 1, which had been estimated at 576–640 pages,[222] actually ran an astonishing 898 pages of text. The total cost to the DFG was the large sum of 25,000 RM.[223] Among these costs were 4,000 RM shared between Bieberbach and Vahlen, whereas the publishers of no other journal put out by the DFG had such honoraria, and 3,000 RM total paid out as honoraria to contributors, a practice common to no other mathematical journal.[224] *Deutsche Mathematik* was also very expensive to produce, involving, for example, multicolored anaglyphs and tipped-in spectacles to view them so as to create a three-dimensional illusion.

In addition, internal politics in the DFG had resulted in the replacement of the radical romantic Nazi Johannes Stark with the opportunistic pragmatic Nazi Rudolf Mentzel. The first-year costs to the DFG had gone well beyond those foreseen, and consequently on January 28, 1937, Mentzel wrote Bieberbach and Vahlen, saying costs had to be brought down—what would they suggest? (He suggested fewer pages and elimination of expensive inserts for two measures.)[225] Bieberbach's reaction three weeks later "in agreement with Prof. Vahlen" is remarkable. While claiming (as is often the practice in cost overruns) that he had technically not gone over cost, he also foresaw a smaller size for volume 2, which would not be "substantially" over about 640 pages. Additionally, he emphasized the community-building nature of *Deutsche Mathematik* as counterweight to the other mathematical journals. In his view, these had the following deficiencies. The *Mathematische Annalen* had a Jewish editor [Otto Blumenthal[226]], the *Mathematische Zeitschrift* contained articles dedicated to Jewish communists [Emmy Noether[227]], *Crelle* contained papers by emigrés,[228] and the German journal in the history of mathematics (*Sources and Studies in the History of Mathematics*) was directed by a Jew [Otto Toeplitz] and a mixed-race emigré [Otto Neugebauer[229]]. If *Deutsche Mathematik* were to turn down a good paper by a *Volksgenosse*,[230] that would force its publication in one of the more established but suspect journals, as well as inhibiting *Deutsche Mathematik*'s community-

[221] See cover letter sent to these principals in BAK R73 15934.

[222] BAK R73 15934, Verlag S. Hirzel to Griewank, July 2, 1935; Bieberbach to DFG, Feb. 17, 1937. In this letter, the "normal size" of each volume is estimated at about 640 pages.

[223] BAK R73 15934, Mentzel to Theodor Vahlen, Apr. 8, 1937.

[224] Ibid., and BAK R73 15934, Verlag S. Hirzel to Griewank, July 2, 1935.

[225] BAK R73 15934, Mentzel to Bieberbach and Vahlen, Jan. 28, 1937. The letter was written by Karl Griewank.

[226] For Blumenthal, see above, chapter 6, "The Case of Otto Blumenthal."

[227] B. L. Van der Waerden published an obituary of Emmy Noether in this journal, a courageous act for even (or perhaps especially) a foreigner.

[228] Among others, Kurt Mahler, Stefan Bergmann, Otto Toeplitz, and Richard von Mises. Bieberbach does not seem to have observed that still listed on the editorial page, though thoroughly Jewish by Nazi standards, was Kurt Hensel (his mother was a baptized Russian Jew; his paternal grandmother the equally baptized Fanny Mendelssohn).

[229] Neugebauer was, to the best of my knowledge, not Jewish, even by Nazi standards.

[230] Literally "folk (or national) comrade."

building nature. Bieberbach stressed that his journal was not just a collection of papers but had a broader educative function as well. He further insisted on continuing to give honoraria to contributors (instead of offprints) as another community-building measure. He insisted that Erhard Tornier, Werner Weber, and Ernst August Weiss, principal helpers on the journal, should be recompensed, as well as the occasional external referee—though he would be frugal in such matters.

Furthermore, the idea of recompensing him and Vahlen was consistent with other scientific journals, was unsolicited, had been initiated by the previous president Stark, and was implicitly recognized by the authorities. Also, now that Vahlen had retired, he had even less reason to work without pay, and his energy and name were important.[231]

But Mentzel held the purse strings, and in his view they needed to be drawn tighter. He would only contribute 12,000 RM to the journal, slightly less than half the cost to him of volume 1 (which had had advertising costs as well). On March 2, 1937, Bieberbach met with Mentzel and discussed how to bring down the costs, with Griewank taking official notes. The upshot was that Bieberbach and Vahlen relinquished their honoraria. Honoraria for authors were also relinquished. To judge by the figures in the correspondence, these two measures alone saved 7,000 RM, or more than half of Mentzel's demanded saving. Mentzel, however, did declare himself ready to see to suitable stipends for coworkers (presumably this meant Tornier, Weber, and Weiss). The second volume of *Deutsche Mathematik* would be kept in the compass of roughly 640 pages (it actually contained 734 pages of text, and volume 3 [1938] was 730 pages). In addition, logistic work (Scholz) would continue to be published, but only as separata of about forty pages, and not within the journal as well (as had happened with the Hermes-Scholz article). Also, advertising expenses redounding to the DFG would be eliminated, and students (presumably unpaid) would be used more heavily in bringing out the journal. Finally, Bieberbach's "community-building" was expressly recognized in the hope that a mathematical organization would come into being as a result of cooperative work on the journal.[232]

Herbert Mehrtens says of Mentzel's funds reduction: "Obviously the representatives of the state did not expect much of Bieberbach's *Deutsche Mathematik*."[233] This seems to me to be mistaken. Mentzel's need to reduce funds seems genuine, and, as has been noted, substantial sums for volume 1 had gone for what might be termed unusual honoraria. Mentzel also seems genuine in his wish that a future *völkisch* mathematics community would come into being. Further, he had no intention of ever asking that the DFG be reimbursed.[234] In succeeding years he maintained support at the new level of 12,000 RM per annum and the journal occasionally grew again beyond the specified size.

[231] BAK R73 15934, Bieberbach to DFG, Feb. 17, 1937.

[232] BAK R73 15934, Griewank's summary of the agreement.

[233] Mehrtens 1987: 223.

[234] BAK R73 15934, Mentzel to Vahlen, Apr. 8, 1937. Mentzel pleaded the pressures of his obligations under the four-year plan as a further reason for reducing funds.

Bieberbach spoke to Vahlen about the loss of honorarium; however, he also advised that Mentzel do so, presumably as "ministerial director" to (retired) ministerial director so that Vahlen would be hearing from a hierachical "equal," and so not be insulted. While Mehrtens seems to be wrong, and some accomplishment was expected of the journal, clearly it was only one small matter among many for Mentzel to attend to, and Griewank had to remind him at least twice to write Vahlen before he did so on April 6.[235]

Bieberbach made one more attempt to increase the size of volume 2, arguing in June that already the first two issues had occupied 376 pages and he had material for about another 128 pages ready, so circa 640 pages for six issues was much too few, as well as ringing the usual changes on the journal's political role.[236] Mentzel, however, stood firm: the funds available to the DFG did not "at the time" permit a greater subsidy. Griewank wrote Bieberbach suggesting he either reduce the number of contributions, delay the appearance of the issues, or get by with a smaller number of issues. Bieberbach had no choice but to comply. In fact, issues 4 and 5 for 1937 each contained only eighty-one pages, and issue 6 did not appear until January 1938 and contained only seventy-four pages. Thus the actual number of published pages during 1937 was 660, roughly conforming to Mentzel's prescription.[237] However volume 3 (1938) began in March and comprised within 1938 six issues totaling 730 pages; in January, total published pages again approach the size of volume 1; so Mentzel must have relented somewhat.[238] Bieberbach even took Mentzel's 640 as a *minimum* number of publishable pages in advertising the journal.[239] Volume 4 had only 656 pages in five issues (the last in September 1939)—it is unclear whether the DFG or the incipient war prevented issue 6 from appearing, but I would guess only the latter. Volume 5 was spread over two years and managed 588 pages during a period when Germany was generally triumphant.[240] Given wartime conditions, the two remaining volumes were naturally smaller. Volume 6 began appearing in September 1941, had effectively only four issues, and ended in September 1942, but still comprised 586 pages. The final volume 7 had only three issues, the first appearing in November 1942, the second in June 1943, and the third, promised for autumn 1943, not until June 1944. Volume 7 contained 608 pages. However, Mentzel's goodwill is perhaps demonstrated by his still authorizing on June 29, 1944, for fiscal year 1944–45, a contribution of "up to" 12,000 RM for *Deutsche Mathematik*.[241] Perhaps this also is one more example of the German academic establishment's unwillingness to contemplate the possibility of German defeat in the war.

[235] BAK R73 15934, Griewank's report to Mentzel of Bieberbach's conversation with Vahlen, Mar. 5, 1937, with handwritten note dated Mar. 24, 1937.

[236] BAK R73 15934, Bieberbach to DFG, June 21, 1937.

[237] See page prior to page 1 in *DM* 2 (1937).

[238] See page prior to page 1 in *DM* 3 (1938).

[239] See inside front cover of unbound issues of the journal.

[240] March 1940–May 1941.

[241] BDC, Bieberbach file, Mentzel to Bieberbach, June 29, 1944.

Was *Deutsche Mathematik* successful? In three volumes published during 1936–38, *Deutsche Mathematik* managed a total of 2,360 pages, *Mathematische Annalen* a total of 2,335 pages, and *Mathematische Zeitschrift* a total of 2,358 pages. When one considers that the size of *Deutsche Mathematik*'s pages was much larger than those of the other two journals and consequently the space allotted to print was over one and one third times that of the other two journals, it would seem that from the point of view of attracting contributions, *Deutsche Mathematik* was initially quite successful.[242]

With respect to attracting subscribers, the matter is less clear. Hirzel Verlag originally made what it considered a conservative estimate for the number of subscribers at 500, especially since it recognized that *Deutsche Mathematik* had to overcome opposition in the German mathematics community, and no German mathematics journal in 1935 had a circulation of more than 725.[243] It therefore seems respectable that the first issue of 1938 (volume 3) sold 533 copies.[244] Certainly *Deutsche Mathematik* did not achieve the dominant position Bieberbach had hoped for, and the *völkisch* mathematics community envisioned by him, Mentzel, and Hirzel Verlag did not come into being, but the journal does seem to have had a readership. The initial printing of volume 1, no. 1, was 6,500,[245] not from hypersanguinity as to its attractiveness, but for advertising purposes—recall, for example, that every secondary school in Germany was to receive a copy. But the amount of issues necessary for these purposes was overestimated. In 1942 the Hirzel Verlag representative suggested pulping numbers of residual copies (paper, in fact, was becoming dear) and made the sarcastic comment that it was unlikely that after the war ended, Bieberbach intended greater propaganda for his journal for which he required trial copies. But still he only suggested pulping, for example, 1,500 out of 2,110 remaining copies of volume 1, no. 1.[246] Thus, from the point of view of attracting contributions, *Deutsche Mathematik* was initially a success; from the point of view of attracting subscribers, its performance was adequate compared to other mathematics journals. However, given that it was being sold at a much-reduced price, this might be accounted a partial failure. Presumably the anticipated "resistance" in the German mathematical community was somewhat balanced by the low price. What about Bieberbach's more general aspirations?

Here, failure was complete. Not only did a *völkisch* mathematics community

[242] The *Deutsche Mathematik* page was 18.5×27 cm, with 14×21.5 cm allotted to print. The other two journals had pages that measured 15×23 cm, with 11.5×19 cm, allotted to print. Thus the margins in *Deutsche Mathematik* were larger as well, giving an impressive appearance.

[243] BAK R73 15934, S. Hirzel Verlag to DFG, July 2, 1935.

[244] The number of copies printed of each issue appears in small print at the bottom of the last page of each issue. On March 21, 1942, Hirzel Verlag wrote the DFG concerning the pulping of unsold copies, giving amounts remaining. From volume 1, no. 3 through the last issue of volume 4 in 1939, subtraction provides usually a surprising constant 533 copies sold, exceptions being 543 for vol. 4, no. 3, and 534 for vol. 4, no. 4.

[245] Mehrtens 1987: 223.

[246] BAK R73 15934, O. Carlsohn for Hirzel Verlag to DFG, Mar. 21, 1942. A curious slip occurs at one point in this letter, in which *kalkulieren*, "to calculate," is typed for *makulieren*, "to pulp."

fail to materialize, but the journal lost the exhortatory tone of volume 1 and settled down to being just more or less another mathematics journal with the occasional racist article or *völkisch* gibe—articles that were carried in no other mathematics journal.

Aside from the dwindling success in maintaining a *völkisch* tone, the journal was also distinguished by its continuing concern for mathematical pedagogy. Nevertheless, the student involvement that Bieberbach desired for these reasons, and Mentzel for economic ones, also seemed to fade. One feature of *Deutsche Mathematik* was summary reports on the mathematical curricula and activity at universities across the country. Judging by these, most voluntary mathematical activity among students in *Arbeitsgemeinschaften* ("study groups") or *Fachschaftsarbeit* ("disciplinary work") devoted to some mathematical topic of mutual interest became strictly mathematical. There were, of course, exceptions, and some of these, which appeared prominently in volume 1, were discussed earlier. The university in Berlin was always an exception—not only because of Bieberbach (seconded among the faculty by Tornier, Werner Weber, and the mathematically marginally competent Klose), but also because Karl-Heinz Boseck[247] was the student leader. For example, in 1936–37, various groups devoted themselves to Bieberbach's *völkisch* view of mathematical and physical creativity, and one even delved into the National Socialistically respectable (but hardly mathematical) book, Alfred Rosenberg's *Myth of the Twentieth Century*. However, they also dealt with serious and suitable mathematical topics by any standard.[248] In 1937, there was again a division between the *sachlich*, or "mathematical-factual," and the *weltanschaulich*. The latter contained study groups working on the politically respectable but bizarre pseudo-scientific "World Ice Theory" and "Hollow World Theory."[249] However, here they actually corrected scientific errors in a "scientifically false" film promoting the "World Ice Theory." [250] Such activities seem the exception among student mathematical activities of the time; more prosaic ones like partial differential equations, geodesy, meteorology, Nevanlinna theory, and so forth were the usual fare. Even some of the clearly nationalistically oriented studies seemed to have had solid mathematical-physical content (e.g., the study of the life and work of Copernicus at Königsberg),[251] though there were also the occasional political/*völkisch* activities, like the "World Ice Theory" at Rostock,[252] or "Racial Questions in Physics" at Freiburg.[253] Even Berlin became less stridently *völkisch*. After Christmas 1937, the "World Ice Theory" at Berlin was dissolved, and a new group under Boseck's leadership was formed to study Madame Curie and the discovery of ra-

[247] See above, chapter 6, "Mathematics in the Concentration Camps."

[248] *DM* 2 (1937): 349.

[249] *DM* 2 (1937): 641–642. Heinrich Himmler was an especial devotee of the "World Ice Theory" and commanded the *Ahnenerbe* to investigate its validity. For the theory itself, see Gardner 1957.

[250] *DM* 2 (1937): 641; *DM* 3 (1938): 476.

[251] *DM* 3 (1938): 490.

[252] *DM* 2 (1937): 653; 3 (1938): 496.

[253] *DM* 2 (1937): 359.

dium. While the intention was to discover precisely her achievement and "to investigate its *völkisch* connections (*Gebundenheit*)," still, as in the case of the various Copernicus study groups, or the one in Berlin that soon also appeared on "Cauchy, Riemann, Weierstrass and the Beginnings of Function Theory" (reported by Oswald Teichmüller), here genuine scientific content was wrapped in *völkisch* paper.[254] Reports of instruction or of voluntary *Fachschaftsarbeit* appeared in *Deutsche Mathematik* for the last time in 1939, presumably because of the further decline in student numbers, as well as the temporary closing of the universities with the onset of war.

The war also seemed to contribute to a further decline in *völkisch* pronouncements in the journal. Although excrescences like the paper "Mathematics and Race" by Max Draeger might appear occasionally,[255] it was the exception rather than the rule. Volumes 2, 3, and 4 of *Deutsche Mathematik* are for the most part unexceptional, though the first issue of volume 2 of *Deutsche Mathematik* seems rather in the spirit of volume 1: the first three articles in it are by Alfons Bühl, Udo Wegner, and Bieberbach. Bühl emphasized, Bieberbach-style, that science and political will spring from the same roots, and the question is where those roots are: National Socialism contains a suitably German ideology within it. History and political thought, law, and even medicine have understood this.

> Only in the natural sciences is it first of all always still a small circle, which, completely enlisted, laboriously seeks the German way. A sad fact if one considers that it was precisely two Nobel laureates in physics [Phillip Lenard and Johannes Stark] who belong among the oldest fellow-combatants of the *Führer*.

Bühl said that liberalism "revels" in wanting to understand everything—Jews make use of this. There is an attack on Einsteinian space-time (propagandized by the "Jewish press") and acausality in physics and on those who want to introduce such acausal notions into biology. For Bühl, this is shocking. The Jew should do research according to his type, but, for Germans, space, time, and causality stem from contemplation of nature—of course, by the time of Bühl's address to natural-science students in August 1936, Jews had been almost entirely expelled from German academic life, and the "Jewish-related" were soon to follow. However, Bühl declared that many "Aryan researchers" were influenced by foreign Jewish thought, and the interesting comparison was given of Weimar statesmen who were Marxist but Aryan.[256]

Bühl's exhortation that science is not objective and that natural scientists had better understand this was followed by a less original piece on research and teachers by Udo Wegner that attacked the misuse of mathematics to lend inappropriately apodictic certainty (which he claimed comes from a "mechanistic ideology," now overcome). There is a demurrer about statistics, but only insofar as it collects and describes data, not insofar as attempts may be made to use it

[254] *DM* 3 (1938): 476; 4 (1939): 143; 4 (1939): 115–116 (summary of a report given by Teichmüller); 4 (1939): 121 (summary of report on Curie by W. Jahn).

[255] *DM* 6 (1941–42): 566–575.

[256] *DM* 2 (1937): 3–5.

as an excuse for action. The issue is fitting mathematics into the new organic world-picture. Here again is the worry about the survival of mathematics in the new state. Wegner stressed that mathematics has epistemological value and is not just a playing with symbols. Mathematics is linked to art and music as equal expressions of *völkisch* qualities. German thought is linked to Greek thought in mathematics,[257] as is the German relationship to "nature." Citations range from Plato and Xenophon to the famous French historian Hippolyte Taine, and on to the "classic" racist writer Houston Stewart Chamberlain, and the contemporaries Vahlen and Lenard. Bieberbach has clearly also been a source, though he is not explicitly cited. Thus mathematics is validated both for its applications and for its educative value for "right thinking" and apprehension of truth. Subtracting the *völkisch* or Nazi twist placed on such justifications by Wegner or E. A. Weiss, these are ancient justifications of mathematics subject to some debate.[258]

The article by Bieberbach takes up such pedagogical issues as a practical matter within the Nazi state, rather than just theoretically. The "radical" ideas of Bruno Kerst[259] that issues central to the *Volk* should be the issues motivating mathematics instruction are promoted, as they had been in Bieberbach's paean of a review of his pamphlet "Breakthrough in Mathematical Instruction."[260] There needs to be a revolution in mathematics instruction to bring it "nearer to life." Specific instructional ideas are sketched. Mathematics explicitly is to be justified not as a cultural good, or general education, or formal learning, but as one of those things that "tested in the life of our people has been revealed as important." In addition to promotion of Germanicism in education through learning about great German mathematicians, an expected theme, Bieberbach's printed talk also promoted his journal. In addition, like the articles of Bühl and Wegner, delivered orally at a Heidelberg student camp, Bieberbach promoted two pedagogical ideas having nothing to do with ideology and with value in themselves. One was the creation of a mathematics dictionary. The other was loosening the formality of university instruction and bringing student and instructor closer for a more "personal education." Bieberbach ended with the (given the location) obligatory reference to Lenard, hoping his ideas would contribute to Lenard's having "a friendlier judgment of the value and wishes of our German mathematical science."[261]

The rest of volume 2 is, however, unexceptional. The occasional racist article appears, but on the whole, the content is largely mathematical. Teichmüller was far from the only person to publish "real mathematics" in *Deutsche Mathematik*. Sometimes pedagogical articles that had a stated National Socialist motivation were in fact generally useful, such as E. A. Weiss's selection of student letters

[257] A famous book on the influence of Greek thought on German intellectual life in general is E[liza] M. Butler, *The Tyranny of Greece over Germany* (London: Macmillan, 1935).

[258] *DM* 2 (1936): 6–10.

[259] Bruno Kerst, *Umbruch im mathematischen Unterricht* (Berlin: S. G. Grote, 1935). (This is only forty-seven pages long.)

[260] *DM* 1 (1936): 110.

[261] *DM* 2 (1937): 11–16.

describing "nontraditional" occupations for mathematics students, like meteorology, insurance, the optical industry, ballistic studies, or office computation. Nor did the writers in *Deutsche Mathematik* have any qualms about citing Jews: among others, Richard von Mises, Emmy Noether, Richard Brauer, Friedrich Levi, A. A. Albert, Paul Bernays, Issai Schur, Peter Scherk, Moritz Cantor, and Heinrich Liebmann. The statement by Helmut Lindner that Jewish mathematicians went uncited in Nazi Germany seems therefore untrue, and at best needs serious qualification.[262] Mathematics is a peculiar discipline, and the proponents of a *Deutsche Mathematik* seem to have really believed what they said: a mathematical fact was a mathematical fact, whoever had found it. However, certain styles of approach, certain subdisciplinary subject matters, certain pedagogical attitudes, certain beliefs about the nature of mathematics, were "racially" determined, and were corrupting to those not of the same ethnic background.

Issue 3 of volume 4 was dedicated to Theodor Vahlen on the occasion of his seventieth birthday. Yet, aside from the full-page portrait of Vahlen wearing Nazi insignia on his lapel and the encomium by Friedrich Engel, there seems nothing particularly nationalistic or *völkisch* about the papers. They include some by authors not usually represented in the pages of *Deutsche Mathematik*, such as Wilhelm Süss (Bieberbach's doctoral student) and Hellmuth Kneser (Vahlen's old friend from Greifswald days). But papers by regular contributors who were "Deutsche Mathematiker," such as Werner Weber, E. A. Weiss, H. J. Fischer, Udo Wegner, or associates like Alfred Klose, or Bieberbach himself, were completely mathematical in content. Quite interestingly, when giving lectures in an ideologically supported atmosphere, such authors would drop *völkisch* remarks, which, if nothing else, signified their *völkisch bona fides*. When *writing* as opposed to *lecturing* about mathematics, such remarks seem to have been eschewed, even in so supposedly receptive a medium as *Deutsche Mathematik*, even in an issue dedicated to Vahlen. Thus at the first national mathematics camp for students and faculty, effectively from June 29 to July 3, 1938, Werner Weber might ask rhetorically whether it was significant that a German (David Hilbert) had first solved "Waring's problem" in number theory by showing a solution existed, but "Jews and foreigners" were mostly involved in exact computations of related constants.[263] Similarly, Klose at the same meeting distinguished between the nature of the contributions of "the Irishman" William Rowan Hamilton and "the Jew" Carl Gustav Jacobi to so-called Hamilton-Jacobi theory (of course, Hamilton was insightful and physically inspired, whereas Jacobi was formal) and spoke of the racial relationships of different mathematical thought forms.[264] And Wegner took as a starting point for his lecture ideas of the Nazi educational theorist Ernst Krieck, managing along the way to praise

[262] Helmut Lindner, "'Deutsche' und 'gegentypische' Mathematik Zur Begründung einer 'arteigenen' Mathematik in 'Dritten Reich' durch Ludwig Bieberbach," in Mehrtens and Richter 1980: 107.

[263] *DM* 4 (1939): 127.

[264] Ibid.: 111–115. This is a summary report of a Berlin *Arbeitsgemeinschaft*; citation is from 114–115.

intuitionism, talk about the Germans as culture-bearers, and refer to Bieberbach's 1934 lectures.[265] Even Teichmüller managed to speak of Weierstrass's many Jewish students, cite the well-known passage in his letter to Sonja Kowalewsky about mathematical imagination,[266] and question whether Bernhard Riemann could be rightly considered a forerunner of Einstein and Hermann Minkowski (also Jewish).[267] However, to judge from the summaries, even at this camp, where the keynote address was by a Berlin biologist and entitled "Racially Bound Thinking and Creativity in the Natural Sciences," purely mathematical talks (e.g., Willi Rinow on the four-color problem, or Teichmüller on partial derivatives) or ones dealing with educational problems showed not a breath of *völkisch* air. Indeed, the well-known physicist C. F. von Weizsäcker seems to have cautiously but firmly defended contemporary physics.[268]

Some mathematics departments even openly ignored all pretense at being *völkisch*, and this was reported without negative comment in *Deutsche Mathematik*. Thus, at the Münster of Behnke and Scholz, 1938 saw two lectures by the famous French mathematician Henri Cartan, one by the American Marston Morse, and two by the Polish logician Stefan Lesniewski.[269]

Although in volume 1, Tornier inveighed against "jugglers of definitions," axiomatic articles were not foreign to *Deutsche Mathematik*. For example, Fritz Klein-Barmen wrote on lattice theory, and Ernst Foradori on his variant approach to some of the same material, which he called "Part Theory" (*Teiltheorie*).[270] The support given Heinrich Scholz's logistic school has already been discussed. A brief book (fifty-eight pages) on geometric axiomatics by Eugen Roth (from the Scholz school) was very favorably reviewed.[271] Max Steck and Baron Freytag-Löringhoff had an exchange on the philosophy of mathematics and the meaning of axiomatics in which nothing was ever spoken of as the German way of thinking.[272] Although Bieberbach may have opined that foundational disputes in mathematics could be attributed to racial interests, *Deutsche Mathematik* published a serious survey article by Gerhard Gentzen on the state of foundational research that was free of ideology.[273] In fact, this, together with a new version of Gentzen's famous proof of the consistency of number theory, appeared in Scholz's series of separata (issue no. 4), encouraged by Bieberbach and supported by the DFG. Similarly, the later distinguished historian of mathematics,

[265] Ibid.: 130–131.

[266] Edited by G. Mittag-Leffler in "Weierstrass et Sonya Kowaleskaya," *Acta Mathematica* 37 (1923): 133–198, p. 191.

[267] *DM* 4 (1939): 115–116.

[268] Summaries of all talks at the camp appear in *DM* 4 (1939): 109–140. This report was edited by Johannes Juilfs, a student in Berlin. A second camp was held in May 1939.

[269] *DM* 4 (1939): 154–155.

[270] *DM* 5 (1940): 37–43. Books by Foradori and the well-known logicians Karl Schröter and Wilhelm Ackermann were also advertised in *Deutsche Mathematik*; see, for example, the back covers for volume 6 (1941–42). These last two were in the series edited by Heinrich Scholz.

[271] *DM* 3 (1938): 347.

[272] *DM* 3 (1938): 467–473; 4 (1939): 238–240.

[273] *DM* 3 (1938): 255–268. For more about Gentzen, see below, chapter 8.

Joseph Hoffman, published considerable early work in *Deutsche Mathematik*, and none of it seems to have been tendentious in a *völkisch* manner.

For the adherent of "Deutsche Mathematik," mathematics and biology were of course an inciteful mixture, and a report in the journal on the first joint gathering (in 1938) of the subgroups of the NSLB[274] devoted respectively to biology and to mathematics and natural science contains some of the expected material on ideology and education, or mathematics as an aid in the study of racial inheritance, or biology and ideology. However, most of the talks seem to have been scientifically substantive, dealing, among other matters, with the optical qualities of polymers, fungal symbiosis, a new approach to integral calculus, the mathematics and physics of aviation, and similar topics. A former president of the German Mathematical Society, Richard Baldus of Munich, spoke on the topic "Axiomatics in Science and School"; the well-known Munich topologist Heinrich Tietze discussed knot theory. Among nonmathematicians, two world-famous scientists, Peter Debye and Karl von Frisch, gave purely scientific talks: Debye on approaches to absolute zero, and von Frisch on using bees for biological instruction. Of the purely pedagogical talks, the first was given by the influential mathematics textbook writer Kuno Fladt.[275] According to the summary of his speech:

> Mathematics instruction ought not to be lacking in the new German school, because mathematics is simply indispensable in the life of our people, second, occupation with mathematics schools not only the understanding, but also the will, and finally mathematics is a cultural possession in which every German should have a share in appropriate measure.

Slightly less desperate was the opening statement of the gathering:

> On the basis of our experience with youth, we feel obligated to once more strongly emphasize the value of intellectual work. By this we mean the education to intellectual work and achievement, which is decisive for the affirmation and rise of our people, as well as the much-undervalued education of character and will by intellectual work.

Clearly in 1938 intellectually serious secondary-school instructors felt beleaguered, and mathematics ones especially so. Fladt was discussing the new (state-determined) course of instruction in mathematics, and others spoke about similar decrees in the other sciences. In particular, Fladt remarked that the preference for "intuition" (*Anschauung*)" meant a primacy of geometry over algebra in school instruction.[276]

Similarly, even when reviewing a tendentious book on mathematical methods in biology, particularly with respect to the theory of inheritance and racial science, Bieberbach limited himself to brief biological and mathematical summa-

[274] *Nationalsozialistischer Lehrerbund*, or "Union of National Socialist Teachers," the compulsory organization for nonuniversity teachers.

[275] See above, chapter 5, "Secondary and Elementary Mathematics."

[276] Report of the gathering by August Engel, *DM* 3 (1938): 607–610. The opening statement was by a secondary-school teacher, L. Baumgartner, the executive of the local organizing committee.

ries, the only hint of something *völkisch* being the wish that the author (Friedrich Ringleb) would do original work in the subject.[277]

The same is true for volume 5 (1940–41). Here there are virtually no "work" articles, and the others seem unexceptionable. Even a Nazi (and Bieberbach) sympathizer like Harald Geppert, while praising the "German genius" of Gauss, does not take the available occasion for a Bieberbach-style diatribe against Jacobi.[278] In fact, the book *The Methods of Physics* by the philosopher of science Hugo Dingler, an active supporter of "Deutsche Physik" and virulent opponent of Einstein,[279] is thoroughly savaged by the reviewer for *Deutsche Mathematik*.[280] This is even more striking when one realizes that the preceding volume of *Deutsche Mathematik* contained a praiseworthy review of the same book by Bruno Thüring, a young astronomer and ardent supporter of "Deutsche Physik." Thüring's review calls it one of the most important books in German science in a long time and speaks of the "two and one half thousand years of work of Aryan natural science research".[281] The condemnatory review by Adolf Kratzer of Dingler's book appeared in the very next issue and was no doubt intended as a corrective—what is significant is that Bieberbach had it printed, and so gave credence to such a corrective. Presumably after the war's start, physics (of whatever sort) was considered more important than *völkisch* respectability (Kratzer states that Dingler's ideas can in no way further physics). A brief article by Hellmuth Kneser in volume 5, written in a lightly sarcastic style, appeals for German words (rather than Germanicized Latin "monsters") for new mathematical concepts, but explicitly refrains from insisting on Germanification of old and fixed words like "determinant" (stemming from Gauss, writing in Latin). It also contains implicit criticism of Wilhelm Blaschke as arrogant, and its penultimate paragraph might be read as a covert statement of the internationalism of science.[282] With respect to the Germanification of words, and the use of good German, Bieberbach himself took a similar position, insisting that words like *radius*, *orthogonal*, and *konform* could not be usefully Germanicized.[283] Kneser's and Bieberbach's remarks make more sense when it is known that in 1938 a ministerial decree placed emphasis on such Germanification. In fact, a complete list of such "foreign words" and "German equivalents" appeared in 1941 divided into five categories: (1) words that should no longer be mentioned in school; (2) words that may be mentioned but not used operationally in schools; (3) words whose replacement by (given) German expressions is desirable; (4) words whose Germanification is recommended; and (5) words for which sug-

[277] *DM* 2 (1937): 733.

[278] "Wie Gauss zur elliptischen Modulfunktion kam," *DM* 5 (1940–41): 158–175, esp. 175.

[279] See, e.g., Mehrtens 1987.

[280] *DM* 5 (1940–41): 83–84.

[281] *DM* 4 (1939): 654–655.

[282] Hellmuth Kneser, "Quatérnion oder Quaternión. Ein Wort über Fachfremdwörter," *DM* 5 (1940–41) 259–261.

[283] Review of Walther Lietzmann, *Mathematik in Erziehung und Unterricht* (Quelle and Meyer, 1941), *DM* 6 (1941–42): 505–506.

gested Germanifications are given. Altogether, well over 300 mathematical words, mostly from Latin or Greek roots in a German form, were to have German equivalents whose use ranged from mandatory to suggested possibilities. A few terms, like *Mathematik*, *Arithmetik*, *transzendente Funktion*, *Logarithmus*, had no suggested equivalents.[284]

Volume 6 contained Draeger's mentioned article on "Mathematics and Race," but even here there was a difference. Draeger's opening apologetic sentences would have been unheard of in *Deutsche Mathematik* only five years previously:[285]

> First of all, a relevant statement seems to me necessary. The following arguments have nothing to do with the cheap exploitation of an existing trend. My investigations of this theme go back far before 1933 and have always been realized in instruction. Each year of secondary-school students to whom I had to teach mathematics was made aware of racial conditionedness (*Bedingtheit*).

A few other aspects of this secondary-school teacher's remarks are worth brief examination. The structural differences between Greek and modern mathematics are elucidated as an example of the methodological differences in different cultural groups' conceptions of mathematics. Oswald Spengler is quoted with approval on the importance of mathematics as a culture clue and the changes it underwent, though his philosophy of history is rejected and "the publicity-seeking title of his work" (*The Decline of the West*) is deplored. Two "demonstrations" of the abstract relationship between mathematics and race are given: one is that mathematics is a function of culture, and culture is a function of race; the other is that philosophy is racially conditioned, and philosophy is intimately bound up with mathematics (Spengler). Thus Draeger comes around to proposing (like Bieberbach) an investigation of how closely the dispute over the foundations of mathematics is racially conditioned. Bieberbach and Vahlen are of course cited, as is the "known fact" that "creative Jewish mathematicians" were always analysts and never geometers. This Draeger connected to the arithmetic nature of Babylonian mathematics, though demurring that appropriate judgments are made difficult by the contribution of the non-Semitic Sumerians. Needless to say, Draeger's perceived three fundaments of modern mathematics—the concept of function, the concept of geometric relationships, and the concept of the infinitesimal leading to the creation of calculus—were for him "spirit of our spirit and blood of our blood," and the last was hypothesized as "most strongly related" to the "Nordic soul." This sort of language is familiar and need not be pursued further. Two things, however, are striking. One is a comprehensive summary of the connection between mathematics and race that (of course) does not affect the validity of mathematical facts. This is threefold:[286]

[284] Walther Lietzmann was on the committee deciding on appropriate Germanifications, and his book (preceding note) contains a complete list of them (pp. 135–140). I am also indebted to Prof. Karl Stein (interview, Feb. 26, 1988) for a discussion of this topic.

[285] Ibid.: 566.

[286] *DM* 6 (1941–42): 572.

1. Materially, which has to do with the building up of subspecialties and their mutual relationship (e.g., Algebra-Geometry)
2. Formally, which concerns the methodology in the sense of the conception underlying the whole investigation
3. philosophically, and indeed,
   a) epistemologically with respect to the fundamental conception and the type of concept formation, and
   b) metaphysically, insofar as mathematics is to be addressed as a cultural symbol of the first rank.

The second striking thing about Draeger's article is that, even in 1941, he emphasized that the justification of mathematics is *not* its applications. Justification through applications is rejected as "Americanism," and in no way a Germanic sort of behavior. The importance of number theory and the Germanic number-theoretic tradition is stressed. Bieberbach's statement in 1934—"As proof of the national necessity of mathematics, one mostly calls upon applications. It seems to me that it suffices to refer to the fact that in mathematical creativity, national (*völkische*) originality powerfully reveals itself"[287]—is cited and emphatically endorsed.

But Draeger's article seems to be isolated in volume 6. Not only are the articles overwhelmingly devoted to research, with *völkisch* remarks, as usual, absent from these, but also the book reviews, pedagogy, history, and other "work" avoid such remarks. This last category is in this volume mostly concerned with issues around the teaching of descriptive geometry or applied mathematics. Similar observations apply to volume 7: even the obituary of Ernst August Weiss in this volume, although it naturally mentions his SA activity, the unfortunate outcome of World War I, and so on, is devoid of *völkisch* commentary. Volume 7 also contains the long article by Heinrich Scholz on formalized studies in the foundations of mathematics mentioned earlier, which, *en passant*, severely takes Max Steck to task.[288]

The mathematical content itself of *Deutsche Mathematik* was, however, peculiar in two respects. The first was the publication of mathematical, particularly statistical, applications to biology. This seems to have been *the* journal for such publications. These were stimulated partly by a tendentious racism (as in the article by Drenckhahn discussed earlier), and probably partly by the National Socialist view that biology should be the ultimate basis for all thought. The second was the far greater preponderance of articles on geometric subjects than in the three leading German mathematical journals—probably no single journal has published more proofs of Morley's trisector theorem.[289] However, much

[287] *DM* 6 (1941–42): 575.

[288] "Was will die formalisierte Grundlagenforschung," *DM* 7 (1942–44): 206–248.

[289] Morley's trisector theorem says that, for an arbitrary triangle, the trisectors of the angles meet in the vertices of an equilateral triangle. It was first discovered around 1900; the lateness of discovery probably is because, in general, the angle trisectors cannot be constructed with unmarked straightedge and compass (the classical Euclidean tools).

more substantial geometry also appeared. This was likely a result of the ideological emphasis on *Anschauung*, and hence geometry, as particularly German (see appendix).

Despite the decline in the *völkisch* content and National Socialist rhetoric in *Deutsche Mathematik*, the DFG under Mentzel, as noted still supported the journal, and at the same 12,000 RM per year.[290] Supplements to *Deutsche Mathematik* were also authorized (though these would be sold rather than given away). These actually amounted to unremunerated expenses for the DFG.[291] Three such came out, in 1939, 1940, and 1941, and were respectively a book by Weber on Pell's Equation in number theory (151 pages); a reprint of Vahlen's *Abstract Geometry* of 1905 in honor of his seventieth birthday (114 pages); and a thirty-one-page monograph by Vahlen on the paradoxes of relative mechanics. Although the three issues comprising volume 7 stretched from November 1942 to June 1944, in that very month Mentzel authorized another 12,000 RM.[292] There were two obvious reasons for the reduced number of articles: first, as both Bieberbach and Behnke remarked for their respective journals, the war meant a reduction in numbers of submitted articles;[293] second, the war meant the increasing scarcity of paper. It is almost symbolic that the last article ever printed by *Deutsche Mathematik* was a partially negative review by Bieberbach of the book *Jewish and German Physics* by Johannes Stark and Wilhelm Müller, which he condemned as too simplistic.[294] For Nazis like Stark and Müller, the issues were simple, and then dressed up in appropriate political and racist language. For Bieberbach, no less given to racist posturing, these were serious intellectual issues.

The Case of Herbert Knothe

The pedagogical element of the inspiration for *Deutsche Mathematik* seems clear, even if that pedagogy was only partly mathematical, and partly intended to rear the new German Nazi youth through mathematics. Bieberbach himself was seriously interested in students and was considered a stimulating teacher, even though his mathematical exposition was not always precisely correct.[295] This was clear to all: Oswald Teichmüller spends a whole page of a seven-page article to correct errors in the appropriate citation in a book by Bieberbach;[296] Werner Fenchel, who was an emigré in 1933 and had been Bieberbach's doctoral student in 1928, spoke of how much one could learn from him provided one avoided mentioning sloppinesses in his books and lectures. But Bieber-

[290] BAK R73 15934, Mentzel to Bieberbach, Apr. 29, 1938. The journal's fiscal year apparently began in April.

[291] BAK R73 15934, Griewank to Bieberbach, June 28, 1938; Griewank to Mentzel, Apr. 21, 1939.

[292] Above, note 241.

[293] BAK R73 15934, Bieberbach to DFG, Feb. 25, 1940. For Behnke, see Behnke 1978: 134.

[294] *DM* 7 (1942–44): 608.

[295] Biermann 1988: 193–194, 211–220.

[296] *DM* 4 (1939): 455–461, p. 461.

bach's lectures were not popular and sometimes evoked unserious attitudes in the students.[297] Seventeen students (including three women) wrote dissertations under his supervision—including, besides Süss and Fenchel, Helmut Grunsky and Hubert Cremer. Grunsky and Fenchel in particular had distinguished mathematical careers. This concern for students seems to be what led Bieberbach into a lengthy contretemps with Wilhelm Blaschke that reflects on both men as well as on official government attitudes toward internal academic mathematical matters like plagiarism.

Herbert Knothe had been Blaschke's doctoral student, receiving his degree on February 10, 1933.[298] He then became Bieberbach's *Assistent* in Berlin. In February 1937 he sent Blaschke a novel proof of the so-called isoperimetric inequality in three dimensions, remarking that it held for any dimension. This is an inequality relating the volume and surface area of any ovoid three-dimensional figure, with equality holding only for the sphere. Knothe's proof, however, turns out to be incorrect in every dimension but two (where the inequality relates perimeter to arc, equality holding only for a circle). Blaschke acknowledged Knothe's communication (February 22, 1937), inviting him to give a report on the argument in the *Hamburger Abhandlungen*, a journal founded by Blaschke. Knothe declined because he had already promised something to Bieberbach for *Deutsche Mathematik*. Three days after his first card, Blaschke sent Knothe another, pointing out his error in dimension three and all higher dimensions. In March, Blaschke published in an Italian journal in Italian the two-dimensional case using essentially Knothe's proof, but without mentioning him by name.[299] On May 2, Knothe wrote the Berlin Dekan (Bieberbach), essentially complaining that Blaschke had stolen his intellectual property—it seems that, in fact, Knothe had lectured that January on the two-dimensional case to Bieberbach and Erhard Schmidt.[300] One can have some sympathy for Knothe; the note was brief, and Blaschke was already famous. In fact, it was his 140th publication, whereas Knothe was just beginning a career. Bieberbach immediately wrote the education ministry via the Rektor requesting that a disciplinary procedure be commenced against Blaschke, whose behavior was "simple theft" and scientifically dishonorable.[301] In June, Blaschke wrote Knothe, incensed; in his nearly thirty years of teaching he and his fortunately many students had used each others' preliminary results freely, and never had a charge of plagiarism been made.[302] Nevertheless, the same day he sent an addendum to the journal editor where his note had appeared, crediting the idea to Knothe,

[297] Fenchel 1980: 161, 162. Fenchel was one of Bieberbach's distinguished students; he also was forced as a Jew to leave Germany in 1933.

[298] Blaschke, *Gesammelte Werke* (1982), vol. 6 contains on pp. 363–365 a list of all of Blaschke's doctoral students, with dates.

[299] Wilhelm Blaschke, "Sulla proprietà isoperimetrica del cerchio," *Rendiconti di Mat. Roma* 4 (1937): 233–234, reprinted in Blaschke 1982, 2:329–330.

[300] BL, Knothe to Dekan (Bieberbach), May 12, 1937; Blaschke to educational authorities via Rektor Hamburg, July 3, 1937. All material on Knothe case below is from BL.

[301] BL, Bieberbach to national educational authorities via Berlin Rektor, May 14, 1937.

[302] BL, Blaschke to Knothe, June 18, 1937.

and the fact that it worked without error in the plane to his student Ta Ten Wu (also previously unmentioned).[303] According to Blaschke's letter of defense to the educational authorities, it was around February 27 that he and Ta Ten Wu had noticed that the proof worked for dimension two. In mid-March (apparently without notifying Knothe that his idea was satisfactory in dimension two), Blaschke gave three lectures in Rome, during which he mentioned Knothe's name and his method. Asked for a little something for the Rome mathematical institute reports, he gave the two-dimensional proof (covering slightly more than a page); but his earlier oral mention of Knothe was not succeeded by a written one. His excuse for this was complicated, acceptable, and yet somewhat devious, having to do with a failed attempt to avoid mention of Knothe's mistake while yet allowing him some credit. Devious because, after all, he need only have said, "Following an idea of Herbert Knothe . . ." or something similar. His last paragraph said that Knothe was anyway not a very good student and suggested that long-standing difficulties between him and Bieberbach were the source of the accusation.[304]

Two days later, following an oral discussion with Blaschke, the Hamburg authorities exonerated him.[305] By the end of August, this decision had gone through the hierarchical pipeline and back to Bieberbach. He was not satisfied. In mid-September he sent a five-page reply to the national educational authorities together with a five-page comparative analysis of Knothe's and Blaschke's work. Knothe himself prepared such a comparison on September 9, but whether this was sent independently or used by Bieberbach is unclear. In the course of this complaint, he rejected Blaschke's suggestion that the difficulties in the German Mathematical Society in 1934–35, when he and Blaschke were on opposite sides, had anything to do with his accusation—in fact, he had even praised Blaschke in an article published in 1936.[306] The ministry did not reply until three months later—as in the case of Rudolf Weyrich and Ernst Weinel's promotion discussed in chapter 4, this seems to have again been a Nazi application of the Parkinsonian principle: "Delay is the deadliest form of Denial."[307] When the reply came, however, it not only denied Bieberbach's accusation, but it chastised him severely:[308]

> With more careful examination you must have come to the same judgment of the material [that is, Blaschke's innocence of plagiarism]. Your heavy-weighing accusation of a deserving university teacher, which rested on one-sided information, was therefore thoroughly unjustified. I express to you therefore my serious disapproval and expect that in the future you will observe the proper care in formation of your judgments.

[303] BL, Blaschke to Scorza, June 18, 1937.

[304] BL, Blaschke to educational authorities via Hamburg Rektor, July 3, 1937.

[305] BL, Statement by Karl Witt dated July 5, 1937.

[306] BL, Bieberbach to ministry, Sept. 14, 1937. See Bieberbach's article in *Deutsche Saat in fremder Erde*, discussed above under "Efforts to Ideologize Mathematics," note 145.

[307] C. Northcote Parkinson, *The Law of Delay* (Boston: Houghton Mifflin, 1971).

[308] BL, ministry (Bojunga) to Bieberbach, Jan. 3, 1938.

While this silenced Bieberbach officially, it did not prevent him from grousing to others. He sent the whole documentation to Wilhelm Süss, who was by then president of the German Mathematical Society, as well as Bieberbach's former student. Süss (a geometer) agreed with Bieberbach, and indeed was willing to talk to others (and did so, at least with Georg Feigl) about the matter. As he told Bieberbach, he was also willing, if possible, to use his office to help, though that possibility might be dubious, simply for the justice of the matter. Vahlen advised Bieberbach against lodging a complaint with the ministry. Bieberbach himself was reluctant to make the whole matter public because he did not want to bring foreign disrepute on German mathematics.[309] In 1950, Blaschke published *Introduction to Differential Geometry*. Here, several different proofs are given of the isoperimetric inequality in the plane, including the one based on Knothe's idea, which is ascribed entirely to Knothe.[310]

Weighing this incident is difficult. Knothe's proof, which used an idea from integral geometry, a subject created by Blaschke, was false as presented to Blaschke. Such elegant proofs for the known three-dimensional case were unknown, and Knothe may have been motivated by this to present the three-dimensional case to Blaschke. Unfortunately, his idea could not be correctly generalized from two dimensions. Blaschke (or Ta Ten Wu) noticed that it worked in the plane, and when pressed in Rome for a little something, provided it as novel and interesting. For Blaschke it may have been, as he wrote Knothe, simply a discussion within his school, whose correct version he published when pressed for a contribution. Also, the traditional quasi-monarchical role of the Ordinarius may have been a tacit influence on his behavior. Bieberbach's assertion of Knothe's intellectual rights (and Knothe could and should certainly have been mentioned in the paper) might simply have been an Ordinarius zealously pursuing the cause of his *Assistent*, since Knothe could scarcely have been successful himself. The ill-feeling of 1934–35 between Blaschke and Bieberbach probably did not help Bieberbach think the best of Blaschke, but it seems unlikely to have been a primary motivation for pursuing Knothe's case. Genuine pedagogical concern for Knothe, and righteous anger at Blaschke's behavior, seem much more prominent. *Prima facie*, the idea *was* Knothe's, as Blaschke's 1950 text acknowledges. Furthermore, while Knothe was a young student just starting out, Blaschke was already internationally known.

And yet, on the one hand, this does not seem to be an isolated instance reflecting Blaschke's personality and academic ethics; on the other, Bieberbach seems to have become a lightning rod for complaints about Blaschke's academic behavior. In December 1940, Gerhard Kowalewski wrote Bieberbach from Prague complaining about Blaschke's treatment of another former student, a certain Otto Varga. At the time, Kowalewski was still recovering from a serious

[309] BL, Süss to Bieberbach, Mar. 31, 1938; Bieberbach to Süss, Apr. 4, 1938.

[310] Wilhelm Blaschke, *Einführung in die Differential Geometrie* (1950), 33. In a note in 1937 stemming from a mathematical congress in Florence, Blaschke attributed the proof (in a footnote) to an idea of Knothe. ("Sulla Geometria Integrale," in Blaschke 1982, 2:331–333).

operation. Born in Hungary in 1909, Varga had proceeded via a *Realgymnasium* in Zips (in Slovakia) to the technical university in Vienna. From there he went to Prague, where he eventually completed his doctorate in 1934 at the German university. Between 1934 and 1936 he studied in Hamburg, wrote a paper together with Blaschke, and is mentioned a number of times in Blaschke's papers of this period. He then returned to Prague and "habilitated" at the German university in 1937. He had a differential geometry paper accepted in *Deutsche Mathematik*, which appeared in 1941.[311] His published work was entirely in Blaschke's sort of differential or integral geometry. According to Kowalewski, Varga gave a quite original talk in integral geometry at Baden-Baden and, naturally enough, apparently sent Blaschke a manuscript of his talk. Blaschke asked him for a manuscript publishable in the Hamburger *Abhandlungen*. Varga replied that he had already sent it elsewhere. Many weeks later Varga received his manuscript back "with the dry remark" that in Hamburg similar results were already known. Then Blaschke published the same result, even using Varga's notation. Varga's paper appeared in an obscure journal published in Pressburg (modern Bratislava) with a footnote referring to his Baden-Baden lecture. Not only did Blaschke not cite Varga, but the two publications are said to be virtually identical.[312] Varga naturally complained—but this only earned him Blaschke's dislike (and, after all, he was only a Prague *Privatdozent*, whereas Blaschke was an internationally known Hamburg professor). In fact, when Kowalewski enthusiastically recommended Varga for an open position at Braunschweig, and Blaschke heard of it, he apparently remarked to the Braunschweig authorities that "he did not know whether Varga stood 100% in agreement with the new state"—enough to prevent Varga from getting the position. Blaschke also tried (unsuccessfully) to prevent Varga from advancing academically within Germany itself, as he pressured Varga to accept a job in Pressburg, a suggestion Kowalewski called "rather shabby." It was after this last, in early December, that Kowalewski wrote Bieberbach—writing earlier seems to have been delayed by illness.

The parallels with Knothe's case are striking, including Blaschke's reaction to assertions of intellectual independence by publication other than in his journal. In fact, Kowalewski said that, according to rumor, Blaschke had behaved similarly previously; apparently he was *not* talking about Knothe, as Bieberbach, in his reply, reviewed that case and its outcome, suggesting that Kowalewski write the ministry about this new evidence of Blaschke's behavior.

Bieberbach's Standing with Colleagues

Bieberbach's mathematical colleagues who did not share his political aims seem to have ostracized him to some extent after his resignation as secretary of the German Mathematical Society. He felt this deeply and wrote Süss:[313]

[311] *DM* 6 (1941–42) 192–212.

[312] BL, Kowalewski to Bieberbach, Dec. 23, 1940; a copy of Varga's *Lebenslauf*; copy of Blaschke to Petersson (continuing a letter from Blaschke to Varga), Dec. 5, 1940.

[313] BL, Bieberbach to Süss, Apr. 14, 1938.

> At present it is indeed customary to strike people from lists for position recommendations if it is suspected that they are more closely acquainted with me, or in Pyrmont had been on my side. I have recently heard that in some out-of-town faculty meeting this was brought against [Werner] Weber, who was never even in Pyrmont.

Similarly, his attempts to get Ernst August Weiss a position in Berlin were frustrated by Erhard Schmidt. Schmidt accused Weiss of being someone who sought academic advancement through following political fashion. Rather interestingly and somewhat pitifully in praise of Weiss, Bieberbach adduced Weiss's SA service and advancement; his point was the effort taken by Weiss.[314] As mentioned above, Schmidt had similarly blocked the *Habilitation* of another Nazi-oriented student he considered incompetent.

Not only was Bieberbach isolated among his colleagues, with just a few supporters like Kubach, Boseck, Vahlen, Klose, Weber, Wegner, Weiss, and Tornier (who was losing his mind), but the pedagogical effort represented by *Deutsche Mathematik* to build new Nazi-oriented mathematicians who were highly responsive to students, after a promising beginning, proved in many ways a failure. Aside from the large number of geometrically oriented papers, *Deutsche Mathematik* became more or less like the other mathematical journals, and the student contribution dwindled to nothing.

The final blow, of course, was that, far from becoming a pope of mathematics, Bieberbach had no standing at all with the education ministry, especially after Vahlen's resignation therefrom in 1937. This is shown by the decisive way in which he was slapped down in the Knothe case. The plain fact was that the ministry did not care about internal academic disputes; however respectably *völkisch* the source might be. Support for a journal like *Deutsche Mathematik*, which attempted an ideological orientation to academic subject matter, was clearly in order. Roiling academic waters was not. *Völkisch* ideology, at least as it pertained to academe, seems to have become a sort of "bread and circuses" in the view of Germany's rulers. It was all right for those who believed in it, and it kept them as supporters, but the really important issues were elsewhere and involved material utility. Similar remarks might be made about physics,[315] pharmacy,[316] and no doubt other disciplines.

After World War II, Bieberbach apparently maintained to an Allied interrogator (who actually happened to know some mathematics) the validity of his views on ethnic personality types and mathematics—he was a true believer to this extent.[317] In his interview with Herbert Mehrtens on September 21, 1981, the ninety-five-year-old Bieberbach maintained that his distinction between a Jewish style and an Aryan style was valid but implied no comparative valuation

[314] Ibid.

[315] Steffen Richter, "Die Deutsche Physik," in Mehrtens and Richter 1980: 116–141; Beyerchen 1977.

[316] Gerald Schröder, "Die Wiedergeburt der Pharmazie—1933–1934," in Mehrtens and Richter 1980: 166–188.

[317] The interrogator was Adolph Grünbaum, who would become a distinguished philosopher of science. Personal communication from Prof. Grünbaum.

of the styles.[318] This may have been true for Felix Klein in 1892, and in fact explicitly was so,[319] but forty-plus years later the atmosphere was rather different, and the aged Bieberbach's recollections after another forty or so years seem at best disingenuous.

The Case of Richard Rado

Though Bieberbach may not have acknowledged the existence of concentration camps until the 1950s, there is at least one instance thereafter in which he attempted to obtain reparations for a mathematician forced to leave Germany by the Nazis.

Richard Rado was a brilliant student at Berlin, a pupil of Issai Schur, who was awarded a doctorate on May 31, 1933 (though his oral examination on his thesis had been eighteen months prior).[320] Unable to "habilitate" in Germany after the April 7, 1933 law regulating the civil service, Erhard Schmidt (who had obtained a research fellowship for him in February) and Schur helped him leave in August for England, where he had a distinguished mathematical career.[321] On June 2, 1933, Bieberbach wrote the following general letter of recommendation for Rado:

> Richard Rado has studied and obtained his doctorate here in Berlin. In my opinion he is one of the most gifted young mathematicians who have come from the Berlin school in recent years. His excellent papers are distinguished by their acuity and analytic-arithmetic cleverness. His personal manner serves thoroughly to recommend him.

Such a letter in June 1933 from Bieberbach for someone of Jewish background is startling, especially since six weeks later Bieberbach would publicly discuss mathematics and race for the first time. Several possibilities suggest themselves: perhaps Bieberbach was unaware of Rado's background, or perhaps Bieberbach only became virulently pro-Nazi between June 1933 and April 1934. Neither of these seems likely, and, in fact, the last line of Bieberbach's recommendation letter might be a covert reference to Rado's ethnic ancestry. What seems most likely is that, given the early Nazi emphasis on emigration of Jews, Bieberbach was willing to help Rado leave Germany, and the letter was intended for British authorities. Perhaps Schmidt was responsible for getting Bieberbach to write it.

In any case, some thirty-three years later, Maximilan Pinl, the mathematician who had insisted on bringing to his colleagues' consciousness the number, quality, and fate of their Nazi-persecuted fellow mathematicians, wrote Bieberbach reminding him of his June 2, 1933, letter (of which Rado had kept a copy) and asking him for help with Rado's present situation. At sixty, Rado was ap-

[318] BI, partial transcript of interview with Prof. Mehrtens (above, note 3).

[319] Above, chapter 6, "The Bieberbach-Bohr Exchange and the 1934 Meeting of the DMV."

[320] Biermann 1988: 228, 362. While such delay was rare, it was not unique to Rado's case.

[321] Pinl 1969: 190.

proaching retirement, his only child had kidney problems that required regular and extremely expensive dialysis, and his prospective pension was slight. Rado had asked the West German government for money in the general restitution process carried on on behalf of Nazi-persecuted Germans (*Wiedergutmachung*). This would have gone through except that, having failed to "habilitate" in Germany, Rado could not say he had been deprived of occupation by the Nazis. Occasionally, payments were made in cases like his, and Bieberbach (then seventy-nine) was asked to bolster his case. Bieberbach replied favorably and made some suggestions. Two years later, the West German government still had not acted, and the situation was much the same. The mathematician Klaus Wagner at Köln wrote Bieberbach for help—perhaps he could certify that Rado would have "habilitated" were it not for the racial laws. Wagner also mentioned that he recalled as a young student dining with Bieberbach in Bad Pyrmont. Bieberbach responded apparently as he had to Pinl, and Wagner in turn asked him for a new letter of recommendation. Bieberbach responded immediately, his recommendation deploring that, after World War II, no systematic effort was made to recall German emigrants to repair the "bloodletting" experienced by German science after 1933. Indeed, he spoke of "lip service" paid to such efforts that only resulted in new posts in a few cases. He praised Rado (and his teacher Schur) as a mathematician, yet he clearly was unaware that Schur had died (in Tel Aviv) in 1941.

Over five years later, the situation still had not changed. The West German government still had not acted, and neither Bieberbach nor anyone else had provided the proof required that Rado's *Habilitation* had been politically prevented. This time Horst Tietz, the "half-Jew" who had secretly studied mathematics in Hamburg, and to whom Bieberbach some twenty years earlier had tearfully acknowledged the conditions that had existed in the Nazi concentration camps, took up Rado's cause. By now Rado was sixty-seven, and Tietz's letter was sent one month before Bieberbach's eighty-seventh birthday. Bieberbach wrote two letters in reply to Tietz, and although, as might be expected, his detailed memory of events forty years in the past was not precise as to dates, he did provide the explicit statement that Rado's *Habilitation* was politically prevented. The second letter suggested connection with a West German parliament committee dealing with such issues on the advice of one of his sons, who was a lawyer. In February 1974 the matter was still (after seven years) hanging fire. Rado died. Presumably the West German government never did honor his request.[322] Rado was not unique. Von Mises, among numerous others, also had trouble (though ultimately was successful) in this regard, because he had left Germany "voluntarily."[323]

[322] The above Rado case is drawn from BL: copy of Bieberbach recommendation, June 2, 1933; Pinl to Bieberbach, June 7, 1966 and June 28, 1966; Wagner to Bieberbach, Feb. 28, 1968, Mar. 14, 1968, and Apr. 23, 1968; Bieberbach recommendation for Rado, Apr. 26, 1968; Tietz to Bieberbach, Nov. 26, 1973, Feb. 15, 1974; Bieberbach to Tietz, Nov. 28, 1973, Jan. 24, 1974.

[323] Bieberbach was also involved in trying to help von Mises and his wife (and former *Assistent*), Hilda Geiringer, obtain postwar reparations. See Jochen Brüning, Dick Ferus, and Reinhard Sieg-

The Rado case shows us Bieberbach trying to aid reparations in one particular case, but failing to acknowledge his own role in creating such a situation, a role that was substantial—and that he apparently at one time wished could have been even more substantive. Bieberbach seems to have always maintained that the distinction between Jewish mathematics (at least in style and pedagogy) and Aryan mathematics was significant. Though he might deplore an individual's suffering, or forced expulsion (at a time when it was politically appropriate), he never seems to have rejected the consequences of carrying out those beliefs politically.

---

mund-Schultze, *Terror and Exile* (Deutsche Mathematiker Vereinigung, 1998), 56, 16. This is the book of an exposition held on the occasion of the International Mathematical Congress in Berlin, 1998. In this case, the ministry approached Bieberbach (at Hilda von Mises's suggestion). The restitution that came to von Mises was posthumous.

CHAPTER EIGHT

Germans and Jews

THE preceding chapters have shown that mathematics as a subject matter was in a difficult position when Hitler came to power.[1] Not only was the Nazi ideology centered on nonrational process as ideal, not only was it involved in a biologistic reductionism of all learning to a fundament of "racial genetics," but the general lay opinion of mathematics at the time was none too favorable. The mathematicians had formed an organization (the *Mathematische Reichsverband*) to encourage better public understanding of their discipline, but whatever headway it had been making was slow. Given its purpose, it is not surprising how quickly and smoothly it became integrated into the Nazi regime. Pedagogues like Walther Lietzmann might harp consistently on the military importance of mathematics, but this was a minor theme—even in the openly rearming Germany of 1935 and later. The initial Nazi view of German academe was a mixture of contempt and suspicion. The latter was somewhat overcome by the speed with which the universities—with, in Fritz Kubach's words, "students in front"—became integrated into the Nazi state. The German Mathematical Society as an organization reacted initially with indifference to the new regime, provided the doing of mathematics by those remaining was not affected. Its crisis in 1934 was caused by a desire on the part of some, led by Ludwig Bieberbach, to make the society an active participant in, rather than a passive recipient of, Nazi structures and regulations. Some who initially supported Bieberbach, like Hellmuth Kneser and Wilhelm Süss, grew disillusioned with, and became opposed to, the Nazi government and its behavior. No doubt Kneser's friendship with Theodor Vahlen and the fact that Bieberbach had been Süss' mentor had something to do with their initial reaction, however much they may have come to change their minds. Bieberbach himself seems to have been a committed believer who carried on the ideological struggle on intellectual grounds, but his efforts "faded into nothingness." *Deutsche Mathematik* as a journal became increasingly and almost totally mathematical (and less student oriented), whatever its founder's original purpose.

Some mathematicians "biologized" in defense of their rational subject; others "militarized"; but most simply wanted to be left alone to do their mathematics. They would try to protect their subject from ideological incursions, such as they might be, but otherwise would be "unpolitical" or perhaps even tacit opponents of the Nazis. This, of course, was not what the regime would have liked. Indeed, as will be seen, so successful was Hans Petersson in his dissimulation of

[1] The title of this chapter is also the title of a well-known book by George Mosse (1970); it nevertheless seems the most appropriate one for this chapter.

Nazi ardor that political officials praised him (to his good fortune) as an exception among mathematicians.

Yet the plain fact is that a number of prominent mathematicians, while not Bieberbachs or Teichmüllers, were also not overly distressed with the regime—they were not Heckes either. Whatever the differing motivations and attitudes of men like Helmut Hasse, Georg Hamel, and Wilhelm Blaschke, they undeniably aided the regime and made propaganda for it.

Many mathematicians, presumably like many academics, though, were able to retreat into their mathematics, fending off as much as possible attempted ideological incursions. Thus the Bonn faculty rejected Erhard Tornier and repeatedly suggested Erich Kamke for a professorship, while Max Steck was prevented from succeeding Constantin Carathéodory at Munich. Faculty action in the complicated case of Ernst Weinel at Jena may have been similar. Many comparable stories no doubt could be told in other mathematics departments. If the mathematicians as a society were acquiescent in the Nazi regime, as professionals they resisted interference. This seems completely in keeping with the Humboldtian tradition of the unpolitical professor; how appropriate such a response was in 1933–34 is another matter that can produce as unending an argument as any hypothetical historical event. However, the only reason such professional resistance could be at all successful was that basically the regime did not care what happened in the academic world, provided Jews and active "troublemakers" were expelled from their positions. Though someone like Max Zorn, though not a Jew, nevertheless a communist, could not hope to "habilitate" in Nazi Germany, someone as established as Kurt Reidemeister was could find another job when he lost his first one in the initial wave of enthusiastic expulsion of the academic opposition, and someone like Erich Hecke could maintain his distinguished position despite his apparent refusal to say "Heil Hitler."

The education ministry only insisted on superficial ideological conformity, such as membership in a National Socialist organization (the SA, which had very little real power after June 30, 1934, became particularly popular). The education minister, Bernhard Rust, seems to have been not only an *alter Kämpfer* but also an alcoholic with a severe head wound from World War I, which led to rumors of mental instability (shades of Johann Weniger!). Originally a secondary-school teacher, Rust was dismissed in 1930 because of allegations of child molestation (though the charges were hushed up, and the same year he entered the *Reichstag* as a Nazi). Hitler's contempt for formal learning is well known, and with the, at best, semi-competent Rust heading the education ministry, that contempt was translated into action. The ministry seems itself to have been regarded contemptuously by those at the top of the government, and Rust, though a *völkisch* adherent, usually lost the internecine struggles among the Nazi hierarchy to more competent people like Goebbels, Göring, Himmler, and even Alfred Rosenberg, who, himself more an Aryan ideologue than a political infighter, also often "lost." An underground joke of the time may reveal the true state of the education ministry: "One Rust equals the smallest unit of time between a decree and its reversal."[2] Whatever one thought "innerly," at least superficial adherence to Nazi rules and regulations and participation was required.

There does seem to have been more of an effort to present a positive Nazi face to the world in the mathematics departments in Berlin and Göttingen than elsewhere. This is undoubtedly because in 1933, they were not only the two most distinguished departments in Germany, but two of the most distinguished in the world. However, it is also because of the attitudes and personalities of the leadership in those departments. In Berlin, the conservative Erhard Schmidt had little interest in administration. Though he twice served as president of the German Mathematical Society, mathematics, not the administration of it, was his concern. The chair of Issai Schur was never refilled after his forced retirement and emigration, though Erhard Tornier, who had succeeded to Landau's chair in Göttingen, carried it with him on his transferral to Berlin in 1936.[3] Richard von Mises, as another forced emigré, was replaced first by Theodor Vahlen and then by the astronomer Alfred Klose, whose main credentials were political. However, Ludwig Bieberbach proved a driving pro-Nazi force and gave mathematics in the capital the desired "Nazi face."

Helmut Hasse's "development," in Werner Weber's sense—his gradual accommodation to being a spokesman for the regime—came to do the same for Göttingen, though Hasse, whatever his politics, always cared about mathematics first.

In both Berlin and Göttingen, there was an enthusiastic Nazi ideologue who was an *Assistent* in the department. In Berlin, it was Karl Heinz Boseck, who has been discussed. In Göttingen, it was a protegé of Erhard Tornier named Paul Ziegenbein. It is certainly possible that Hasse and Bieberbach were actually dominated by these young men who played the role of "political controller" in their respective departments, but this seems not true. In Bieberbach's case, this is patent; in Alexander Dinghas' words, the distinguished mathematician "had respect for" Boseck.[4] In Hasse's case, his correspondence with Harold Davenport, his efforts to join the Nazi party, and, as discussed below, his refusal to countenance Alfred Stöhr as Carl Ludwig Siegel's *Assistent* (though Ziegenbein was involved in this last) all show his "development" in Werner Weber's sense. In addition, neither the Hamburg department, which was nonexistent in 1919 but had risen rapidly under the leadership of Blaschke and Hecke, arguably to "third place," nor the highly respected department at the University of Munich, promoted this "Nazi face." In the case of Hamburg, as will be seen, perhaps Blaschke was enough of a publicist, both for himself and as an agent of the government, to obviate this need. It can be argued that this was one of his intentions, but it surely was not his only one. As for Munich, it seems as though the *Ordinarien* in mathematics, Constantin Carathéodory, the topologist Hein-

[2] Hans-Jochen Gamm, *Der Flusterwitz im Dritten Reich* (1963), 59.

[3] Scharlau et al. 1990: 37.

[4] Above, chapter 6, "Mathematics in the Concentration Camps," note 464.

rich Tietze, and Oskar Perron, none of them sympathetic to the Nazis, managed to keep Nazi ideology at bay.[5]

These matters are further elucidated by more detailed consideration of personal histories. Some of these refer to people already considered above, but context precluded more adequate delineation of attitude and behavior. Other personal histories illuminate aspects as yet unmentioned. This chapter is devoted to filling some of those blank spaces. Among others, there will be consideration of Wilhelm Blaschke, Erich Hecke, and Heinrich Behnke. Blaschke and Hecke were the founding and premier members of the Hamburg department (and personal enemies); Behnke was Hecke's student, became himself a prominent mathematician, and was among Blaschke's principal accusers during the postwar denazification. Also to be examined in more detail are Oswald Teichmüller and his friend Ernst Witt, both brilliant mathematicians (in different fields) and both dynamic young Nazis involved with the events at Göttingen in 1933–34. Then there were Jews (though with little or no religious practice) like Richard Courant, Edmund Landau, and Felix Hausdorff, who at first could not believe what was happening in the country they loved and felt a part of. All suffered some form of tragedy—in Hausdorff's case, fatally. Ernst Peschl and Hans Petersson are examples of mathematicians who successfully (and dangerously) dissembled their true beliefs, one as a consequence of his sincere Catholicism, the other because he had made a marriage outside the accepted Nazi limits. Paul Riebesell is an example of a non academic mathematician, originally elevated to a position of prominence under the Nazis, who fell into disfavor because of an honest adherence to general academic standards. One small indication of the Munich professors' attitude (and the lack of concern of Rust's education ministry) was their giving Paul Riebesell, when he was in official disgrace in 1938, an adjunct professorial postion. Helmut Ulm and Alfred Stöhr represent among mathematicians the situation of non-Jews who at the time had too many Jewish friends. Gerhard Gentzen represents the "innocent Nazi," the famous mathematician who was a party member because that was how one got ahead, but who was apparently too unworldly to realize when party membership became a disability, and consequently died tragically, despite the fact that he never had any known "Nazi activity." Erich Kähler, on the other hand, is an antipodal figure. Apparently never a party member, he is an example of the extreme right-wing nationalist who, even at an advanced age, excused Nazi "excesses." The case of Ernst Zermelo shows that seemingly trivial behavior could have serious consequences under the Nazis, as well as denunciation at work among mathematicians. Finally, it is useful to consider even more detail about Wilhelm Süss, who led mathematics during much of the Third Reich; while on the one hand, he benefited materially from his status, on the other, he was in a position to save dozens of his non-Jewish colleagues from tragic fates.

[5] See, for example, S. L. Segal, "Topologists in Hitler's Germany," chap. 30 in I. M. James, ed., *History of Topology* (Amsterdam: Elsevier, 1999), 849–861, esp. pp. 857–860.

Wilhelm Blaschke

A typically complicated case that has already been met several times is that of the highly creative and justifiably famous mathematician Wilhelm Blaschke. Indeed, the late Hans Zassenhaus told me that understanding Blaschke (whom he had known as a senior colleague) was the key to understanding mathematicians and mathematics in Nazi Germany, at least in Hamburg.[6]

Blaschke's early career has already been sketched.[7] His career under the Nazis seems to have been that of a traditionally "unpolitical" German academic—one who used the politics that were for self-advancement without trying to influence them; in short, an opportunist. But in Blaschke, a well-traveled and cosmopolitan man of great mathematical ability, and some vanity, who was a German nationalist, opportunism for personal advancement seems a driving force. In 1938 Blaschke, who was certainly anti-Semitic long before Hitler came to power,[8] wrote a review of the *Semicentennial History of the American Mathematical Society* (edited by R. C. Archibald) in which he speculated that the exact sciences might be a substitute for religion in "Dollarland" and commented that the book contained interesting material on the ancestry of American mathematicians, "with careful and shameful silence about only one race." This latter could lead to mistakes because of the "old American preference for old Biblical names." Later he said, "Most astonishing is the large mathematical undertaking in the small niggertown (*Negerdorf*) of Princeton." He also praised Johns Hopkins and Chicago (where he had visited in 1932).[9]

Nor were Blaschke's politics hidden from his mathematical readers. The preface to his *Geometry of Webs* (with his student Gerrit Bol), dated Messina (Italy), March 13, 1938, stated, "Whereas earlier volumes [of mine on differential geometry] appeared in murky times, this book was completed as a dream of my youth was fulfilled, the union of my more narrowly seen homeland, Austria, with my larger homeland, Germany."[10] In Weimar days, the 1923 preface to the second volume of his lectures on differential geometry (written with Kurt Reidemeister) had ended, "Most especial thanks are due the publisher, who had not lost his energy while the cousin of the famous French geometer performs his war dances on the corpse of the German nation (*Reiches*)."[11] In 1935, Blaschke published a little book on integral geometry (in German) in the well-known French series *Actualités scientifiques et industrielles*. The preface to this book alluded disparagingly to Bieberbach's theories of ethnic mathematics and its relation to abstraction or intuition.[12] This was written less than two months after

[6] Hans Zassenhaus, interview, Nov. 17, 1987.

[7] Above, chapter 7, "The Frankfurt Succession."

[8] Ibid.

[9] *JDMV* 49 (1939): Abteilung 2, pp. 80–81.

[10] Wilhelm Blaschke and Gerrit Bol, *Geometrie der Gewebe* (1938), vi.

[11] Wilhelm Blaschke, *Vorlesungen über Differentialgeometrie II* (1923), vi. The reference is to Raymond Poincaré, the French president, who was a cousin of Henri Poincaré.

[12] Wilhelm Blaschke, *Integralgeometrie* (1935), 3.

the conclusion of his conflict with Bieberbach described in chapter 6, which resulted in Blaschke's loss of the presidency of the German Mathematical Society. Yet, a few years later, Blaschke could write: "The Romans thereby seem to offer a clear example of the dependence of mathematical aptitude on race, a question that naturally is abundantly difficult and complicated." This was said in the context of a remark that the sole Roman contribution to mathematics was the negative one of murdering Archimedes.[13]

Thus Blaschke was a German nationalist who welcomed *Anschluss* (like many of his fellow Austrians), but apparently was no ethnic ideologue about the value and provenance of different mathematical styles. He maintained flexibility about such matters, perhaps depending on his audience.

Blaschke was also quite a proud man, and a sampling of material about him from various sources will serve to indicate how very difficult it can be to judge the attitudes and actions of some during the Nazi period. With the end of World War II, Blaschke was removed by the Allied authorities from office. In fact, in a protocol of the Hamburg university senate from June 26, 1945, Blaschke is described as a member of the party since 1937 and a member of the Nazi *Dozentenbund* who had been Dekan for many years, until 1943. "Although he cannot be described as a typical National Socialist, he used political pressure in issues of appointments. The Rektor indicates Blaschke's old sympathies with fascism." The Rektor in question was the philosopher Emil Wolff. The description is Erich Hecke's as then-Dekan. Both Hecke and Wolff were among those who had never been involved with the party and after the war were placed in positions of authority with British approval. They consequently also served on the denazification commissions set up by the military government in mid-November 1945, though Hecke had to be replaced because of illness in July 1946.[14]

Blaschke, found culpable, fought for reinstatement. On November 4, 1946, he wrote Oswald Veblen, also a well-known geometer, that he had been successful:[15]

> It gives me great pleasure to be able to tell you that I have again been placed in office without conditions. I wish heartily to thank you and other colleagues for your readiness to help in this business. Unfortunately the conditions here are in many respects very difficult, in particular the food supply has again gotten worse.

[13] Wilhelm Blaschke, *Mathematik und Leben* (1941). This was a lecture celebrating the 250th anniversary of the Hamburg Mathematical Society. Although there was no university in Hamburg prior to 1919, the Hamburg Mathematical Society was perhaps the oldest scientific society in the world devoted to a single discipline (see Scharlau et al. 1990: 145). The quoted remark is omitted in the revised version published as part of *Reden und Reisen eines Geometers* (Berlin, 1957, 1961).

[14] I am grateful to Christoph Maass and Karin Reich for copies of the denazification materials used above. In 1932, Wolff had protested the arrogant attitude of Nazi students who wished to replace the intellectual autonomy of the university with National Socialist ideas. See Geoffrey J. Giles, *Students and National Socialism in Germany* (1985), 97.

[15] VP, Blaschke to Veblen (filed under B).

Veblen in Princeton was nonplussed. He had never written anyone on Blaschke's behalf, and so told Carl Ludwig Siegel, who was visiting in Germany. Siegel replied [in English]:[16]

> If you were here, you would not be astonished at Blaschke's letter. Everybody receives letters from Blaschke. He wants to prove that he is an influential person again. Hecke proposed that he should retire and get a pension; this would have been the only satisfactory solution.

Yet Blaschke had presumably been asked for the names of people the authorities might write to on his behalf; at first it seems he may not have known who was written and so simply thanked Veblen as an irreproachable name he might well have put down as a fellow geometer and long-time mathematical acquaintance. However, it turns out that Blaschke did know at least some of the people who wrote letters about him, since at least one of them incensed him.[17]

On September 22, 1945, in response to an inquiry two weeks earlier, Heinrich Behnke wrote Erich Hecke, acting as Dekan under the British military government, about three instances to his knowledge of "political activity of a National Socialist stamp" by Blaschke. All three involved Behnke personally. Behnke claimed that at the time of the crisis in the management of the *Zentralblatt* in 1938, when a number of non-Germans on its advisory board and non-German reviewers withdrew their support because of the enforced nazification of the journal's policies, he asked Blaschke for his advice. Behnke (and Ferdinand Springer, the publisher) were concerned that Americans would be stimulated to start a competing reviewing journal (as in fact happened—the first year of *Mathematical Reviews* was 1940). According to Behnke, Blaschke's reply was that he knew "his Americans," and that they were not capable of mounting a mathematical reviewing journal of international stature. Furthermore, Behnke claimed that Blaschke proceeded to write Springer with a political denunciation in an attempt to push Behnke off the *Zentralblatt*. Similarly, Behnke claims that Blaschke denounced him politically to university academics at Munich in 1943, when the question of a successor to Carathéodory's chair arose. Actually, though the chair had been vacated in 1938, there was no successor until 1944, when Eberhard Hopf filled the post. In 1942 Behnke traveled on a lecture trip to Romania. Sometime later Blaschke traveled to Romania as well (Blaschke was friendly with the Romanian geometer D. Barbilan, who was sympathetic to the Germans). It came to Behnke's ears that Blaschke had claimed that he had heard that Behnke had disparaged the German political

[16] VP, Siegel to Veblen (filed under S). The letter was sent on January 28, 1947, and received by Veblen one month later.

[17] According to Christoph Maass ("Das Mathematisch Seminar der Hamburger Universität, 1933–1945," in E. Krause, L. Huber, and H. Fischer, eds., *Hochschulalltag im "Dritten Reich"* [1991]), Heinrich Behnke's letter, about which discussion follows, was forwarded to Blaschke on November 2, 1945. The letter itself is in the Hamburg Staatsarchiv, Personalakten Blaschke (PAB), as well as in UAG, Nachlass Herglotz. All information below about Blaschke, unless otherwise annotated, is from PAB. For Blaschke's seeing the letter, cf. PAB, Wolff to Landahl, Dec. 3, 1945.

situation, and had arranged that Berlin should hear about it so that Behnke would never again be allowed foreign travel. On the face of it, while Behnke may have been insulted, and even felt threatened, in the first two instances he was perhaps fortunate that Blaschke only wrote to Ferdinand Springer or professors of mathematics at Munich (none of whom had any party associations), rather than to government officials. That is, the purported denunciatory letters would injure Behnke, but not destroy him. The third instance mentioned above seems to have been handled similarly.

Needless to say, Blaschke protested when he discovered the sort of information Behnke had provided. On November 7, he defended himself to the authorities. With respect to the *Zentralblatt*, he said that he was in agreement with Otto Neugebauer, and later Harald Geppert, that it needed to be an international journal, and that this had roughly been managed until 1938–39, despite threats to the journal from "hotspurs" (*Heissporne*) like Bieberbach. He then talked about his leadership in the struggle against Bieberbach, which had been not without personal danger. Blaschke also claimed to have been responsible for keeping the famous Polish mathematician Waclaw Sierpinski as an editor of the *Zentralblatt*. He maintained that he was in fact worried about American competition, though he was named the "Hamburg representative" of *Mathematical Reviews*. Blaschke also mentioned that he had saved the Polish mathematician Stanislaw Gołab from a concentration camp[18] and that he and Harald Geppert (who became chief editor of the *Zentralblatt* in 1940) had made "partially successful" efforts to help French prisoners.[19] Behnke's "tactless" and "ambitious" mixing into international situations had been unwanted. Furthermore, Behnke had given a Nazi-oriented political talk on the occasion of Blaschke's visit to Münster (when, Behnke claimed he had inquired of Blaschke about the *Zentralblatt* problems). This gives a feeling for Blaschke's defense on this "*Zentralblatt* issue," which did not fail also to mention the emigré Emil Artin ("absolutely no Nazi") and his putative difficulties in enduring Behnke as a student. This defense almost screams for annotation and reveals why Zassenhaus found Blaschke so problematic.

Blaschke was a cosmopolitan and basically internationally inclined. He was also a conservative German nationalist, who had had no trouble singing the Nazi tune when they were giving him his bread. Now he was singing British tunes. The "struggle against Bieberbach" has been discussed in chapter 6, and can in fact be seen as Blaschke's attempt to retain the power vested in him by the German Mathematical Society in its attempt to avoid the *Führerprinzip* versus Bieberbach's attempt to impose that principle. In that struggle, Blaschke lost personal power because of a "cosmopolitan" mistake. The accusation that Behnke made Nazi noises seems scarcely in the character of the Behnke already discussed—an anxious man, unlike Blaschke never a party member, and with a

[18] On Golab, see M. Kucharzewski, "Stanislaw Golab—Life and Work," *Aequationes Mathematicae* 24 (1982): 1–18. Golab's internment in concentration camps is mentioned on p. 3.

[19] Blaschke mentions no specific names.

half-Jewish son, who did, in the last years of the Third Reich, speak out about the pedagogical difficulties caused by Nazi policies, but who basically retreated into his mathematics while claiming that he was threatened by his international and Jewish collegial connections.

The crisis at the *Zentralblatt* was occasioned by the demand that the Italian Jewish mathematician Tullio Levi-Civita be dismissed from the editorial board. Not only the name of Levi-Civita, but also that of Richard Courant had appeared on the title page through volume 17 (1938), and Courant, as a Jew, albeit without religious practice, had long since emigrated to the United States. Neugebauer was chief editor, and he was on the title page of volume 18 (1938), which was identical but for the omission of Levi-Civita's name as well as that of the Russian mathematician Paul Alexandroff. For the next volume, however, Neugebauer's name was gone, as were those of Courant and four other foreign editors (Harald Bohr, G. H. Hardy, Oswald Veblen, and J. D. Tamarkin). Neugebauer was replaced by Egon Ullrich, and the others by a Swiss, the Pole Sierpinski, two "acceptable" Italians, Helmut Hasse, and two other Germans, the latter colleagues of Ullrich at Giessen. One of these was Harald Geppert. The second issue for 1939, volume 20, added the well-known Japanese mathematician T. Takagi. For volume 21, van der Waerden's name was omitted (as noted above, van der Waerden was now a citizen of an enemy country, and in danger of losing his livelihood); in the succeeding 1940 volume, Geppert replaced his colleague Ullrich as chief editor. Blaschke seems to have made no protest; his claim that he advised Geppert not to replace Sierpinski's name is and was unverifiable, as were most of his other claims, including especially that he and Geppert tried to help some French prisoners.[20]

At the war's conclusion, Geppert committed suicide together with his family. Geppert had been a leading exponent of German cultural imperialism, Nazi-style. In 1940 he obtained, through Bieberbach's influence, the position of Ordinarius in statistics (as successor to the mentally incapacitated Erhard Tornier[21]) at Berlin. However, Geppert was no statistician originally. His doctoral dissertation was written at Breslau and had to do with function-theoretic expansions of arbitrary functions; his *Habilitationsschrift* at Giessen was on a topic in differential equations and, he said, was stimulated by suggestions of and conversations with Richard Courant.[22] Thus Geppert also falls into that frequently mentioned curious category of the Nazi mathematician with a personally important "Jewish" teacher. It was in 1938 that Geppert became prominent as a statistician through the publication with Siegfried Koller of the book *The Mathematics of Inheritance* (*Erbmathematik*), which brought him to Bieberbach's attention as a thinker

[20] The name of the well-known French mathematician Gaston Julia, reputedly a German collaborator, continued on the title page of the *Zentralblatt* from the first volume in 1931 until the end of World War II.

[21] Above, chapter 4, "Hasse's Appointment at Göttingen."

[22] See *Mathematische Annalen* 95 (1926): 519–543, p. 519. For the doctoral dissertation, see *Mathematische Zeitschrift* 20 (1924): 29–94.

about "problems important for the future."[23] Since Tornier had worked in probability theory, Geppert was now no doubt a suitable replacement both in Nazi dedication (he had joined the SA in 1933) and in mathematical interests. In 1940, Geppert became editor of both the *Jahrbuch* and the *Zentralblatt*, thus obviating the competition between the two in his own person. He also became a professor in Berlin and undertook a trip to German-occupied Paris in December. The main purpose of this reconaissance trip (*Erkundungsreise*) was to ascertain whether the Institut Poincaré desired to carry on further the organization of international congresses of mathematics under the auspices of the Mathematical Union, hated by German nationalists, or whether it would be possible to reorganize them as centrally German. In any case, Geppert and Blaschke were likely friends. Geppert also spoke Italian (his mother was Italian) and was the sort of man who knew as an aid to academic advancement how to ingratiate himself politically during the Nazi era. He was a competent mathematician, as was his sister, despite Landau's accurate savaging of one of her papers.[24] Blaschke was already a famous mathematician, who added Nazi luster to his fame through political ingratiation. Geppert was a political climber. Blaschke survived the suicide Geppert and was thus able to use him as an unavailable reference for his own *bona fides*.[25]

Similar remarks can be made about Blaschke's defense of the other two points made by Behnke. He used Behnke's failure to be successor to Carathéodory at Munich to say that clearly Behnke's attack on him was motivated by disappointed vanity and a desire for personal vengeance, whereas, claimed Blaschke, Erich Hecke was united with him in not seeing Behnke as a suitable person to succeed to Carathédory's chair. Behnke, in fact, had been Hecke's student, and had written a doctoral dissertation in analytic number theory, but then went off in an entirely different direction to found the modern theory of several complex variables and a distinguished mathematical school. Carathédory had strongly encouraged him in this effort.[26] Both Hecke (with primary interests in algebra and number theory) and Blaschke (as a geometer) may well have initially undervalued Behnke's mathematical prowess. However, this does not excuse a politically denunciatory letter by Blaschke, if Behnke's claim is correct. Blaschke wrote Munich asking whether such a letter from him could be found in the faculty archives. No such letter could be found. The postwar Rektor K.

[23] Koller was a student of Felix Bernstein (who emigrated because he was both a Jew and an active political figure) in 1933. Koller's dissertation involved the statistics of blood groups (1931). For his involvement in the Nazi era and rehabilitation in West Germany, see Götz Aly and Karl Heinz Roth, *Die restlose Erfassung* (1984). Koller was one of those people whose research could be easily turned to Nazi purposes, even though it could also be pursued in a purely scientific fashion (as Bernstein notably did).

[24] Above, chapter 7, "Bieberbach and Landau." She was, however, a statistician of some note who after the war had a distinguished career in West Germany. See Hermann Witting, "Mathematische Statistik," in *Ein Jahrhundert Mathematik* 1990: 781–815, esp. 802–805. The paper Landau complained about was in a very different area.

[25] On Geppert and the Nazi plan for the reorganization of science as respected mathematics, see Siegmund-Schultze 1986: 1–17.

[26] Behnke 1978: 71.

Clusius consulted Oskar Perron on the matter. Perron replied that Blaschke was an outstanding mathematician, but unfortunately not such an outstanding character, and he had no doubt that such a denunciatory letter was written by him "to some unclear (*dunklen*) address" influencing attitudes toward Behnke. Perron continued, "Naturally we cannot research the cloacas of the *Dozentenschaft*, and Blaschke depends on that." Thus he advised a reply that no such letter could be found in the faculty archives.[27] Blaschke reported as much to the Hamburg authorities.[28] Blaschke's mention of Hecke, whom he accused earlier in the letter of "collecting material against me," further noting, "I have had a tense relationship with Hecke for twenty-five years," was a backhand way of justifying his negative opinion of Behnke's suitability as successor to Carathéodory. If the man whom Blaschke accused of carrying on a postwar vendetta against him also agreed about Behnke, what more justification of his opinion could be asked for? Blaschke even suggested as an item in his defense that he had helped Hecke out with the Nazi authorities. This seems unlikely. The implication was that Hecke was an ingrate.[29]

As to the Romanian trip, Blaschke countercharged that Behnke had agitated against him in Switzerland, and that in May 1943 he learned in Bucharest of Behnke's further fulminations against him the previous December in Romania. Blaschke asserted his failure to understand how in wartime someone could carry personal quarrels with a colleague into a foreign country. He reported Behnke's purported behavior to Geppert and Wilhelm Süss, and commented how, in contrast to Behnke's trip, Süss' trip to Romania went just fine. Blaschke added unverifiable evidence of his *bona fides* with respect to Romania as well: he claimed that in 1936 in the Romanian city of Jassy (Iasi) he gave a popular lecture in German before about 1,500 students "of whom about half were Jews and a good portion of the remainder belonged to the Iron Guard"; the lecture took place "without brawling, although the Rektor there earlier had had one of his ears cut off by radical students." The claim seems unlikely, but again is unverifiable. The Iron Guard (Legion of the Archangel Michael) and its predecessors were virulently anti-Semitic, and its leader Corneliu Zelea Codreanu even plotted in 1923 to murder those involved in granting Romanian citizenship rights to Jews. The law faculty at the university in Iasi, where Codreanu was a student, was dominated by the ardent anti-Semite, A. I. Cuza, Codreanu's early mentor, for whom the university would eventually be named.[30]

Although Blaschke's rebuttal of Behnke seems unsubstantiated and unsub-

[27] University Archives, Munich O-N-14 (Oskar Perron), Clusius to Perron, May 25, 1946; Perron to Clusius, May 29, 1946. Perron's response is handwritten.

[28] PAB, Blaschke to Paul Harteck, June 1, 1946. Harteck was a well-known physical chemist who replaced the mortally ill Hecke as Dekan. He had been denounced (unsuccessfully) to the Gestapo in 1943. For his substantial role in German nuclear power projects during World War II, see Walker 1989: passim. For the denunciation, ibid.: 127–128; and Angela Bottin, *Enge Zeit* (1992), 98.

[29] PAB, Blaschke to military government, Sept. 17, 1945, and Hecke to Wolff, Nov. 25, 1945

[30] For the Romanian situation, see Eugen Weber, "Romania," in Hans Rogger and Eugen Weber, eds., *The European Right* (1965), 501–574.

stantiable, presumably, *prima facie* acceptance of it helped him win reinstatement.

Apparently the *Dozentenführer* in June 1934 characterized Blaschke as "pacifistic"[31]—but this seems a typical exaggeration of an enthusiastic Nazi, rather like the characterization of Hellmuth Kneser as leftist mentioned earlier. Blaschke, after all, had inserted aggressive political statements into a mathematics book as early as 1923. Others judged Blaschke eminently acceptable—prior to consideration of him for the Ernst Abbé memorial prize (in mathematics or physics) for 1937, a check was made with the Hamburg Rektor Adolf Rein as to Blaschke's political acceptability. The response was that though Blaschke had had no political activity prior to 1933, there were no second thoughts necessary about him, that he worked along in university politics and was occupied also with work in foreign countries.[32]

Blaschke certainly was occupied with work in foreign countries. Not only did he visit Italy (whose language he spoke) almost every year, as well as Romania, Bulgaria, Greece, and Slovakia, but also, in the course of a long career, the United States, India, Japan, China, The Netherlands, Spain, and German-occupied France, garnering numerous honors (and students) along the way. In 1938 he turned down an invitation to Moscow (for political reasons?). All these visits were prior to 1945, many prior to 1933, some in the course of a "round-the-world tour." However, Blaschke's love of travel continued in the 1950s, which saw professional visits to, among other places, Turkey, Argentina, and Chile.

Blaschke's reports on those trips which took place between 1933 and 1945 seldom failed to have a political dimension, sometimes with unanticipated results. In autumn 1941 Blaschke gave lectures for the German air force in occupied France and in Italy, as well as visiting colleagues and lecturing at universities. The last paragraph of his brief report is: "I might also indicate that it is unsuitable to send professors without or with all too little knowledge of Italian to Italy. In particular it is now truly unfortunate that if in Italy, German professors attempt to lecture in French." The response of the ministry was prompt; in a note to the Rektor, it asked Blaschke for the names of German professors who sought to use French for guest lectures in Italy. Blaschke backpedalled and said Italian colleagues had not named names, and he knew none; they had only spoken of the linguistic difficulties during lectures by German scholars and the undesirability of lectures in French.[33]

Blaschke was someone who felt able to complain to the Nazi authorities about his travel difficulties—which were generally minor compared to those of persons considered more dubious by Nazi officials, like Hecke and Behnke. When he made such complaints in connection with the wartime "international"

[31] Maass 1991: 4 and n. 12.

[32] PAB, Ministerialrat Stier to Rein, Dec. 29, 1936; together with a handwritten annotation, obviously the content of the response dated Dec. 31 in PAB.

[33] PAB, Blaschke's report dated Dec. 5, 1941; ministry to Rektor Universität Hamburg, Jan. 19, 1942; Blaschke to Rektor, Feb. 4, 1942. The Rektor had not asked him for names of offending professors until February 2.

mathematical congress held in Rome in November 1942, the ministry demanded chapter and verse within two weeks, which he gave. His complaint about the size of the permitted German delegation was, however, rejected on Foreign Office grounds.[34] This report also again gave Blaschke opportunity for political commentary that might well be personally advantageous:

> One hears today also from German officials in Italy very unfavorable judgments concerning the Italian mood about the war. I consider these judgments one-sided. Businessmen and financiers there tremble frequently over their fortunes. However, it would be difficult to object much to the sentiments of a number of universities such as Florence and Bologna. Perhaps this is connected with the fact that one can only seldom characterize the faculty of Italian universities as capitalists.

This was followed by the suggestion of the official furtherance of cultural exchange with Italy. The same report contrasted the fact that Hasse and Carathéodory had spoken in German, which was understood with difficulty, whereas he had spoken in Italian.[35] Further self-promotion as an official cultural ambassador to Italy can perhaps be seen in the report on a 1938 lecture tour to Greece and Italy, where, he observed: "The increase in the anti-Semitic movement at the universities in Italy since the preceding year is remarkable."[36]

It is in Blaschke's report of May 1943 on his trip to Bucharest and Sofia that Behnke is mentioned. Here again, the emphasis in the report is on Blaschke as potentially useful cultural ambassador. So far as *this* report goes, the only mention of other German mathematicians reads:

> Of the numerous German mathematicians invited there [to Bucharest] in recent years, the lectures of [Wilhelm] Süss, Freiburg, were especially mentioned to me appreciatively. In contrast the lectures of [Hans] Petersson, Strassburg, were plainly too difficult, and the stay of Behnke, Münster, seems to have given rise to some political difficulties.[37]

But 1943 would prove a disastrous year for Blaschke, though he was Dekan for the second time. On July 28 he and his wife went on holiday; July 29–30 saw a devastating air raid on Hamburg in which Blaschke's house was completely bombed out. On August 31, the Rektor wrote Blaschke at the Tyrolean village where he was staying, urging him to return so he could both obtain a new dwelling and take up his duties as Dekan. This letter crossed one from Blaschke (dated August 28) that complained that he had not yet received his August pay, said his wife and children were going to Graz (presumably much safer than Hamburg) on September 2, and mentioned that he was going on the

[34] PAB, Blaschke's report dated Dec. 1942; ministry to Rektor, Jan. 26, 1943; Blaschke to ministry Feb. 8, 1943.

[35] The Germans occupied all of France (including Vichy) in November 1942. November 1942 is a fair date for the apogee of Hitler's *imperium*. With the foreseen Axis victory, Blaschke no doubt saw himself as the ideal cultural (mathematical) ambassador between the partners Germany and Italy.

[36] PAB, report dated Apr. 28, 1938.

[37] Blaschke's trip was May 5–May 25, 1943. The report is dated May 31, 1943.

same date to the university at Heidelberg. The Rektor replied again urging his return, telling him in response to his query that many colleagues like him had lost all they owned, but only one had been killed in the raid. Blaschke, however, had other plans, and by return mail said that Udo Wegner[38] had invited him to stay in Heidelberg, where he was sorely needed and could also participate in some military research. Though Blaschke did not mention it, both he and the Rektor, Eduard Keeser, understood that a romantic university town by the Neckar, like Heidelberg, was much safer than an urban metropolis and production center, like Hamburg. Keeser immediately asked the acting Dekan (in Blaschke's absence) whether Blaschke was dispensable at Hamburg; the Dekan similarly addressed Hecke. The reply was as might be anticipated—not because of Hecke's and Blaschke's mutual dislike for one another, but simply because Blaschke, a senior member of the faculty, appeared to be trying to evade a danger the others were inevitably subject to. All the present mathematics faculty and assistants, with the sole exception of Hecke, had been drafted. The situation was such that Blaschke's disappearance from Hamburg would produce the same situation as already existed in Heidelberg. So, says Hecke, "Looking back on the instructional needs of the previous war-semester, I must therefore designate *as indispensable the collaboration of Blaschke* in representing the discipline of mathematics here" (emphasis in original). Back through Dekan Klatt to Rektor Keeser went the declaration of indispensability, and on to Blaschke on September 23.[39] In the meantime, Blaschke had rented rooms for himself in Hamburg, had asked to put off his return there until lectures began (in mid-October), had spoken of spending the intervening time in Graz with his family, *and* had asked for a special financial emolument to cover his future separation from his family.

Keeser's letter of September 23 said that while he understood Blaschke's situation, and his concern for his family's safety, his absence from Hamburg was inconsistent with his position as Dekan, and consequently he was being removed from that position. Furthermore, there had been a decree that all faculty who stayed away from Hamburg without explicit permission beyond August 10 (in consequence of the air raid) would have their pay blocked.[40]

Receiving this in Graz, Blaschke sent a telegram of protest announcing his immediate return, and was coolly answered by a telegram saying that Keeser was traveling and not expected to return until October 11, when Blaschke could speak to him. Blaschke did return immediately, discussed the matter with Keeser's deputy, Rudolf Sieverts, and wrote Keeser a sarcastic letter about it. Shortly thereafter, Blaschke's request for the "separation emolument" was rejected as unjustified.

Soon after Blaschke returned to Hamburg, the Rektor suggested to him a

[38] For Wegner, see in particular chapter 4, "Hasse's Appointment at Göttingen."

[39] PAB, Keeser to Blaschke, Sept. 23, 1943. Though 45,000 people lost their lives in "Operation Gomorrah," this "terror attack" did nothing to reduce the size of the Hamburg student body. See Giles 1985: 297–298.

[40] PAB, Keeser to Blaschke, Sept. 23, 1943.

draft of an announcement of his resignation from the position of Dekan that was formal and correct and clearly dictated out of collegial consideration. Blaschke rejected it. Consequently the Rektor felt obliged to return to his September letter, and in November wrote the ministry that "Herr Professor Blaschke, in the view of many of his colleagues, in the period after the catastrophe that befell Hamburg, had not behaved as one must expect of a Dekan," and laid out the whole story.

A full six months later, *Dozentenführer* Anschütz, writing on party stationery, put in a good word for Blaschke, who "did not feel himself completely correctly understood." Especially "at the present time" (May 1, 1944), said Anschütz, comradely disturbances were to be avoided. He went on to praise Blaschke as teacher and scholar. A month later Anschütz endorsed Blaschke's receiving 20,000–30,000 RM to go to Paris for a week to buy books for the mathematics department. The Rektor approved, and on July 5, so did the ministry. Whether Blaschke actually went is unknown to me as, by the time he received permission, the Allies were in Caen, also pushing toward St. Lô, and would be in Paris in seven weeks.

Dismissed by the British military on September 3, 1945, Blaschke wrote a seven-page defense of his behavior on September 17, five days before Behnke wrote the letter about him already cited. Here Veblen was mentioned as a reference, and Levi-Civita as a Jew invited to speak at Hamburg. Here also was a long list of cases he claimed attested to his help for others in political trouble. These are not all worth individual comment—some are irrelevant to the Nazi situation, and in some his role seems somewhat exaggerated. One that is not irrelevant is a story about helping Hans Zassenhaus, who had made unfortunate remarks about a picture of Göring.[41] Somewhat startling is his claim that he protected Hecke from *Dozentenführer* Anschütz. The struggle with Bieberbach in 1934–35 of course is also used to attest to an anti-Nazi fight. The issue of the München professorial succession is mentioned to his credit, as he claimed that he had prevented, not without danger to himself, Max Steck from being Carathéodory's successor.[42] Blaschke emphasized his genuine cosmopolitanism, and excused his involvement with an organization known as the "Foreign Office of the *Dozentenschaft*" on these grounds. Blaschke also managed to twist the truth about his loss of the position of Dekan in 1943, blaming it on Anschütz. Given acknowledged "devils" like Bieberbach and Anschütz, it was reasonable for Blaschke to set himself up in apparent opposition to them as much as possible. The argument made to me by Christoph Maass that Blaschke's real concern was with what the Nazi regime could do for him rather than what he could do for them neglects the fact that a certain *quid pro quo* was involved. Obtaining what Blaschke wanted in prestige and position from the Nazis involved aiding them.

A number of people wrote letters for Blaschke, some more positive than

[41] This story is corroborated by a personal conversation with Hans Zassenhaus, interview, Nov. 17, 1987.

[42] See above, chapter 6, "Max Steck and the 'Lambert Project.'"

others. For many German academics in 1945–46, it was quite rational to view the dismissal of less egregious colleagues by the Allies as a political counterpoint to the prior dismissals by the Nazis—one more instance of the politicization of the "apolitical" university. And Blaschke was a difficult case. Despite all his personal opportunistic self-aggrandizement under the Nazis, Blaschke had helped the likes of Zassenhaus, Reidemeister,[43] and Zorn when they were in official difficulty; he also seems to have helped Gołab. According to a letter written by Zassenhaus, he allowed Emil Artin to refrain from the Hitler salute when beginning his classes, he worked to prevent the *Gleichschaltung* of the mathematics department, and only hung a picture of Hitler when it was "demanded by special order"; he never requested political activity of department members and ignored even the anti-Nazi political beliefs of department members. Yet Hans Petersson's widow says that they never felt comfortable with Blaschke, never knew what his true feelings might be, and suspected those feelings might be pro-Nazi.[44]

This unease, this feeling that Blaschke was, in Hecke's cited words, "not the typical National Socialist," though one with "old sympathies for fascism," was reflected in a second letter of Behnke to Hecke, sent a week after the one already cited, in which he attempted to assess Blaschke: "The German mathematician most influential in appointments during the past fifteen years." Behnke acknowledged Blaschke's considerable mathematical abilities (even if he found the estimation of his mathematical work somewhat inflated) and Blaschke's skill in fostering younger mathematicians. Yet he also found Blaschke lacking in patriotic love (*Heimatliebe*) for Hamburg or in any sense of responsibility arising from having worked there for so many years.[45] In fact, according to Behnke, the only motives that concerned Blaschke were purely egocentric ones—people behaving in other, less self-centered ways he found incalculable or naive; people speaking of other motives he considered "literati",[46] a class he despised. Academic politics, however, were more important to Blaschke than national politics, despite his fascist and National Socialist sympathies—the nickname "Mussolinetto" he had already acquired in 1925 referred more to his character, said Behnke, than to his convictions (and no doubt to his great love for Italy as well).

According to Behnke, what changed the Blaschke who had signally helped Reidemeister keep a German professorship in 1933, and opposed Bieberbach in 1934, to a more pro-Nazi stance in academic politics was not just his "will to power" and unending appetite for it, but friendship with Harald Geppert and Udo Wegner. This rings true. Wegner and Geppert both acquired enormous

[43] PAB, referred to in a letter, Blaschke to Emil Wolff, Sept. 28, 1945.

[44] Interview, Apr. 14, 1988. For Petersson, see below, "Hans Petersson."

[45] PAB, Behnke to Hecke, Sept. 29, 1945, p. 2. Cf. Behnke 1978: 120, for the attitude he found lacking in Blaschke.

[46] One might recall that this phrase of contempt was common among people of a conservative disposition. One well-known example is Thomas Mann's stigmatization in 1918 of his brother Heinrich as *Zivilisationsliterat* in *Betrachtungen eines Unpolitischen* (Berlin: S. Fischer Verlag, 1918).

cachet well beyond their status as mathematicians through political correctness. For a man as given to status as Blaschke, their example must have been as attractive as Behnke has suggested. Hecke also found Blaschke's relationship with Wegner symptomatic of his Nazi fellow-traveling and willingness to achieve his aims through Nazi methods.[47] Although Blaschke came to support the "unclean doings of Naziism"[48] within mathematics, Behnke said that Blaschke was never dangerous as teacher or researcher, and would not be in the future. On the other hand, for any leadership or advisory academic activity, he was far too unreliable, "indeed expressly dangerous."

Giving credence to Behnke's opinions and accusations, these being only the most serious ones, against Blaschke, Emil Wolff as Rektor thought the appropriate treatment of Blaschke was not dismissal but a somewhat early retirement (Blaschke was sixty) with pension, which would allow him to continue to be scientifically active. Hecke, Blaschke's old personal enemy, who nonetheless recognized that Blaschke was highly valuable as a mathematician and fosterer of mathematical talent as well as that he had not been a "genuine" National Socialist, agreed. The man Blaschke was thus characterized in Wolff's thoughtful words:[49]

> Herr Blaschke was by no means a politically active or intellectually convinced National Socialist. He faced these things with a cool superiority and skeptical indifference. The reproaches that could be raised against him were rather more of a moral than a political nature. They come down to that he was not very conscientious in the choice of means for the utilization of his strong influence and had adapted to the existing political situation with flexible opportunism.
>
> The problem set by him and his inwardly consistent character, always undisguisedly expressed by him, is very correctly summed up by Professor Reidemeister[50] in the following way: "Blaschke is a case *sui generis* and is only to be measured with his measures. In the effect of his activity, however, according to my impression, thoroughly preponderates the furtherance of ideas that he indeed did not make his own, but to whose representatives he rendered singular service."

The Vicar of Bray (or Charles Maurice de Talleyrand-Périgord) could perhaps not have been better described.

Wolff had suggested a way for Blaschke to remain mathematically active and receive income, yet suffer some punishment for his power-hungry Nazi fellow-traveling. Blaschke, however, insisted on full reinstatement and was unwilling to admit he had done anything wrong. One strategy he pursued repeatedly was blaming his present troubles on his old feud with Hecke. In the middle of June 1945, Hecke had mentioned Behnke's trip to Romania and his potential successorship to Carathéodory's Munich chair as items on Blaschke's account—Behnke had told him of them years earlier. However, on June 27, Hecke en-

[47] PAB, Hecke to Wolff, Nov. 25, 1945.

[48] The phrase is Behnke's, PAB, Behnke to Hecke, Sept. 29, 1945.

[49] PAB, Wolff memoir, Oct. 25, 1945.

[50] Reidemeister in fact had suggested that Blaschke receive a research position.

tered the hospital for an operation—the cancer that would kill him less than two years later had already been diagnosed before the war was over. Consequently, until October 15, the role of Dekan was assumed by the botanist Gustav Bredemann. Thus Hecke was not functioning as Dekan when Blaschke was dismissed, nor when Behnke's two letters were received. Furthermore, on October 18, Hecke asked Wolff to work on Blaschke's case without him, given his ancient and unresolvable quarrel with Blaschke.[51] In fact, Hecke had argued against Blaschke's dismissal and for his pensioning off, and never failed to acknowledge Blaschke's standing as mathematician and teacher, though he thought Blaschke dangerous if allowed any political influence in university matters.[52]

Nevertheless, Blaschke so conducted his defense and accused Hecke of playing a malicious role that Wolff felt obligated to break off all relations with him, asking that any further communication between them be through Bredemann. Indeed, neither Wolff nor Hecke (nor Behnke) thought Blaschke a Nazi by conviction. Wolff, however, felt compelled to deny explicitly that he knew anything of the background of Blaschke's relationship with Hecke, and to assert that it would in no way influence the disposition of Blaschke's case.[53]

The case dragged on. Open questions that remained included how closely and in what capacities Blaschke had worked with Wegner, and the purported denunciation of Behnke to Munich authorities. Among others putting in a good word for Blaschke was the well-known British mathematician W.V.D. Hodge. On October 23, 1946, Blaschke was completely reinstated by the British.

On September 30, 1953, Blaschke retired at Hamburg. The following year he received the national prize of the DDR (East Germany) and three years later became an honorary member of its academy of sciences. He died March 17, 1962.

Blaschke was a "special case," neither the committed believing ideologue like Bieberbach or Vahlen, nor the committed anti-Nazi like Kamke or Hecke. His sole interest seems to have been personal self-aggrandizement—in Lady Macbeth's advice, looking like the time to beguile it. "Concentered all in self," he could never have understood how his friend Reidemeister, whom he had helped in a bad political situation, and his life-long enemy Hecke, could judge him similarly and agree on what his postwar academic fate should be.

This detailed consideration of Blaschke is not to pillory his memory, but to realize that perhaps what made him a "special case" was that he continued one variant of the traditional apolitical stance of the German professor throughout the Nazi period. He worked to promote Hamburg mathematics, but in part because he was its leader. What was most important to him seems to have been

[51] PAB, Hecke to Wolff, Nov. 25, 1945, and *Hamburger Abhandlungen* 16 (1949): 6, where a sketch of Hecke's life is appended to a reprinting of the memorial address for Hecke given by Wilhelm Maak.

[52] PAB, Hecke to Wolff, Nov. 25, 1945.

[53] PAB, Wolff to Landahl, Dec. 3, 1945, and Wolff to Blaschke, Dec. 3, 1945. Cf. Wolff to Schulverwaltung, Hamburg, Mar. 25, 1946.

the traditional prestige and rank of the German Ordinarius. During the Nazi hegemony, it was easy for him to make the necessary political representations, as he was sympathetic to both the anti-Semitic and the Greater German emphases of the regime. Also, his mathematics, being highly geometric, was the sort that even enemies like Bieberbach ideologically approved. He made these representations to foster his personal interests—political position outside of mathematics did not interest him, since politics itself had no appeal for him. Academic political position did interest him. A positive consequence (since Blaschke was a first-rate mathematician for whom mathematics was his primary interest) was the relatively undisturbed continuation of Hamburg mathematics, and his geometric school in particular, during the Third Reich. This does not condone Blaschke's crass opportunism, nor his willingness to use Nazi-tinted political levers as available, but it does raise the question that perhaps a reason to condemn his behavior is that he adapted to the politics without looking more closely at what they meant. He simply did not care. He did not have the genuine conservative principles that seem to have activated men like Heinrich Scholz, Süss, Prandtl, Hellmuth Kneser, and Erhard Schmidt (though each in a somewhat different way). Inclined to cosmopolitanism and conviviality, a genuinely fine mathematician and mentor to students, in a different political time he would have perhaps been just as opportunistic and less negatively judged by posterity for it. The basic principles in Blaschke's life seem to have been selfish—which did not hinder him from helping others when it incurred no personal disadvantage. Thus he was a puzzle, both to his defenders, like Zassenhaus, and to his accusers, like Behnke. Many mathematicians, many academics, in Germany during the Nazi period were probably like him, only less so. Rather than trying to decide whether Blaschke was more a "realistic" hero, who preserved the Hamburg mathematics department from political intrusion, or a cynical opportunist who subscribed publicly to Nazi beliefs solely for purposes of self-aggrandizement, perhaps it is best to admit that under the Nazis he was both: like Sir Lancelot, "and faith unfaithful kept him falsely true."

The Development of Heinrich Behnke's Attitudes

Behnke may have been adamant that Blaschke was dangerous as an administrator of influence, though not as a teacher or researcher, yet he was quite able to overlook similar flaws among practitioners of other disciplines. Just as the Allied occupation of Germany saw Hecke installed as Dekan for mathematics and science at Hamburg, Behnke became similarly installed at Münster, a position he held until 1949. There were, as with Blaschke, the usual arguments as to who was deserving of punishment and for what (among those apparently writing completely positively for Blaschke, in addition to Zassenhaus, were Konrad Knopp and Gustav Herglotz). One case at Münster concerned the pharmaceutical chemist and, since 1931, director of the Münster Pharmaceutical Institute, H. P. Kaufmann. Though it does not deal with a mathematician, brief delineation of this case serves as a counterpoint to Blaschke's case, not least because both situations involved Behnke.

After the war, Kaufmann was reinstated in his position (in 1943 he had gone to Berlin, presumably in connection with scientific military activity), despite the fact that he had, as a Nazi party member, written a denunciatory letter about a colleague. This letter is cited below because it actually is extant and accurately gives the flavor of what was offensive in 1935, whereas Blaschke's purported letter to Munich seems no longer to exist, if it ever did. The denounced colleague was the physicist Adolf Kratzer. In 1935 Kratzer was proposed as interim Dekan at Münster, and on January 9 of that year Kaufmann wrote the Dekan still in office with the following complaint ("c/o party member Prof. Trier"):

> At the beginning of the last semester, Professor Kratzer, in the presence of a large number of colleagues, declared: "60% of those who entered the party after January 30 [1933] are rascals." Professor Kratzer, who sat diagonally across from me, must have known that I first entered the party in 1933. For your orientation I note that in 1919–1920 I belonged to the German People's Party (*Deutsche Volkspartei*), and since then have been partyless. . . .
>
> How unjustified the criticism cited above is from the mouth of a nonmember of the party, I scarcely need to tell you, since you certainly entered at the same time as I. I would have been pleased if all members of the faculty had joined the party from innermost conviction after they had an opportunity to get to know Adolf Hitler as Leader of the German people. You have surely also frequently heard the opinion that one can be a National Socialist even without belonging to the party. According to my conviction, however, only he who is a party member can completely grasp the sense of the movement and also entirely advocate it. Today the party has a very high percentage of young members, whose entrance took place with the agreement of, and after screening by, the local organizations. The reproach of Professor Kratzer thus means an attack on their honor.

According to Kaufmann, not only was Kratzer, the nonparty member, unsuited to such a position, but Kaufmann explicitly denied any desire to have the position for himself. He had only stood for it at the urging of the leader of the National Socialist students, and had come one vote behind Kratzer, but he did not desire academic office:

> My activity for the party lies outside the university in the National Socialist education of nearly 3,000 fellow Germans (*Volksgenossen*) and making them able to defend themselves (*Wehrhaftmachung*). I was called to this task, which gives me great pleasure, by the trust of the upper leadership of my corps, as an old soldier who served in the front lines, and for the past two years I devote all my free time to it.

Blaschke never wrote anything so patently ideologically committed, nor, so far as is documentable, so potentially deleterious to a colleague. Yet Behnke as Dekan apparently "protected" Kaufmann from postwar recrimination, though Kratzer was clearly a friendly acquaintance with whom he had collaborated on a book under the aegis of Süss' book project, and who was as distressed as he in

1933 when they found Nazi leaflets within the university. This "protection" enraged the organic chemist Fritz Micheel, who became the postwar head of chemistry at Münster. Indeed, when Micheel wrote an angry letter to Behnke (not his first, and itself apparently in reply to an angry letter) reminding him without mentioning names that as Dekan he admitted protecting a denunciant, Behnke telephoned Kaufmann, saying he guessed this accusation referred to him. Kaufmann asked for confirmation; should it then be positive, he "would be in a position to take the necessary steps [presumably legal] against Micheel." In Behnke's 1978 memoir, Kaufmann is mentioned once in a friendly manner in connection with helping Dekan Behnke deal with the unusual postwar socio-economic pressures then bedevilling pharmaceutical study—indeed, "he helped me immediately and soon also took over all responsibility." Micheel is mentioned not at all by Behnke. Kaufmann was about forty-four when he became a party member; Kratzer was four years younger.[54]

This little drama took place in 1949. The Behnke of 1945–46, so condemnatory of Blaschke, by whom he felt personally injured, by 1949 seems to have accepted men like Kaufmann as bygones who could be extremely helpful. It is unclear what, if anything, Kratzer suffered as a result of Kaufmann's letter.

By 1978, Behnke's increasing and understandable tendency to let the dead past bury itself—after all, German academic mathematics would not move forward if it were preoccupied with past recrimination—seems to have led to the somewhat evasive style of his 1978 memoir, *Semesterberichte*, when dealing with the Nazi period. While Behnke accurately conveys its atmosphere, both personalities and the various countervailing pressures on them during the period are largely avoided. The scant mention of van der Waerden has already been noted—Blaschke is mentioned only as the remarkable department administrator and teacher he undeniably was, and Bieberbach once as a lecturer and once as the doctoral supervisor of the forced emigré (Jewish) Stefan Bergmann. While this is true, standing alone, it is certainly somewhat misleading. While Behnke discusses his Romanian trip, he nowhere mentions Blaschke in this connection.

Erich Hecke

It is interesting to see the attitude of a mathematician like Behnke, whose school of mathematics was becoming dominant in postwar West Germany, toward another figure who was important in the continuation of mathematical activity and traditions through the Nazi period, his own doctoral supervisor, Erich Hecke. Furthermore, though Behnke turned in directions far removed from his

[54] For the "Kaufmann case" as mentioned here, see *Nachlass* W. Herrmann (Münster): Kaufmann to Dekan, Jan. 9, 1935; Micheel to Behnke, Aug. 10, 1949, and Nov. 3, 1949. For Kaufmann in Berlin and Micheel as head of chemistry, see Poggendorf (1960), vol. 7A. For Behnke and Kratzer finding Nazi leaflets, Behnke 1978: 116. On this occasion, Heinrich Scholz was apparently less offended. For their joint book, see *Nachlass* W. Herrmann, Wilhelm Süss to Mevius, Apr. 21, 1943. For Kaufmann's postwar helpfulness, Behnke 1978: 183.

dissertation, he and Hecke worked closely together on the publication of the *Mathematische Annalen* throughout the Hitlertime.[55]

Hecke was certainly associated with tacit resistance groups and apparently helped the resistance efforts of Hiltgunt Zassenhaus, who illegally brought food and medicine to Scandinavian prisoners while acting as an interpreter.[56] On Hecke's character, Behnke says (speaking from his seven years' experience as his student and *Assistent*):[57]

> In Hamburg I had many long talks with him during walks. His radical skepticism got more and more on my nerves. Probably he had had it very difficult with himself. He found no midpoint. He had had his great success too early in his life. Although I knew of anti-Semitic remarks by him from earlier days, in the time of the brown garden [i.e., the Nazi period] he was very brave, indeed already dangerously without religion [or "nondenominational"].

Behnke's picture of Hecke may be contrasted with the view of another of his students, Wilhelm Maak, who opened a memorial address for his teacher with the words, "Only a truly outstanding human being (*Mensch*) can be a truly good teacher, as Hecke was," and ended with the thought that he cannot better express his thanks to Hecke "than by repeating my promise that I will make the effort later on to be a similarly good teacher and fatherly friend for others as Hecke was for me." In between, he spoke of how Hecke encouraged students to try to understand the mathematical thought processes that led great mathematicians in various directions, of Hecke's love of Anton Bruckner's symphonies and Richard Strauss's operas (and antipathy to Bach) and his genuine interest in Goethe. Even with all allowances for the exuberances of a heartfelt eulogy, Maak's picture of Hecke as intellectual is very different from Behnke's. Perhaps the difference is between the 1920s, when Behnke was Hecke's student, and the late 1930s.[58]

In those late 1930s, Hecke was someone whose opposition to the Nazis was known, even if it could be safely expressed only by a look. Horst Tietz was, in Nazi jargon, a "Jewish miscegenant of the first degree" (as was Behnke's son), that is, he had exactly one Jewish parent. However, because his non-Aryan father had been on the front lines in World War I, in 1939 he was allowed to become a university student. In the event, this dispensation lasted only a year, and in December 1940, a secret decree forced the university to expel him. Tietz described Hecke as the central figure for new mathematics matriculants, and "certainly the most fascinating personality that I have been permitted to experience in universities!" According to Tietz, when Hecke entered the lecture hall, he always greeted his students with a silent nod of the head, even though the

[55] Above, chapter 6.

[56] Behnke 1978: 80. For Hiltgunt Zassenhaus, see her autobiographical memoir, *Ein Baum blüht im November* (Hamburg: Hoffman und Campe, 1974, 1976), and Bottin 1992: 70. She was the sister of the mathematician Hans Zassenhaus.

[57] Behnke 1978: 79.

[58] W. Maak, "Erich Hecke als Lehrer," *Hamburger Abhandlungen* 16 (1949): 1–6.

"Germanic greeting" of "Heil Hitler" with accompanying salute was prescribed. Tietz told the following personal anecdote:

> While coming from the mathematics department . . . we [Tietz with some other students] overtook Professor Hecke; there was a bone-rattling chill in the air and we were all heavily wrapped up; with a smart "Heil Hitler, Herr Professor" and quickly raised right arm my fellow students passed the old gentleman [Hecke was fifty-two]; with an astonished-indulgent look Hecke glanced beside him, raised his hat, lightly bowed, and said, "Good morning, ladies and gentlemen!" Since then I have further heard the "Germanic greeting" from none of my fellow students.

Tietz further said that at first there was no political activity among the students, though clearly every variety of opinion toward National Socialism was maintained in the student body. In particular, a "happily astonished looking up in view of Hecke's behavior quickly bound together [those] in like-minded opposition," while the convinced National Socialists revealed themselves by uncomprehending head-shaking.

When expelled, Tietz sought Hecke's advice. Hecke said that naturally Tietz should come secretly to his mathematics lectures, also those of Hans Zassenhaus, and likely the theoretical physicist Wilhelm Lenz, with whom Hecke would speak. Tietz was somewhat afraid of Zassenhaus, who always wore the insignia of a Nazi organization, but Hecke calmed him by saying that Zassenhaus was their trusted agent who behaved in a politically correct manner so as to protect them. This in fact happened in 1942, when somehow Tietz's presence as a secret student at the university became known and a denunciation was threatened: Zassenhaus warned Tietz, and Hecke immediately cancelled the class, giving back the student fees.[59] In fact, Tietz only discovered much later that Zassenhaus was connected with a resistance effort that helped hide endangered people.[60] Blaschke's ambiguous political stance also appears in Tietz's story. Zassenhaus, in his letter supporting Blaschke's reinstatement, cited to his credit that he knew about Tietz and did nothing.[61] Tietz himself has said he was always afraid of Blaschke.[62]

After the devastating air raid on Hamburg in late July 1943, Tietz and his parents fled to Marburg. Hecke arranged contact with Kurt Reidemeister, who was willing to help Tietz with private study (as Zassenhaus had done at Hamburg after it was no longer possible for Tietz to hear lectures secretly). However, on Christmas Eve, Tietz and his parents were arrested by the Gestapo and sent to the concentration camp at Buchenwald—only Horst Tietz survived.[63]

[59] This story was told to me independently by Hans Zassenhaus (interview, Nov. 17, 1987), and also appeared in Tietz's unpublished recollections entitled "Studium mit Hindernissen" (March 1984). Prof. Tietz repeated the story in an interview (Apr. 5, 1988). I am grateful to Prof. Tietz for a copy of his recollections.

[60] Tietz n.d.

[61] PAB, Zassenhaus memorandum, "Ref. Dismissal of Professor W. Blaschke," Sept. 18, 1945.

[62] Horst Tietz, interview, Apr. 5, 1988.

[63] Tietz n.d.

There was little ambiguity about Hecke's position, and many positive stories about his behavior under the Nazis exist, leading to his postwar membership on the denazification commission, as well as his *Dekanat*. However, as noted, his illness forced his replacement on the former by Zassenhaus, and Bredeman substituted for him as Dekan. Nevertheless, as in Blaschke's case, he seems to have been determined to be just while unforgiving of those who promoted the Nazi cause.[64] Yet, over thirty years after his teacher's death, Behnke said:[65]

> After the war Hecke obtained a great deal of power. He was, however, too inexperienced in the ways of the world and too unsteady to use it. At that time he caused much calamity. Some of this is to be pardoned by his serious illness. As early as 1947 he died in Copenhagen—an enemy to the world and himself. I, however, will remain eternally grateful to him.

Perhaps the difference between Behnke and Hecke (or the Behnke of 1945 and the Behnke of 1978) can be found in what Behnke said were their different attitudes in 1933 toward academic Nazi "fellow travelers"—for Behnke they were fools, for Hecke they lacked character.[66]

It is, of course, impossible to say whether mathematics would have continued as respectably as it did at Hamburg had Blaschke's attitude toward the Nazi regime been more like Hecke's. The example of Tübingen, though, would seem to indicate that the results would not have been disastrous. Recall that at Tübingen, when Karl Kommerell retired in 1937, the *Dozentenführer*, complaining about the attitudes of Konrad Knopp and Erich Kamke toward the new order, said a real Nazi was needed to provide a different perspective. The mention of Kamke in this context is testimony to his importance as a mathematical personality in the department, as he was only an associate professor, his career having been delayed by his World War I service. In the event, Hellmuth Kneser was hired, but he turned out to be a man of conservative rather than Nazi principles. Coincidentally, Hamburg and Tübingen each had a prominent mathematical figure whose marriage met Nazi disapproval, Kamke in the case of Tübingen, and Emil Artin in the case of Hamburg.

Oswald Teichmüller

If Blaschke reveals the well-known mathematician as self-aggrandizing opportunist, whom some believe gave necessary aid to his department by his actions, the well-known mathematician as dynamic believer in the Nazi cause is presented by Oswald Teichmüller. Teichmüller's behavior in the matter of Helmut Hasse's Göttingen appointment and his spurring on of Werner Weber was dis-

[64] Promoting the cause as distinct from joining the party. Gustav Bredemann and Wilhelm Lenz, both mentioned above, are examples of professors at Hamburg who became party members for self-protection but did nothing to promote the Nazi cause. Compare also Hecke's student Hans Petersson, below. For Lenz, see also Tietz n.d.

[65] Behnke 1978: 80.

[66] Ibid.: 126.

cussed in chapter 4. It was Teichmüller who encouraged Weber in the attempt to get Udo Wegner to design a pedagogically Nazi-oriented mathematics program. It was Teichmüller who came to Edmund Landau's office when his classes were boycotted. At the time, he was twenty.

(Paul Julius) Oswald Teichmüller was born on June 18, 1913, in Nordhausen, a town in the Harz mountains. However, his home was in the even smaller village of St. Andreasberg in the Harz, to which mother and baby returned after a few days. His father, (Adolf Julius) Paul Teichmüller, was a weaver, and thirty-three at his son's birth; his mother Gertrude, née Dinse, was six years older, and the couple had no further children.[67] His father was wounded during World War I and died when Oswald was twelve, though whether these events were related is unclear. In any event, his father's business was shut down in 1915 when he went to war. Presumably the father on his return in 1918 took up this trade again, since the family lived in St. Andreasberg until his death.[68] The family was poor, but the son, with perhaps the sense of moving up in social class as a student, always listed his father's occupation as factory owner rather than employee.[69]

According to his mother, when Teichmüller was three-and-a-half, she discovered that he knew how to count, and he also learned to read on his own, his first self-instruction being from labels on tin cans. When his father returned in 1918, young Oswald read to him fluently and recited a poem previously unknown to his mother. On his father's death, she took him from St. Andreasberg, "whose school he had long outgrown," to Nordhausen, where he lived with an aunt[70] and attended the *Gymnasium*.

Teichmüller entered Göttingen in the summer semester 1931 as a brilliant but lonely student from the hinterlands. He was not yet eighteen. Peter Scherk, a student at the time, and Hans Lewy, a young instructor, both of whom would become well-known mathematicians (and both forced emigrés in 1933), told anecdotes of the ungainly student's brilliance.[71] After one semester, he joined the NSDAP and the SA as well. While it is tempting to assign at least part of the reason for this to Teichmüller's provincial background and the declining material fortunes of his petit bourgeois family, it should not be forgotten that the town of Göttingen was a cauldron of right-wing sympathies.[72] That this was a declaration of political idealism is perhaps indicated by the fact that Teich-

[67] The above facts are the same in Erhard Scholz's biography of Teichmüller in N. Schappacher and E. Scholz, eds., "Oswald Teichmüller—Leben und Werk," *JDMV* 94 (1992): 1–39, p. 3; William Abikoff, "Oswald Teichmüller," *Mathematical Intelligencer* 8 (1986): 8–16, 33; and Teichmüller's *Lebenslauf* dated July 17, 1935, in UAG. The birthdate given in Teichmüller's *Collected Works*, ed. L. Ahlfors and F. W. Gehring (1982), is incorrect. Material on Teichmüller below is from either Schappacher and Scholz 1992 or Abikoff 1986, unless otherwise cited.

[68] Abikoff 1986, citing letters of Gertrud Teichmüller to H. P. Künzi.

[69] Schappacher and Scholz 1992: 3, n. 3.

[70] Ibid.: 3.

[71] Abikoff 1986: 10.

[72] The well-known mathematician Werner Fenchel, himself an emigré from Germany, ascribed a significant role to Teichmüller's provinciality (letter from Abikoff to author, Sept. 6, 1988). For the political atmosphere of Göttingen, see Marshall 1972.

müller joined the party first (with no. 587,724) and the SA three weeks later, presumably to add activism to his beliefs.[73] Party activity seems to have provided friends for Teichmüller, and he became the deputy leader of the Nazi organization of mathematics and natural-science students. By autumn 1933, with two years' service in party and SA, Teichmüller gave his beliefs explicit expression by leading the November 2 boycott of Edmund Landau's calculus class, described in chapter 4. He later met Landau in his office to talk over the situation, and, at Landau's request, put down his view of the boycott in a letter. This is the letter that Landau, after deleting Teichmüller's name, passed on to the Göttingen authorities with the request to retire.[74]

Though it is unlikely that Teichmüller and Ludwig Bieberbach had any contact prior to the Landau boycott, Teichmüller's letter to Landau seems the earliest setting down of ideas thereafter associated with Bieberbach. The "manly rejection" of Landau by the Göttingen students was taken by Bieberbach as the frontispiece of his Easter Tuesday address that became something of a *cause célèbre*. Rather than the Göttingen boycott confirming Bieberbach's ideas, it seems more than possible that, at least in part, it inspired them. Thus, when Teichmüller's mother wrote about the Landau boycott in 1948, ". . . And Oswald? He had blown in Bieberbach's horn, and louder than he,"[75] she did not realize that "Oswald" had perhaps helped construct "Bieberbach's horn." Teichmüller did not have the elaborate intellectual typological rationale Bieberbach borrowed from Jaensch for his opinions, but that does not alter their similarity.

Teichmüller explained to Landau that external forces could alter the students' temper and inspire them to change unsatisfactory situations previously considered unchangeable. A failure by a teacher to care about or understand the majority student mentality could also lead to student disruption. As to Teichmüller's beliefs about the student action:[76]

> You [Landau] expressed the assumption yesterday [in our conversation] that it had been an anti-Semitic demonstration. I stood and stand by the view (*Standpunkt*) that a special action inimical to Jews should be directed against almost anyone else before you. It was, for me, not about making difficulties for you as a Jew, but solely about protecting German students in their second semester from being instructed by a teacher of a completely foreign race precisely in differential and integral calculus, while sparing as much as possible all others therefrom. I dare as little as any other person to doubt your capability for pure international-mathematical-scientific teaching of suitable students of whatever heritage. However, I also know that many academic lectures, especially also differential and integral calculus, at the same time have educa-

[73] UAG, Werner Blume (*Dozentenführer* at Göttingen) to Kurator Göttingen (Justus Valentiner), Oct. 22, 1935. For those members of both, usually SA membership preceded party membership. The dates given agree with this in Teichmüller's *Lebenslauf* of July 17, 1935.

[74] That this student was Teichmüller has been proved by Schappacher and Scholz (1992). Previously Werner Weber (presumably as Landau's *Assistent* at the time) was suggested.

[75] Schappacher and Scholz 1992: 5 n. 14.

[76] Cited and translated from ibid.: 28–30. Teichmüller's "own experience" refers either to Werner Fenchel or to Richard Courant.

> tional value and lead the student not only into a new conceptual world, but also to a different mental viewpoint (*geistige Einstellung*). Again, since the mental viewpoint of an individual depends on his mentality (*Geist*); which thus should become transformed; this mentality, again, according to fundamental rules, not only contemporary ones, but already long recognized, depends completely substantially on the racial composition of an individual; allowing Aryan students to be educated by a Jewish teacher, for example, ought not in general be recommended. I can here speak from my own experience. For the student [taught by a teacher of foreign race] remains really only two paths: perhaps (*entweder*) he draws out of the teacher's lecture only the international-mathematical skeleton and clothes it with his own flesh. That is mathematically-philosophically productive work, to which only the fewest have grown. . . . The third path, to take over the material in its foreign form, leads to a spiritual (*geistigen*) degeneration that you could not well expect of a student today and also do not wish. The possibility, however, that you transmit to your hearers the mathematical kernel without your own national coloration is so small as it is certain that a skeleton without flesh does not run but falls in a heap and disintegrates.
>
> From this, my view, also follows that there were little to argue against it if you wish to hold more advanced lectures, building on the already present mental viewpoint worked out for application or knowledge of important mathematical facts, now as before in the best relationship with the students in our university. This is a view that only a few of my comrades have joined.

Teichmüller went on to say that he could only think of the student majority opinion that Landau should never again lecture as anti-Semitism, but that the distinction between his and their view was at the moment irrelevant. He emphasized that all students were united and there was no question of a division into radicals and moderates; they were all good comrades and only differed "over the purely theoretical question of whether yesterday's action [the boycott] had an anti-Semitic or a pro-German character."[77] Teichmüller's letter might be termed "bizarre," but in it are found the same arguments that Bieberbach and Eva Manger were to make within several months, the same disclaimer of anti-Semitism, the same emphasis on the virtues of apartheid. Perhaps the differences between Bieberbach and "P.S." also have the same character: Bieberbach was "pro-German," while "P.S." was explicitly "anti-Semitic," though Bieberbach, a much older and more sophisticated man than Teichmüller, must have realized that the dividing line in fact barely existed in the circumstances.[78] Such ideas of the distinctions between "Aryan" and "non-Aryan" thought were of course commonplaces at the time, but with a difference. Though Jaensch and Bieberbach might point to the well-known passage in Felix Klein's "Evanston Colloquium," or even a letter from Karl Weierstrass to Sofya Kowalewskaya, as evidence of the nineteenth-century currency of such ideas, in fact both Klein and Weierstrass, while suggesting that different ethnic groups, Jews in particular, thought about mathematics differently, never suggested that therefore some

[77] Ibid.

[78] Teichmüller's letter is characterized as "bizarre" by Schappacher and Scholz (1992: 28).

were unsuitable to teach "German" youth.[79] Weierstrass did say that Jews lack the *Phantasie* necessary for the most significant research, but he also cited a non-Jew so lacking in his opinion, and never suggested that Jews are inadequate teachers. In fact, one wonders to what extent his remark to Kowalewskaya was motivated by Weierstrass's well-known antipathy to Kronecker.[80] The Nazi education ministry dismissed Jews under the April 7, 1933 law because they were Jews, not because of any intellectual rationale about Jews being unfit teachers for Germans. The law was for "reforming the civil service"; university teachers were civil servants; Jews *per definitionem* could not wholeheartedly support the new state; hence they had to go. In fact, one paragraph of the law addressed dismissal of non-Jews who failed to be sufficiently enthusiastic. However, there were exceptions (apparently inserted at von Hindenburg's insistence), and both Landau and Courant fell under these exceptions. Hence Courant initially was only "furloughed" and not dismissed. Hence also, Landau, no doubt naively, and perhaps with the self-assurance of an important professor, thought he could resume his lectures in the autumn of 1933.

Teichmüller's letter seems the earliest extant expression applying these ideas to the suitability of university mathematics teachers. There is no reason, though, to think that Bieberbach explicitly borrowed from him. What seems more probable is that Bieberbach, disliking Landau, philosophically attuned to the National Socialist program of renewal, and always eager to advance his hierachical standing in mathematics, saw the Göttingen student action as an opportunity to become the leader of mathematics by way of being the older senior guide of the mathematics students' revolution. This speaks both to his use of the Landau boycott in his 1934 lectures/papers and to the explicit pedagogical aims of *Deutsche Mathematik*, as well as to the high initial involvement of Nazi students under the leadership of Fritz Kubach in that effort.[81]

In October 1936, Teichmüller prepared the ground for transfer to Berlin and Bieberbach, with whom he "habilitated" in March 1938. Prior to this he took his doctorate in Göttingen. In the academic year 1933–34, Franz Rellich had held a Göttingen seminar on operator theory. One of Teichmüller's fellow students (and SA comrade) Hermann Wachs had the idea of generalizing the Hilbert spaces treated in the seminar.[82] Teichmüller independently developed the theory of operators in what he called "Wachs Space" and proposed it as a doctoral dissertation. At the time, Hasse (and Tornier) had come to Göttingen

[79] Cf. chapters 2 and 7 above.

[80] *Acta Mathematica* 39 (1923): 191, extract from a letter of Weierstrass to Kowalewskaya, Aug. 27, 1883. The letter criticizes Kronecker explicitly. Sofya Kowalewskaya died February 10, 1891; Kronecker, December 29, 1891. Since the given name of Kowalewskaya appears in the literature as both Sofya and Sonya, it should be made clear that the first was her name at christening, the second what all her friends called her (ibid.: 134, 147).

[81] See above, chapter 7, passim. Cf. also BDC file for Kubach, letter from Bruno Baron von Freytag-Loringhof (student mathematics leader in Greifswald) to Kubach, Oct. 11, 1934.

[82] His idea was to use the quaternions for scalars. See Norbert Schappacher in Schappacher and Scholz 1992: 14–15.

only a few months previous, and the only other Ordinarius left in Göttingen mathematics was Gustav Herglotz. Teichmüller's choice of dissertation supervisor and approver is curious. Rellich, whose lectures the draft dissertation followed closely, would have been the most suitable supervisor, but he was suspect as Courant's *Assistent*. Herglotz, whose wide-ranging mathematical interests and knowledge were well known, kept himself removed from anything political. Teichmüller gave the draft to Hasse, the only member of the faculty of mathematical standing who had had any positive connection with the Nazi authorities at the time, if in no other way than by his official appointment. Furthermore, it would appear that the Nazi students soon realized that Hasse was ready to cooperate fully with them.[83] The subject matter was far removed from the algebraist Hasse's expertise or knowledge, and so Teichmüller's choice seems rather more a political than a mathematical one. Hasse did send it to an expert, Gottfried Köthe in Münster, who guided it to final form in early 1935,[84] and Teichmüller's final examining committee was Hasse, Herglotz, and the physicist Robert Pohl. Teichmüller passed this examination on June 26, 1935, and was officially awarded the doctorate some five months later.

Teichmüller's dissertation was his only paper in the area of functional analysis. His next few papers showed Hasse's mathematical influence, and were algebraic. Politically, however, Teichmüller thought little more of Hasse than he had originally. In a letter to Bieberbach expressing his desire to transfer to Berlin, he apparently expressed the thought that complete *Gleichschaltung* of the Göttingen Institut would not happen in the foreseeable future under Hasse.[85] Before he left Göttingen, however, he heard Rolf Nevanlinna's lectures as a visiting faculty member—these inspired work in complex analysis, work in an area familiar to Bieberbach, and his *Habilitationsschrift*.[86] Bieberbach, in his referee's report on this *Habilitationsschrift*, in fact included a dig at the "less original" work in algebra that preceded it. While Teichmüller in 1936 made four mathematical contributions to the first volume of *Deutsche Mathematik*, three of them algebraic, thereafter he published nothing algebraic in Bieberbach's journal (with the exception of one singular paper in 1940) and much analytic work, while his algebraic work was largely reserved for "Crelle's Journal," of which Hasse was the editor.[87]

[83] Heinrich Kleinsorge (leader of the Nazi organization of mathematics students at the time) in an interview with Erhard Scholz, Mar. 2, 1985, as cited in Schappacher and Scholz 1992: 6 n. 8.

[84] Norbert Schappacher in Schappacher and Scholz 1992: 14.

[85] UAG, Werner Blume to Kurator Göttingen, Oct. 22, 1935. Teichmüller was also friendly with Werner Weber, who had caused Hasse so much trouble at his Göttingen appointment. According to Peter Scherk (see Abikoff 1986), Teichmüller converted Weber to Naziism. Weber did not join the party until 1933 (no. 3,118,177). See BDC file, Werner Weber.

[86] Rolf Nevanlinna was a distinguished Finnish mathematician who created the active area of complex analysis now known as "Nevanlinna Theory." Nevanlinna taught at Göttingen in 1936–1937 and in the negotiations leading to this position indicated his "sympathy for Germany," MI, Hasse to Göttingen Academy of Sciences, Nov. 5, 1936. Nevanlinna's first wife was "a great admirer of Hitler," see Weil 1992: 130. See also HK, Bieberbach to Kneser, Jan. 21, 1938.

[87] See Teichmüller 1982, or the list on pp. 33–34 of Schappacher and Scholz 1992.

Teichmüller was a politically committed Nazi, but his first love was mathematics. Shortly after he passed his final doctoral examination, Hasse obtained a position as *Assistent* for him, and consequently Teichmüller also joined the *Dozentenschaft*. In proposing Teichmüller, Hasse spoke of his "extraordinary mathematical gifts," and remarked that he "promises to become a mathematician of importance" and that his lecturing style was of a "painfully exact, in high degree suggestive, and impressive sort."[88] At the time Teichmüller had the position of *Rottenführer* in his SA troop, as he explicitly remarked in his attached curriculum vitae. Yet mathematics took precedence over university political activity for him. Asked on the occasion for approval of Teichmüller's departure to Berlin, *Dozentenführer* Blume remarked that Teichmüller had had ample opportunity to be active in university politics:

> [Concerning such activity] I have, however, until now noticed only very little or better said none at all. Teichmüller also refused to participate in the first roll-call of the newly called to life *Dozentenbund*, which was drawn up as a required gathering in the assembly hall, since, on that evening, he did not wish to miss a scientific mathematical lecture.

Thus, though Teichmüller, according to reports of his fellows, as well as a reference from Tornier, had outstanding scientific qualifications, if "he wished to leave Göttingen, I am of the opinion that one should calmly let him go and ought to put no difficulties in the way of a planned change of atmosphere." This report also mentioned Teichmüller's activity *contra* Hasse as a "disagreeable affair," and remarked, concerning his character, that Teichmüller made the impression of someone inexperienced in the ways of the world; even Tornier characterized him as eccentric.[89]

So Teichmüller went to Berlin and Bieberbach. Erhard Scholz, in his Teichmüller biography, argues that Teichmüller's version of "Deutsche Mathematik" and Bieberbach's were different.[90] This seems rather convoluted. It was precisely support of what came to be Bieberbach's ideological program that Teichmüller's letter to Landau reveals. If, as Erhard Scholz claims, Teichmüller understood by "Deutsche Mathematik" some general political program for the direction of mathematical institutes by Nazi-minded mathematicians, then it hardly seems he would have found Hasse so unsuitable. Men like Teichmüller, Kubach, and Weber saw themselves as the youthful dynamic vanguard of a new German regeneration under the Nazi aegis. Bieberbach, as a sort of mathematical Phillipe Égalité, joined in promoting that dynamism, presumably for his own mathematical-political ends. That Teichmüller was a mathematical genius, whereas Weber was run-of-the-mill and Kubach even less competent as a mathematician,[91] does not affect their sharing of political attitudes. It is true that Teichmüller, in a letter to his former fellow student Adolf Bruns, spoke of Berlin as a

[88] UAG, Hasse to Dekan, July 16, 1935.

[89] UAG, Werner Blume to Kurator Göttingen, Tornier's estimate is cited by Blume.

[90] Schappacher and Scholz 1992: 8–9.

[91] Kubach's doctoral dissertation was a historical one on Kepler.

"foreign city" and wondered "what could be going on in Göttingen." However, that should perhaps be seen not as personal isolation within the Berlin mathematics department, but rather as the reaction of someone rurally raised to the metropolis of Berlin in contrast to the much smaller university city of Göttingen.[92]

Scholz also seems to think that the conflict between Berlin and Göttingen only became inimical with the concentration of *Deutsche Mathematiker* in Berlin.[93] While it is certainly true that the eventual presence of Werner Weber and Tornier (as well as Teichmüller) in Berlin would hardly induce cordial relations with a Göttingen led by Hasse, it is also true that the enmity went back a long way—one need only think of the feelings of Weierstrass and especially Frobenius about Klein, or, somewhat later, the contrasting attitudes of Bieberbach and Erhard Schmidt on the one hand, with Hilbert on the other, toward the Bologna congress of 1928, as discussed in the preceding chapter. In fact, it almost seems as if Bieberbach, rather than causing the division of Berlin with Göttingen, was more likely stimulated by the already extant enmity. Teichmüller's letter to Bruns was written one month after *Kristallnacht*, yet he does not seem to mention it. Hans Wittich, however, recalled speaking to Teichmüller about these events around the same time, to which he replied "in his free and easy tone, 'You are a reactionary bourgeois (*Spieser* [*sic*]) and don't comprehend the Führer's ideas.'"[94]

Teichmüller's dedication to the Nazi cause and ideology seems complete, and he seems to have shared Bieberbach's version of its application to mathematics. Far from being isolated mathematically in Berlin prior to the war, beginning with the *Habilitationsschrift*, Teichmüller published seven papers and his epochal monograph (197 pages) on "quasiconformal mappings and quadratic differentials." All this work was done in the two Berlin years (April 1937–July 1939). All the papers appeared in *Deutsche Mathematik*, six dealt with analytic equations, some in areas cultivated by Bieberbach, and one of these was an explicit improvement of a (partly fallacious) argument of Bieberbach—so much for isolation! Throughout this period Teichmüller was supported by a stipend provided by the education ministry chief for science, Theodor Vahlen.

On July 18, 1939, Teichmüller, just twenty-six, was drafted. By then, war was in the air. March had seen the German annexation of the non-Sudeten Czech lands as the "Protectorate of Bohemia and Moravia"; Slovakia became an "independent" puppet state. In April, Mussolini annexed Albania. Britain guaranteed the Polish borders on August 25, 1939—something it had refused to do fourteen years earlier at Locarno. The Nazi-Soviet pact of August 23 was merely Hitler's final step to the Polish invasion. Teichmüller was originally called up for only eight weeks' service, but the war intervened. In April 1940 he took part in

[92] The letter is reproduced in part in Schappacher and Scholz 1992: 30. It was written December 7, 1938.

[93] Schappacher and Scholz 1992: 9.

[94] Letter, Hans Wittich to Lars Ahlfors, Sept. 22, 1982. I thank William Abikoff for providing me with a copy of this letter.

the invasion of Norway; later, however, he was transferred to army headquarters in Berlin for cryptographic work—apparently Werner Weber was similarly employed.[95] As the publication record shows, even while in the army, Teichmüller did not give up mathematics. In academic year 1942–43, Bieberbach managed to get a partial release for him to give lectures at the university on "uniformization theory." However, with the defeat at Stalingrad in February 1943, Teichmüller gave up his relatively safe cryptological position to answer a new call to arms and entered a unit involved in the famous tank battle at Kursk. This began on July 4, 1943, but by mid-July the Germans needed to send reinforcements to the west because of the Sicilian landings, and the Russians had counterattacked. By early August, the Germans were falling back along most of the front. In August Teichmüller received a furlough home; his unit was in the vicinity of Kharkov, which was recaptured by the Russians on August 23. By the beginning of September, Teichmüller's unit had been surrounded and for the most part wiped out. In those same days, Teichmüller attempted to rejoin it, apparently in Poltava, southwest of Kharkov, and still east of the Dnieper. He was lost in the confused and bloody fighting of the German retreat sometime before the Russian advance reached the Dnieper on September 22.[96]

Teichmüller was a committed Nazi and a dedicated mathematician. He believed in "Deutsche Mathematik," yet his papers quote without comment mathematics first found by Jews. He could admire Landau as a mathematician while insisting on his unsuitability to teach elementary classes because he was a Jew. Hans Wittich reported seeing Teichmüller in Göttingen prior to his leaving for the Eastern front, when he was depressed and "unwilling to express his motives for volunteering for the front." Wittich believed that Teichmüller had changed his mind politically, and that his volunteering was a sort of self-imposed punishment for his former opinions. However, Wittich says this in the context of wondering how "an otherwise so sharp and critical thinker" could believe Nazi slogans.[97] That Teichmüller reformed his opinions seems extremely unlikely. Queries like the just cited one posed by Wittich are all too common, and all too unjustified. Nothing necessarily connects a person's brilliance in one area with a particular insight into politics. Wittich's attitude is mentioned because it seems important to reject it, even if he did know Teichmüller personally. Teichmüller *was* a gifted, brilliant, and seminal mathematician; he was also a dedicated Nazi. Nor is this some consequence of a mathematical "unworldliness" or "naïveté." Among many other well-known "brilliant" people who, for some time, albeit in quite different ways, openly promoted the Nazi cause were the ethologist Konrad Lorenz, the psychologist Carl Gustav Jung, the philosopher Martin Heidegger, the poet Gottfried Benn, the art historian Wilhelm Pinder, and the surgeon Ferdinand Sauerbruch. No academic profession was immune from having its Bieberbachs and Teichmüllers.

[95] Schappacher and Scholz 1992: 12, n. 36.

[96] Ibid.: 13–14.

[97] Letter, Wittich to Ahlfors, Sept. 22, 1982, as in note 94.

Ernst Witt

If Teichmüller was dedicated, his friend Ernst Witt, also an extremely gifted mathematician, was perhaps an instance of the young naive devotee of the Nazi cause. Horst Tietz had the impression that Witt was not serious about anything and was blind to everything outside mathematics. This was in 1939.[98] As was seen earlier, as a student at Göttingen, he had the naiveté to appear at Emmy Noether's private seminar in her home in an SA uniform, but, during the same period in the disturbances around Hasse's accession at Göttingen, adopted an attitude that, although pro-Nazi, nevertheless placed mathematical ability above politics (at least for "Aryans"). It was Hans Zassenhaus's opinion that Teichmüller converted Witt when both were students at Göttingen.[99]

In any case, like Teichmüller, Witt grew up far from any German metropolis. His parents were Chinese missionaries, and though he was born in 1911 on an island in the Baltic, he spent the first nine years of his life in China, and consequently spoke Chinese as well as German fluently. In 1920, his parents shipped him to Germany, where he ended up in Müllheim, a small town in Southern Baden, under the care of an uncle also in the missionary movement. Both his father and, later, a teacher in the "high school" he attended in Freiburg recognized Witt's mathematical gifts. The subject of his doctoral dissertation at Göttingen was suggested by Emmy Noether, but as she, by the time of his examination, had been dismissed, and in any case had never been allowed to be the principal official supervisor of a dissertation, that role was taken by Gustav Herglotz. Herglotz, together with Hermann Weyl and the physicist Robert Pohl, formed his examining committee. His "habilitation" was done with Hasse in 1936. In spring 1933, like so many others, Witt joined the Nazi party (no. 1,903,092) and the SA. Effectively in 1938, but officially on September 1, 1939, he succeeded to the chair of Emil Artin in Hamburg, Artin having emigrated because of his half-Jewish wife. Witt seems to have ceased SA activity after he went to Hamburg, though whether he ever formally resigned is unclear.[100] Witt never left the party—it has been said that the latter would have had heavy consequences, presumably the loss of his position. Though there were certainly any number of *Ordinarien* who were not party members, a newly appointed professor under the new Nazi dispensation would probably have had difficulty had he suddenly become apparently "less Nazi." The same *Dozentenführer* Blume who in 1936 found Teichmüller unfortunately too engaged in his mathematics to be a really active party comrade, in the following year found Witt a "comrade whom it is always good to see" in *Dozentenbund* meetings, and who

[98] Horst Tietz, interview, Apr. 5, 1988. Ernst Witt succeeded to Emil Artin's chair in mathematics at Hamburg in 1938.

[99] Hans Zassenhaus, interview, Nov. 17, 1987.

[100] According to an obituary by his student Ina Kersten, *JDMV* 95 (1993): 166–180, pp. 167–168, citing a statement of the astronomer Otto Heckmann. Cf. BDC file on Witt. Material on Witt not otherwise annotated can be found in Kersten's obituary.

had participated in an SA camp.[101] Nevertheless, thirty-one months earlier, *Kreisleiter* (roughly "cell leader") Thomas Gengler (by profession an astronomer) complained that Witt, despite being a party member, was "politically colorless" and had yet to understand that mathematics and natural science were also racially connected; he mistakenly thought them international.[102] This conforms with Witt's attitudes as seen earlier in the Hasse matter; yet clearly his activity in party organizations was independent of Teichmüller. In 1936 also, Witt would be supported in vain by F. K. Schmidt and Gustav Herglotz as Otto Toeplitz's successor at Bonn.[103] The naiveté that seemed to characterize Witt as a student in Göttingen and a young professor in Hamburg apparently persisted. Manfred Knebusch found Witt, with whom he studied after the war, truly "unpolitical," a "loner," and someone who didn't pay overly much attention to others' opinions, and so, for example, wore braids in his hair for a while.[104]

In 1941, Witt was drafted for the Russian campaign, but for reasons of health he was shipped back to Germany the following year, where he entered the cryptographic service—he and Teichmüller may have become reacquainted at this time. Witt remained there until the war's end, then returned to Hamburg, where the British promptly suspended him. In 1947, however, he was fully reinstated. Witt was beloved by students (and supervised thirteen Ph.D. dissertations—all but three postwar), but on his retirement in 1979 he withdrew from social engagement with his colleagues.

Ernst Witt seems to have actually suited a usual caricature of a mathematician—both heedless and ignorant of the world, somewhat naive, self-absorbed in his mathematical universe, truly unpolitical. This is also a caricature sometimes used to explain much academic reaction to the Nazis. In both cases it is almost always false. Witt is worth consideration because his life seems to show that the caricatures could, in fact, both be true.

Richard Courant

Witt may have been a naive "Aryan" as Nazi supporter and mathematician. However, many German Jewish mathematicians, like most successful German Jews, could not anticipate their loss of prestige and rank, let alone the holocaust to come. Were mathematicians like Richard Courant, Edmund Landau, and Felix Hausdorff naive? Courant has had an extremely readable book devoted to him, already cited several times. To discuss him in detail here would therefore be both otiose and inappropriate. Suffice to mention, though, that at the time of the planned Nazi boycott of Jewish stores on April 1, 1933, Courant thought the whole trouble Einstein's fault, and a recent speech of Hitler had made "a

[101] BDC file on Witt, Blume to the Göttingen branch of the NSDAP, Sept. 9, 1937.
[102] BDC file on Witt, Gengler memorandum dated Feb. 4, 1935.
[103] Personalakten Toeplitz, Bonn.
[104] Manfred Knebusch, interview, Feb. 25, 1988.

quite positive impression on him."[105] But if Courant was naive, it was the naiveté of a man who had been wounded as a front-line soldier during World War I, who had had an interest in socialism, and who was active in the Silesian plebiscite.[106] His service in World War I made Courant technically exempt from the April 7 law "for reform of the civil service." Thus, instead of being dismissed outright, he was legalistically put on a forced leave of absence on April 13. If Courant was naive, so was Justus Valentiner, who, as Kurator, was the Nazi education ministry's delegate in Göttingen during the early Nazi years,[107] since he fully supported the petitions to the ministry to reinstate Courant.[108] Courant originally had been interested in getting along with the government, and several petitions were organized in his favor. These included a letter from Hermann Weyl accompanying an explanatory letter from Courant detailing his front-line service, excusing his six-months-long membership in the Socialist party by his interest in its extreme right wing, and ending with his "readiness anytime to work for the national state without reservation" (a condition of future employment under the April 7 law), as well as a petition organized by Kurt Friedrichs and Otto Neugebauer, and a letter from Harald Bohr in Copenhagen, speaking also for his brother Niels and G. H. Hardy in England. In addition, David Hilbert wrote directly to the ministry; another direct communication was a brief letter from Hellmuth Kneser, Kurt Friedrichs, and Ludwig Prandtl, asking to be heard in the case. Kneser and Friedrichs had both been Courant's *Assistenten*, and originally a third such, Udo Wegner, was to have signed the letter, but he demurred. As seen in chapter 4, Wegner would become the Nazi students' candidate to succeed to Courant's chair.[109] At this time Kneser was very sympathetic to what he thought were Nazi aims, and was a friend of Theodor Vahlen (twenty-nine years his senior), who had become chief of the Nazi education ministry's scientific office.[110] Thus those sympathetic to Courant were by no means necessarily liberal; indeed, they included both Kneser and Blaschke. There were also those who refused to sign the Friedrichs-Neugebauer petition out of fear. One of the few honest, perceptive, and not "naive" people was Erich Hecke, who refused to sign the Friedrichs-Neugebauer petition because he said it would do no good—and it didn't.[111]

[105] Reid 1976: 139–140. For Courant, see also Schappacher 1987. The speech was probably Hitler's rather temperate address on March 23 in Potsdam (preparatory to the passage of the *Ermächtigungsgesetz*, or "Enabling Law").

[106] Reid 1976: 74, 85.

[107] Dahms, in Becker et al. 1987: 25.

[108] Schappacher 1987: 350–351.

[109] Ibid.: 350–351, and n. 61 on p. 364. Also Reid 1976: 149–152. The Friedrichs-Neugebauer petition had twenty-eight signatures. For the relation between Courant and Neugebauer, see Reid 1967: 108. For Kneser's attitude, ibid.: 151.

[110] For Kneser's attitudes at the time, HK, Kneser to Courant, May 20, 1933, and Reid 1976: 144. For friendship with Vahlen, HK, Kneser to Vahlen, Mar. 10, 1933, May 7, 1933, and other letters. Courant apparently attributed Kneser's political attitudes at this time to Vahlen's influence, see Reid 1976: 148. The letter cited by Reid (pp. 144 and 148) is the cited Kneser to Courant, May 20, 1933.

[111] Reid 1976: 152. A copy of the petition is in HK.

EDMUND LANDAU

What of Edmund Landau? Landau was the son of a well-to-do Berlin gynecologist (who invented a myomectomy operation). His mother was from the banking family of Jacoby, and Landau grew up in a Jacoby house amid other Berlin banking families. His father Leopold was both a nationalist patriot and someone politically engaged in Jewish issues—not a contradiction in the first decade of the century. In 1905 Landau married Marianne Ehrlich, daughter of Paul Ehrlich, who would share the 1908 Nobel Prize for Medicine; Ehrlich had been a fellow student with Landau's father. Thus Landau grew up a very well-connected and well-to-do person (he employed six servants in his household) who saw no contradiction between being Jewish and being German. According to Landau's son Matthias, he was the only ostensibly Jewish member of the Göttingen faculty who had some Jewish practice and a connection to a synagogue.[112] From some time in the 1920s, Landau added the Hebrew name Yechezkel to his given ones of Edmund Georg Hermann—in honor of his relationship to a famous Prague rabbi.[113] In 1925 he was invited to give a talk at the newly opened Hebrew University in Jerusalem; instead of giving such a talk in German, he quickly learned Hebrew with the aid of the mathematician Jacob Levitzki.[114] In 1927–28 he was guest professor for a year in Jerusalem.

Landau was not only wealthy, well connected, and a Jew with some small religious practice and Zionist leanings, he was also something of a prodigy. Legend has it that at age three, when his mother forgot her umbrella in a carriage, he replied, "It was number 354," and the umbrella was quickly reacquired. He achieved his doctorate in 1899 at age twenty-two; the next year he wrote Hilbert about his ideas for a proof of the prime ideal theorem for algebraic number fields, an outstanding unsolved problem at the time.[115] In 1901 he "habilitated" at Berlin with a paper on Dirichlet series. By 1909 he had published nearly seventy papers. In that year he went directly from *Privatdozent* at Berlin to Ordinarius at Göttingen. In that same year, his epoch-making book on prime number theory appeared. Formally both doctorate and *Habilitation* were with Georg Frobenius, who, however, underestimated his student repeatedly.[116]

Landau was also something of a cynical snob. The story is well known that he used to tell people who would ask for his address in Göttingen, "You'll find it easily; it's the most splendid house in the city."[117] Landau also apparently would only travel in trains with first-class compartments. He was an old friend of Fritz Rathenau, who worked in the Prussian interior ministry and was a cousin of the assassinated foreign minister Walther Rathenau. Is it any wonder

[112] Matthias Landau, interview, Düsseldorf, Jan. 26, 1988.

[113] Fraenkel 1967: 162.

[114] A reproduction (in Hebrew) is in Landau 1984, 8: 289–297.

[115] Reproduced in Kluge 1983: 12–26. This letter was written from Paris. Cf. also ibid., Landau to Hilbert, June 9, 1900, pp. 27–28. Landau would solve this problem along the sketched path.

[116] Biermann 1988: 163, 182–183, 328.

[117] Kluge 1983: 88. Reid 1970: 118.

that this wealthy, well-established, well-connected, brilliant man whose mathematical work had been more than once groundbreaking, and whose students included some of the best-known mathematicians of the next generation,[118] should have underestimated the Nazis? There is a story that reveals both Landau's cynicism and this underestimation. Once, on a visit in 1932, Fritz Rathenau told Landau that he didn't know whether the Nazis would come to power, but if they did, he had heard they planned to construct a concentration camp for Jews on the Lüneburg heath. Landau's response is supposed to have been, "In that case I should immediately reserve for myself a room with a balcony with a southern exposure."[119]

Teichmüller recognized Landau's genius and would have gladly studied advanced material from him (and Landau would have gladly taught him). Bieberbach saw the wealthy, brilliant, somewhat arrogant, and not entirely pleasant man with whom he had already had some mathematical differences, and whom he made a convenient archetype of "Jewish mathematics." Teichmüller had a leading role in expelling Landau from his position, Bieberbach in exploiting that expulsion. Landau died in retirement in Berlin, five days after his sixty-first birthday in 1938.

Felix Hausdorff

Neither Landau nor Courant saw the "transvaluation of all values" that the Nazi hegemony meant, at least for academics. Even more poignant is the story of Felix Hausdorff, without question one of the great mathematicians of the early twentieth century—a seminal codifier and creator in both set theory and general topology. Hausdorff's disastrous fate ironically may have stemmed in part from the fact that, unlike Courant or Landau, he did not suffer the immediate loss of his job. Like Courant and Landau, Hausdorff fell under the exceptions to the April 7 law; however, perhaps because Bonn was not Göttingen, either in mathematical prominence or in the rabid right-wing orientation of the town or student population, Hausdorff remained a full professor until he retired on grounds of age in March 1935.

Felix Hausdorff was born in Breslau (modern Wrocław) on November 8, 1868; however, when he was still young, his well-to-do family moved to Leipzig, where he grew up.[120] He showed such great musical ability as a youth that he wanted to be a composer, and only the persistent pressure of his father made him give this idea up. So he studied mathematics, achieving his doctorate

[118] Among others, Konrad Knopp, Dunham Jackson, Henry Blumberg, Paul Bernays, Erich Kamke, Gustav Doetsch, Carl Ludwig Siegel, Alexander Ostrowski, Arnold Walfisz, Werner Rogosinski, and Hans Heilbronn. There were thirty-three doctorands (including four in which Landau acted with [for] Emmy Noether) in all.

[119] Kluge 1983: 93. All otherwise unannotated biographical information above about Landau can be found in Kluge.

[120] Information about Hausdorff not otherwise footnoted comes from Magda Dierksmann et al., "Felix Hausdorff zum Gedächtnis," *JDMV* 69 (1967): 51–76, and the article "Felix Hausdorff," in *Dictionary of Scientific Biography* 1970: vol. 6, 176–177.

in 1891, and "habilitating" in 1895. It is perhaps surprising, since Hausdorff became famous for his work in set theory and topology, that both of these efforts were in astronomy, and Hausdorff's first four papers dealt with astronomical and optical matters. Actually, at the time, Hausdorff was developing a career as a litterateur. Under the pseudonym Paul Mongré, he published poems and books with a Nietzschean flavor—indeed, the first of these, *Sant' Ilario*, had 378 pages and appeared in 1897 with the subtitle, *Thoughts from Zarathustra's Country*. Thus the young "mathematician" Hausdorff was primarily interested in literature and philosophy and mixed in those circles. As the scion of a wealthy family, he did not have to worry about making a career as a mathematician; for him, mathematics, both as research and as a subject to teach, was more an avocation than anything else. However, around 1904, his literary production as Paul Mongré slacked off, though a satirical play he wrote in that year was produced with substantial success in 1912. Nevertheless, he seems around this time to have become primarily a mathematician. Hausdorff's most productive mathematical period began around 1909 (when he was forty-one). For all his seminal importance in set theory, one should not forget that he made distinctive and distinguished contributions to other areas of mathematics as well (e.g., Hausdorff measure, Hausdorff matrices, and the Baker-Campbell-Hausdorff formula in noncommutative algebra).

Hausdorff was also an extremely modest man—in fact, so modest, and so unworried about career, that, in 1902, comfortably esconced as an "associate professor" in his hometown of Leipzig, he turned down a comparable offer at Göttingen. Admittedly, Göttingen was not yet the premier center it would become when, within a few years, Klein and Hilbert would be joined by Minkowski and Runge, and the young Prandtl would begin his aerodynamical work (as a member of the technical physics faculty). It seems to have been the geometer Eduard Study who convinced Hausdorff to abandon Leipzig for Bonn in 1910, but after three years as an "associate professor," he left to be Ordinarius at Greifswald, returning to Bonn in 1921. One gets a feeling for Study's help by noting that at Greifswald, Hausdorff succeeded the algebraist Friedrich Engel, who himself had been Study's successor there. This apparent friendship between the two men throws additional light on Hausdorff's support for Ernst August Weiss, discussed earlier.[121] Of course, by the time he returned to Bonn, Hausdorff was a famous mathematician—his groundbreaking book *Foundations of Set Theory* had appeared in 1914, as had a number of significant papers.

In 1928 the Nazis had won twelve parliamentary seats, in 1930, 107. When Heinrich Brüning, who had been governing by emergency decree, resigned in June 1932, the Nazi vote the following month yielded 230 parliamentary seats, making them by far the largest single party in the Reichstag, though they did not have a majority. Thus it was small wonder that when Hausdorff's student Magda Dierksmann that same year said good-bye and promised to see him

[121] Above, chapter 5, "Mathematical Camps."

on his seventieth birthday in 1938, he replied: "By then everything will be different."[122]

Hausdorff had never denied his Jewish origins, nor had he ever opted for baptism. In a Nazi civil-service formulary that he apparently filled out in early 1935, he listed in his distinctive small hand his religion as *israelitisch* and his racial status as non-Aryan. His wife, the former Charlotte Goldschmidt, though likewise non-Aryan, had long before converted to Lutheranism, as had her sister. One day before his sixty-sixth birthday, Hausdorff took the new civil-service oath sworn to Adolf Hitler.[123] However, the following January, a new law was decreed enforcing the retirement of all civil-service faculty who had passed their sixty-fifth birthday, unless there were "supervening university interests." Accordingly, Hausdorff, who had continued teaching since he fell under the exceptions to the April 7 law, was retired in March 1935.[124] The situation for Jews in Germany grew progressively bleaker. Apparently in early 1939 Richard Courant received a "very touching" letter from the seventy-one-year-old Hausdorff inquiring about the possibility of a research fellowship;[125] but this attempt at emigration failed.

The only Bonn mathematician who maintained contact with Hausdorff after this forced emeritization, ostensibly on account of age, was Erich Bessel-Hagen.[126] The story of Hausdorff's last year appears in papers collected by Bessel-Hagen as well as in his personal correspondence.[127] In April 1941, he wrote Elisabeth Hagemann:[128]

> Things go tolerably well with the Hausdorffs, even if they can't escape from the vexation and the agitation over continual new anti-Semitic chicanery. The tax burden and the monetary subtractions that are imposed on them are so high that he can no longer live from his income alone and must use his savings (*Vermögen*); it's good that he still has these reserves. Besides, they have been compelled to give up a part of their house, whereby their space is very crowded. However, I am glad that there are more people who worry about the Hausdorffs, as I occasionally verify when during a visit I meet one or another. Recently I met, for example, a musician who had just played together

[122] Dierksmann et al. 1967: 54.

[123] Both oath and formulary in Personalakten Hausdorff, Universität Bonn.

[124] Personalakten Hausdorff, Universität Bonn, Rust to Kurator, Mar. 5, 1935.

[125] See Reinhard Siegmund-Schultze, *Mathematiker auf der Flucht vor Hitler*, Dokumente zur Geschichte der Mathematik no. 10 (Braunschweig and Wiesbaden: Friedr. Vieweg, 1998), 121. Hausdorff's letter is apparently not extant.

[126] *Nachlass* Zermelo, Universität Freiburg, Bessel-Hagen to Zermelo, Dec. 23, 1940; Hans Bonnet in Dierksmann et al. 1967: 76.

[127] The historian of mathematics Erwin Neuenschwander devotes a preprint of the *Fachbereich Mathematics*, Technische Hochschule Darmstadt, dated January 1992 (for the fiftieth anniversary of Hausdorff's death), to this: "Felix Hausdorffs letzte Lebensjahre nach Dokumenten aus dem Bessel-Hagen Nachlass." As Prof. Neuenschwander managed to seal Bessel-Hagen's *Nachlass* against the use of others, the material below is cited from this preprint. Material cited earlier in chapter 5 from Bessel-Hagen is from the original sources, as it was seen prior to Prof. Neuenschwander's action.

[128] Neuenschwander, ibid.: 5–6.

with Hausdorff. That is really lovely, that in this way some joy will be brought to them in the house.

However, less than four months later, he would write the same correspondent:

> I often had great anxiety (*Sorge*) about the Hausdorffs. Mrs. Hausdorff was for a long time seriously ill from an old ailment—I don't know what it is. Scarcely was she over the worst than there came the agitation about the intended internment of the Jews. Here the procedure was mad. In the early part of the year, old nuns were forcibly driven out of a cloister on the Kreuzberg; these poor old women who never harmed anyone and only carried on a retiring life devoted to their pious usages, and who naturally are completely estranged from the machinery (*Getriebe*) of the outer world. Now all the Jews still living in Bonn will be compulsorily interned in this stolen building; they must either auction their things, or place them for preservation in "faithful" hands.

In Hausdorff's case, the university ("for once," says Bessel-Hagen) behaved decently and stood up for Hausdorff, saying that he should be allowed to remain in his house.[129] Hausdorff's sister-in-law hoped to go to her daughter in America, but the American authorities made difficulties, and few neutral boats sailed the Atlantic in mid-war. In October, the Hausdorffs were forced to wear the "yellow star," among other indignities. Toward the end of 1941 they were threatened with deportation to Cologne, but as Bessel-Hagen realized, this would only be "a preliminary to deportation to Poland. And what one hears concerning the accommodation and treatment of Jews there is completely unimaginable."[130] By New Year's, however, the threat had vanished. In mid-January came a new order that the Hausdorffs were to be interned in Endenich (a suburb of Bonn); Bessel-Hagen learned of this during a visit on Thursday, January 22, and was somewhat surprised to hear Hausdorff say definitively that the family would no longer live to see the day when things got better. On the next Sunday, Hausdorff wrote the following letter to the Jewish lawyer Hans Wollstein:[131]

> Dear Friend Wollstein:
>
> By the time you receive these lines, we three will have solved the problem in another way—in the way from which you have continually attempted to dissuade us. Whenever we first may have overcome the difficulties of removal, the feelings of security that you have predicted for us totally do not wish to appear, on the contrary,
>
> Also Endenich is still perhaps not the end![132]

[129] Neuenschwander, ibid. The same was done for Alfred Philippson, an emerited Jewish geography professor in Bonn. Phillippson was eventually deported to Theresienstadt, but managed to survive through influential help. For Philippson, see also Hannah Arendt, *Eichmann in Jerusalem* (1994), 134.

[130] Neuenschwander (as in note 127), 7.

[131] First published ibid.: 12–13, together with a facsimile. A copy of the original is also in my possession.

[132] In German, this involves a pun: *Auch Endenich ist noch vielleicht das Ende nich(t)*. Hausdorff omits the *t* in *nicht* to make clear he is punning.

What has been done against the Jews in recent months arouses well-founded anxiety that we will no longer be allowed to experience a bearable situation.

Tell the Philippsons[133] whatever you think is good, along with thanks for their friendship (which, however, above all you deserve). Also give Herr Mayer[134] our heartfelt thanks for everything that he has done for us and if need be would have still done, we have sincerely wondered at his organizational achievements and successes, and, had we not that anxiety, would have gladly given ourselves over to his care which indeed would have brought along a feeling of relative security—unfortunately only a relative one.

With a will dated October 10, 1941, we have made our son-in-law, Dr. Arthur König, Reichardstieg 14, Jena, our heir. Help him insofar as you can, dear friend! Help also our live-in Minna Nickol or whoever else asks you; our thanks must we take with us to the grave. Perhaps the furniture, books, etc. can still remain in the house beyond January 29 (our removal date); perhaps also Frau Nickol can also still remain in order to wind up the running obligations (bills for city services, etc.)—tax records, bank correspondence and the like that Arthur needs are in my study.

If it is possible we wish to be cremated, and enclose for you three declarations to this effect. If not, then either Herr Mayer or Herr Goldschmidt must arrange what is necessary.* We will take care of defrayal of costs insofar as possible; besides my wife is a member of a Lutheran burial fund—the documents are in her bedroom. What monetarily still is lacking to cover costs, either our heir or Nora[135] will undertake.

Forgive us, that we still cause you trouble beyond death; I am convinced that you will do what you *are able* to do (and which perhaps is not very much). Forgive us also our desertion! We wish you and all our friends will experience better times.

Yours faithfully,
Felix Hausdorff

* My wife and sister-in-law, however, are of the Lutheran religion.

That evening they took lethal doses of barbiturates (veronal); in the morning Hausdorff and his wife were dead; her sister lingered for a few days in a coma. After some struggle, a place was found in the Poppelsdorf cemetery for the urns with their ashes.[136] In 1948, Hindenburgstrasse, on which the Hausdorffs had lived, was renamed Hausdorffstrasse.

A note is in order about the succession to Hausdorff's chair. Following his enforced retirement, the faculty's first list of suggestions on May 31, 1935, was Helmut Hasse; Eberhard Hopf (still on leave from Berlin at MIT) and F. K. Schmidt (tied in second place); and Emanuel Sperner and Erich Kamke (tied for third). This was clearly an impossible list of fine mathematicians, none of whom (except Sperner) had at the time pro-Nazi ideological connections: Hasse had just gone to Göttingen, Hopf was in America and had already turned down a

[133] See note 129.

[134] Dr. Siegmund Mayer was a lawyer. He died in Auschwitz (Neuenschwander, as in note 127, n. 14).

[135] I.e., Hausdorff's daughter Lenore, married to Arthur König.

[136] Neuenschwander, as in note 127, p. 9, Erich Bessel-Hagen to Elisabeth Hagemann, Mar. 26, 1942.

call from Heidelberg (though he would later return to Leipzig), and Schmidt had just gone (from substituting for Emmy Noether in Göttingen) to Jena. Sperner had been a foreign group leader for the NSDAP in Peking in 1933–34 and had just gone to Königsberg, while Kamke was the oldest of the five at forty-five, and was known as a non-Nazi (the faculty's recommendation avoided mentioning his wife).[137]

On December 3, Rudolf Mentzel at the DFG suggested Erhard Tornier's name—the first truly ideological suggestion.[138]

On December 18, Hans Beck, a good Nazi, and the only mathematics Ordinarius remaining in Bonn (Otto Toeplitz had been forcibly placed on leave in 1935), traveled to Berlin and was told that the first four on the above list were impossible for the reasons already indicated (it is unclear whether Kamke was discussed).

The university was asked to consider the Hausdorff and Toeplitz successions together; on February 26, 1936, they submitted the following names for the Hausdorff succession: Ernst Kähler, Konrad Knopp, and Karl Dörge; for the Toeplitz succession: Eberhard Hopf (again), Egon Ullrich, and Kamke (again). Kamke was described accurately as "a man who stands fast in storm and bad weather, . . . a war veteran, a Lutheran, married, and with three children." The faculty rejected the possibility of appointing the resident E. A. Weiss or Fritz Rehbock on the grounds that they were geometers; there already was a geometric Ordinarius (Beck), and what was needed were two analysts.

A month earlier, the then-Dekan of the philosophical faculty had written a letter to the ministry rejecting the Nazi Tornier as an appointment at Bonn on grounds of the narrowness of his research, and as someone who, at forty-two, was unlikely to broaden his interests; furthermore, all evidence pointed to his being a poor teacher, which Bonn could ill afford. It is unclear whether there was malice in Dekan Oertel's mentioning that Tornier had collaborated with Willy Feller (who was Jewish), or in saying (given the atmosphere):[139]

> He [Tornier] conceives, as does Hausdorff, of probability as an additive set function, and specializes this in a very particular way. From this he develops then a mathematical-logical theory, which is spoken of with appreciation. That can also not alter the fact that certain not-insignificant particular points originate, as Tornier himself admits, with Kaluza and Feller.

These lists (and other suggestions) also led to nothing, but both lists and the letter about Tornier make it clear that in 1935–36, the Bonn faculty was attempting to keep the ideological wolf as far from the door as possible, while

[137] Personalakten Hausdorff, Universität Bonn, Dekan Rothacker to education ministry via Rektor and Kurator, Bonn, May 31, 1935.

[138] Personalakten Hausdorff, Mentzel to education ministry, Dec. 3, 1935; copied to Dekan Oertel, and by him to the chemist Paul Pfeiffer (who was a sort of subdean of the mathematics and natural sciences division of the philosophical faculty).

[139] Oertel to ministry, Jan. 23, 1936, in Personalakten Hausdorff, Universität Bonn.

obtaining solid analysts for the faculty.[140] In the end, the Hausdorff chair was converted into that of an associate professor, and filled from 1937 on by Ernst Peschl (who had only "habilitated" in 1935), while Toeplitz was not succeeded until 1939, and then by the well-known algebraist Wolfgang Krull.

Ernst Peschl

Ernst Peschl is a good example of the young mathematician who, in order to begin a career, was forced into at least a nominal political stance in which he apparently did not believe. Born in 1906, he obtained his doctorate under Carathéodory at Munich in 1931, and then went for two years to Jena to work with Robert König, followed by eighteen months with Behnke at Münster; finally "habilitating" in 1935 back at Jena, where he remained König's *Assistent.* In 1937 he went to Bonn, and in 1941 was called up to be an army interpreter (Peschl spoke fluent French). Peschl had been a leader in a Roman Catholic youth group and was in Jena (in Thuringia, which "went Nazi" before Hitler's appointment as chancellor). His youth organization was dissolved by the Gestapo, just as he was about to "habilitate"; as a consequence, he yielded to the increasing pressure and joined both the SA and the NSDAP; however, he avoided all activity in either organization and did not hold any office. After about a year, he was able to "shake off" service in the SA. Among others who could testify to his inner anti-Nazi attitudes, he cited Behnke and Max Pinl.

From June 1934, he no longer did any SA service. As an *Assistent*, he willy-nilly became a compulsory member of the *Dozentenschaft* and the NSLB[141] without any action on his part. In fact, Peschl resisted attempts to get him to become an individual instead of a corporative member of the NSLB, and around 1938 managed to drop out of it entirely by not paying his corporative dues. He refused to fill out various formularies for membership in the *Dozentenschaft*, despite warnings. Thus Peschl was the young Catholic, former activist, who determined to make a career in Germany believing, he says, that the fanaticism of the Nazis would be tempered by the realities of governing. Thus he suppressed his beliefs and "went along," but only as far as was *pro forma* necessary.[142] In this he might be contrasted with the former Catholic activist Peter Thullen, who decided to leave Germany, and the considerably older former Catholic activist Gustav Doetsch, who became a "110% Nazi."[143] Indeed, Doetsch and Peschl had something of a run-in.

[140] A combined list for succession to both Hausdorff and Toeplitz is in Personalakten Toeplitz, Universität Bonn, Pfeiffer to Oertel, Oct. 26, 1935. See also in the Personalakten Toeplitz, Oertel to ministry, Mar. 18, 1936; Bachér (at ministry) to Kurator, Bonn, Apr. 15, 1936; and Oertel to Kurator, May 25, 1936.

[141] *Nationalsozialistischer Lehrerbund*, the Union of National Socialist Teachers.

[142] The above information is from Personalakten Peschl, Universität Bonn, Peschl's *Lebenslauf*, prepared Oct. 18, 1946, and *Fragebogen* replies, prepared May 25, 1946, for Allied authorities.

[143] Thullen had completed his studies. See above, chapter 5, "Students and Faculty before and during Wartime." For Doetsch, see above, chapter 4, "The Süss Book Project."

Peschl worked in the Gröbner-Doetsch industrial mathematics institute[144] (which exempted him from military service) from March 1943 until March 1945, when he was let go without prior notice nor explanation. This made him immediately subject to the draft, and Doetsch tried to get him drafted; however, the Osenberg organization[145] came to his rescue to prevent that calamity, and he spent the next weeks, until the occupation of Braunschweig by the Allies on April 11, working at the technical university there. Peschl apparently had, at least in the latter part of his service at the institute, offered passive resistance to Doetsch's rigid militaristic conceptions (recall that Doetsch was a major as well as a professor). Consequently, as Peschl learned from the Osenberg group, Doetsch had had him fired (on March 29, 1945) because his "attitude towards work did not correspond to the necessities of the war at the moment."[146]

Paul Riebesell

Paul Riebesell seems to be an example of the nonacademic mathematician, for whom carrying out adjunct academic duties honestly proved deleterious under the Nazis, but whose establishment position during the Nazi period and early party membership nevertheless threatened him after the Allied victory in World War II. Indeed, it was hard to tell in many cases who was a true believer in the Nazi faith and who was not. In the case of Riebesell, a mathematical statistician and expert in actuarial mathematics, an action he took in 1937 described below brought him into sufficient disfavor with the party that he was given a severe reprimand and dismissed from his then-prominent position. In 1937 such an action could only have been taken as a matter of principle.

Paul Riebesell was born in 1883, studied in Munich and Kiel, and in 1909 became a secondary-school teacher.[147] He served in World War I, and from 1917 until 1922 was a director of the public youth welfare organization in Hamburg. Though his early work had been in mathematical physics (ranging from direct current generators to relativity theory), he became interested in statistics, and with the establishment of a university in Hamburg in 1919, applied for a paid adjunct position in "practical mathematics." The faculty had difficulty approving this, presumably because Riebesell had yet to "habilitate."[148] This he did "without the usual formalities"[149] and on February 14, 1920, gave his public inaugural lecture, "The Importance of Mathematical Statistics for the Sciences." Thereafter he gave lectures in statistics and actuarial mathematics at

[144] For Doetsch and Gröbner's institute, see above as cited in note 143.

[145] Above, chapter 5, "The Wartime Drafting of Scientists."

[146] Personalakten Peschl, Universität Bonn, Peschl replies to *Fragebogen*.

[147] Material not otherwise footnoted on Riebesell and cited below comes from his Personalakten in the Hamburg Staatsarchiv.

[148] Ibid., Riebesell to Senate (of university), Dec. 14, 1919, and Dekan Rabe to Senate, Jan. 19, 1920.

[149] Scharlau et al. 1990: 147.

the university. In addition to academics, Riebesell was also active in Hamburg insurance circles and in 1923 became director of the municipal fire insurance. In thanks for his rebuilding and expanding it after the 1923 inflation, in 1931 he was named president, partly in honor of twenty-five years of municipal service. Not quite fifty when Hitler came to power, Riebesell joined the Nazi party on April 27, 1933 (counted as of May 1); he later said on the Allied *Fragebogen* (denazification questionnaire) that this was to insure against his removal from this post. In October he published a well-received "Handbook of Insurance," and on the twenty-sixth of that month wrote an article for a Hamburg newspaper entitled "Mathematics and Natural Science in the Service of National Education." This underscored the need to support the National Socialist ideology, emphasizing that the best educational values were the eternal ones of German culture and science. It cited Kant on science being true insofar as it was mathematical, and found in mathematical education the same national values Weiss would—indeed, emphasized irresponsible facile intellect as the opposite of the "strict mathematical spirit," while denying that mathematics led to materialism. Along the way, the value of mathematics for racial investigations and the theory of genetic inheritance was mentioned. There was also the usual Germanic name-dropping. Thus far, Riebesell seems like a middle-aged man of some position, at least protecting himself through ideological lip-service, and possibly possessed of a more fundamental ideological belief. And the results were positive. In 1934 Riebesell was called to Berlin to be president, and thus *Führer*, of the National Union of Public Statutory Insurance Companies.[150]

However, in 1937 Riebesell did a small thing that revealed his true attitudes, at least by then. A Jewish author in Vienna published a book on life insurance. Riebesell praised the book, and this praise was used by the publisher in advertising the book. In fact, Riebesell said that a book of this sort had not previously existed in the German language, and that the prior German-language texts were outmoded. But this was not all. Riebesell also wrote German insurance journals to say:

> I gave this judgment to the author who came to me in my capacity as a university teacher and through the mediation of an irreproachable Aryan personality with the request for a scientific evaluation of his book, with the express indication thereupon given that under no circumstances would I recommend it to a publisher in, or the circulation of the book in, Germany. I have communicated exactly this to two German publishers when they asked.

On October 21, 1937, an article in *Schwarze Korps*, the official SS journal, denounced Riebesell for these errors, saying that had he but known who the author and publisher (identified also as Jewish) were, he surely would not have provided the encomium he did. The statement to professional journals only compounded his error,[151] in the article's view. A denunciation in *Schwarze Korps*

[150] Including Fire, Life, Hail, Accident, and Liability companies.

[151] The German word used is *Panne*, which is the word for an auto breakdown or tire blowout.

was not to be taken lightly, and on December 21, 1937, a chamber of the NSDAP Supreme Court decided that party member no. 3,030,498, Paul Riebesell, had acted contrary to the efforts of the party, and he was punished "with a warning." Werner Zschintzsch, writing for the education ministry, which also examined the affair, told Riebesell:

> By your evaluation, praising the scientific work of a Jew and through the characterization of the work as worthy of circulation while simultaneously declaring the impossibility of a publication of the book in Germany because of Germany's fundamental attitude toward the Jewish question, you have demonstrated an attitude (*Auffassung*) that I must most sharply censure, given your position as a university instructor.[152] At least as negligence I must further lay to your charge that the Jewish author and his publishing house made use of your declaration for advertising purposes, and have thereby damaged the reputation of the German government abroad. With respect to your behavior otherwise, I plainly am convinced that you have not consciously acted against German interests, but your mode of procedure is to be explained by a lack of insight into the Jewish problem as a whole (*an sich*) and the questions connected with it.
>
> I declare to you the serious disapproval [of the ministry].

What "warning" and "censure" and "disapproval" meant for Riebesell was a loss of all his national public positions, but maintenance of his party membership. He left Berlin and went to Munich, where he became the director of a private life insurance company. In 1941 he planned to publish a little book on mathematics in daily life, but, at the insistence of his publisher, needed to obtain a release from membership in and regulation by the *Reichsschrifttumskammer* (the Nazi body that regulated professional writers). This went smoothly, and in 1942 and 1944 Riebesell published two small works (of sixty-one and fifty-one pages, respectively).[153]

What leads to the thought that Riebesell was acting on principle, rather than merely being naive, as Zschintzsch thought, are the circumstances of his actually obtaining his national role. It seems as though party member Riebesell was elected to his national role by a unanimous vote (as the Nazis would have desired) in order to forestall the assumption of that position by Hans Goebbels (brother of Joseph) or the local *Gauleiter* Schwede-Coburg. Indeed, after Riebesell's dismissal, Schwede-Coburg called him an enemy of the state and sharply criticized a colleague who had written Riebesell a friendly letter of farewell.[154] The story of the "Aryan mediator" was perhaps invented by Riebesell in order to give him a plausible excuse for his positive criticism; furthermore, in 1937 Austrians were not yet German. In fact, as national superintendent of insurance, Riebesell and some colleagues had prevented gross excesses desired

[152] Riebesell had transferred his adjunct position at Hamburg to a similar one in Berlin. The Jewish author was named Werner Levi; whether he was a former pupil of Riebesell is unclear.

[153] BDC file for Riebesell.

[154] Personalakten Riebesell, Hamburg Staatsarchiv, Hans Otto Schmitt to Allied authorities, Oct. 26, 1945; Alfred Neuschler to Riebesell, July 1, 1946; Fehrmann to Riebesell, Nov. 29, 1945.

by Nazi ideologues, such as the insurance of Nazi party premises without the payment of a premium. The private insurance company in Munich to which he went from Berlin seems to have been the only one in Munich that did not receive a medal for excellent Nazi party performance (less than 10 percent of its employees belonged to the party). What seems likely is that Riebesell, despite some care (such as refusing a recommendation to a German publisher), overreached himself in the matter of the insurance text, and it was used by the likes of Schwede-Coburg to launch the anonymous attack in the *Schwarze Korps* leading to Riebesell's dismissal.

With the Allied victory, party member Riebesell was arrested as a "leading industrialist." In defense of his real attitudes (formally in answer to questions on the Allied *Fragebogen*), Riebesell obtained numerous letters, all full of praise for his quiet professional actions against Nazi ideological domination of his profession, including those mentioned above. Two were from former concentration-camp internees. One was from a communist to whom Riebesell gave a job while ignoring (unlike others) his political views, which he knew about. When this man was drafted, Riebesell continued his salary in support of his family.

Paul Riebesell seems to have been a man of conservative nationalist attitudes, initially somewhat attracted to Hitler, but also a personally decent man who became disillusioned with the political leadership and tried to prevent ideological idiocy from infecting his profession. He was certainly not the only such. On June 27, 1946, the military authorities found Riebesell unburdened by his activities during the Nazi era, and on July 6 he was released from house arrest; in August he was reinstated in his old Hamburg position. In 1948 he was reappointed at Hamburg to an adjunct professor's position for actuarial mathematics. He died suddenly on March 16, 1950, mourned by family and colleagues alike.

Helmut Ulm and Alfred Stöhr

For a man like Riebesell, already of some position and prestige, remote associations with Jews could lead to denigration and dismissal. For younger people they could be even more disastrous. Helmut Ulm, perhaps best known today for a theorem in the theory of infinite groups, considered himself an applied mathematician who also worked at the intersection of algebra and topology.[155] Ulm had studied with Otto Toeplitz and Felix Hausdorff in Bonn, and his "interest in algebra was awakened" by Toeplitz, who arranged for Ulm to enter Emmy Noether's circle at Göttingen. He also became an assistant to Richard Courant. Thus, *all* the mathematicians influential on his career were Jews, and they not only were teachers but became friends. Furthermore, Ulm did not make a secret of this, nor did he make National Socialist pronouncements. Thus

[155] Material on Helmut Ulm not otherwise footnoted is from Personalakten Ulm, Universität Münster.

he could not "habilitate" in the Göttingen of Hasse and Tornier, and in 1935 went to Behnke in Münster. But the political difficulties pursued him, and he did not successfully "habilitate" until 1945, fifteen years after his doctorate. Nevertheless, he was successful at Münster as a low-paid *Dozent*, and even (given his applied interests) managed to be seconded to the Foreign Office during the war, which gave him the enviable "indispensable" status, preserving him from the threat of the draft.

Similarly, in Berlin, Alfred Stöhr, as a young student who had not yet been awarded his doctorate, suffered career misfortunes attendant on associating with Jews. Stöhr's dissertation was approved by Werner Weber and Erhard Schmidt. However, Issai Schur was also a strong influence on him.[156] Stöhr's oral defense of the dissertation took place on November 23, 1938, and he was awarded his doctorate on January 25, 1939. According to a letter from Carl Ludwig Siegel to Hecke dated November 15, 1938, Stöhr had gotten into political trouble in Berlin because he had occasionally spoken to the mathematician Robert Remak and had made two visits to Schur after he was finally forcibly emerited in 1935.[157] There are also letters from Hasse to Hecke about Stöhr in which Hasse explains how Stöhr's political trouble at Berlin made it impossible to hire him as an *Assistent* at Göttingen.[158] Needless to say, Hasse and Siegel differed sharply on this. In the eventuality, through Siegel's mediation, Stöhr went to work with Hecke at the *Annalen*. This seems rather like Ulm finding a haven with Behnke. Yet Erich Bessel-Hagen apparently did not suffer at Bonn because of his visits to the Hausdorffs. The reason for this difference may be (if there is a reason to be found other than the local variability of Nazi practice) that Berlin and Göttingen were so prominent mathematically that more attention was paid to the politics of probationers (and being not so junior may also have shielded Bessel-Hagen somewhat). Furthermore, Bieberbach at Berlin and Hasse at Göttingen were both inclined (though perhaps for somewhat different reasons) to exercise such political judgments.

The situation of Stöhr is also interesting because it is mentioned by Oswald Teichmüller in a letter dated December 7, 1938, to his close friend Adolf Bruns in Göttingen.[159]

> However, to return to more serious things: you indeed have had Stöhr there a while.[160] That he did not become a junior assistant[161] there pleases me: indeed there is no basis

[156] See the obituary by Erich Härtter in *JDMV* 83 (1981): 159–168. In 1956 Stöhr wrote a paper "dedicated to the memory of my honored teacher I. Schur." Another of Stöhr's teachers cited by Härtter was (somewhat ironically) Werner Weber.

[157] Schur was originally (technically illegally) dismissed in 1933, but efforts, primarily by Erhard Schmidt, kept him in office until 1935.

[158] *Nachlass* Hecke, Universität Hamburg, Hasse to Hecke, Nov. 19, 1938. For Hasse's attitudes, see above, chapter 4. I am indebted to Prof. E. Bredendiek for copies of Siegel's and Hasse's letters.

[159] Partially reproduced in Schappacher and Scholz 1992: 30–32.

[160] Stöhr was originally scheduled to begin at Göttingen on November 1, if his doctorate were completed by mid-November as planned. See *Nachlass* Hecke, Universität Hamburg, Hasse to Hecke, Nov. 19, 1938.

[161] *Hilfsassistent*. Stöhr made a trip to Göttingen in vain for November 1, and as a consequence of

> for hatred of Stöhr, however the development of the affair ought to prove that with respect to your institute some optimism is warranted, that Göttingen does not allow itself to be degraded to Berlin's garbage pail. For a brief while, indeed as a result of the brisk exchange, almost a line of combat was introduced: it seemed that the inner unity of both institutes would be bought with hostility against one another. This separation was, however, thoroughly undesirable, it would have come to simple irrelevant rivalry without any deeper sense. Now Stöhr, whose mathematical ability is certain, did not become a junior assistant in Göttingen, even though he was rejected by Berlin, anticipatorially indeed for that reason, or respectively on the same grounds.

Thus, for Teichmüller, Stöhr's situation was nothing but a political symbol; even when mathematical ability was certain, political considerations should rule, and not least because of the opportunity for a unity on political grounds of Berlin and Göttingen. While it is unclear how much Teichmüller knew about the deep roots of this antagonism,[162] he certainly saw it being bypassed by political means.

Ernst Zermelo

Ernst Zermelo provides another case of what in Nazi times was reckoned a bold act with damaging results. While Zermelo is known to most mathematicians for his work in set theory and mathematical logic, his doctoral dissertation at Berlin was devoted to the calculus of variations and was written under H. A. Schwarz's supervision;[163] his *Habilitationsschrift* at Göttingen five years later in 1899 was a contribution to kinetic gas theory ("Hydrodynamic Investigations of Vortices [*Wirbelbewegungen*] on the Surface of a Sphere"). In 1901 he wrote his first paper in set theory, and in 1904 and 1908 the famous studies in set theory and mathematical logic briefly discussed earlier. But he never lost his interest in calculus of variations and similar areas, publishing relevant papers as late as 1930 and 1931, when most of his work had been of a quite different sort.[164]

Despite this variety, he considered David Hilbert "his first and only teacher in science."[165] Thus he was a mathematician of considerable breadth. He was also, to judge from his surviving correspondence and tales about him, a man of considerable humor, often sardonic. He seems to have inspired people to friendship. Among well-known stories about him are that once, when asked the origin of his rather unusual German name, he said: "It was really 'Walzer-

---

his rejection by Hasse, Siegel felt obligated to reimburse Stöhr's travel costs from his own pocket. See *Nachlass* Hecke, Universität Hamburg, Siegel to Hasse, Nov. 15, 1938.

[162] See above, chapter 7, and Biermann 1988: passim. The disagreements between Hasse and Bieberbach, e.g., at Bad Pyrmont in 1934 (above, chapter 6, "The Bieberbach-Bohr Exchange and the 1934 meeting of the DMV"), no doubt continued this "tradition."

[163] Biermann 1988: 161.

[164] Pinl 1969: 221–222.

[165] Zermelo *Nachlass*, Universität Freiburg (hereafter cited as ZN). This is the source of unannotated material on Zermelo below. The remark about Hilbert is in Ernst Zermelo to Richard Courant, Feb. 4, 1932.

melodic' [waltz tune], but they had to drop the first and last syllables." Another was his opinion (expressed around 1905) that it would be impossible to reach the North Pole because the amount of whiskey necessary to reach a certain latitude was proportional to the tangent to the latitude and hence approached infinity as one got nearer and nearer the pole. Similarly, his student and friend Arnold Scholz always addressed him as "Dear Zero," and other correspondents remember the pleasure of his humor. In 1910 Zermelo went as Ordinarius to Zürich, taking the young Bieberbach with him. However, in 1916 he had to be emerited (at age forty-five) because of a lung ailment.[166] He retired to the Black Forest, where he spent the next ten years as a private teacher; but in 1926 he was made an "honorary professor" at the University of Freiburg (which allowed him to teach classes as he wished). In 1935 Zermelo (then sixty-three) refused to give the "German greeting" (Hitler salute) and was reported to have made disparaging remarks about Hitler. The occasion was the return of the Saar to Germany on January 15. The denunciant was Eugen Schlotter, the same *Assistent* who made trouble over Heinrich Kapferer's failure to use "Heil Hitler" as a morning greeting. At the time, Schlotter was Gustav Doetsch's *Assistent*. A formal university disciplinary procedure was initiated, aimed at expelling Zermelo from the university and forbidding him from ever again teaching. At the hearing, Doetsch declared that Zermelo's continued refusal to give the Hitler salute was, he thought, "dangerous for young people."[167] To forestall this, on March 2, 1935, Zermelo renounced all further teaching activity at Freiburg in a letter to the Dekan. By March 13 he still had not had any response, and he wrote again to the Rektor redeclaring his decision and suggesting plausible excuses why he had not yet had a reply. The close of the letter reads, "With all due respect" (*In geziemender Hochachtung*). Zermelo never published after 1935, though he seemed to have maintained a lively correspondence with mathematicians at all levels. His Freiburg colleague Wilhelm Süss in particular seems to have remained friendly after 1935, inviting him, for example, to colloquia, though on one occasion he sent a letter explaining his disinvitation:[168]

> On next Tuesday a lecture will take place in [our] mathematical society. Since Herr Doetsch is giving the talk, I have not had an invitation sent you. On the one hand, I did not believe you would gladly participate. On the other, I wish at present to avoid everything that could cause a quarrel within the society and could have as a conse-

[166] This is according to Pinl (1969: 221–222) and rings truer than either Fraenkel 1967: 149, where Zermelo is accused of having made a sarcastic joke at the expense of Switzerland, or the persistent gossip in the mathematical community about putative mental illness.

[167] See above, chapter 5, "*Dozentenschaft* Reports". The papers referring to this event have been collected from the Personalakten Zermelo in the Universitätsarchiv Freiburg under the title "Ernst Zermelo Diffamierung in Freiburg 1935." This includes statements by Schlotter, Doetsch, Süss, Heffter, and others. I am grateful to Volker Peckhaus for providing me with a copy of this collection. See also HK, Süss to Kneser, Oct. 9, 1945. This last is handwritten and clearly in haste. Here the date of Zermelo's trouble is erroneously given as 1934.

[168] ZN, Süss to Zermelo, July 11, 1936.

> quence its falling asunder. I hope therefore to have met your understanding if I therefore ask you this time to stay away from the presentation.

This note continues in a very friendly fashion.

As seen in chapter 6, Süss and Doetsch had been quite friendly a year previous and associated together with extreme nationalist attitudes. It is tempting to speculate that the denunciation of Zermelo had something to do with the break between the two men.

Thus, Ernst Zermelo, arguably the founder of axiomatic set theory, and a member of the "older generation," seems to have felt about the Nazis the way Bieberbach felt about axiomatics. However, it would be a mistake to categorize mathematical logic in general in any such way. There was, for example, the curious relationship between Heinrich Scholz, the doyen of a "German school of logic," and Bieberbach and Mentzel's DFG. But Scholz seems to have been initially a conservative nationalist who came to reject the Nazis, even if that could only be expressed tacitly.

Gerhard Gentzen

Quite different was one of the authors in the Scholz series of monographs, perhaps the most famous one: Gerhard Gentzen.[169] Gentzen was born in 1909 in Greifswald, though he spent his youth on the somewhat isolated Baltic island of Rügen. He took his doctorate under Hermann Weyl's supervision in 1933 with a dissertation in mathematical logic. The following year he became Hilbert's *Assistent* at Göttingen. Gentzen joined the SA on June 14, 1938.[170] In September 1939 he was drafted, but his *Habilitationsschrift* was apparently sufficiently far along that he "habilitated" in February 1940, though he did not receive the *venia legendi*, or right to teach, at that time, presumably because he was a soldier.[171] He was a wireless operator in an air intelligence unit (*Luftnachrichtenregiment*). The frictions of his duties apparently brought on a nervous collapse, and he was released from the army as unable to function in June 1942. It appears that he had already had a nervous breakdown prior to his military service.[172] After some time in a sanatorium, Gentzen was able to function as mathematician and lecturer, and in February 1943 received the right to be a

[169] Material on Gentzen not otherwise cited can be found in either the BDC file for Gentzen or Manfred Szabo's introduction to Gentzen's *Collected Papers* (Amsterdam: North Holland, 1969).

[170] M. Pinl, "Kollegen in einer dunklen Zeit IV," *JDMV* 75 (1973–74): 173–174.

[171] A copy of the referees' judgments, dated Feb. 9, 1943, is in Gentzen's BDC file. The referees were Wilhelm Ackermann and Heinrich Scholz, experts in the field, but not at Göttingen. Kaluza, Herglotz, and Hasse assented. The paper was published in *Mathematische Annalen* 119 (1943): 149–161. It might be noted that in 1940, in writing their praise of Gentzen's work, neither Ackermann nor Scholz avoids the name of a Jewish emigré like Bernays.

[172] HK, Gentzen to Kneser, Aug. 28, 1937; Gentzen to Kneser, Apr. 23, 1942; Gentzen to Kneser, May 11, 1941; Kneser to Gentzen, May 14, 1941. Hellmuth Kneser befriended Gentzen when he was a student in Greifswald in 1928–29. The Kneser *Nachlass* also contains correspondence between Kneser and Szabo about Gentzen.

university teacher.[173] In May of that year he gave a trial lecture in Prague, choosing Kepler's laws and their relation to both the Ptolemaic theory of epicycles and Newton's gravitational theory as a subject.[174] These lectures apparently showed "great didactic skill." He was appointed at Prague on October 5, 1943, though his pay was still coming through his previous assistantship at Göttingen. In November a paid position for Gentzen at Prague was requested. By the time this made its way up through the hierarchy (Dekan, Rektor, Kurator, the Reichsprotektor of Bohemia and Moravia, the education ministry in Berlin), it was February 1944, but the official appointment was made.[175] Gentzen taught at Prague from autumn 1943 through April 1945.

As the Russians advanced on Prague, against the recommendation of Maximilian Pinl, Gentzen refused to give up his university position voluntarily. He was placed by Czech partisans in a forced labor camp along with all other university personnel, and he died there on August 4, 1945. One of his fellow prisoners apparently tried to tell their guards that Gentzen was a world-famous scholar with American and Russian letters on his person, but this provoked no interest. The labor camp conditions were the usual ones of unimaginable overcrowding and frequent beating. After a few weeks the prisoners were taken to do forced labor in the city, and therewith acquired some extra food.

> After a week's work, a hysterical Czech female threw a paving stone on Dr. Gentzen's hand, which squashed two fingers. As a consequence he could no longer go to work. In the beginning I [Dr. Franz Krammer] could bring him some bread from my work place, but later we were strictly searched . . . we were soon buggy and full of lice. The worst, however, was the inconceivable hunger. I had never believed that hunger could be so painful. Dr. G[entzen], who, unlike us others, did not have the slight improvements during work, was the first in [our] cell to succumb to this hunger . . . on August 4. . . . Dr. Gentzen did not die of typhus, but was truly starved.[176]

Gentzen's political attitudes are extremely unclear. Was his publishing a purely mathematical paper in *Deutsche Mathematik* an act of solidarity or an act of convenience?[177] The logician Georg Kreisel hinted that colleagues thought him a Nazi sympathizer.[178] His refusal to flee Prague in the face of the Russian advance may be evidence of this, or may just reflect a conviction that, with his

[173] BDC file for Gentzen, Hans Rohrbach (as director of the mathematical institute in Prague) to ministry of education, Nov. 13, 1943.

[174] Prague was where Kepler, working as court astronomer to the Holy Roman emperor Rudolph II, devised his laws of planetary motion (1609, 1619); thus this was an almost ideal subject for a German lecturer.

[175] BDC file for Gentzen. In November 1943, Ernst Mohr had been drafted (for Mohr, see chapter 4, "Hasse's Appointment at Göttingen"), and the institute director, Hans Rohrbach, had had his own "indispensable" status lifted, with the result that he could teach only every second week.

[176] Pinl 1973–74, for Gentzen's refusal. The story of Gentzen's death and the citation is from a letter from Dr. F. Krammer to H. [sic] Pinl, Nov. 23, 1946; a copy is in HK.

[177] *DM* 3 (1938): 255–268.

[178] Georg Kreisel, "Review of Manfred Szabo, *The Collected Papers of Gerhard Gentzen*," *Journal of Philosophy* 68 (1971): 255–256.

history of mental illness, he would not get as good a job ever again—his illness left his speech somewhat halting, though apparently this did not affect his pedagogical ability.[179] Gentzen's joining the SA may have been an act of political commitment, or an attempt to obtain necessary political credentials for an academic appointment. He had been determined to be a mathematician since childhood, and the introduction to his *Collected Works* represents him as the caricature of the naive idealist highly strung mathematician. All in all, it is difficult to judge, but that he never personally persecuted anyone seems assured.

Hans Petersson

Hans Petersson was a somewhat older mathematician than Gentzen who spent the early war years in Prague. Petersson reveals the situation of a talented mathematician whose career had just substantially started in 1933 (doctorate in 1925, *Habilitation* in 1929, both with Hecke), and who was a man of principle. Petersson's story during the Third Reich also gives added perspective on some of the personalities, attitudes, and situations we have already seen.[180] On June 8, 1934, Hecke, Blaschke, and Emil Artin jointly signed a request for Petersson's promotion to associate professor (without civil-service status). This was approved by the Dekan, who said nothing about politics, but only discussed Petersson's mathematical and pedagogical ability. In particular, he was good at elementary courses and had worked out lectures of a "very unusual and important sort that are scarcely held otherwise in Germany." Also, "in personal relationships he had thoroughly proved himself." However, it would appear that Hecke drafted the Dekan's letter. The promotion proposal was rejected by Hamburg officials on the technical grounds that one must have at least six years' teaching experience "at German universities" before such a promotion, and Petersson had had only five. The petitioners were told to reapply in July 1935. Thus Petersson had no immediate hope of enhancing his meager salary as a *Privatdozent* in this official way.

Consequently he turned to an unofficial method. He had been born in Bentschen in the province of Posen in 1902. His father had been a local court official (*Amtsgerichtsrat*) and, presumably in this capacity, had met the jurist Hans-Heinrich Lammers, who was twelve years his junior. At the conclusion of World War I, Bentschen became Polish (in the "corridor"), and then or earlier the Petersson family had moved west. With Hitler's accession, Lammers became "State Secretary and Chief of the State Chancellery."[181] On November 11, Hans

[179] Hans Rohrbach, *Gutachten* (Evaluation) of Gentzen's probationary lectures, July 1, 1943, in BDC file for Gentzen.

[180] All material on Hans Petersson not otherwise annotated can be found in the Personalakten Petersson, Hamburg Staatsarchiv.

[181] Hans-Heinrich Lammers, born in 1879, was head of the Reichschancellery in the Nazi government, and a close legal advisor of Hitler. After November 26, 1937, he was minister without

Petersson, saying he had heard there were several vacant chairs of mathematics in Germany, turned to his father's former friend for help in obtaining one of them. Lammers, saying he knew only the father, not the son, forwarded the request to the education ministry.[182] Such applications for academic jobs were certainly not usual, but on January 23, 1935, a ministry official, Dr. Franz Bachér, sent letters asking for estimations of Petersson. These went to Riebesell (by then in Berlin), Tornier, Bieberbach, and the *Dozentenführer* in Hamburg. It should be noted that none of the three mathematicians had anything to do with Petersson's field of research, though they were asked to comment on his "scientific qualifications." However, all three (including Riebesell at the time) could be counted on for political judgments. The replies are interesting, especially for what Tornier and Bieberbach have to say about mathematics as well as about Petersson. Riebesell sent two sentences saying that so far as he knew Petersson was qualified on all three accounts (scientific ability, teaching ability, personality). Tornier refused to judge personality or pedagogical ability, but:

> His scientific accomplishments lie in the area of theoretical algebraic number theory and certainly qualify him scientifically for an associate professorship, in case he should negotiate for it.
>
> For a full professorship I would not unconditionally say so, since there are certainly more proficient people still available, and personally to me it also does not seem desirable that the purely algebraic number-theoretic direction (therefore completely foreign to natural science) in mathematics now once more receives the upper hand, for it is only an ornament in the edifice of mathematics.

Bieberbach accounted Petersson "diligent, one-sided, talented, energetic, however less original in his papers." He also said that someone who knew Petersson's area of expertise better might judge him better. Petersson made "neither a particularly clever, nor a particularly reliable impression." According to Bieberbach, he was an inferior product of the Hamburg School.[183]

Tornier, of course, had just several months previously had his struggle with the algebraic number theorist Helmut Hasse.[184] Bieberbach was in the midst of the contretemps that had started the previous September in Bad Pyrmont, where Hecke had proved no friend to his ideas. In fact, his ultimatum obtained through Vahlen deposing Blaschke as head of the German Mathematical Society

---

portfolio. His job was coordinating ministries and overseeing personnel. He gradually lost influence to Martin Bormann, who ordered his arrest on April 25, 1945. He escaped being shot by falling into the hands of the Americans, and at the "Wilhelmstrasse trial" in 1949 was sentenced to twenty years in prison. This was later reduced by half, but he was in fact released on December 16, 1954, and died in 1962.

[182] BDC file, Petersson, Lammers' adjutant to Petersson, Nov. 27, 1934; Lammers to ministry, Nov. 27, 1934.

[183] BDC file, Petersson, Bachér to those named, Jan. 23, 1935, Tornier to Bachér, Jan. 25, 1935; Riebesell to Bachér, Jan. 25, 1935; Bieberbach to Bachér, Jan. 27, 1935. The *Dozentenführer* (below) replied on Feb. 4, 1935.

[184] Above, chapter 4, "Hasse's Appointment at Göttingen."

had been sent only eight days earlier. That Bieberbach would fail to support a student of Hecke from the Hamburg of Hecke and Blaschke does not seem surprising.[185] Yet Tornier's reply certainly, and Bieberbach's perhaps, reveal that already in early 1935, the *Deutsche Mathematiker* had decided that some large areas of mathematics were better not pursued. Bieberbach's remarks about Petersson's field might be a genuine expression of lack of knowledge, but also, somewhat ambiguously, might reflect on the field itself. Certainly the word "reliable" (*zuverlässig*) with respect to personality carried a political code.

As to the *Dozentenschaft*, it reported that Petersson had talent as scientist and teacher. However, he was personally very retiring, so "that his personality is very difficult to evaluate." This led to a nonrecommendation. The ministry on April 4 told Lammers (after he had inquired in March about his receiving no reply) rather vaguely that Petersson was not sufficiently well recommended. Privately they were furious. A memorandum accompanying the handwritten draft declared: "We must put an end to this sort of request under any circumstances." It also worried about inappropriate people seeing the ministry files on academics.[186]

The ministry saw no reason to communicate with Petersson, and in June 1935, as had been earlier suggested, the Dekan renewed the request mentioning explicitly the fulfillment of six years' time since *Habilitation*. Hecke was asked for an accompanying scientific evaluation of Petersson, but his absence from Hamburg delayed this at least until September. At the end of October the Hamburg Rektor Adolf Rein[187] sent Hecke's glowing scientific and pedagogical evaluation along with his endorsement of Petersson, and cited a report from the *Dozentenschaft*, which said:

> Dr. Petersson belongs to the naval SA as an active member (*Sturmmann*), so that hardly any second thoughts of any sort need be held concerning his political orientation. In other matters Dr. Petersson enjoys no sympathies in the circles of the Hamburg *Dozentenschaft*.

The startling last sentence is glossed by Rein to say that "Dr. Petersson is personally not particularly valued by some of his colleagues."[188]

Rudolf Mentzel was in charge of the case at the ministry this time. He asked only Hasse and Bieberbach for opinions. Hasse at least was knowledgeable about Petersson's subfield of mathematics and spent eight lines full of positive adjectives about Petersson as colleague and researcher. Bieberbach was far more positive this time, and called the promotion warranted, but could not avoid gibing that "nevertheless one misses in [Petersson's] papers the correct aim. Also

[185] Above, chapter 6, "The Bieberbach-Bohr Exchange and the 1934 Meeting of the DMV."

[186] BDC file, Petersson, Lammers to ministry, Mar. 7, 1935; ministerial memorandum (handwritten) dated Apr. 4, 1935, on draft of reply to Lammers.

[187] For more information about Rein, see Giles 1985. Despite its title, this book deals mostly with the situation in Hamburg.

[188] BDC file, Petersson, Rein to Hamburg educational authorities, Oct. 5, 1935, also in Staatsarchiv Hamburg, Personalakten Petersson.

a certain one-sidedness is apparent." Here again the phrases are ambiguous, and it is easy to suspect the *Deutsche Mathematiker*'s evaluation of the field instead of the person.[189]

Petersson was asked on December 4 for the usual religious birth and marriage certificates for himself and his wife going back two generations. He told the Dekan that he could not get either his or his father's birth certificates because Bentschen was now Polish; however, a baptismal certificate for his father and a copy of a 1917 copy of Petersson's own birth certificate were permitted as substitutes.

While Gentzen's reasons for joining the SA are unclear, Petersson's are obvious. He had married on September 30, 1933, and needed more money. On October 15, he joined the SA.[190] Hecke gave a statement to the postwar denazification committee that he had advised Petersson, like many young people, to join the SA so he could advance. Petersson was in Hamburg until 1939, and Hecke said that during this time he was never political, mostly avoided party service, and was "busied successfully and blamelessly with his mathematical work."[191] But Petersson's SA membership, to which he added Nazi party membership in 1937, had a deeper root. Copies of the proofs of Aryan descent for him and his wife, the former Margarete Ehlers, exist in both the Hamburg Staatsarchiv and the Berlin Document Center. In the BDC copies, all religious affiliations are suitably Protestant, but in the Hamburg copies the religious affiliations of his wife's maternal grandmother and grandfather are blank. As the German denazification committee declared:[192]

> The principal objection against Prof. Petersson is his having joined the SA in December [sic] 1933. He explains the necessity of this by the fact that his wife, whom he had married in September 1933, had a Jewish grandparent[193] and had been a member of the Socialist party, both facts he had not made known to the authorities. He hoped to get SA protection from the possible consequences of his marriage by his membership. Professor Hecke, who is a well-known opponent of National Socialism, had advised him to join the SA. His membership in the party and the NSDB follow, as he plausibly explains, from his belonging to the SA. That he was not very active in the SA is shown by the fact that during twelve years of membership he did not reach a rank higher than private first class.

It also commented that Petersson's party membership had nothing to do with his later positions at the universities of Prague and Strassburg. In March 1947 he was reinstated.

No wonder, then, that Petersson, in 1934, engaged in running a seminar with

[189] BDC file, Petersson, Mentzel to Hasse and Bieberbach, Feb. 21, 1936; Hasse *Gutachten*, Feb. 24, 1936; Bieberbach to Mentzel, Feb. 24, 1936.

[190] BDC file, Petersson, *Fragebogen* filled out Jan. 27, 1936.

[191] Staatsarchiv Hamburg, Personalakten Petersson, Hecke to authorities, Oct. 19, 1945.

[192] Staatsarchiv Hamburg, Personalakten Petersson, report of Denazification Committee (E. Wolff, Bredemann, Zassenhaus), Aug. 15, 1946.

[193] Her maternal grandfather was Karl Lampe; her maternal grandmother, Anna Jacobi.

Blaschke on ballistics, a subject far from either's research interest, or that he applied himself dutifully to the *Fachschaftarbeit* ("disciplinary work") reported in *Deutsche Mathematik*. More than Peschl, he had reasons beyond a simple distaste for the regime to hide his true feelings so he could advance. He scrupulously explained in his proof of Aryan descent when he joined the SA why his surname was spelled in the Swedish way instead of the German way (Peterssohn). It seems his father changed his name back to the original Swedish after Petersson's great-grandfather, a military physician, had changed to a German spelling.[194] Indeed, in 1939, in a handwritten response to the usual questionnaire about ancestry, Petersson spelled his father's and grandfather's names with an "h," as opposed to his earlier practice.[195]

In fact, Petersson, who Horst Tietz said was "very antifascist,"[196] was so successful at dissembling his true feelings that in April 1938 the Hamburg division of the Nazi party could write,[197]

> The information requested from [his SA] naval group (*Sturm*) about P[etersson] has turned out very good. Besides P[etersson] has been a party member since May 1, 1937. P[etersson] in the meantime has likewise completed an eight-week military exercise.
>
> Prof. Petersson represents the rarely occurring case of a mathematician who stands without reservation on the ground of the National Socialist *Weltanschauung* and besides is active in the SA with a will (*mit Lust und Liebe*).

Consequently an academic promotion for Petersson was justified. Similarly, when, in early 1939, Petersson requested extra money to live on, the *Dozentenschaft* leader could approve, not only because of his low pay and high scientific ability, but because "politically he stands absolutely positive."[198]

Petersson's going to Prague was not the first attempt to find him the deservedly better job than the one he had at Hamburg. In 1935 Blaschke tried to obtain a professorship for him in Calcutta. In 1938 there were negotiations about the possibility of Petersson's having an associate professorship in Halle. These negotiations were somewhat lengthened by the lack of timeliness of a political report on Petersson, and it was March 24, 1939, before the official appointment was made by the ministry.[199] By this time Prague was German. On a trip to Sofia, Bulgaria, in May, Blaschke ran into the Prague Rektor, and recommended Petersson for the far better job of Ordinarius at Prague (the job that Rudolf Weyrich[200] also wanted). On September 9, Petersson received that appointment. But apparently he now wanted to stay in Hamburg. Perhaps the

[194] Hamburg Staatsarchiv, Personalakten Petersson, Petersson to Dekan, Dec. 17, 1935.

[195] BDC file, Petersson, official questionnaire dated June 15, 1939.

[196] Horst Tietz, interview, Apr. 5, 1988.

[197] BDC file, Petersson, Hamburg *Dozentenbundführer* to national leadership of the *Dozentenbund*, Apr. 5, 1938.

[198] BDC file, Petersson, Anschütz to *Reichsdozentenführung*, Jan. 18, 1939.

[199] For Calcutta, see Hamburg Staatsarchiv, Personalakten Petersson. For Halle, see both this and the BDC file on Petersson.

[200] Above, chapter 4, "The Winkelmann Succession."

beginning of World War II had something to do with this reluctance; though Prague, even if further East, might seem safe enough, it was nevertheless in a certain sense "enemy territory." At the time Petersson's family contained an infant son as well as a three-year-old. In any case, on October 7, 1940, he was ordered immediately (*mit sofortiger Wirkung angeordnet*) to Prague to be codirector of the mathematical institute at the university and at the technical university—but only as an associate professor.[201] In 1941 he went to Strassburg, at the opposite end of "Greater Germany," where the Nazi government had established a "national university." In 1944, in the face of the advancing Allies, this university was closed, and on October 20, Blaschke, acting for the Dekan,[202] requested that Petersson and the mathematician Emanuel Sperner be transferred from the closed university to Hamburg. Not only were there not enough instructors in Hamburg, but Sperner and Petersson both had war contracts with the Hamburg naval observatory. Petersson was approved provided he simultaneously took over for Gerrit Bol in Greifswald; however, this was later rescinded.[203] Petersson left Strassburg for Hamburg on November 23, 1944[204]—Allied troops entered Strassburg the same day. Petersson's rank was unclear; following his reinstatement he wrote a letter claiming to be *de jure* a full professor. Nevertheless, he was apparently kept on from year to year at Hamburg—even being refused a multi-year contract. Petersson's problem was that even though he had been a full professor in Strassburg, his position at Hamburg had no civil-service status. In 1951 (he was then forty-nine), he declined an offer of an associate professorship at Jena, as well as a full professorship at Rostock (then both in East Germany). As one academic commented on his case in 1951: "The tragedy of the overfilling of West German universities with professors who have fled the East will make situations like Petersson's unavoidable for an even longer time." Finally, in 1953, Petersson received the position of Ordinarius at Münster, where his long-time full-professor colleague was Heinrich Behnke, a fellow student of Hecke who had finished his doctorate two years before him.[205]

Hans Petersson had the courage to marry the woman he loved despite her "familial taint," to hide it successfully, and to convincingly go through the motions of being an ardent Nazi supporter when he was not. Despite these strains, his mathematical activity flourished. As opposed to the much more senior Wilhelm Blaschke, for Petersson personal survival and human relations, rather than personal self-aggrandizement, were involved. Indeed, the Peterssons were always a little afraid of Blaschke, and unclear as to where he really stood.[206]

[201] Hamburg Staatsarchiv, Personalakten Petersson.

[202] BDC file, Petersson, Blaschke to Rektor writing as acting Dekan, Oct. 20, 1944.

[203] BDC file, Petersson, "Vermerk," Nov. 20, 1944. Sperner went to Freiburg at Wilhelm Süss's instance. Bol was remaining there. Both were at Süss's newly formed national mathematics institute. The distance from Hamburg to Greifswald is about 140 miles.

[204] BDC file, Petersson, Kurator Strassburg to ministry, Dec. 27, 1944.

[205] Staatsarchiv Hamburg, Personalakten Petersson.

[206] Interview with Margarete Petersson (in the presence of Holger Petersson), Apr. 14, 1988.

Erich Kähler

A direct contrast with Petersson is provided by another young Hamburg mathematician who made a name for himself, and who "habilitated" the year after Petersson: Erich Kähler. Kähler's extreme nationalism and defense of events of the 1930s and 1940s was so pronounced, even in 1988, that, setting out to interview him, I was warned that he had been a Nazi party member. In fact, he seems not to have been. Not only did he insist on this fact, but there seems to be no BDC file indicating he was.[207] One of the reasons for interest in Kähler is that he personally knew at different times, briefly or at length, a number of the people appearing in these pages—the view of a mathematician still unashamedly (but somewhat mystically) on the far right in 1988 provides, therefore, a different and perhaps valuable perspective. But he is also extremely interesting in himself as a type not yet examined. He seemed to be a man self-educated to Nietszchean ideas of leadership who thought and acted as he did out of the deep values of loyalty and service: "Theirs not to question why. Theirs but to do or die." He seemed to display none of the ulterior motives, good or bad, that color many of the people in these pages. Neither did he seem to have had any developed system of values, interests, and experience by which to judge (when Hitler came to power he had just turned twenty-seven—his most important mathematical papers appeared in the next two years). It is hard not to respect Kähler as principled while abhorring his principles. In 1935, Kähler went to Königsberg, where he became Ordinarius the following year.[208] Thus far information provided by Kähler is as for his listing in the membership of the German Mathematical Society. However, in 1935 also, he told me, he volunteered for military service;[209] he was in the navy in 1937, and on August 24, 1939, in the army. He spent the whole of World War II in the German military and 1945–47 as a prisoner of war. Thus, what arguably might have been his most productive mathematical years were spent in military service. In 1948, when he was over forty-two, he again became an Ordinarius, this time in the city of his birth and early student days, Leipzig, then in East Germany. In 1964 he returned to Hamburg, where he had obtained his doctorate under Blaschke, and later he retired to a Hamburg suburb.

Politically, Kähler said that he was inclined toward German nationalism (like many of his colleagues), and Hitler awakened in him the feeling of a greater Germany (which he apparently responded to by desiring military service, rather than by joining the party). The *Führerprinzip* was, he felt, not bad in itself: for

[207] Erich Kähler, interview, Jan. 30, 1988. All information about Kähler below and not otherwise cited is from this interview.

[208] There seems to be an error in Scharlau et al. 1990: 198. Kähler succeeded Gabor Szegö (forced out as Jewish) in 1935 in Königsberg.

[209] In January 1935, the scheduled plebiscite in the Saar resulted in its rejoining Germany. In March 1935, Hitler openly began to rearm in repudiation of the Versailles Treaty.

to be *Führer* meant to be responsible.[210] His oath to Hitler (as a civil servant) was very important to him: he thought of Hitler as his Kaiser. There was a "cult of genius" around Hitler; hence, in Kähler's view, there are no neo-Nazis because one can't be a Nazi without Hitler, whose real aim was the nullification of the Versailles Treaty. He thought Hitler's high politics could not be widely understood, as they were supranational, like Roman politics. No one had dug so deeply into history as Hitler.[211] No doubt there were criminals in high places in Hitler's Germany, but the leadership of the country was not criminal. Similarly, Kähler thought the blaming of the whole German people, a *Kulturvolk* (cultural nation), for the criminal acts of some was not only inappropriate, but resulted from an intentional desire to make Germany politically impotent.

As to the "Jewish question," Hitler, thought Kähler, had the insight to see that there *was* a "Jewish question."[212] How Hitler handled the Jewish question may have been wrong—and Kähler thought *Kristallnacht* was wrong—but this did not negate Hitler's insight. Having said this, he immediately attempted to mitigate the wrong by commenting that much worse things were happening in Lebanon [in 1988]. Indeed, there were only two things that made Kähler doubt Hitler's politics: his handling of the Jewish question, and his marching into Prague in March 1939 (because it meant he had lied when he said the Czech Sudetenland was all he wanted—Kähler acknowledged the Bible as a strong early influence in his life). However, he made the point, a familiar one from the German right, that none of the extermination camps was on German soil, but all were in the East, as if to argue that the German people would not have tolerated them. Furthermore, he pointed out that Auschwitz was liberated by Russians; news of Auschwitz came from Russians and was intended to defame the Germans. For Kähler, the destruction of the Jews (which he seemed to admit was wrong) is used internationally as a "wooden hammer" (*Holzhammer*) to end any serious analysis or discussion of the German question (this, of course, was said in 1988, prior to the reunification of Germany).

As to World War II, Kähler thought it mostly had to do with three peoples (*Völker*): Jews, Russians, and Germans, all of which had the "insightful intellectual spirit" (*zuschauenden Geist*). The other peoples in the war were completely secondary. When the conversation shortly thereafter shifted to his admiration of Dostoyevski as a "Russian Nietzsche," and I mentioned that Moeller van den Bruck[213] had translated Dostoyevsky into German, Kähler recalled that Gregor Strasser had led a wing of the NSDAP that had wanted to stay open to the Russians (i.e., truly socialist), and then added, somewhat surprisingly "I believe he was later killed." Since Gregor Strasser was one of the victims on June 30, 1934, "The Night of the Long Knives" or *Röhmputsch*, this may reflect either a

[210] This is, of course, a reading of Friedrich Nietzsche's *Also sprach Zarathustra*.

[211] Alan Bullock, in the conclusion of his well-known biography *Hitler, A Study in Tyranny* (1962; original publication 1952), remarks that Hitler made the modern world.

[212] So did some conservative opponents of Hitler, like Carl Goerdeler; cf. above, chapter 3.

[213] Moeller van den Bruck originated the term "Third Reich." For more about him, see Stern 1961.

selective loss of memory or the fact that many of the victims were not widely known at the time.

It would seem as though the same sort of nationalism that motivated Erich Kähler in 1935 continued to motivate him over fifty years later. A former student told me that when Kähler was active as a professor in Hamburg, he used to keep a Nazi naval flag in his office. Kähler's thinking the Jews were an important people of genius like the Germans because they created a nation *de novo* seems rather like the disturbing fact that much of Adolf Eichmann's knowledge of Jews was garnered from reading Theodor Herzl's *Judenstaat*, with which Eichmann was positively impressed.[214]

What did this man who thought that it was nonsense to speak of twelve years of injustice think of his mathematical colleagues during those twelve years? All professors, he said, had to be inclined a bit toward the Nazi party (*NS-parteilich*) in order to remain in office. Like himself, Blaschke believed in a Greater Germany and was the "protective angel" of the Hamburg department. Hecke was "not political enough," but Blaschke said to leave Hecke alone and he would care for him. Blaschke was sophisticated, Hecke just the opposite (*weltfremd*); but Blaschke protected Hecke and Artin (until his emigration in 1937). As to "Deutsche Mathematik," as an idea in itself it had nothing against it. However, Bieberbach's mistake was to mix philosophy and mathematics—one could be interested in both (as Kähler was), but they should not be mixed. Teichmüller was not naive, and neither was the philosopher Martin Heidegger; in Kähler's view, they wanted and believed in a Greater Germany. The tragedy, in the Greek sense of the word, was that one had to do something immoral because one was compelled (presumably in aiming toward a greater good as an end).

Spending time on Kähler's right-wing philosophical-political perspective may seem irrelevant to mathematics, but in a rather interesting way this is not so. For some time he had been interested in philosophically mathematizing, as it were, the world and human existence. His philosophical hero beside Nietzsche was Leibniz. Thus, in a paper published in 1986 (when he was eighty), Kähler spoke (in English) of a "mathematical monadology offered by a philosophic transposition of the local algebra, about which I have reported elsewhere," and stated that "dynamics of monads find their best representation in arithmetic and purely algebraic relations." The point is not what meaning such phrases may have nor their relevance to the attendant mathematics, but the effort at a "philosophical mobilization of mathematics."[215] Kähler believed that mathematics is called to develop Nietszche's thought in the same way as Maxwell was called to develop Faraday's.[216] In our 1988 discussion, Kähler remarked that we do not

[214] Arendt 1994: 40–41, 57, 209. For parallels between Imperial German nationalism and Jewish nationalism, see Mosse 1970.

[215] Erich Kähler, "The Poincaré Group," in J.S.R. Chisholm and A. K. Common, eds., *Clifford Algebras and Their Applications in Mathematical Physics* (1986), 265–272. Somewhat curiously, given the present interest in "string theory" in ten dimensions, this paper, which is about the mathematics of cosmology and relativity, *inter alia* discusses a purportedly relevant ten-dimensional Lie group.

[216] Nietzsche was by training and occupation a classical philologist. Faraday, essentially self-edu-

live in states founded on law and justice (*Rechtstaat*) because an appropriate analysis of history is not possible, but also that his "people" (*Volk*) was now humanity, and the global separations in the world are the real problem (in 1988, Kähler was referring to the Cold War). There will never be a world peace if one does not think globally and create a world-nation. Treaties will never bring world peace. What is the core? According to Kähler, it is Germanic thought (*Deutsches Denken*) or, equivalently, Roman thought, which would see the world as potentially a single *imperium*. Mathematical analysis is the philosophical route to that "Brave New World." As a reviewer of a recent paper by Kähler (published when he was eighty-six) remarked: "The main thesis of the paper is that algebraic geometry is a prolegomenon to a mathematical theory of monads."[217]

To assure the reader of Kähler's solid mathematical credentials (at least fifty years ago), it should be noted that in 1944, when he was effectively a prisoner at St. Nazaire (a fortress at the mouth of the Loire), he claimed that no less notable French mathematicians than Elie Cartan and André Weil sent him mathematics books.

Erich Kähler and his mathematical philosophy are certainly *sui generis*. His blending of expertise in sophisticated mathematics, German idealist philosophy, extreme nationalism, Leibniz, and Nietzsche provides a strange mixture. He is certainly an outlier among the mathematicians of this period. He died in 2001.

Wilhelm Süss

Far from an outlier, perhaps the central mathematical figure during the Third Reich, certainly the political spokesman for mathematics from 1937 to 1945, was Wilhelm Süss. Corresponding to this critical role are the several times he has already appeared earlier in these pages. An attempt at further understanding his attitudes, however, seems crucial to understanding the German mathematical community during this critical period. The fact that both the German Mathematical Society and the faculty at Freiburg chose him as their leader during those years reflects the values they saw in him. It was the way in which Süss acquired the trust of the mathematicians and brought them through this crucial time that made Alexander Ostrowski say in a memorial address:[218] "Certainly no

cated, knew no mathematics, but invented the concept of a force field, later exploited mathematically by Maxwell.

[217] Doru Stefanescu, reviewing Kähler's paper "Also sprach Ariadne," in *Mathematical Reviews* 956 (1995): 6.

[218] Alexander Ostrowski, "Wilhelm Süss, 1895–1958," *Freiburger Universitätsreden, Neue Folge* no. 28 (1958): 12. I am indebted to Richard Ellis for a copy of this talk at a memorial service for Süss. Biographical material below about Süss, not otherwise annotated, comes from the following sources, sometimes redundantly: Süss's *Lebenslauf* in Personalakten Süss in the archive at the Universität Freiburg (hereafter PAS); an article in the *Freiburg Wochenspiegel* 19/20 (Dec. 1957), and an obituary in the *Freiburger Studentenzeitung* 4 (June 1958), both also in PAS; as well as an obituary by Helmuth Gericke in *JDMV* 69 (1968): 161–183, and Ostrowski's eulogy.

one since Felix Klein has done so much for German mathematics as Wilhelm Süss."

Wilhelm Süss was born in Frankfurt-am-Main on March 7, 1895. His father was a teacher, and the whole family had musical gifts, including young Wilhelm. His paternal uncles were professional musicians. His mother came from a family that had produced "Bürgermeisters" (mayors) of her locality for generations. Brilliant in all subjects at the *Gymnasium*, Süss decided to study mathematics, and began his university work in Freiburg, where, among others, he heard lectures by Alfred Loewy, whom many years later he would succeed as Ordinarius.[219] As was the fashion at the time, he also studied in Göttingen and Frankfurt, but in October 1915 was drafted and served on the front lines in World War I until demobilized in late November 1918, except for an eight-week period when he had malaria. One brief humorous episode at this time was that when drafted, Süss had been erroneously listed as "stud. med." instead of "stud. math." As a consequence he was for a time delegated as assistant to the regiment's veterinarian, and learned a great deal about the care of horses and their ailments. He later was transferred to a (more appropriate) sound-ranging group. On returning from war, Süss went back to Frankfurt to study. In March 1920 he finished his doctorate under the supervision of Ludwig Bieberbach. It dealt with problems concerning the definition of volume. During this period of little more than a year he also met his future wife, another mathematics student named Irmgard Deckert. Her father was an academic geographer, and her maternal grandfather was Charles Goodyear.[220] When Bieberbach went to Berlin in 1921, Süss went with him as his *Assistent*. However, he seems to have spent most of his time actually working for the *Notgemeinschaft der Deutschen Wissenschaft*, the predecessor of the DFG.[221] In January 1923 Süss accepted a position in Kagoshima, Japan, where he spent almost exactly five years from March of that year. In Japan, Süss's duties were not primarily mathematical, but more concerned with the study of German language and literature. However, he also continued mathematical work and publication.

Homeward winds were blowing. Karl Reinhardt, who had been Bieberbach's first student,[222] had also been a childhood acquaintance. By now Reinhardt was an Ordinarius in Greifswald, and Süss wrote him. The upshot of this correspondence was assurance of *Habilitation* at Greifswald (on the basis of Süss's publications) and a position there. At Greifswald he met also Hellmuth Kneser, who would become his life-long, very close friend.

Politics in late Weimar was often the politics of extremes. Süss, a veteran of the war from a solidly bourgeois background, returning from five years abroad to be met by economic bad times and the still-unfulfillable demands of Versailles, was understandably, like many with front-line war experiences, of a

[219] Süss 1967.

[220] I. Süss, interview, Mar. 25, 1988.

[221] Compare Biermann 1988: 215 n. 4; Ostrowski 1958: 7; and Süss *Lebenslauf*, in PAS.

[222] His geometrical work of 1918 lay more or less dormant for fifty years, but its subject matter of tiling has recently had a revival of interest. See, e.g., Doris Schattschneider, "In Praise of Amateurs," in David Klarner, ed., *The Mathematical Gardner* (1981).

conservative nationalist frame of mind. This does not mean that he was any sort of proto-National Socialist. On the contrary, both he and his wife had been DDP[223] voters, and in the last parliamentary election in 1932 (November 6) voted social-democratic as a stronger way of voting anti-NSDAP. In the mathematical contretemps at Bad Pyrmont and afterward, Süss (like Doetsch and Hellmuth Kneser) adopted an ultra-nationalist stance. They really did think that German national pride was in question, especially with respect to Blaschke's actions at the time. But while Doetsch put on his uniform again and became an inspired Nazi follower, and Kneser toyed with the idea of joining the NSDAP to demonstrate his approval of its nationalist actions and assertion of nationalist feelings,[224] Süss seems to have resisted close association with the Nazi party, though almost willy-nilly he became a member.

It is probably fair to say that during the Weimar years, most Germans were committed nationalists, whatever their internal politics. Süss had resisted association with the NSDAP, and was shocked when in March 1933 all teachers at the university in Greifswald were ordered to join the SA the next day. Learning that only people already members of the veterans' organization, the *Stahlhelm*, were exempted, and that the university was under political compulsion, late that night he joined the *Stahlhelm*. On July 3, 1933, however, the *Stahlhelm* was integrated into the SA.[225] When Süss later was appointed Ordinarius at Freiburg, he bought a Nazi flag for outward show so his opinions would not be questioned.[226]

Nevertheless, according to his wife, when Süss was negotiating for the Ordinarius position with the ministry, he began the discussion by announcing his nonmembership in the party.[227] He was appointed to succeed Alfred Loewy, who, Jewish, blind, and sixty-one, had been compulsorily removed from office, though technically he fell under the exceptions clause of the April 7 law, and apparently he had continued to teach successfully. Süss had a friendly visit with his old teacher. In 1937, the NSDAP rolls were again opened, and the then-Freiburg Rektor Metz, unbeknownst to Süss, recommended him for party membership. Süss found out when, to his astonishment, he received a letter telling him time and place to enroll. Süss resisted. However, Metz and other colleagues told him that as a member he could hope to combat inimical tendencies of the party from within,[228] and so, in early 1938, he became a member. Süss had already been acting Dekan in 1936, and for 1939–40 he was Dekan; in 1940 he

[223] This was a small center-left party, in which the (Jewish) mathematician Felix Bernstein had been a leader. Süss n.d.: "Vor," p. 3; PAS, Süss to Dekan, Oct. 8, 1945.

[224] HK, Kneser to Courant, May 20, 1933; Reid 1976: 148–149. However, Kneser did not do so.

[225] For dates, see Broszat 1981: 88, 204, though Süss (PAS, Süss to Dekan, Oct. 8, 1945) says this happened in 1934. Integration may have been later in Greifswald, or Süss may have misremembered.

[226] Süss n.d.: "Nach," p. 4. Earlier, in Greifswald, the family had seen their house stoned because it failed to display a swastika.

[227] Ibid.: p. 7.

[228] Metz was a fairly early party member. Whether he really believed what he seems to have told Süss, or this was just an argument to enforce Süss's acquiescence, is unclear.

succeeded to the position of Rektor.[229] As already discussed, in 1937 he became president of the German Mathematical Society. He held both posts until the end of World War II. While his continuous presidency of the German Mathematical Society made him its effective *Führer*, as discussed in chapter 6, Süss apparently did not aspire to such a position—in fact, he tried to get the consent of his friend Hellmuth Kneser to stand for office and replace him.[230] As Rektor he was "Master of the Art of the Possible" and was able to steer the university through the chaotic dangers of the Nazi period.[231]

With the war's end, Süss was, as a Nazi university official, suspended from office. Letters poured in in his defense. The Allied-installed Rektor at Freiburg spoke of Süss having made a "sacrificium intellectus" by joining the party, and all non–National Socialists understood his act and were confident of his inner beliefs.[232]

In his reply to his suspension, Süss himself not only mentioned the above facts, but also his standing up "against party, SD, and Gestapo" for numerous colleagues and students. Furthermore, he was struck from the rolls of the SA for lack of activity and not even given an "honorable discharge."[233]

The letters urging Süss's reinstatement mentioned his speech at the Rektor's conference, his use of his connections "to work good," his difference from other party members, and his active helping of others in desparate straits. Kamke wrote of his situation during the Third Reich and how Süss (and Gerlach) had saved him from a "labor camp" in 1944 when all "Jewish-related" persons were to be so interned.[234] A letter composed by Threlfall, Seifert, and Hermann Boerner (at Munich) mentioned a list of twenty-five individual cases of Süss's rescue of colleagues threatened with political persecution. This was subscribed to by twelve other colleagues, including all the Munich *Ordinarien* (Carathéodory [emer.], Heinrich Tietze, E. Hopf, Oskar Perron).[235] Another letter from Carathéodory mentioned Süss's rescue of a Serbian mathematician, Nikolas Saltykow, who, though aged and in declining health, was being held hostage in a prison. Süss secured his release when half a dozen others whom Carathéodory had approached refused to help.[236] Behnke wrote of Süss's continual support of him despite his being "Jewish-related" (through his son), espe-

[229] This could never have happened had he not been a party member. Süss was not eager for this job. See HK, Süss to Kneser, June 27, 1940. This brief handwritten letter says in reference to the Rektorship, "I hope this chalice passes me by; I no longer have any ambition [thereto]." It also speaks of the principal difficulty for a Freiburg Rektor being the attempts of the NS *Dozentenschaft* and *Studentenschaft* to participate in running the university.

[230] HK, Kneser to Süss, Apr. 7, 1939 (Blatt 2).

[231] Ostrowski 1958: 11.

[232] PAS, Rektor to education ministry (Freiburg), Oct. 12, 1946.

[233] PAS, Süss to Dekan, Oct. 8, 1945.

[234] PAS, Kamke to Rektor, Oct. 24, 1945.

[235] PAS, undated, but apparently sent in mid-October 1945. The Munich *Ordinarien* in mathematics subscribed in a separate letter (stating that Perron agreed with the other three). This letter was also successfully circulated at the Oberwolfach Institute for further signatures.

[236] PAS, Carathéodory to University, Oct. 24, 1945.

cially during 1944–45 when the Gestapo was continually checking on him and he was threatened with (at the least) expulsion (because Baden was proximate to Switzerland). Also in 1941, Süss aided Behnke in preserving manuscript material of the famous French mathematician Henri Cartan that he had left behind in Strassburg.[237] Ludwig Prandtl and Max Planck sent a lengthy letter, half of which was devoted to a quotation from Süss's address to the Rektors' conference in 1943.[238] This was also mentioned by Gustav Herglotz,[239] while Gerrit Bol, then at Oberwolfach, spoke of Süss's support of foreigners in Nazi Germany.[240] Süss's attempt to save Ernst Mohr's life when Hans Rohrbach made him aware of Mohr's imminent execution has already been discussed (though, while Mohr was not executed, this had other sources);[241] however, a few more cases of how Süss, as a politically influential mathematician, helped some of his colleagues will round out the picture of him and of the situations of mathematicians.

Gerhard Grüss was a mathematician at the mining school in Freiberg. Of all the *Ordinarien* there, he was the only one who had not been a soldier. Once (apparently before 1933) he had even told a colleague that he was a pacifist. He was not a party member, and when asked why not, he told a colleague that he especially could not agree with the standpoint taken by Hitler toward the Jewish question. He also had forbidden his son to receive the Hitler Youth magazine *Der Pimpf* (The cub [scout]) and explained (in writing) that he did not consider it so necessary that, given the contemporary lack of paper, he could be responsible for a subscription when valuable scientific publications were failing to appear. Finally, on April 20, 1944, he had stood in front of his house, which was near the Freiberg stadium where a political ceremony was taking place, and refused to give the Hitler salute during the singing of the national anthem, though others near him had done so.[242]

These actions also have to be seen in the context of the mining academy at Freiberg.[243] Grüss's predecessor had been Friedrich Willers, who had come there in 1928, five years after "habilitating" at the technical university in Berlin at the rather advanced age of forty (he had spent considerable time as a secondary-school teacher). Willers, who was certainly "Aryan," attempted to raise the stan-

[237] PAS, Behnke *Gutachten*, Nov. 19, 1945.

[238] PAS, Prandtl and Planck to "Betrifft Stellungnahme . . . ," Oct. 24, 1945. See above, chapter 6, for a discussion of this speech.

[239] PAS, Herglotz to university, Oct. 21, 1945.

[240] PAS, Bol to Rektor Freiburg, Nov. 20, 1945. Gerrit Bol was Dutch.

[241] Above, chapter 4, "Hasse's Appointment at Göttingen." Cf. BAK R73 12976, Fischer to Süss, July 28, 1944, where Fischer mentions a visit from Rohrbach on behalf of Mohr (and when his penalty was unknown but thought to be some sort of imprisonment).

[242] BAK R73 12976, Fischer to Süss, July 28, 1944, details the charges. April 20 was Hitler's birthday.

[243] Freiberg, which is in Saxony near Dresden, should not be confused with Freiburg (im Breisgau), at the other end of Germany in the Black Forest.

dards of the students. This was met with strong resistance by them, resulting in the National Socialists compelling his retirement in 1934.[244]

Grüss's failure to give the Hitler salute was the immediate cause for the complaint against him. The other material was then assembled. An action against Grüss was contemplated by the Nazi leadership in Saxony but not taken.[245] Instead, the leadership applied to the education ministry that Grüss be punished for political unreliability. In addition, Gustave Scheel, who in 1936 had been the Nazi national student leader, became the national *Dozentenführer* in 1944 and demanded Grüss's removal from the university. The Reich Chancellery also became involved. Dr. Fischer, the ministry official in charge, believed that the strictest rules had to be applied to university personnel, who were, after all, role models for academic youth, and so nothing could be done for Grüss, despite the desperate need (in 1944) for people with his knowledge and skills. Fischer was sure Süss agreed.[246]

Süss had tried, when he first heard of Grüss's difficulties, to suggest a positive solution: perhaps a position at another institution. Süss had already applied in June to Werner Zschintzsch, the executive secretary (*Staatssekretär*) of the ministry, who, once Süss was in contact with Fischer, allowed standard ministry procedures to proceed. The circumstances of Süss telling Zschintzsch about Grüss are interesting. As Rektor, Süss had apparently held a celebration in early June for the seventieth birthday of Eugen Fischer, the anthropologist, geneticist, and prominent racial hygienist, who always had felt his home to be Freiburg.[247] Zschintzsch had been invited and had had a good time—so much so that he inquired after the possibility of acquiring some Baden wine. This was the opportunity Süss took.[248]

Though Wilhelm Fischer considered Grüss's case hopeless, Süss repeatedly suggested Grüss's removal to Aachen as a solution to his difficulties. Fischer had said he was sure Süss agreed with him, and Süss did, insofar as he would not use Grüss (who was unknown to him) at the new national mathematical institute (which became situated at Oberwolfach).[249] However, Süss did think Grüss suitable for a grant for military research (provided Fischer had no reservations about it). He considered that Grüss's actions resulted from short-sightedness, lack of adequate reflection, and a childish spirit of contradiction, and it was "more than probable" that he was loyal and had no really inimical political

[244] For Willers, see Pinl 1969: 216–217; Scharlau et al. 1990: 104; and the obituary (with publications list) by Robert Sauer and Helmut Heinrich in *Zeitschrift für Angewandte Mathematik und Mechanik* 40 (1960): 1–8.

[245] BAK R73 12976, Süss to Fischer, July 1, 1944.

[246] BAK R73 12976, Fischer to Suss, July 28, 1944.

[247] For Eugen Fischer, see Proctor 1988: passim, and Müller-Hill 1984: passim. An interview of Fischer's daughter by Müller-Hill (ibid.: 119–121) revealed, among other things, Fischer's love for Freiburg.

[248] BAK R73 12976, Süss to Fischer, July 1, 1944.

[249] For the founding of the Oberwolfach Institute, see above, chapter 6.

position. One of the reasons for urging a position for Grüss at Aachen (presumably as successor to Robert Sauer) was not only that it was a center for applied mathematics, but that in July 1944 no one went there gladly; thus there was little competition for the open position, and Grüss had told Süss he was open to *any* position in Germany. The fighting in northern France made Aachen (the French Aix-la-Chapelle) an unwelcome voluntary location, and it would fall into Allied hands on October 21.[250] Süss managed in this way to validate Fischer's official view of Grüss while suggesting a procedure that would rescue Grüss from the Nazi bureaucracy in Saxony. Moreover, from a Nazi point of view, Fischer's placing Grüss "on the front lines," as it were, would be a validatable action. In the eventuality, Grüss stayed at the Berg Akademie in Freiberg until his death in 1950—indeed, in the postwar rebuilding of German universities, he became its Rektor. It is quite probable that Süss's intervention on Gruss's behalf slowed the Nazi bureaucratic machinery until, as "a man who always stood up for his convictions even when it was connected with danger to his existence," Grüss was safe.[251]

Süss helped or rescued many other mathematicians; for example, he rescued Georg Lorentz with wife and child from a resettlement camp to work on "militarily important" projects with Erich Kamke in late 1943.[252]

Further detailing of Süss's actions on behalf of mathematicians threatened by state or party in one way or another would become repetitious, even otiose. However, one peculiar situation deserves a final mention as an example of his activity, since it is of a different sort. Charles Pisot[253] was an Alsatian mathematician trained in France.[254] He spent his early career as a secondary-school teacher, but by 1938 was already teaching at a university. With the fall of France he was offered a French associate professorship but chose instead to go over to the Germans. Apparently there were difficulties with Alsatians obtaining German citizenship, and without German citizenship Pisot could not possibly get a university job in Germany. Efforts seem to have been made on his behalf as early as 1941 or 1942, to no avail. In April 1943 an official at the education ministry wrote that a rectification of Pisot's situation in Germany was called for on grounds of both justice and politics. The official died, and in October 1944 Pisot was still not a German citizen. While his civil-service status was thus technically that of Alsatian secondary-school teacher, he was *de facto* a univer-

[250] BAK R73 12976, Süss to Fischer, July 1, 1944, July 13, 1944, and Oct. 12, 1944. For most of July, the Allies were, however, contained in Normandy.

[251] The citation is from an obituary of Grüss by Friedrich Willers, *Zeitschrift für Angewandte Mathematik und Mechanik* 30 (1950): 232.

[252] BAK R73 12976, Süss to Fischer, Nov. 23, 1943. In 1992, this saving of Lorentz was recalled personally to me by one of his students at a garden party in Duisburg.

[253] Charles Pisot was a well-known number theorist, and his name is enshrined in the important class of algebraic numbers known as Pisot-Vijayarhagavan (or P.V.) numbers. Süss always called him Karl instead of Charles. In 1979, the fiftieth anniversary of his degree was celebrated in France with an extensive symposium.

[254] BAK R73 13854, Süss to Mentzel, Feb. 22, 1945; Süss to Präsidum RFR, Feb. 22, 1945; Süss to Fischer, Feb. 22, 1945. BAK R73 12972, Süss to Fischer, Oct. 12, 1944.

sity teacher "until recalled." As early as 1940 he had, in fact, had a temporary position in Freiburg, and in 1941 one in Greifswald.[255] In 1944 he was doing "militarily relevant" research in Freiburg, and later in the national mathematical institute established at Oberwolfach. Süss agitated throughout early 1945 for a rectification of the situation so he could appoint Pisot a regular associate professor either in the national mathematics institute, or at Freiburg, or both. In Süss's opinion, this needed to be done as quickly as possible, but would not make up for "the other possibilities that he had relinquished as a German in France."[256]

This seems to be just a case of Süss pushing the Nazi bureaucracy to do what was right. Also, perhaps Pisot considered that German citizenship might protect him from charges of collaboration—but this is pure speculation. What happened with Süss's application on Pisot's behalf is unclear. In any case, Pisot came through the war and its aftermath whole and continued his significant mathematical career after it as a (politically appropriate) Frenchman.[257] The case of Pisot illustrates how devotion to his country (and the mathematics done in it) was always a primary motivation for Süss. The Nazis were Germany's extreme misfortune, to be gotten around, or made to see nonideological reality, as much as possible. His address to the Rektors is in the same spirit. For Süss, Pisot, no ideologue, seems to have been neither collaborator with the Nazis nor French pariah, but simply a (very good) mathematician of ambiguous national heritage, trying to persevere in his discipline despite the extreme dislocations of war. Helping such a man helped mathematics and helped human beings worth helping.

This is not a biography of Süss, and throughout this book enough has been said to give some view of the man responsible for guiding the fortunes of mathematics (and many non-Nazi mathematicians) for the last two-thirds of the Third Reich. Though originally a conservative nationalist, he was always anti-Nazi, and it was as a conservative humanist who genuinely cared for people that he fulfilled his various positions. Süss's later career or the later fortunes of the German Mathematical Society are irrelevant to this book except as already briefly touched upon. Suffice to say here that Süss was reinstated; though there was some hesitation about doing so for a "Nazi Rektor," the Allies, even the French (who occupied Freiburg), understood his great service acting "from within" to defuse as much as possible destructive Nazi actions.[258] Very few party members had even contemplated such action. His university showed its confidence in Süss by again choosing him as Rektor for the academic year beginning

[255] For Freiburg, see HK, Süss to Kneser, Oct. 14, 1940. For Greifswald, Scharlau et al. 1990: 133.

[256] The citation is from BAK R73 13854, Süss to Mentzel, Feb. 22, 1945.

[257] While his fellow mathematicians seem to have reaccepted Pisot, he apparently did suffer recriminations from other French (personal communication from Liliane Beaulieu, reporting on a conversation with Pisot's widow).

[258] PAS, Rektor Janssen *Aktennotiz (Auszug)*, Nov. 20, 1945.

April 15, 1958. However, ill with liver cancer,[259] he died on May 21, 1958. At his death, among other expressions of sympathy came one from an advisor to the British embassy in Bonn with the history-drenched name of Robert Cecil, who spoke of Süss's importance in the postwar organization of Anglo-German university conferences.[260] Süss cared most deeply about the preservation of his discipline and its capable practitioners who may have been endangered by their anti-Nazi convictions, and did this often effectively in the worst of times. This was the task that came to be given him. In words that Clement Attlee applied to Winston Churchill: "He was the horse for the course."

The Positions of German Mathematicians

It is not surprising to find a contempt for the academic world in Hitler's Germany, as well as a general feeling that advanced "pure mathematics" was a frivolous excess with no practical relation to the *Volk* (similar attitudes toward "pure mathematics" are well known in nontotalitarian societies). Ludwig Bieberbach, Ernst August Weiss, Erich Jaensch, and Claus Hinrich Tietjen are examples of different people who tried to give pure mathematics *völkisch* importance instrumental to raising Nazi youth, but their effort in this direction seems to have been largely in vain. However, the neglect of most applied mathematics, even during the war, does seem surprising. The aerodynamics of Ludwig Prandtl is one exception, no doubt because of its obvious immediate applicability and the support of Göring. Another exception was Alwin Walther's computational institute at Darmstadt, which was continually well-funded and did calculations of immediate military importance, as, for example, for the rocket installations at Peenemünde. But in many other places, applied mathematics was allowed to wither: in Aachen, Theodor von Kármán was succeeded by the proponent of Aryan physics, Wilhelm Müller; in Berlin, Bieberbach promoted the astronomer Alfred Klose as Richard von Mises's replacement, despite his inadequacy as an applied mathematician, because of his political credentials; in Jena, despite the importance of the *Zeisswerke*, the appointment of a competent applied mathematician stumbled not only against international politics, but against a great variety of personal political squabbling. Wilhelm Süss's book project was extricated from the personal back-biting of Gustav Doetsch only with difficulty; Doetsch, himself, though much of his own work on Laplace Transforms was highly applicable, and though he was a *Luftwaffe* officer, produced little in this direction. Even in the area of machine computation, creative innovators like Konrad Zuse could scarcely get a hearing. The not-so-invisible hand of political machination in the name of ideology colored many, perhaps most, academic transactions. Pedagogues like Walther Lietzmann had been stressing the military importance of mathematics since before World War I, but

[259] Ostrowski 1958: 11.

[260] PAS, Robert Cecil to Rektor, June 4, 1958.

not until 1943 and the severe pressures of World War II was the importance of applications of academic work in all areas, including mathematics, recognized or addressed officially with Goebbels's Heidelberg speech about the importance of the intellectual worker and Süss's address to the Rektors.

Exactly this lack of concern, either of the Nazi academic educational establishment for mathematics, or of those Nazis who had real political power for academe, is what helped enable a man like Wilhelm Süss to carry out the protective and defensive aspects of his intermediary role for mathematics and mathematicians. Absent Süss, or someone of his skill, convictions, and intentions, the political and ideological forces of Nazi rule would have done even more damage than they did to both mathematics and mathematicians. That his efforts were as successful as they were, however, is partly because meaningful political-ideological powers found mathematics unimportant.

Was there any difference in the mathematics done inside the Third Reich? Was any significant or even reasonable mathematics done? After all, a prejudice at least as old as David Hume says that only in democratic-republican societies can science thrive. A comprehensive survey of *Mathematische Annalen*, *Mathematische Zeitschrift*, *Crelle*, and *Deutsche Mathematik* (see appendix) during the years 1933–45 does show varying biases in the content of the articles published (with *Deutsche Mathematik* beginning in 1936). While it is sometimes difficult to categorize an article, the accompanying table shows the percentage of articles and pages in "pure geometry" and "geometry +" (i.e., geometry with a significant admixture of analysis, topology, or algebra).

	Pure geometry articles	*Pure geometry pages*	*Geometry + articles*	*Geometry + pages*
Annalen	10.8%	10.1%	6.4%	6.0%
Crelle	15.0%	15.0%	5.0%	10.5%
Zeitschrift	6.8%	4.6%	11.1%	16.2%
Deutsche Mathematik	29.0%	31.3%	2.4%	2.9%

It seems fair to say that the density of geometry articles in *Deutsche Mathematik* is roughly twice that of the other journals, and the geometry is far more "pure."

As observed, men like Tornier and, perhaps more surprisingly, Bieberbach had a tendency to denigrate nongeometric mathematics, especially algebra, as insufficiently *völkisch*. E. A. Weiss was fortunate in that his Nazi political-cultural predilections and his mathematical ones coincided in his geometry. A similar remark might perhaps be made about Wilhelm Blaschke. However, such effects were only marginally, if at all, significant. Major prewar work was done in Germany by, among many others, Heinrich Behnke, Wilhelm Blaschke, Erich Hecke, Helmut Hasse, Erich Kamke, Hellmuth Kneser, Wolfgang Krull, Hans Petersson, Franz Rellich, F. K. Schmidt, and Hans Zassenhaus—not to mention, on the one hand, Oswald Teichmüller, and on the other, Carl Ludwig Siegel (who left Germany in 1940). There was significant work even in abstract logic (namely Heinrich Scholz's school, and particularly Gerhard Gentzen). Although the war caused increasing disturbance to academic activity, especially from 1942 onward, especially by the drafting of young people, academic mathematics continued in all areas, and some of the young people were rescued by the "Osenberg action."

Although German mathematics suffered tremendous losses through the expulsions, and these wounds could not be healed (even if the government had tried harder to do so), and although Nazi academic and pedagogical practices caused further weakening, nevertheless, Germany under the Nazis was far from a Sahara of mathematics.

The Nazi period in German history was not some horrible excrescence of German national character; rather, it was the product of sociohistorical events. Various social crises during the Weimar period each produced branching paths of action; the paths chosen, perhaps for purely idiosyncratic reasons on occasion, led to January 30, 1933. Hitler was not inevitable; neither was he purely accidental. But this has not been yet another study of how the Nazis came to power, or how they exercised it. Rather, it has been an attempt to examine how a particular academic discipline, mathematics, reacted to political and ideological pressures, and how its practitioners comported themselves when faced with those pressures, whether welcome or not. Here issues of "national character," or, preferably, socialization, may play a role. Most German mathematicians of the 1920s came from middle- to upper-middle-class families. The denouement of World War I in the Versailles Treaty resulted for many Germans in feelings of unfair degradation to the advantage of the Allies, a theme about which many books have been written at the time and since. For many academics, to this was added the "republican" leveling of their prestige under the Weimar Republic compared to a Wihelminian existence. Also, the lost war, the catastrophic inflation, and then the world-wide depression threatened to destroy bourgeois existence. In addition, many among the younger, established academics had served in the war—in vain. Thus it was easy for many to see the Nazi movement, for all its "plebeian excess," aimed primarily at reestablishing Germany's place among the nations.

Whatever the initial reaction among mathematicians, undeniably there was also for many an evolution of attitudes in several different directions, though for some prominent figures, of course, any evolution was illusory. Men like Wilhelm Süss and Ludwig Prandtl "supped with the devil" and attempted to alleviate his effects on their fellow human beings. Wilhelm Blaschke was too much of a pan-German nationalist cynic to have changed much during the twelve-year Reich; under the Allies, he behaved as cynically as he had under the Nazis. Ludwig Bieberbach seems to have been committed to his own mixture of racial science, psychological typology, and mathematical philosophy without much change from 1933 onward, for whatever reasons. There were also, of course, the solid examples of unchanging opponents of Nazism like Erich Kamke or Erich Hecke, even if their expressions necessarily remained somewhat muted.

But many attitudes did evolve. Gustav Doetsch, the former Catholic activist and sometime defender of academic rights, became a "110% Nazi," while of the equally active young Catholics, Ernst Peschl and Peter Thullen, one retreated into a disguised participation in the Nazi state while the other emigrated voluntarily, the difference in part probably being their career positions at the time. There was also the sort of evolution that moved Hellmuth Kneser from initial nationalist sympathy with the Nazis to gradual rejection of them, guided, it seems, by a basic set of honest conservative principles. Similar principles seem to have led Erhard Schmidt from a nationalist appreciation of what Hitler had achieved in foreign policy to an upright rejection of what the Nazis did. Helmut Hasse, as he became increasingly involved as a representative of the Nazi state in the international community, seems to have evolved in the other direction, becoming someone who was sympathetic to the Nazi state as well as an apologist for it. There were the dynamic "true believers" like Werner Weber and Oswald Teichmüller, whose dynamism sought even more significant expression, and those who acquired protective coloration for necessary reasons, like Hans Petersson. Ernst Mohr went from an impecunious and ambiguously pro-Nazi student (having denounced Hasse as anti-Hitler) to one who listened to forbidden radio broadcasts, ended in a concentration camp, and narrowly escaped a death-sentence decree.

While many other people who lived under the twelve-year Reich evolved in their attitudes, for mathematicians it shows that the mathematical notion of fixed truth logically arrived at played—unsurprisingly—no role in their nonmathematical, political, and cultural lives. Nevertheless, the notion that training as a mathematician permits better analytical and logical thinking is a common belief held, as shown, by people ranging from E. A. Weiss to the lay respondents to Heinz George's questionnaire discussed in chapter 5.

Readers will make their own judgments about these and other characters, their motivations, and the state of mathematics during the Third Reich. They will also make their own decisions about Thomas Reissinger's thesis mentioned in chapter 1: on closer examination, does mathematical training in fact incline toward misjudgment in politically and ideologically charged situations? I have tried to present the evidence, which is often not unambiguous, in a fair manner.

An enduring question, however, for any historical period in which issues of ethical behavior abound, is what one's own behavior might have been in that period. For the Nazi period, and for most mathematicians who were not automatically victims, perhaps an answer is provided by a sensitive letter written after the war by Hubert Cremer. Cremer was a student of Bieberbach and made a significant contribution to a (still not completely resolved) question in number theory. He also was apparently a tacit anti-Nazi.[261] He was best known to his colleagues, however, as the author of *Carmina Mathematica*, a collection of humorous poems on mathematical themes. On March 7, 1947, he wrote the fol-

[261] See Litten 1996: 199

lowing letter to his old acquaintance, Friedrich Levi, a well-known algebraist who, as a forced emigré, had found a position in Calcutta, India.[262]

> I send you my heartiest thanks for your friendly letter, which I received yesterday. It moved me deeply through the terrible news it contained. As I so lightly sent you "cordial greetings," I had no idea that also your poor mother and your sister were killed by those murderers' hands which have disgraced the German name forever. If this possibility had been known to me, I would never have dared to write in such a harmless fashion to you. I did not know the dead, however I had a mother, and have a sister, and it is possible for me to have some feeling what such a loss in such a way must mean. And I feel myself guilty as well. To be sure, we knew nothing of these horrors, however, indeed, we dimly suspected them; we had immeasurable and pitiable fear and because of this fear for our own lives, we were silent. Today I feel that we should have stood up and spoken out, even in the certainty of being murdered ourselves. At the time I silenced my conscience with the impoverished objection that such a sacrifice were indeed senseless, and with secret gifts of money to Jews in need. Even such actions were always in the fear that someone would find out and it would land us also in a concentration camp.
>
> This is not the first horrible news that I have received. The mother, sister, brother-in-law, and parents-in-law of my friend, Alfred Brauer, were also murdered, and indeed I am aware of the number who were murdered; however, every time it is a new blow when I find again verified in a particular case what to me is still and forever incomprehensible. May I at least express to you and your wife my honestly acknowledged participation in this terrible occurrence, which weighs on me heavily? And may I thank you that after such horrors you have not transferred to me only too justified feelings of hate; on the contrary, have written me such a cordial letter! It did me a great deal of good.
>
> In honestly acknowledged participation, I am yours,
> Hubert Cremer

[262] The letter may be found in the Levi *Nachlass* at the Universität Freiburg.

A P P E N D I X

A survey of mathematical literature in the years 1933–44 in Germany in four journals was compiled by Beata Smarczynska at the University of Rochester under my supervision. I have a detailed copy, article by article, of this survey, which roughly classified each mathematical article as to general subject matter. A summary follows. In this summary, year by year, both the number and the percentage of pages devoted to a particular subject matter are indicated.

In addition to the earlier remarks about the weight of geometry articles in *Deutsche Mathematik* as opposed to the other journals, this appendix reveals several things.

First, whereas in the *Annalen*, *Zeitschrift*, and *Deutsche Mathematik*, the number of articles in algebra was half or less than half of those in analysis (in the *Zeitschrift*, about a third), and a fifth or less of the total (in *Deutsche Mathematik*, about a ninth), in "Crelle" it was twice the amount of analysis and nearly half the total. The editor at "Crelle" was Helmut Hasse, a well-known "pure" algebraist; at the *Annalen*, the editor was Erich Hecke, and at the *Zeitschrift* it was Konrad Knopp. Knopp was an analyst, and Hecke's work was essentially in analytic number theory. "Crelle" therefore became the place for algebraic work in Germany, despite the opinions of the likes of Bieberbach and Tornier, and the expulsions of Emmy Noether, Issai Schur, and Richard Brauer, among others. For example, in addition to Hasse, Wolfgang Krull, Max Deuring, Bartel van der Waerden, Ernst Witt, and Hans Zassenhaus, as well as other algebraists, remained in Germany throughout the Nazi period.

One can also note that while the amount of topology published was always small, it was primarily published in the *Annalen*, and it fell off considerably after 1935.

Again, "Crelle" and *Deutsche Mathematik* each published about half as many physics-related mathematics pages as the *Annalen* or the *Zeitschrift*. Also, *Deutsche Mathematik* published a surprising amount of probability, though presumably often with a tendentious Nazi slant. (In making the tables here, probability and statistics were grouped together.) On the other hand, "Crelle" published almost no probability.

Readers with specific interests will make their own discernments about the information in these tables.

Percentages do not sum to 100 for the usual reasons: because the percentage of pages devoted to an article was carried to only two decimal points, and because articles do not usually fill a last page.

TABLE A.1
Mathematische Zeitschrift, by Article Subject, 1933–44

	Year(s) (Volume no.)						
	1933 (1) (vol. 36)	*1933 (2) (vol. 37)*	*1933–34 (vol. 38)*	*1934–35 (vol. 39)*	*1935–36 (vol. 40)*	*1936 (vol. 41)*	*1936–37 (vol. 42)*
Algebra							
No. of articles	7	10	10	13	14	18	10
No. of pages	134	176	204	268	145	218	147
% of total pages	13.09	20.42	26.19	32.64	16.88	27.42	10.53
Analysis							
No. of articles	30	39	18	31	26	22	22
No. of pages	528	431	276	354	381	368	367
% of total pages	51.56	50	35.43	43.12	44.35	46.29	57.89
Geometry							
No. of articles	0	3	4	3	8	2	5
No. of pages	0	24	43	33	57	44	41
% of total pages	0	2.78	5.52	4.02	6.64	5.53	10.53
Logic							
No. of articles	0	2	0	4	0	2	0
No. of pages	0	28	0	89	0	30	0
% of total pages	0	3.25	0	10.84	0	3.77	2.63
Topology							
No. of articles	2	1	0	0	3	4	1
No. of pages	33	17	0	0	44	41	17
% of total pages	4	1.97	0	0	5.12	5.16	2.63
Probability							
No. of articles	0	2	0	1	2	1	2
No. of pages	0	38	0	9	72	20	37
% of total pages	0	4.41	0	1.1	8.38	2.52	2.63
Misc.							
No. of articles	0	1	1	0	0	0	0
No. of pages	0	16	2	0	0	0	0
% of total pages	0	1.86	0.26	0	0	0	0
Physics							
No. of articles	8	4	2	2	3	0	0
No. of pages	201	68	45	29	34	0	0
% of total pages	19.63	7.89	5.78	3.53	3.96	0	0
Biology	0	0	0	0	0	0	0
A/G[1]							
No. of articles	0	1	0	0	0	0	1
No. of pages	0	11	0	0	0	0	24
% of total pages	0	1.28	0	0	0	0	2.63

Year(s)
(Volume no.)

1937–38 (vol. 43)	*1938–39 (vol. 44)*	*1939 (vol. 45)*	*1940 (vol. 46)*	*1940–42 (vol. 47)*	*1942–43 (vol. 48)*	*1943–44 (vol. 49)*	*1944 (vol. 50)*
8	6	9	3	10	12	4	0
156	77	158	38	113	210	55	0
18.2	9.47	17.79	4.76	14.56	30.26	6.92	0
19	23	30	25	25	21	22	3
400	352	444	371	368	342	428	68
46.67	43.3	50	46.43	47.42	49.28	53.84	42.5
2	3	4	5	3	2	4	0
60	26	59	65	31	29	30	0
7	3.2	6.64	8.14	3.99	4.18	3.77	0
0	1	2	1	2	0	1	0
0	9	34	16	42	0	15	0
0	1.11	3.83	2	5.41	0	1.89	0
1	1	0	0	0	0	1	0
21	14	0	0	0	0	19	0
2.45	1.72	0	0	0	0	2.39	0
0	1	0	0	0	0	1	0
0	7	0	0	0	0	3	0
0	0.86	0	0	0	0	0.38	0
0	0	0	0	0	0	0	0
0	0	0	0	0	0	0	0
0	0	0	0	0	0	0	0
2	0	3	0	3	1	0	0
65	0	64	0	41	4	0	0
7.58	0	7.21	0	5.28	0.58	0	0
0	0	0	0	0	0	0	0
0	1	2	4	2	1	1	0
0	32	55	92	47	3	27	0
0	3.94	6.19	11.51	6.06	0.43	3.4	0

TABLE A.1
(*continued*)

	Year(s) (Volume no.)						
	1933 (1) (vol. 36)	*1933 (2) (vol. 37)*	*1933–34 (vol. 38)*	*1934–35 (vol. 39)*	*1935–36 (vol. 40)*	*1936 (vol. 41)*	*1936–37 (vol. 42)*
An/Al[2]							
No. of articles	3	2	4	2	2	1	0
No. of pages	128	17	59	39	22	4	0
% of total pages	12.5	1.97	7.57	4.75	2.56	0.5	0
An/G[3]							
No. of articles	0	3	9	0	7	6	8
No. of pages	0	36	150	0	104	70	149
% of total pages	0	4.18	19.28	0	12.11	8.81	7.89
T/G[4]	0	0	0	0	0	0	0
Total							
No. of articles	50	68	48	56	65	56	49
No. of pages	1024	862	779	821	859	795	782
% of total pages	100.78	100.01	100	100	100	100	99.99

[1]A/G = a mixture of algebra and geometry.
[2]An/Al = a mixture of analysis and algebra.
[3]An/G = a mixture of analysis and geometry.
[4]T/G = a mixture of topology and geometry.

Year(s) (Volume no.)							
1937–38 (vol. 43)	*1938–39 (vol. 44)*	*1939 (vol. 45)*	*1940 (vol. 46)*	*1940–42 (vol. 47)*	*1942–43 (vol. 48)*	*1943–44 (vol. 49)*	*1944 (vol. 50)*
1	4	1	2	1	1	1	0
7	94	13	37	16	3	28	0
0.82	11.56	1.46	4.63	2.06	0.43	3.52	0
7	5	2	6	5	4	3	1
148	202	61	180	118	103	190	92
17.27	24.85	6.87	21.53	15.21	14.84	23.9	57.5
0	0	0	0	0	0	0	0
40	45	53	46	51	42	38	4
857	813	888	799	776	694	795	160
99.99	100.01	99.99	100	99.99	100	100.01	100

TABLE A.2
Mathematische Annalen, by Article Subject, 1932–44

	Year(s) (Volume no.)					
	1932–33 (vol. 107)	*1933 (vol. 108)*	*1933–34 (vol. 109)*	*1934–35 (vol. 110)*	*1935 (vol. 111)*	*1935–36 (vol. 112)*
Algebra						
No. of articles	15	7	11	10	22	12
No. of pages	301	102	83	171	311	177
% of total pages	35.56	12.94	9.71	21.67	38.25	21.72
Analysis						
No. of articles	18	18	22	20	20	16
No. of pages	307	365	428	306	342	350
% of total pages	34.23	46.32	50.06	38.78	42.07	42.94
Geometry						
No. of articles	5	6	6	4	8	4
No. of pages	95	101	89	103	52	65
% of total pages	10.59	12.82	10.41	13.05	6.4	7.98
Logic						
No. of articles	2	1	2	3	2	3
No. of pages	27	19	62	79	22	103
% of total pages	3.01	2.41	7.25	10.01	2.71	12.64
Topology						
No. of articles	3	3	5	1	3	2
No. of pages	95	85	105	22	17	26
% of total pages	10.59	10.79	12.28	2.79	2.09	3.19
Probability						
No. of articles	2	3	2	0	0	2
No. of pages	59	30	22	0	0	12
% of total pages	6.58	3.81	2.57	0	0	1.47
Misc.						
No. of articles	0	0	1	0	2	1
No. of pages	0	0	7	0	10	11
% of total pages	0	0	0.82	0	1.23	1.35
Physics						
No. of articles	1	0	2	1	4	2
No. of pages	4	0	43	9	32	29
% of total pages	0.45	0	5.03	1.14	3.94	3.56
Biology		0	0	0	0	0
Al/G[1]						
No. of articles	0	4	3	3	2	0
No. of pages	0	40	12	52	27	0
% of total pages	0	5.08	1.4	6.59	3.32	0

Year(s) (Volume no.)						
1936–37 (vol. 113)	*1937 (vol. 114)*	*1937–38 (vol. 115)*	*1938–39 (vol. 116)*	*1939–41 (vol. 117)*	*1941–43 (vol. 118)*	*1943–44 (vol. 119)*
7	16	8	12	4	16	0
105	219	98	224	57	257	0
12.73	24.89	11.33	26.05	6.83	33.33	0
20	17	20	21	23	14	11
324	278	473	377	483	263	252
39.27	31.59	54.68	43.84	57.84	34.11	74.34
4	5	6	4	5	5	2
84	120	63	40	85	98	45
10/18	13.64	7.28	4.65	10.18	12.71	13.27
1	1	2	0	2	2	1
39	11	44	0	54	16	22
4.73	1.25	5.09	0	6.47	2.08	6.49
0	0	1	2	1	2	0
0	0	23	17	13	53	0
0	0	2.66	1.98	1.56	6.87	0
2	0	0	0	1	2	0
55	0	0	0	19	59	0
6.67	0	0	0	2.28	7.65	0
0	0	0	1	1	0	0
0	0	0	9	21	0	0
0	0	0	1.05	2.51	0	0
3	5	3	4	2	0	1
35	100	81	79	20	0	20
4.24	11.36	9.36	9.19	2.4	0	5.9
0	0	0	0	0	0	0
6	4	4	0	2	0	0
122	58	83	0	83	0	0
14.79	6.59	9.6	0	9.94	0	0

Table A.2
(continued)

	Year(s) (Volume no.)					
	1932–33 (vol. 107)	*1933 (vol. 108)*	*1933–34 (vol. 109)*	*1934–35 (vol. 110)*	*1935 (vol. 111)*	*1935–36 (vol. 112)*
An/Al[2]						
No. of articles	0	3	0	3	0	0
No. of pages	0	46	0	38	0	0
% of total pages	0	5.84	0	4.82	0	0
An/G[3]						
No. of articles	1	0	0	0	0	1
No. of pages	9	0	0	0	0	36
% of total pages	1	0	0	0	0	4.42
T/G[4]						
No. of articles	0	0	1	1	0	1
No. of pages	0	0	4	9	0	6
% of total pages	0	0	0.47	1.14	0	0.74
Total						
No. of articles	47	45	55	46	63	44
No. of pages	897	788	855	789	813	815
% of total pages	100.01	100.01	100	99.99	100.01	100.01

[1]A/G = a mixture of algebra and geometry.
[2]An/Al = a mixture of analysis and algebra.
[3]An/G = a mixture of analysis and geometry.
[4]T/G = a mixture of topology and geometry.

Year(s) (Volume no.)						
1936–37 (vol. 113)	*1937 (vol. 114)*	*1937–38 (vol. 115)*	*1938–39 (vol. 116)*	*1939–41 (vol. 117)*	*1941–43 (vol. 118)*	*1943–44 (vol. 119)*
2	3	0	3	0	2	0
25	76	0	90	0	19	0
3.03	8.64	0	10.47	0	2.46	0
1	0	0	1	0	1	0
36	0	0	24	0	6	0
4.36	0	0	2.79	0	0.78	0
0	0	0	0	0	0	0
0	0	0	0	0	0	0
0	0	0	0	0	0	0
46	52	44	48	41	44	15
825	880	865	860	835	771	339
100	100.01	100	100.02	100.01	99.99	100

TABLE A.3
Journal Fuer Reine Und Eingewandte Mathematik, by Article Subject, 1932–49

	Year(s) (Volume no.)							
	1932–33 (vol. 169)	*1933–34 (vol. 170)*	*1934 (vol. 171)*	*1934–35 (vol. 172)*	*1935 (vol. 173)*	*1935–36 (vol. 174)*	*1936 (vol. 175)*	*1936–37 (vol. 176)*
Algebra								
No. of articles	7	7	12	7	9	9	12	18
No. of pages	68	89	153	77	117	143	106	158
% of total pages	27.42	35.32	60.71	27.8	46.06	55.64	42.4	62.95
Analysis								
No. of articles	8	8	3	7	4	5	4	4
No. of pages	96	113	66	95	52	91	71	69
% of total pages	38.71	44.84	26.19	34.3	20.47	35.41	28.4	27.49
Geometry								
No. of articles	1	3	0	2	5	2	3	1
No. of pages	8	32	0	48	54	15	46	7
% of total pages	3.23	12.7	0	17.33	21.26	5.84	18.4	2.79
Logic								
No. of articles	0	1	0	0	0	0	0	0
No. of pages	0	6	0	0	0	0	0	0
% of total pages	0	2.38	0	0	0	0	0	0
Topology								
No. of articles	1	0	0	0	3	0	0	0
No. of pages	8	0	0	0	31	0	0	0
% of total pages	4.76	0	0	0	12.2	0	0	0
Probability								
No. of articles	0	0	0	0	0	0	0	0
No. of pages	0	0	0	0	0	0	0	0
% of total pages	0	0	0	0	0	0	0	0
Misc.								
No. of articles	0	1	0	0	0	0	1	0
No. of pages	0	3	0	0	0	0	16	0
% of total pages	0	1.19	0	0	0	0	6.4	0
Physics								
No. of articles	0	0	0	2	0	1	0	0
No. of pages	0	0	0	25	0	5	0	0
% of total pages	0	0	0	9.03	0	1.95	0	0
Biology	0	0	0	0	0	0	0	0
Al/G[1]								
No. of articles	1	1	2	0	0	0	0	0
No. of pages	25	9	33	0	0	0	0	0
% of total pages	10.08	3.57	13.1	0	0	0	0	0

Year(s)
(Volume no.)

1937 (vol. 177)	1937–38 (vol. 178)	1938 (vol. 179)	1939 (vol. 180)	1939–40 (vol. 181)	1940 (vol. 182)	1940–41 (vol. 183)	1942 (vol. 184)	1943 (vol. 185)	1944–49 (vol. 186)
11	6	8	5	4	18	9	9	8	5
122	67	128	111	74	144	92	133	60	59
49.59	27.8	49.81	43.53	17.62	50.35	30.98	52.57	30.83	22.1
2	5	2	4	4	2	3	0	3	6
41	52	15	80	98	46	54	0	92	72
16.67	21.58	5.84	31.37	23.33	16.08	18.18	0	31.94	26.97
5	5	3	2	2	1	4	5	2	6
58	122	36	35	35	7	55	61	28	82
23.58	50.62	14.01	13.73	8.33	2.45	18.52	24.11	9.72	30.71
0	0	0	1	0	1	0	1	0	0
0	0	0	6	0	1	0	20	0	0
0	0	0	2.35	0	0.35	0	7.91	0	0
1	0	0	1	0	1	0	1	1	0
18	0	0	12	0	17	0	26	13	0
7.32	0	0	4.71	0	5.94	0	10.28	4.51	0
0	0	0	0	0	0	0	0	0	1
0	0	0	0	0	0	0	0	0	9
0	0	0	0	0	0	0	0	0	3.37
0	0	0	0	0	0	0	0	0	0
0	0	0	0	0	0	0	0	0	0
0	0	0	0	0	0	0	0	0	0
0	0	1	1	0	1	2	0	1	2
0	0	9	11	0	40	46	0	38	14
0	0	3.5	4.31	0	13.99	15.49	0	13.19	5.24
0	0	0	0	0	0	0	0	0	0
0	0	1	0	2	1	0	1	0	1
0	0	45	0	213	31	0	6	0	11
0	0	17.51	0	50.71	10.84	0	2.37	0	4.12

TABLE A.3
(continued)

	Year(s) (Volume no.)							
	1932–33 (vol. 169)	*1933–34 (vol. 170)*	*1934 (vol. 171)*	*1934–35 (vol. 172)*	*1935 (vol. 173)*	*1935–36 (vol. 174)*	*1936 (vol. 175)*	*1936–37 (vol. 176)*
An/Al[2]								
No. of articles	3	0	0	2	0	0	1	0
No. of pages	43	0	0	32	0	0	11	0
% of total pages	17.34	0	0	11.55	0	0	4.4	0
An/G[3]								
No. of articles	0	0	0	0	0	1	0	1
No. of pages	0	0	0	0	0	3	0	17
% of total pages	0	0	0	0	0	1.17	0	6.77
T/G[4]								
No. of articles	0	0	0	0	0	0	0	0
No. of pages	0	0	0	0	0	0	0	0
% of total pages	0	0	0	0	0	0	0	0
Total								
No. of articles	21	21	17	46	21	18	21	24
No. of pages	248	252	252	277	254	257	250	251
% of total pages	101.54	100	100	100	99.99	100.01	100	100

[1]A/G = a mixture of algebra and geometry.
[2]An/Al = a mixture of analysis and algebra.
[3]An/G = a mixture of analysis and geometry.
[4]T/G = a mixture of topology and geometry.

Year(s)
(Volume no.)

1937 (vol. 177)	1937–38 (vol. 178)	1938 (vol. 179)	1939 (vol. 180)	1939–40 (vol. 181)	1940 (vol. 182)	1940–41 (vol. 183)	1942 (vol. 184)	1943 (vol. 185)	1944–49 (vol. 186)
1	0	1	0	0	0	1	1	0	1
7	0	9	0	0	0	19	7	0	7
2.85	0	0	0	0	0	6.4	2.77	0	2.62
0	0	1	0	0	0	0	0	1	1
0	0	15	0	0	0	0	0	47	13
0	0	5.84	0	0	0	0	0	16.32	4.87
0	0	0	0	0	0	1	0	1	0
0	0	0	0	0	0	31	0	10	0
0	0	0	0	0	0	10.44	0	3.47	0
20	16	17	14	12	25	20	18	17	23
246	241	257	255	420	286	297	253	288	267
100.1	100	95.51	100	99.99	100	100.01	100.01	99.98	100

TABLE A.4
Deutsche Mathematik, by Article Subject, 1936–44

	Year(s) (Volume no.)						
	1936 (vol. 1)	*1937 (vol. 2)*	*1938 (vol. 3)*	*1939 (vol. 4)*	*1940–41 (vol. 5)*	*1941–42 (vol. 6)*	*1942–44 (vol. 7)*
Algebra							
No. of articles	15	8	5	11	6	8	8
No. of pages	171	88	45	255	71	55	77
% of total pages	20.19	13.08	7.29	28.94	7.64	9.15	12.24
Analysis							
No. of articles	14	23	14	15	18	19	14
No. of pages	124	221	241	277	239	190	159
% of total pages	14.64	32.84	39.06	31.44	25.73	31.61	25.28
Geometry							
No. of articles	30	17	14	15	15	14	18
No. of pages	297	149	218	167	367	206	214
% of total pages	35.06	22.14	35.33	18.96	39.5	31.61	34.02
Logic							
No. of articles	1	0	2	2	0	0	1
No. of pages	40	0	21	14	0	0	43
% of total pages	4.72	0	3.4	1.59	0	0	6.84
Topology							
No. of articles	0	0	0	1	0	0	0
No. of pages	0	0	0	15	0	0	0
% of total pages	0	0	0	1.7	0	0	0
Probability							
No. of articles	10	7	1	1	4	1	3
No. of pages	95	82	3	23	26	6	79
% of total pages	12.05	12.18	0.49	2.61	2.8	1	12.56
Misc.							
No. of articles	4	5	4	3	11	10	2
No. of pages	53	51	37	37	137	86	8
% of total pages	6.26	7.58	6	4.2	14.75	14.31	1.27
Physics							
No. of articles	3	4	1	3	3	1	3
No. of pages	24	35	3	57	50	3	49
% of total pages	2.83	5.2	0.49	6.47	5.38	0.5	7.79
Biology							
No. of articles	6	2	0	1	2	2	0
No. of pages	43	26	0	23	13	9	0
% of total pages	5.08	3.86	0	2.61	1.4	1.5	0

TABLE A.4
(continued)

	Year(s) (Volume no.)						
	1936 (vol. 1)	*1937 (vol. 2)*	*1938 (vol. 3)*	*1939 (vol. 4)*	*1940–41 (vol. 5)*	*1941–42 (vol. 6)*	*1942–44 (vol. 7)*
Al/G[1]	0	0	0	0	0	0	0
An/Al[2]							
No. of articles	0	0	0	0	2	0	0
No. of pages	0	0	0	0	6	0	0
% of total pages	0	0	0	0	0.65	0	0
An/G[3]							
No. of articles	0	1	3	2	1	3	0
No. of pages	0	21	49	13	20	46	0
% of total pages	0	3.12	7.94	1.48	2.15	7.65	0
T/G[4]	0	0	0	0	0	0	0
Total							
No. of articles	83	67	44	54	62	58	49
No. of pages	847	673	617	881	929	601	629
% of total pages	100.83	100	100	100	100	100	100

[1]A/G = a mixture of algebra and geometry.
[2]An/Al = a mixture of analysis and algebra.
[3]An/G = a mixture of analysis and geometry.
[4]T/G = a mixture of topology and geometry.

BIBLIOGRAPHY

This bibliography contains works used in the preparation of this book, though not all of them are cited explicitly within it, and not all explicit citations are herein included.

Books, Articles, and Manuscripts

Abendroth, W. 1966. "Das Unpolitische als Wesensmerkmal der deutschen Universität." In *Universitätstage* 1966: 189–208.

Abikoff, W. 1986. "Oswald Teichmüller." *Mathematical Intelligencer* 8:8–16, 33.

Ackerman, N., and M. Jahoda. 1950. *Antisemitism and Emotional Disorder*. American Jewish Committee. New York: Harper.

Adam, U. 1977. *Hochschule und Nationalsozialismus, Die Universität Tübingen im Dritten Reich*. Tübingen: Mohr.

Alexanderson, G. L., ed. 1987. *The Pólya Picture Album: Encounters of a Mathematician*. Boston and Basel: Birkhauser.

Alexandroff, P. 1969. "Die Topologie in und um Holland in der Jahren, 1920–1930." *Nieuw Archief voor Wiskunde* 17:109–127.

Aly, G., and K. H. Roth. 1984. *Die restlose Erfassung*. Berlin: Rotbuch Verlag.

Arendt, H. 1980. *The Origins of Totalitarianism*. 1 vol. revised ed. Orlando, Fla.: Harcourt Brace. [Original ed., 1950.]

———. 1994. *Eichmann in Jerusalem*. New York: Penguin Books. [Original ed., 1964.]

Atti del Congress Internazionale del Matematici. 1928. Bologna.

Ayçoberry, P. 1981. *The Nazi Question*. Translated by Robert Hurley. New York: Pantheon.

Bahr, H. 1894. *Der Antisemitismus. Ein Internationales Interview*. Berlin: S. Fischer Verlag.

Bahrdt, H. P. 1966. "Soziologische Reflexionen über die gesellschaftlichen Voraussetzungen des Antisemitismus in Deutschland." In W. Mosse, ed., *Entscheidungsjahr, 1932*, 135–155. Tübingen: Mohr.

Baillaud, B., and H. Bourget. 1905. *Correspondence d'Hermite et de Stieltjes*. 2 vols. Paris: Gauthier-Villars.

Barbu, Z. 1956. *Democracy and Dictatorship*. New York: Grove.

Baynes, N. 1942. *The Speeches of Adolf Hitler, April 1922–August 1939*. 2 vols. continuously paginated. Vol. 1. London: Oxford.

Becker, H., H.-T. Dahms, and C. Wegeler, eds. 1987. *Die Universität Göttingen unter dem Nationalsozialismus*. Munich: Saur.

Begehr, H., ed. 1998. *Mathematik in Berlin*. Aachen: Shaker Verlag.

Behnke, H. 1978. *Semesterberichte*. Göttingen: Vandenhoeck und Ruprecht.

———. N.d. "An einer deutschen Universität im braunen Sturm." Unpublished manuscript; probably a draft version of Behnke 1978.

Bekenntnis der Professoren an den deutschen Universitäten und Hochschulen zu Adolf Hitler. 1933. Dresden: W. Limpert.

Bell, E. T. 1940. *The Development of Mathematics*. New York and London: McGraw-Hill.

Benaceraff. P., and H. Putnam. 1964. *Philosophy of Mathematics*. Englewood Cliffs, N.J.: Prentice-Hall.

Benda, J. 1969. *La trahison des clercs (The Treason of the Intellectuals)*. Translated by R. Aldington. New York: Norton.

Benze, R. 1939. *Erziehung im Grossdeutschen Reich*. Frankfurt am Main: Moritz Diesterweg.

Berben, P. 1975. *Dachau, 1935–45, The Official History*. London: Norfolk.

Beyerchen, A. 1977. *Scientists under Hitler*. New Haven: Yale University Press.

Bieberbach, L. 1914. "Über die Grundlagen der Moderne Mathematik." *Die Geisteswissenschaften* 1, no. 33: 896–901.

———. 1934a. "Persönlichkeitsstruktur und Mathematisches Schaffen." *Unterrichtsblätter für Mathematik und Naturwissenschaften* 40:236–243.

———. 1934b. "Stilarten mathematischen Schaffens." *Sitzungsberichte Akademie Berlin* 351–360.

———. 1940. "Die völkische Verwurzelung der Wissenschaft." *Sitzungsberichte Heidelberger Akad. d. Wiss.*

Biermann, K. R. 1988. *Die Mathematik und Ihre Dozenten an der Berliner Universität, 1810–1933*. Berlin: Akademie Verlag.

Bigalke, H.-G. 1988. *Heinrich Heesch*. Boston and Basel: Birkhäuser.

Birkhoff, G. D. 1938. "Fifty Years of American Mathematics." In *A Semicentennial History of the American Mathematical Society, 1888–1938*, vol. 2 (Addresses), 270–315. New York: Arno.

Blaschke, W. 1923. *Vorlesungen über Differentialgeometrie II*. Berlin: Springer.

———. 1935. *Integralgeometrie*. Paris: Hermann and Co.

———. 1937. "Sulla proprietà isoperimetrica del cerchio." *Rendiconti di Mat. Roma* 4:233–234. Reprinted in Blaschke 1992, 2:329–330.

———. 1941. *Mathematik und Leben*. Leipzig and Berlin: Teubner.

———. 1950. *Einführung in die Differential Geometrie*. Berlin: Springer.

———. 1982. *Gesammelte Werke*. 6 vols. Essen: Thales Verlag.

Blaschke, W., and G. Bol. 1938. *Geometrie der Gewebe*. Berlin: Springer.

Blau, B. 1952. *Das Ausnahmerecht für die Juden in den europäischen Ländern, 1933–1945*. New York.

Bleuel, H. P. 1968. *Deutschlands Bekenner*. Bern: Scherz.

Bleuel, H. P., and Ernst Klinnert. 1967. *Deutsche Studenten auf dem Weg ins Dritte Reich*. Gütersloh: Siegbert Mohn.

Bömer, K., ed. 1936. *Deutsche Saat in fremder Erde*. Berlin: Zeitgeschichte.

Bonjour, E. 1960. *Die Universität Basel von den Anfängen bis zur Gegenwart, 1460–1960*. Basel: Verlag Helbing and Lichtenhahn.

Born, M. 1971. *The Born-Einstein Letters, with commentary by Max Born*. Translated by Irene Born. New York: Walker.

Bottin, A. 1992. *Enge Zeit*. Berlin and Hamburg: Dietrich Reimer Verlag.

Bracher, K. D. 1966. "Die Gleichschaltung der deutschen Universität." In *Universitätstage 1966*: 126–142.

Brewer, J. W., and M. Smith, eds. 1981. *Emmy Noether: A Tribute to Her Life and Work*. New York and Basel: Marcel Dekker.

Broszat, M. 1981. *The Hitler State*. Translated by John Hiden. London and New York: Longmans. [German original *Der Staat Hitlers*, published 1969.]

———. 1984. *Die Machtergreifung*. Munich: Deutscher Taschenbuch Verlag.

Brouwer, L.E.J. 1929. "Mathematik, Wissenschaft und Sprache." *Monatshefte für Mathematik und Physik* 36:133–164.

Bullock, A. 1962. *Hitler, A Study in Tyranny*. Revised ed. Harmondsworth: Pelican. [Original ed., 1952.]

Chapman, J. J. 1914. *Deutschland über Alles or Germany Speaks: A Collection of the Ut-*

terances of Representative Germans—Statesmen, Military Leaders, Scholars, and Poets—in Defense of the War Policies of the Fatherland. New York and London: G. P. Putnam's Sons.

"Cinq lettres sur la Théorie des Ensembles." 1905. *Bullètin de Mathèmatiques* 33:261–273.

Dauben, J. W. 1979. *Georg Cantor, His Mathematics and Philosophy of the Infinite.* Cambridge, Mass.: Harvard University Press.

Deuerlein, E. 1974. *Der Aufsteig der NSDAP in Augenzeugenberichten.* Paperback ed. Munich: Deutscher Taschenbuch Verlag.

Dick, A. 1970. *Emmy Noether, 1882–1935.* Beiheft to Elemente der Mathematik. Basel: Birkhäuser.

Dicks, H. V. 1950. "Personality Traits and National Socialist Ideology." *Human Relations* 3:111–154.

Dictionary of Scientific Biography. 1970. Charles Gillispie, editor-in-chief. 16 vols. plus 2 suppl. vols. American Council of Learned Societies. New York: Scribners.

Die Berliner Akademie der Wissenschaften in der Zeit des Imperialismus. 1975. Vol. 2, *1917–1933*; vol. 3, *1935–1945.* Berlin: Akademie Verlag.

Dierksmann, M., et al. 1967. "Felix Hausdorff zum Gedächtnis." *JDMV* 69:51–76.

Dinghas, A. 1998. "Erinnerungen aus der letzen Jahren des Mathematischen Instituts der Universität Berlin." In Begehr 1998: 183–203.

Dipper, C. 1983. "Der Deutsche Widerstand und die Juden." *Geschichte u. Gesellschaft* 9:349–380.

Dnieprov, A. 1962. "Maxwell's Equations." *Beyond Amaltheia.* Moscow: Foreign Language Publishers.

Dorner, A., et al. 1936. *Mathematik im Dienste der nationalpolitischen Erziehung.* 3d ed. Frankfurt am Main: Moritz Diesterweg.

Ebel, W. 1935. "Über Sinn und Gestaltung des Rechenbuchs." *Deutsches Bildungswesen* 3:761–768.

Ebert, H. N.d. "Mathematiker im KZ 1944/1945." Unpublished manuscript, copied from a copy in the possession of Herbert Mehrtens.

Ein Jahrhundert Mathematik, 1890–1990. 1990. Deutsche Mathematiker-Vereinigung. Braunschweig and Wiesbaden: Vieweg.

Encyclopedia of the Holocaust. 1990. New York: MacMillan.

Erdmann, K. 1980. *Deutschland unter der Herrschaft des Nationalsozialismus, 1933–1939.* Deutscher Taschenbuch Munich: Verlag.

Eschenburg, T. 1965. "Aus dem Universitätsleben vor 1933." In *Deutsches Geistesleben und Nationalsozialismus,* 23–46. Tübingen: Wunderlich.

Eubank, K. 1963. *Munich.* Norman: University of Oklahoma Press.

Faust, A. 1973. *Der Nationalsozialistische Studentenbund.* 2 vols. Düsseldorf: Schwann.

Fenchel, W. 1980. "Erinnerungen aus der Studienzeit." *Überblicke Mathematik*: 155–166.

Festinger, L. 1957. *A Theory of Cognitive Dissonance.* Evanston, Ill.: Row, Peterson.

Festinger, L., H. Riecken, and S. Schlachter. 1959. *When Prophecy Fails.* Reprint. New York: Harper Books. Original ed., Minneapolis: University of Minnesota Press, 1956.

Festschrift Ernst Mohr zum 75 Geburtstag am 20. April 1985. 1985. Ed. K. H. Förster. Berlin: Universitätsbibliothek der Technischen Universität Berlin.

Fischer, H. J. 1984–85. *Erinnerungen.* Vol. 1: *Von der Wissenschatt zum Sicherheitsdienst*; vol. 2: *Feuerwehr für die Forschung.* Ingolstadt: Quellenstudien der Zeitgeschichtlichen Forschungsstelle Ingolstadt.

Fisher, C. 1966. "The Death of a Mathematical Theory." *Archive for the History of the Exact Sciences* 3:137–159.

———. 1967. "The Last Invariant Theorists." *Archives Europeenes de Sociologie* 8:216–244.
Flitner, W. 1986. "Das war unser Weg." *Deutsches Allgemeines Sonntagsblatt* 46 (Nov. 16).
Focke, H., and U. Reimer. 1995. *Alltag Unterm Hakenkreuz*. Reinbek bei Hamburg: Rowohlt.
Forman, P. 1971. "Weimar Culture, Causality, and Quantum Theory, 1918–1927." *Historical Studies in the Physical Sciences* 3:1–116.
———. 1973. "Scientific Internationalism and the Weimar Physicists: The Ideology and Its Manipulation in Germany after World War I." *Isis* 64:150–180.
Fraenkel, A. A. 1967. *Lebenskreise*. Stuttgart: Deutsche Verlags-Anstalt.
Frei, G. 1985. "Helmut Hasse (1898–1979)." *Expositiones Mathematicae* 3 (1985): 55–69.
Frei, N. 1987. *Der Führerstaat*. Munich: Deutscher Taschenbuch Verlag.
Freudenthal, H. 1987. *Berlin, 1923–1930*. Berlin: Walter de Gruyter.
Frewer, M. 1979. "Das Wissenschaftliche Werk Felix Bersteins." Thesis, Göttingen.
Fricke, D. 1960. "Zur Militarisierung des deutschen Geistesleben im wilhelminischen Kaiserreich: Der Fall Leo Arons." *Zeitschrift für Geschichtswissenschaft*.
Fromm, E. 1963. *War within Man*. Philadelphia: American Friends Service Committee.
Gamm, H. 1963. *Der Flusterwitz im Dritten Reich*. Munich: List Verlag.
Gardner, M. 1957. *Fads and Fallacies in the Name of Science*. New York: Dover.
Gasman, D. 1971. *The Scientific Origins of National Socialism*. London: MacDonald.
Gay, P. 1968. *Weimar Culture: The Outsider as Insider*. New York: Harper and Row.
General Inequalities 2. 1980. Basel: Birkhäuser Verlag.
General Inequalities 3. 1983. Basel: Birkhäuser Verlag.
Genuneit, J. 1984. "Mein Rechen-Kampf." In K. Fuchs, ed., *Stuttgart im Dritten Reich, Die Jahre von 1933 bis 1939*, 205–236. Stuttgart: Landeshauptstadt Stuttgart.
George, H. 1937. "Laienurteile über den Lebenswert der Mathematik." *Zeitschrift für angewandte Psychologie* 53:80–112.
Gericke, H. 1972. "50 Jahre GAMM." *Ingenieur-Archiv* 41 (supplement).
Geschichte der Universität Rostock, 1419–1969. 1969. Berlin: VEB Deutscher Verlag der Wissenschaften.
Geuter, U. 1985. "Nationalsozialistische Ideologie und Psychologie." In *Geschichte der deutschen Psychologie im 20. Jahrhundert*, 172–200. Opladen: Westdeutscher Verlag.
Giles, G. J. 1985. *Students and National Socialism in Germany*. Princeton: Princeton University Press.
Gillies, D., ed. 1992. *Revolutions in Mathematics*. Oxford: Clarendon.
Glass, Bentley. 1962. "Liberal Education in a Scientific Age." In Paul Obler and Herman Estrin, eds., *The New Scientist: Essays on the Methods and Values of Modern Science*, 215–238. Garden City, N.Y.: Anchor Books.
Goebbels, J. 1972. *Reden*. Edited by H. Heiber. Vol. 2. Düsseldorf: Droste Verlag.
Gödel, K. 1947. "What Is Cantor's Continuum Problem?" *American Mathematical Monthly* 54:515–525.
Graml, H. 1988. *Reichskristallnacht*. Munich: Deutscher Taschenbuch Verlag.
Grunsky, H. 1986. Obituary of Ludwig Bieberbach. *JDMV* 88:190–205.
Gumbel, E. 1984. *Verschwörer*. Reissue. Foreword by Karen Buselmeier. Frankfurt am Main: Fischer Taschenbuch Verlag. [Original ed., 1924.]
Hadamard, Jacques. 1949. *The Psychology of Invention in the Mathematical Field*. Princeton: Princeton University Press.
Hamel, G. 1928. "Ueber die philosophische Stellung der Mathematik." In *Akademische Schriftenreihe der Technischen Hochschule Charlottenburg*. Berlin: Technical University of Berlin.

Hanson, J. 1981. "Nazi Aesthetics." *Psychohistory Review* 251–281.

Hasse, Helmut. 1952. *Mathematik als Wissenschaft, Kunst, und Macht.* Wiesbaden: Verlag für Angewandte Wissenschaft.

Hawkins, T. 1970. *Lebesgue's Theory of Integration, Its Origin and Development.* Madison: University of Wisconsin.

Heffter, L. 1952. *Beglückte Ruckschau auf Neun Jahrzehnte.* Freiburg: Hans Ferdinand Schultz.

Hegel, G.F.W. 1977. *Phenomenology of the Mind.* Translated by J. B. Baillie. 2d ed. New York: Humanities Press. [Original ed., London, 1931.]

Heiber, H. 1966. *Walter Frank, u. sein Reichsinstitut für Geschichte des Neuen Deutschland.* Stuttgart: Deutsche Verlags-Anstalt.

Hellpach, W. 1949. *Wirken im Wirren.* Vol. 2, *Lebenserinnerungen, 1914–1925.* Hamburg: C. Wegner.

Herf, J. 1984. *Reactionary Modernism: Technology, Culture, and Politics in Weimar and the Third Reich.* Cambridge: Cambridge University Press.

Hertzman, L. 1963. *DNVP: Right-Wing Opposition in the Weimar Republic, 1918–1924.* Lincoln: University of Nebraska Press.

Herwig, H. 1987. "Clio Deceived." *International Security* 12, no. 2:5–44.

Heyting, A. 1966. *Intuitionism: An Introduction.* Amsterdam: North-Holland.

Hilbert, D. 1926. "Über das Unendliche." *Mathematische Annalen* 95:161–190.

———. 1930. "Naturerkennen und Logik." *Naturwissenschaften*: 959–963. Reprinted in Hilbert 1965, 3:378–387.

———. 1965. *Gesammelte Abhandlungen.* 3 vols. Chelsea reprint. New York. [German original, Berlin: Springer, 1935.]

Hitler, A. 1943. *Mein Kampf.* Translated by Ralph Manheim, from first edition of 1925 and 1927. Sentry edition. Boston: Houghton Mifflin.

Hobson, L. 1947. *Gentleman's Agreement.* A novel. New York: Simon and Schuster.

Holborn, H., ed. 1973. *Republic to Reich.* Translated by Ralph Manheim. New York: Vintage.

International Council of Philosophy and Humanistic Studies of UNESCO. 1955. *The Third Reich.* New York.

Jäckel, E. 1983. "Der Machtantritt Hitlers—Versuch einer geschichtliche Erklärung." In *1933, Wie die Republik der Diktatur erlag.* Stuttgart: W. Kohlhammer.

Jaensch, E. R. 1931. *Über die Grundlagen der Menschlichen Erkenntnis.* Leipzig: J. A. Barth.

———. 1938. *Der Gegentypus. Beiheft to Zeitschrift für angewandte Psychologie und Charakterkunde.* Leipzig: J. A. Barth.

Jaensch, E. R., and F. Althoff. 1939. *Mathematisches Denken und Seelenform.* Leipzig: J. A. Barth.

Jentsch, W. 1985. "Emmy Noether's Letters to Heinrich Brandt" (April 8, 1933, and April 26, 1933). *Historia Mathematica* 13:5–12.

Jordan, P. 1935. *Physikalische Denken in der neuen Zeit.* Hamburg: Hanseatische Verlagsanstalt.

Kähler, E. 1986. "The Poincaré Group." In J.S.R. Chisholm and A. K. Common, eds., *Clifford Algebras and Their Applications in Mathematical Physics*, 265–272. Boston: D. Reidel.

Kahle, P. 1945. *Bonn University in Pre-Nazi and Nazi Times, 1923–1939.* Privately printed. London: Portsoken Press. [Prepared 1942.]

Kalkmann, U. 1991. "Die Vertreibung Aachener Hochschullehrer durch die Nationalsozialisten." Aachen Universitätsarchiv.

Kant, I. 1959. *Foundations of the Metaphysics of Morals* and *What Is Enlightenment*. Translated by Lewis Beck. Library of Liberal Arts no. 113. New York: Liberal Arts Press.

Kater, M. 1974. *Das Ahnenerbe der SS, 1935–1945*. Stuttgart: Deutsche Verlags-Anstalt.

Kaufmann, W. 1953. *Monarchism in the Weimar Republic*. New York: Bookman.

Kerst, B. 1935. *Umbruch im mathematischen Unterricht*. Berlin: S. G. Grote.

Kessler, H. 1972. *In the Twenties*. Translated by Charles Kessler (from Harry Graf Kessler, *Tagebücher, 1918–1937*). New York: Holt, Rinehart, Winston.

Klemperer, V. 1987. *LTI*. Röderberg-Taschenbuch no. 35. Köln: Paul-Rugenstein. [Original ed., Halle, 1957.]

Kluge, W. 1983. "Landau, E., Staatsprufungsarbeit." Thesis, University of Duisburg.

König, R. 1941. "Mathematik als biologische Orientierungsfunktion unseres Bewusstseins." *Zeitschrift für mathematischen und naturwissenschaftlichen Unterricht* 2:33–47.

Koenigsberg, R. 1977. *The Psychoanalysis of Racism, Revolution and Nationalism*. New York: The Library of Social Science.

Kosambi, D. D. 1945. "George David Birkhoff, 1884–1944." *Mathematics Student* 12:116–120.

Kreisel, G. 1971. "Review of Manfred Szabo, *The Collected Papers of Gerhard Gentzen*." *Journal of Philosophy* 68:238–265.

Kuhn, H. 1966. "Die Universität vor der Machtergreifung." In *Die Deutsche Universitat im Dritten Reich*, 13–43. Munich: Piper.

Kuhn, T. S. 1977. *The Essential Tension*. Chicago: University of Chicago Press.

Kurcharzewski, M. 1982. "Stanislaw Gołab—Life and Work." *Aequationes Mathematicae* 24:1–18.

Lakatos, I. 1963–64. "Proofs and Refutations." *British Journal for Philosophy of Science* 14. Reprinted as chapter 1 of *Imre Lakatos, Proofs and Refutations* (Cambridge: Cambridge University Press, 1976).

Landau, E. 1984. *Collected Works*. Vols. 7, 8, 9. Essen: Thales Verlag.

Lane, B., and L. Rupp, eds. and trans. 1978. *Nazi Ideology before 1933*. Austin: University of Texas Press.

Laqueur, W. 1962. *Young Germany, A History of the German Youth Movement*. New York: Basic Books.

Lenard, P. 1936. *Deutsche Physik*. 4 vols. Munich: J. F. Lehmann.

———. 1943. *Grosse Naturforscher*. Munich: J. F. Lehmann. [Original ed., 1929.]

Lewis, T. 1990. "Authoritarian Attitudes and Personalities." *Psychohistory Review* 141–168.

Ley, M., and J. Schoeps. 1997. *Der Nationalsozialismus als politische Religion*. Bodenheim b., Mainz: Philo Verlagsgesellschaft.

Lietzmann, W. 1941. *Mathematik in Erziehung und Unterricht*. Leipzig: Quelle and Meyer.

Lilge, F. 1948. *The Abuse of Learning*. New York: Macmillan.

Lindner, H. 1980. "'Deutsche' und 'gegentypische' Mathematik: Zur Begründung einer 'arteigenen' Mathematik im 'Dritten Reich' durch Ludwig Bieberbach." In Mehrtens and Richter 1980.

Litten, F. 1996. "Ernst Mohr—Das Schicksal eines Mathematikers." *JDMV* 98:192–212.

———. 2000. *Mechanik und Antisemitismus, Wilhelm Müller, 1880–1968*. Heft 34 of Algorismus. Munich: Institut für Geschichte der Naturwissenschaften.

Littlewood, J. E. 1986. *A Mathematician's Miscellany*. Reprint. Cambridge: Cambridge University Press. [Original ed., London: Methuen, 1953.]

Maak, W. 1949. "Erich Hecke als Lehrer." *Hamburger Abhandlungen* 16:1–6.

Maass, C. 1991. "Das Mathematische Seminar der Hamburger Universität, 1933–1945."

In Eckart Krause, Ludwig Huber, and Holger Fischer, eds., *Hochschulalltag im "Dritten Reich."* Berlin and Hamburg: Dietrich Reimer Verlag.

Marshall, E. B. 1972. "The Political Development of German University Towns in the Weimar Republic: Göttingen and Münster, 1918–1930." Ph.D. thesis, University of London.

Mazlish, B. 1981. "Leader and Led, Individual and Group." *Psychohistory Review* 214–237.

Mehrtens, H. 1985. "Die 'Gleichschaltung' der mathematischen Gesellschaften in national-sozialistischen Deutschland." *Jahrbuch Überlicke Mathematik*: 83–103.

———. 1986. "Angewandte Mathematik and Anwendungen der Mathematik in national-sozialistischen Deutschland." *Geschichte und Gesellschaft* 12:317–341.

———. 1987. "Ludwig Bieberbach and 'Deutsche Mathematik.'" In Esther Phillips, ed., *Studies in the History of Mathematics*, 195–241. Mathematical Association of America.

———. 1996. "Mathematics and War: Germany, 1900–1945." In Paul Forman and José Sanchez-Ron, *National Military Establishments and the Advancement of Science and Technology*, Boston Studies in the Philosophy of Science 180, 87–134. Dordrecht and Boston: Kluwer.

Mehrtens, H., and S. Richter, eds. 1980. *Naturwissenschaft, Technik, und NS-Ideologie*. Frankfurt-am-Main: Suhrkamp.

Meinecke, F. 1926. *Die Deutsche Universität und der Heutige Staat*. Tübingen: Mohr.

Mendelssohn, K. 1973. *The World of Walther Nernst: The Rise and Fall of German Science, 1864–1941*. Pittsburgh: University of Pittsburgh Press.

Menger, K. 1979. "My Memories of L.E.J. Brouwer." Chapter 21 in *Selected Papers in Logic and Foundations, Didactics, Economics*. Dordrecht, Holland: Reidel.

Molsen, K. 1935–37. "Über spezielle Klassen irreduzibiler Polynome." *Schriften des Mathematischen Seminars* (Berlin).

Monna, A. F. 1972–73. "The Concepts of Function in the Nineteenth and Twentieth Centuries, in Particular with Regard to the Discussions between Baire, Borel, and Lebesgue." *Archive for History of Exact Sciences* 9:57–84.

Mosse, G. 1964. *The Crisis of German Ideology*. New York: Grosset and Dunlap.

———. 1966. "Die Deutsche Rechte und die Juden." In W. Mosse, ed., *Entscheidungsjahr, 1932*. Tübingen: Mohr.

———. 1970. *Germans and Jews*. New York: Grosset and Dunlap.

Müller-Hill, B. 1984. *Tödliche Wissenschaft*. Hamburg: Rowohlt.

Mulkay, M. 1969. "Some Aspects of Cultural Growth in the Natural Sciences." *Social Research* 36, no. 1. Abridged version in Barry Barnes, ed., *Sociology of Science* (Harmondsworth: Penguin Books, 1972).

Muslin, H. 1992. "Adolf Hitler: The Evil Self." *Psychohistory Review*: 251–270.

Neumann, F. 1966. *Behemoth*. New York: Harper. [Original ed., 1942–44.]

Neumann, S. 1965. *Die Parteien der Weimar Republic*. Stuttgart: Kohlhammer. [Original ed., *Die politische Parteien in Deutschland*, Berlin: Junker u. Dünnhaupt, 1932.]

Noether, E. 1911. "Zur Invarianttheorie der Formen von *n* Variabeln." *Journal für die Reine u. Angewandte Mathematik* 139:118–154.

Ostrowski, A. 1958. "Wilhelm Süss, 1895–1958." *Freiburger Universitätsreden, Neue Folge* no. 28: 1–16.

Parshall, K. 1994. "Toward a History of Nineteenth-Century Invariant Theory." In David Rowe and John McCleary, eds., *The History of Mathematics*, vol. 1, 157–206. San Diego: Academic Press.

Pascher, J. 1966. "Das Dritte Reich erlebt an drei deutschen Universitäten." In *Die Deutsche Universität im Dritten Reich*, 45–70. Munich: Piper.

Peano, G. 1902–6. "Super theorema de Cantor-Bernstein et additione." *Rivista de Matematica* 8:136–157.

Pinl, M. 1969. "Kollegen in einer dunklen Zeit I. *JDMV* 71:167–228.

———. 1971. "Kollegen in einer dunklen Zeit II." *JDMV* 72:165–189.

———. 1972. "Kollegen in einer dunklen Zeit III." *JDMV* 73:153–208.

———. 1973–74. "Kollegen in einer dunklen Zeit IV." *JDMV* 75:166–208.

Pinl, M., and L. Furtmüller. 1973. "Mathematicians under Hitler." *Leo Baeck Yearbook* 18:130–181.

Poliakov, L., and J. Wulf. 1959. *Das Dritte Reich und Seine Denker* (Dokumente). Berlin: Arani Verlag.

Proceedings of the International Mathematical Congress, Toronto, 1924. Vol. 1. Toronto: University of Toronto Press.

Proctor, R. 1988. *Racial Hygiene*. Cambridge, Mass.: Harvard University Press.

Pulzer, P. J. 1964. *The Rise of Political Antisemitism in Germany and Austria*. New York: Wiley.

Quätsch, C. 1961. *The Numerical Record of University Attendance in Germany in the Last Fifty Years*. Berlin: Springer-Verlag.

Radó, S. 1932. "Paths of Natural Science in the Light of Pscychoanalysis." *Psychonanalytic Quarterly* 1:683–700.

Rang, B., and W. Thomas. 1981. "Zermelo's Discovery of the 'Russell' Paradox." *Historia Mathematica* 8:5–22.

Rang- und Organisationsliste [der NSDAP]. 1946. Stuttgart: W. Kohlhammer.

Reich, W. 1970. *The Mass Psychology of Fascism* [1944]. Translated by Vincent Carfagno. New York: Farrar, Straus and Giroux.

Reid, C. 1970. *Hilbert*. Berlin and New York: Springer.

———. 1976. *Courant in Göttingen and New York*. Berlin and New York: Springer.

Reingold, N. 1981. "Refugee Mathematicians in the United States of America, 1933–34: Reception and Reaction." *Annals of Science* 38:313–338.

Reissinger, T. N.d. "Die Verführbarkeit der Mathematiker." University of Mannheim, preprint #120.

Richter, S. 1980. "Die Deutsche Physik." In Mehrtens and Richter 1980: 116–141.

Ringer, F. 1960. "The German Universities and the Crisis of Learning, 1918–1932." Ph.D. thesis, Harvard University, Cambridge, Mass.

———. 1969. *The Decline of the German Mandarins*. Cambridge, Mass.: Harvard University Press.

Robinson, A. 1969. "Some Thoughts on the History of Mathematics." *Compositio mathematica* 20:188–193.

Rogger, H., and E. Weber, ed. 1965. *The European Right*. Berkeley: University of California Press.

Rohrbach, H. 1968. "Erhard Schmidt, Ein Lebensbild." *JDMV* 69:209–224.

Rosen, E. 1971. "Biography of Copernicus." In *Three Copernican Treatises*, 3d ed., 313–412. New York: Octagon Books.

Rowe, D. 1986. "'Jewish Mathematics' at Göttingen in the Era of Felix Klein." *Isis* 77:422–449.

———. 1988. "Die Wirkung deutscher Mathematiker auf die amerikanische Mathematik, 1875–1900." *Mitteilungen der Mathematischen Gesellschaft der DDR* 72–96.

Rühle, G. 1935. *Das Dritte Reich, 1934*. Berlin: Hummelverlag.

Schaper, R. 1992. "Mathematiker im Exil." In *Die Künste und die Wissenshaften im Exil, 1933–1945*. Gerlingen: Lambert Schneider.

Schappacher, N. 1987. "Das Mathematische Institut der Universität Göttingen." A condensed version of Schappacher n.d. In Becker et al. 1987: 344–373.

———. N.d. "Das Mathematische Institut der Universität Göttingen." Unpublished manuscript.

Schappacher, N., and M. Kneser. 1990. "Fachverband-Institut-Staat." In *Ein Jahrhundert Mathematik, 1890–1990*, 1–82. Braunschweig and Wiesbaden: Friedr. Vieweg.

Schappacher, N., and E. Scholz. 1992. "Oswald Teichmüller—Leben und Werk." *JDMV* 94:1–39.

Scharlau, W., et al. 1990. *Mathematische Institute in Deutschland, 1800–1945*. Braunschweig: Vieweg.

Schattschneider, D. 1981. "In Praise of Amateurs." In David Klarner, ed., *The Mathematical Gardner*. Boston: Prindle, Weber and Schmidt.

Schlote, K. 1991. "Noether, F.—Opfer zweier Diktaturen." *NTM Schriftenreihe* 28:33–41.

Schröder-Gudehus, B. 1973. "Challenge to Transnational Loyalties: International Scientific Organizations after the First World War." *Science Studies* 3:93–118.

Schwabe, K. 1969. *Wissenschaft und Kriegsmoral*. Göttingen: Musterschmidt.

Segal, S. 1980. "Helmut Hasse in 1934." *Historia Mathematica* 7:46–56.

Seier, H. 1984. "Universität und Hochschulpolitik im Nationalsozialistischen Staat." In K. Malettke, ed., *Der Nationalsozialismus an der Macht*, 143–165. Göttingen: Vandenhoeck und Ruprecht.

Sharpe, E. F. 1935. "Similar and Divergent Unconscious Determinants Underlying the Sublimations of Pure Art and Pure Science." *International Journal of Psychoanalysis* 16:186–202.

Siegel, C. L. 1966. *Gesammelte Werke*. 3 vols. Vol. 3. Heidelberg: Springer.

Siegmund-Schultze, R. 1984a. "Das Ende des Jahrbuchs über die Fortschritte der Mathematik und die Brechung des deutschen Referatenmonopols." *Mitteilungen der Mathematischen Gesellschaft der DDR* 91–102.

———. 1984b. "Einige Probleme der Geschichtsschreibung der Mathematik im faschistischen Deutschland—unter besonderer Berucksichtung des Lebenslaufes des Greifswalder Mathematikers Theodor Vahlen." *Wissenschaftliche Zeitschrift der Ernst-Moritz-Arndt-Universität Greifswald* 32:51–56.

———. 1984c. "Theodore Vahlen—zum Schuldanteil eines deutschen mathematikers am faschisten Missbrauch der Wissenschaft." *NTM Schriftenreihe* 21:17–32.

———. 1986. "Faschistiche Pläne zur 'Neuordnung' der europäischen Wissenschaft, Das Beispiel Mathematik." *NTM Schriftenreihe* 23:1–17.

———. 1988. "Berliner Mathematik zur Zeit des Faschismus." *Mitteilungen der Mathematischen Gesellschaft der DDR*: 61–84.

———. 1989. "Zur Sozialgeschichte der Mathematik an der Berliner Universität im Faschismus." *NTM Schriftenreihe* 26:49–68.

Sommerfeld, A., and F. Krauss. 1951. "Otto Blumenthal zum Gedächtnis." *Jahrbuch der Rheinisch-Westphälischen Technische Hochschule Aachen* 21–25.

Sontheimer, K. 1957. "Antidemokratisches Denken in der Weimarer Republik." *Vierteljahresheft für Zeitgeschichte* 5.

———. 1962. *Antidemokratisches Denken in der Weimarer Republic*. Munich: Nymphenburger Verlag.

———. 1966. "Die Haltung der Universitäten zur Weimarer Republik." In *Universitätstage* 1966: 24–42.

Spengler, O. 1932. *The Decline of the West*. 2 vols. Translated by C. F. Atkinson. New York: Knopf. [Original ed., as *Der Untergang des Abendlandes*, 1926, 1928.]

Spranger, E. 1955. "Mein Konflikt mit der National-Sozialistische Regierung 1933." *Universitas, Zeitschrift für Wissenschaft Kunst und Literatur* 457–473.

———. 1973. "Hochschule und Staat." In Eduard Spranger, *Gesammelte Schriften*, vol. 10, 189–224. Heidelberg: Quelle and Meyer.

Stachura, P. D. 1978. "'Der Fall Strasser': Gregor Strasser, Hitler and National Socialism, 1930–1932." In Peter D. Stachura, ed., *The Shaping of the Nazi State*. London: Croom Helm; New York: Harper and Row.

Steck, M. 1942. *Das Hauptproblem der Mathematik*. Berlin: G. Lüttke.

Steinberg, M. 1977. *Sabers and Brown Shirts*. Chicago· University of Chicago Press.

Stern, F. 1960. *The Political Consequences of the Unpolitical German*. New York: Meridian Books.

———. 1961. *The Politics of Cultural Despair: A Study in the Rise of Germanic Ideology*. Berkeley: University of California Press.

Stern, J. P. 1975. *Hitler: The Führer and the People*. Berkeley: University of California Press.

Stirk, S. D. 1946. *German Universities through English Eyes*. London: Gollancz.

Strubecker, K. 1943. Obituary of E. A. Weiss. *Deutsche Mathematik* 7:254–298.

Suhling, L. 1980. "Deutsche Baukunst, Technologie und Ideologie im Industriebau des Dritten Reiches." In Mehrtens and Richter 1980: 243–281.

Süss, I. 1967. *Beginnings of the Mathematical Research Institute Oberwolfach at the Country House "Lorenzenhof."* Translated by P. L. Butzer. Mathematisches Institut Oberwolfach (privately printed).

———. N.d. (but not later than 1988). "Erinnerungen." Copy obtained through Prof. M. Barner.

Synnott, M. 1979. *The Half-Opened Door: Discrimination and Admissions at Harvard, Yale and Princeton, 1900–1970*. Westport, Conn.: Greenwood.

Teichmüller, O. 1982. *Collected Works*. Edited by L. Ahlfors and F. W. Gehring. Berlin: Springer.

Thomsen, G. 1934. "Uber die Gefahr der Zurückdrängung der Exakten Naturwissenschataften an der Schulen und Hochschulen." *Neue Jahrbücher für Wissenschaft und Jugendbildung*: 164–175.

Tietjen, Cl. 1936. *Raum oder Zahl*. Leipzig: Friedrich Brandstetter.

Tietz, H. N.d. "Studium mit Hindernissen." Unpublished recollections, manuscript dated March 1984.

Titze, H. 1987. *Das Hochschulstudium in Preussen und Deutschland, 1820–1944*. Göttingen: Vandenhoeck und Ruprecht.

Tobies, R. 1981. *Felix Klein*. Biographien hervorragender Naturwissenschaftler, Techniker, und Mediziner, Band 50. Leipzig: Teubner.

———. 1984. "Untersuchungen zur Rolle der Carl-Zeiss Stiftung für die Entwicklung der Mathematik an der Universität Jena." *NTM Schriftenreihe* 21:33–43.

———. 1986. "Die 'Gesellschaft für angewandte Mathematik und Mechanik' im Gefüge imperialistischer Wissenschaftsorganisation." *NTM Schriftenreihe* 19:16–28.

Todd, J. 1983. "Oberwolfach 1945." In E. F. Beckenbach and W. Walter, eds., *General Inequalities* 3: 19–22. Basel: Birkhäuser Verlag.

Tollmien, C. 1987. "Das Kaiser-Wilhelm-Institut für Strömungsforschung verbunden mit der Aerodynamischen Versuchsanstalt." In Becker et al. 1987: 317–347.

Universitätstage 1966, Nationalsozialismus und die Deutsche Universität. 1966. Berlin: Walter de Gruyter.

van der Waerden, B. L. 1931–66. *Algebra, Moderne Algebra*. 7 editions. Berlin and Heidelberg: Springer-Verlag.

———. 1935. "Nachruf auf Emmy Noether." *Mathematische Annalen* 111:469–476.

van Stigt, W. 1979. "The Rejected Parts of Brouwer's Dissertation on the Foundations of Mathematics." *Historia Mathematica* 6.

Vezina, B. 1982. *Die Gleichschaltung der Universität Heidelberg*. Heidelberg: Carl Winter Universitftsverlag.

Voegelin, E. 1966. "Die deutsche Universität und die Ordnung der deutschen Gesellschaft." In *Die deutsche Universität im Dritten Reich*. Munich: Piper Verlag.

Von Ferber, C. 1956. *Die Entwicklung des Lehrkörpers der deutschen Universitäten und Hochschulen, 1864–1954*. Göttingen: Vandenhoeck und Ruprecht.

von Humboldt, W. 1854. *Ideen zu einen Versuch, die Graenzen* [sic] *der Wirksamkeit des Staats zu bestimmen*. Translated as *The Sphere and Duties of Government* by Joseph Coulthard, Jr. London: Teubner.

von Kármán, T. 1967. *The Wind and Beyond*. Boston: Little Brown.

von Krockow, C. N.d. *Scheiterhaufen*. Severin und Siedler, n.p., n.d. (c. 1988).

von Laue, M. 1948. "The Wartime Activities of German Scientists." *Bulletin of Atomic Scientists* 4.

von Mises, R. 1964. *Mathematical Theory of Probability and Statistics*. Edited and complemented by Hilda Geiringer. New York: Academic Press.

Walker, M. 1989. *German National Socialism and the Quest for Nuclear Power, 1939–1949*. Cambridge; Cambridge University Press.

Weber, M. 1974. *Max Weber on Universities*. Translated and edited by Edward Shils. Chicago: University of Chicago Press.

Weil, A. 1992. *The Apprenticeship of a Mathematician*. Translated by Jennifer Gage. Basel: Birkhauser. [French original, *Souvenirs d'apprentissage*, 1991.]

Weinstein, F. 1980. *The Dynamics of Nazism: Leadership, Ideology and the Holocaust*. New York: Academic Press.

Weiss, E. A. 1933. "Wozu Mathematik." Pamphlet. Bonn.

———. 1935. "Das Mathematische Arbeitslager Kronenburg 3. Lager." Pamphlet. Bonn.

———. 1936. "Das Mathematische Arbeitslager Kronenburg 4. Lager." Pamphlet. Bonn.

———. 1938. "Das Mathematische Arbeitslager Kronenburg 5. Lager." Pamphlet. Bonn.

Weyl, H. 1946. "Mathematics and Logic." *American Mathematical Monthly* 53:2–13.

———. 1959. *Raum, Zeit, Materie*. Dover reprint. Berlin: Springer. [4th ed., Zürich, 1921; English ed., *Space, Time, Matter* (London: Methuen, 1922).]

———. 1968. *Gesammelte Abhandlungen*. 4 vols. Berlin and New York: Springer.

Wiener, N. 1964. *I Am a Mathematician*. Cambridge, Mass.: MIT Press. [Original ed., New York: Doubleday, 1956.]

Wigner, E. 1960. "The Unreasonable Effectiveness of Mathematics in the Natural Sciences." *Communications in Pure and Applied Mathematics* 13:1–14. Reprinted in T. Saaty and F. J. Weyl, eds., *The Spirit and Uses of the Mathematical Sciences* (New York: McGraw-Hill, 1969), 123–140.

Willstätter, R. 1965. *From My Life*. New York: Benjamin. [Original German ed., 1949.]

Winkler, H. 1976. "German Society, Hitler, and the Illusion of Restoration, 1930–33." *Journal of Contemporary History* 11, no. 4:1–16.

Wüssing, H. 1964. *Die Genesis des Abstraken Gruppenbegriffes*. Berlin: VEB Deutscher Verlag der Wissenschaften.

Zacharias, M. 1919. *Zeitschrift für math. -naturwiss. Unterricht* 50:277–280.

Zermelo, E. 1908. "Neuer Beweis für die Moglichkeit einer Wohlordnung." *Mathematische Annalen* 65:107–128.

Zierold, K. 1968. *Forschungsförderung in Drei Epochen*. Wiesbaden: Franz Steiner Verlag.

Archival Sources

The archives of the Bundesarchiv Koblenz, the Berlin Document Center, the U.S. National Archives, and the Mathematisches Institut Göttingen were especially useful in compiling this book. The use of material found in other archives is referenced as appropriate in the text and notes.

Personalakten Hans Beck, Bonn
Personalakten Heinrich Behnke, Münster
Personalakten Erich Bessel-Hagen, Bonn
Personalakten Wilhelm Blaschke, Hamburg
Personalakten Otto Blumenthal, Aachen
Personalakten Constantin Carathéodory, Munich
Personalakten Gustav Doetsch, Freiburg
Personalakten Felix Hausdorff, Bonn
Personalakten Erich Hecke, Hamburg
Personalakten W. Herrmann, Münster
Personalakten Erich Kamke, Tübingen
Personalakten Hellmuth Kneser, Tübingen
Personalakten Friedrich Levi, Freiburg
Personalakten Ludwig Neder, Münster
Personalakten Ernst Peschl, Bonn
Personalakten Hans Petersson, Hamburg
Personalakten Paul Riebesell, Hamburg
Personalakten Heinrich Scholz, Münster
Personalakten Wilhelm Süss, Freiburg
Personalakten Otto Toeplitz, Bonn
Personalakten Helmut Ulm, Münster
Personalakten Theodore von Kármán, Aachen
Personalakten Ernst August Weiss, Bonn
Personalakten Ernst Witt, Hamburg
Personalakten Ernst Zermelo, Freiburg

Correspondence of Ludwig Bieberbach, in Niedersächsisches Staats- und Universitätbibliothek (abbreviated BL).

Correspondence connected with the publication of *Mathematische Annalen*, in the possession of H. Petersson (abbreviated HNMA).

Richard Courant Papers. These are in two collections: one, in the possession of his son, Ernst Courant (at the time I used them they were in the possession of Ernst Courant's mother, Nina Courant), and the other at New York University.

Papers of the Göttingen Mathematical Institute, as arranged by Norbert Schappacher (abbreviated MI).

Nachlass of Hellmuth Kneser, in the possession of his son, Martin Kneser (abbreviated HK).

Papers of Oswald Veblen, in Library of Congress (abbreviated VP).

Weber, W. N.d. "Ursachen und Verlauf meiner Auseinandersetzung mit Prof. Hasse." Unpublished manuscript in the Bundesarchiv Koblenz (abbreviated Weber).

INTERVIEWS

(Interviews were conducted by the author, except as noted.)

Marianne Bernstein-Wiener, numerous telephone conversations, 1989–95.
Ludwig Bieberbach, interviewed by H. Mehrtens, September 21, 1981 (abbreviated BI).
Natascha Artin Brunswick, New York, N.Y., October 1987.
Werner Burau, Hamburg, Germany, January 31, 1988.
Erich Kähler, Hamburg, Germany, January 30, 1988.
Manfred Knebusch, Regensburg, Germany, February 25, 1988.
Martin and Jutta Kneser, Göttingen, Germany, February 18, 1988.
Matthias Landau, Düsseldorf, Germany, January 26, 1988.
Wilhelm Magnus, New Rochelle, N.Y., 1982.
Margarete Petersson, Münster, Germany, April 14, 1988.
Karl Stein, Munich, Germany, February 24, 26, 1988.
Irmgard Süss, Freiburg, Germany, March 25, 1988.
Peter Thullen, Fribourg, Switzerland, March 21–23, 1988.
Horst Tietz, Hannover, Germany, April 5, 1988.
K. H. Weise, Kiel, Germany, March 16, 1988.
Hans Zassenhaus, Columbus, Ohio, November 17, 1987.
Henry Zatskis, Newark, N.J., 1982.
Max Zorn, Bloomington, Ind., March 18, 1991.

INDEX

All mathematicians mentioned, and all people whose commentary on various issues is cited, are included in this name index. It is hoped that this index, together with the detailed delineation in the table of contents of subsections of chapters 4–8, will enable the reader to find efficiently what is wanted. Occasionally, a reference to a page is to a footnote on that page.